ER7QT

AF616104

Building and Solving Mathematical Programming Models in Engineering and Science

PURE AND APPLIED MATHEMATICS

A Wiley-Interscience Series of Texts, Monographs, and Tracts

A complete list of the titles in this series appears at the end of this volume.

Building and Solving Mathematical Programming Models in Engineering and Science

Enrique Castillo
Antonio J. Conejo
Pablo Pedregal
Ricardo Garciá
Natalia Alguacil

A Wiley-Interscience Publication
JOHN WILEY & SONS, INC.

This text is printed on acid-free paper. ♾

Published simultaneously in Canada.

For ordering and customer service, call 1-800-CALL-WILEY.

Library of Congress Cataloging in Publication Data:

Building and solving mathematical programming models in engineering / Enrique Castillo . . . [et al.].
p. cm. — (Pure and applied mathematics)
Includes bibliographical references and index.
ISBN 0-471-15043-6 (cloth : alk. paper)
1. Programming (Mathematics) 2. Engineering models. I. Castillo, Enrique, 1946- II. Pure and applied mathematics (John Wiley & Sons : Unnumbered)
T57.7 .B85 2001
620'.001'5197—dc21 2001026954

Printed in the United States of America

10 9 8 7 6 5 4 3 2 1

DEDICATION

To Alfonso Fernández Canteli
for his constant help and encouragement

Contents

Preface

Modeling is one of the most appealing areas in engineering and the applied sciences. In fact, engineers need to build models to solve real-life problems. The aim of a model consists of reproducing the reality as faithfully as possible, trying to understand how the real world behaves and obtaining the expected responses to given actions or inputs.

Many types of models are used in practice, such as differential equation models, functional equation models, finite difference and finite element models, and mathematical programming models.

Selection of the adequate model reproducing the reality is a crucial step for a satisfactory solution to a real problem. The associated mathematical structures are not arbitrary, but a consequence of the reality itself. In this book we make an effort to connect physical and mathematical realities. We show the reader the reasoning that leads to the analysis of the different structures, models, and concepts. This becomes apparent in the illustrative examples, which show the connection between model and reality.

In this book we deal with mathematical programming models, including linear programming and nonlinear programming.

Mathematical programming problems are particular problems one has to deal with very frequently. One is prepared to solve them using many available tools, procedures, or computer packages. In fact, these problems are studied in detail in undergraduate or graduate courses. However, one may not be prepared to solve other frequently appearing problems such as

1. Linear programming problems with many variables and/or constraints
2. Nonlinear programming problems
3. Decomposition techniques for problems to be solved with mathematical programming tools
4. Transformation rules for other problems to be solved with mathematical programming tools

In this book we give methods that allow the solution of a wide collection of interesting practical problems.

The aim of this book is not to deal with the standard methods of analysis, which are comprehensively covered in many other books. Instead, we try to show the reader the power of these methods for solving practical problems, and then introduce the main concepts and tools.

When analyzing and discussing mathematical programming, one possibility is to give a deep theoretical analysis and a discussion of problems and methods. This approach has some risks. Although initially the treatment seems to be more rigorous and deep, the reader can be led to become too curious and cautious about mathematical details but having no knowledge of what the mathematical procedures lead to or are motivated by. For example, it is not uncommon to give a person who has studied linear programming for years a simple two-dimensional plot in which the feasible set is drawn, and ask him/her to mark the sequence of extreme points the simplex method would follow to find the optimal solution, without having a correct answer. Note that this means a lack of understanding of the basic ideas behind the simplex method.

Alternatively, one can treat this topic with the help of illustrative examples, and try to transmit to the reader the profound and ingenious that underlie these methods, thus making the book more readable and attractive. We do not deal with standard methods or solutions. The reader looking for standard methods or references to published works with this orientation should consult one of the many books on the topic (see Bibliography). On the contrary, in this book the problems above mentioned are discussed from a different point of view.

In addition to obtaining solutions, mathematicians and engineers are interested in analyzing conditions leading to well-defined problems. In this context, the problems of compatibility and uniqueness of solutions play a central role. In fact, they lead to very interesting physical or engineering conclusions, which relate the physical conditions, obtained from reality, to the corresponding conditions behind the mathematical models. The methods to be developed in this book also allow one to conclude whether a set of linear constraints leads to a solution.

The book is organized in four parts. In Part I we deal with models and endeavor to introduce the reader to the attractive field of mathematical programming by means of carefully selected examples. We guide the reader to discover the mathematical statement of the problems after a clear understanding of the physics or engineering problems.

Part II deals with methods and describes the main techniques used to solve linear and nonlinear programming problems. As indicated above, the aim of this part is to give not a rigorous analysis of the methods but a clear understanding of the main ideas.

Since nothing can be practically done without the help of good software, in Part III we have selected GAMS (the general algebraic modeling system) as the main tool to be used in the book. To facilitate its use, a detailed description of how problems need to be stated and the possibilities of GAMS are given. All the problems in Part I and many others are solved using this tool.

Once the reader is familiar with mathematical programming problems and the GAMS software, Part IV is devoted to more dense applications of these techniques to practical problems in many areas of research, including artificial intelligence (AI), computer assisted design (CAD), probability and statistics, economics, engineering, and transportation. All these applications have arisen from real life. In other words, they are mathematical models of physical, economic, or engineering problems. This part also includes a chapter dedicated to some tricks that are useful in dealing with especial cases of problems as, for example, converting some nonlinear problems into linear ones, and it also gives some specific GAMS tricks.

This book can be used as a reference or consulting book and as a textbook in upper-division undergraduate courses or in graduate-level courses. Included in the book are numerous illustrative examples and end-of-chapter exercises. We have also developed the GAMS code to implement the various examples and methodologies presented in this book. The current version of these programs, together with a brief *User's Guide*, can be obtained from the World Wide Web site

`http://www.gams.com/`

This book is addressed to a wide audience, including mathematicians, engineers, and applied scientists. There are few prerequisites for the reader, although a previous knowledge of linear algebra, elementary calculus, and some familiarity with matrices are essential.

Several colleagues and students have read earlier versions of the manuscript for this book and have provided us with valuable comments and suggestions. Their contributions have given rise to the current substantially improved version. In particular, we acknowledge the help of Angel Yustres, José Manuel Vellón, and Marta Serrano, who made a very careful reading of the manuscript and gave many suggestions, and to Gonzalo Echevarría and the Wiley team for their help in dealing with the edition of the book.

Special thanks are given to the University of Castilla-La Mancha, University of Cantabria, and Iberdrola and José Antonio Garrido for their financial support.

Finally, Enrique Castillo wants to express his sincere thanks to Professor Alfonso Fernández Canteli (University of Oviedo) for his constant help and encouragement.

ENRIQUE CASTILLO, ANTONIO J. CONEJO,
PABLO PEDREGAL, RICARDO GARCÍA,
AND NATALIA ALGUACIL

Ciudad Real, Spain
August 2, 2001

Part I

Models

Chapter 1

Linear Programming

1.1 Introduction

Mathematical programming is a modeling technique that is used as a powerful tool in decisionmaking. When dealing with a decision problem, the first step consists of identifying the possible decisions to be made; this leads to identifying the problem variables. Usually, decisions are of a quantitative character and we look for the values that optimize our objective. The second step consists of determining which decisions are admissible; this leads to a set of constraints that are determined according to the nature of the problem under consideration. In the third step, the cost/benefit ratio associated with each decision is calculated; this leads to an objective function, assigning to each set of values for the decision variables a given cost/benefit ratio. The set of all these elements is the data set.

Linear programming (LP), which deals only with linear objective functions and linear constraints, is a part of mathematical programming, and one of the most important areas of applied mathematics. It is used in many fields, as engineering, economics, business administration, and many other areas of science, technology and industry.

In this chapter we introduce linear programming by means of several illustrative examples. To start our exposition we mention that any linear programming problem requires identifying four basic elements:

1. The set of data

2. The set of variables involved in the problem, together with their respective domains of definition

3. The set of problem linear constraints that define the set of feasible solutions

4. The linear function to be optimized (minimized or maximized)

In the following sections we give a list of examples making a special effort in identifying these four sets.

The selected list is only one sample of the huge amount of linear programming problems (LPPs) available in the existing literature. They have been chosen with the aim of illustrating the power of this technique, and allowing the reader to become familiar with the above four elements.

1.2 The Transportation Problem

In this section we introduce and make a detailed discussion of the transportation problem.

Assume that a product is to be shipped in the amounts $u_1, \ldots, u_m$, from each of m shipping origins, and received in amounts $v_1, \ldots, v_n$, by each of n shipping destinations. The problem consists of determining the amounts x_{ij}, to be shipped from origins i to destinations j, to minimize the cost of transportation.

The four main elements of the transportation problem are

1. **Data**

   m: the number of origins

   n: the number of destinations

   u_i: the amount to be shipped from origin i

   v_j: the amount to be received in destination j

   c_{ij}: the cost of sending a unit of product from origin i to destination j

2. **Variables**

   x_{ij}: the amount to be shipped from origin i to destination j.

   It is assumed that these variables are nonnegative:

$$x_{ij} \geq 0; \; i = 1, \ldots, m; \; j = 1, \ldots, n \tag{1.1}$$

   This implies that the direction of the product flow is prefixed by giving the origins and destinations. However, other assumptions can be made. For example, we can use unrestricted real variables, i.e., $x_{ij} \in \mathbb{R}$, if we do not want to prefix what the origins and destinations are.

3. **Constraints.** The constraints of this problem are

$$\begin{array}{rcl} \sum_{j=1}^{n} x_{ij} & = & u_i; \; i = 1, \ldots, m \\ \sum_{i=1}^{m} x_{ij} & = & v_j; \; j = 1, \ldots, n \end{array} \tag{1.2}$$

   The first set of conditions states that the total amount shipped from origin i must be equal to the sum of the amounts going from it to all destinations $j = 1, \ldots, n$.

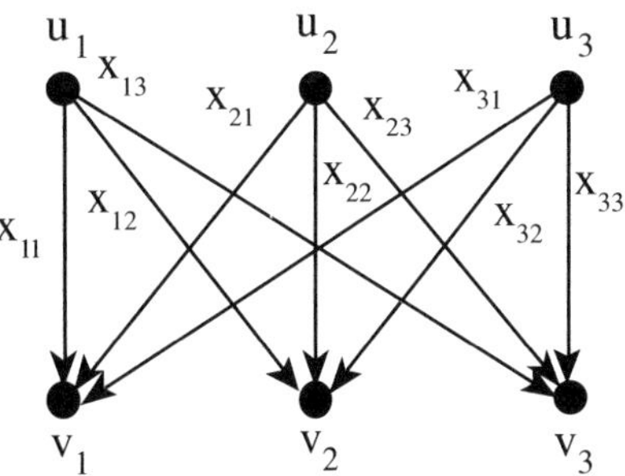

Figure 1.1: Illustration of the transportation problem.

The second set of conditions states that the total amount received at destination j must be equal to the sum of all amounts shipped to that destination from all origins $i = 1, \ldots, m$.

The reader must differentiate between the *variable bounds* (1.1) and the *problem constraints* (1.2).

4. **Function to be optimized.** In the transportation problem we are normally interested in minimizing the total cost of transportation (sum of the unit costs times the amounts being shipped); that is, we minimize

$$Z = \sum_{i=1}^{m} \sum_{j=1}^{n} c_{ij} x_{ij} \tag{1.3}$$

Once these four elements have been identified, we are ready to solve the problem.

Example 1.1 (Transportation problem). Consider the transportation problem in Figure 1.1, with $m = 3$ shipping origins and $n = 3$ destinations, and

$$u_1 = 2, u_2 = 3, u_3 = 4; v_1 = 5, v_2 = 2, v_3 = 2$$

In this case, the system (1.2) becomes

$$\mathbf{Cx} = \begin{pmatrix} 1 & 1 & 1 & 0 & 0 & 0 & 0 & 0 & 0 \\ 0 & 0 & 0 & 1 & 1 & 1 & 0 & 0 & 0 \\ 0 & 0 & 0 & 0 & 0 & 0 & 1 & 1 & 1 \\ 1 & 0 & 0 & 1 & 0 & 0 & 1 & 0 & 0 \\ 0 & 1 & 0 & 0 & 1 & 0 & 0 & 1 & 0 \\ 0 & 0 & 1 & 0 & 0 & 1 & 0 & 0 & 1 \end{pmatrix} \begin{pmatrix} x_{11} \\ x_{12} \\ x_{13} \\ x_{21} \\ x_{22} \\ x_{23} \\ x_{31} \\ x_{32} \\ x_{33} \end{pmatrix} = \begin{pmatrix} 2 \\ 3 \\ 4 \\ 5 \\ 2 \\ 2 \end{pmatrix} \tag{1.4}$$

$$x_{ij} \geq 0; \quad i, j = 1, 2, 3$$

The first three equations correspond to the product balance at the three origins and the last three equations, to the balance at the three destinations.

If we now assume some particular values

$$\mathbf{c} = \begin{pmatrix} 1 & 2 & 3 \\ 2 & 1 & 2 \\ 3 & 2 & 1 \end{pmatrix}$$

for the transportation costs, our problem becomes the minimization of

$$Z = x_{11} + 2x_{12} + 3x_{13} + 2x_{21} + x_{22} + 2x_{23} + 3x_{31} + 2x_{32} + x_{33}. \tag{1.5}$$

Using the GAMS package to solve this problem (see Section 11.2.1), we obtain a minimum value of 14 for the objective function, which implies a minimum cost of

$$Z = 14 \text{ units, at point } \mathbf{x} = \begin{pmatrix} 2 & 0 & 0 \\ 1 & 2 & 0 \\ 2 & 0 & 2 \end{pmatrix} \tag{1.6}$$

Normally, standard packages give solutions to the problems at hand or inform the user about their infeasibility, but they do not inform about the existence of multiple solutions when they exist. In Section A.39, we shall see that this problem has many solutions leading to the same value of the objective function. ■

1.3 The Production Scheduling Problem

We shall deal with two different versions of this problem.

1.3.1 Production Scheduling Problem 1

A manufacturer produces an item, whose demand fluctuates in time, in a form similar to that in Figure 1.2. The manufacturer must always satisfy the monthly requirements established by demand. In general, any scheduling problem has many different schedules that will satisfy the requirements. There are two possible alternatives:

1. **Variable production**. The manufacturer could produce each month the exact number of units required by demand. However, since a fluctuating production schedule is costly to maintain, because of overtime cost in high–production months, and the costs associated with the release of personnel and machinery in low–production months, this type of production schedule is not efficient.

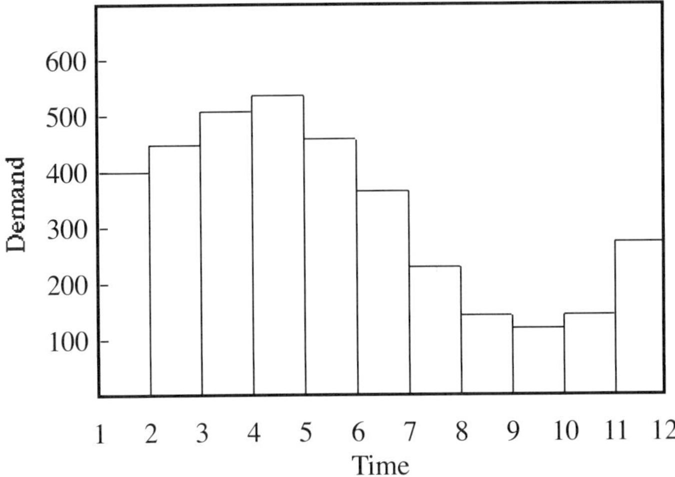

Figure 1.2: Demand function over time.

2. **Constant production**. The manufacturer faced with fluctuating requirements could overproduce in periods of low requirements and store the surplus to be used in periods of high demand. Thus, the production can be made constant, compensating underproduction with past overproduction. However, because of the cost of keeping a manufactured item in storage, such a solution may be undesirable if it yields comparatively large monthly surpluses.

Problems of this nature illustrate the difficulties that arise when conflicting objectives are inherent in a given system. Our objective is to design a production schedule that maximizes benefit after consideration of the costs of production fluctuations and inventories.

The three main elements in the production scheduling problem are:

1. **Data**

   n: the number of months to be considered

   s_0: the amount of an item available in storage at the beginning of the considered period

   d_t: the number of units (demand) required in month t

   s^{max}: the storage capacity

   a_t: the selling price at time t

   b_t: the production cost at time t

   c_t: the storage cost at time t

2. **Variables**

   x_t: the number of units produced in month t

   s_t: the number of units in storage in month t

3. **Constraints.** Since the demand d_t at month t must be equal to the storage change $s_{t-1} - s_t$ plus the production x_t at month t, the storage capacities cannot be exceeded, and demands d_t, storage s_t, and production x_t must be nonnegative, we have the following constraints:

$$\begin{array}{rcl} s_{t-1} + x_t - d_t & = & s_t;\ t = 1, 2, \ldots, n \\ s_t & \leq & s^{max};\ t = 1, 2, \ldots, n \\ s_t, x_t & \geq & 0 \end{array} \tag{1.7}$$

4. **Function to be optimized.** One possibility in the production–scheduling problem consists of maximizing benefit after consideration of the costs of production fluctuations and inventories. This implies maximization of

$$Z = \sum_{t=1}^{n} (a_t d_t - b_t x_t - c_t s_t) \tag{1.8}$$

   If the period is short, a_t, b_t, and c_t can be considered as constant, that is, $a_t = a, b_t = b$ and $c_t = c$. One alternative option consists of minimizing the storage cost:

$$Z = \sum_{t=1}^{n} c_t s_t \tag{1.9}$$

Example 1.2 (Production scheduling problem). In this example we illustrate the production scheduling problem with no storage limit. Consider the demand function in Table 1.1 and assume that the initial storage is $s_0 = 2$. Then, system (1.7) becomes

$$\mathbf{Cx} = \begin{pmatrix} -1 & 0 & 0 & 0 & 1 & 0 & 0 & 0 \\ 1 & -1 & 0 & 0 & 0 & 1 & 0 & 0 \\ 0 & 1 & -1 & 0 & 0 & 0 & 1 & 0 \\ 0 & 0 & 1 & -1 & 0 & 0 & 0 & 1 \end{pmatrix} \begin{pmatrix} s_1 \\ s_2 \\ s_3 \\ s_4 \\ x_1 \\ x_2 \\ x_3 \\ x_4 \end{pmatrix} = \begin{pmatrix} 0 \\ 3 \\ 6 \\ 1 \end{pmatrix}$$

$$s_t, x_t \geq 0;\ t = 1, 2, 3, 4 \tag{1.10}$$

where the 0 in the matrix on the right-hand side comes from substracting the demand for $t = 1$ from the initial storage.

Table 1.1: Demand as a function of time in Example 1.2

Time	Demand
1	2
2	3
3	6
4	1

If we maximize the benefit after consideration of the costs of production fluctuations and inventories, as in (1.8), and assume $a_t = 3, b_t = 1, c_t = 1$, our optimization problem becomes maximization of

$$Z = 36 - x_1 - x_2 - x_3 - x_4 - s_1 - s_2 - s_3 - s_4 \tag{1.11}$$

subject to (1.10).

Using the GAMS package to solve this problem (see Section 11.2.2), we obtain a maximum value of

$$Z = 26 \text{ at point } \mathbf{x} = (s_1, s_2, s_3, s_4, x_1, x_2, x_3, x_4) = (0, 0, 0, 0, 0, 3, 6, 1)^T$$

which implies no storage at all. The reader is suggested to decide a new demand function for the storage to be useful.

1.4 The Diet Problem

The diet problem consists of determining the amounts of different nutrients that must be used to minimize the cost and satisfy some nutrition conditions. More precisely, assume that the nutrient contents of a number of different foods, their costs, and the minimum daily requirements for each nutrient are given. The problem consists of determining the amount of each food to be purchased, such that it satisfies the nutritional requirements and leads to the minimum cost.

The four main elements in the diet problem are

1. **Data**

   m: the number of nutrients

   n: the number of different foods

   a_{ij}: the amount of nutrient i in one unit of food j

   b_i: the minimum amount of nutrient i required

   c_j: the cost of one unit of food j

2. **Variables.** The variables involved in this problem are:

   x_j: the amount of food j to be purchased

Table 1.2: Nutrient contents of five different foods: (DN) digestible nutrients, (DP) digestible protein, (Ca) calcium, and (Ph) phosphorus

Nutrient	Required amount	Corn	Oats	Milo maize	Bran	Linseed meal
DN	74.2	78.6	70.1	80.1	67.2	77.0
DP	14.7	6.50	9.40	8.80	13.7	30.4
Ca	0.14	0.02	0.09	0.03	0.14	0.41
Ph	0.55	0.27	0.34	0.30	1.29	0.86

3. **Constraints.** Since the total amount of a given nutrient i is the sum of the amounts of that nutrient in all foods and the amounts of food must be nonnegative, we have the following constraints:

$$\begin{aligned} \sum_{j=1}^{n} a_{ij}x_j &\geq b_i; \;\; i = 1, \ldots, m \\ x_j &\geq 0; \;\; j = 1, \ldots, n \end{aligned} \tag{1.12}$$

4. **Function to be minimized.** In the diet problem we are interested in minimizing the food cost:

$$Z = \sum_{j=1}^{n} c_j x_j \tag{1.13}$$

where c_j is the unit cost of food j.

■

Example 1.3 (Diet problem). Consider the five nutrients and the minimum requirements of digestible nutrients (DN), digestible protein (DP), calcium (Ca), and phosphorus (Ph) shown in Table 1.2.

Constraints (1.12) become

$$\begin{aligned} \begin{pmatrix} 78.6 & 70.1 & 80.1 & 67.2 & 77.0 \\ 6.50 & 9.40 & 8.80 & 13.7 & 30.4 \\ 0.02 & 0.09 & 0.03 & 0.14 & 0.41 \\ 0.27 & 0.34 & 0.30 & 1.29 & 0.86 \end{pmatrix} \begin{pmatrix} x_1 \\ x_2 \\ x_3 \\ x_4 \\ x_5 \end{pmatrix} &\geq \begin{pmatrix} 74.2 \\ 14.7 \\ 0.14 \\ 0.55 \end{pmatrix} \\ x_1, x_2, x_3, x_4, x_5 &\geq 0 \end{aligned} \tag{1.14}$$

Consider the following unit cost coefficients for the diet problem:

$$c_1 = 1, \;\; c_2 = 0.5, \;\; c_3 = 2, \;\; c_4 = 1.2, c_5 = 3.$$

Thus, we have the following LPP; minimize

$$Z = x_1 + 0.5x_2 + 2x_3 + 1.2x_4 + 3x_5 \tag{1.15}$$

subject to (1.14).

Using some of the existing computer programs to solve the above LPP, as GAMS, we obtain a minimum cost of (see program at end of Section 11.6) $Z = 0.793$, at the point $(0, 1.530, 0, 0.023, 0)$. This means that we must buy only oats and bran.

■

1.5 The Network Flow Problem

Assume a transportation network (pipeline system, railroad, highway, communication, etc.) through which we wish to send a homogeneous commodity (oil, grain, cars, messages, etc.) from particular points of the network, called *source nodes*, to particular destination points, called *sink nodes*. In addition to source or sink nodes, the network can contain intermediate nodes, where no input or output of commodity occurs. Let us denote the flow going from node i to node j by x_{ij} (positive in the direction $i \to j$, and negative otherwise).

The four main elements in the network flow problems are

1. **Data**

   $\mathcal{G}$: the graph $\mathcal{G} = (\mathcal{N}, \mathcal{A})$ defining the transportation network, where $\mathcal{N}$ is the set of nodes, and $\mathcal{A}$ is the set of links

   n: the number of nodes in the network

   f_i: the input (positive) or output (negative) flow at node i

   m_{ij}: the maximum flow capacity of link going from node i to node j

   c_{ij}: the cost of sending one unit from node i to node j

2. **Variables.** The variables involved in this problem are

   x_{ij}: the flow going from node i to node j

3. **Constraints.** By imposing the conservation of flow condition at all nodes, and the capacity constraints at all links, the following constraints are obtained. Conservation constraints are

$$\sum_j (x_{ij} - x_{ji}) = f_i; \quad i = 1, \ldots, n \tag{1.16}$$

   and capacity constraints are

$$-m_{ij} \le x_{ij} \le m_{ij}; \quad \forall i < j \tag{1.17}$$

   where $i < j$ avoids duplication.

4. **Function to be minimized.** The total cost is

$$Z = \sum_{ij} c_{ij} x_{ij} \tag{1.18}$$

   Thus, we minimize (1.18) subject to (1.16) and (1.17).

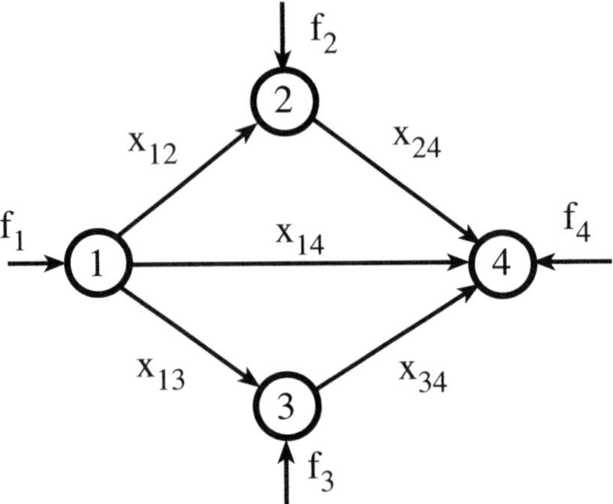

Figure 1.3: Illustration of a network flow problem.

Problems of flow networks are very common in engineering. In fact, the water supply system, the network communications systems, and others lead to network flow problems as those described above. In addition to solving the associated optimization problems one can be interested in analyzing the solution set, and how this set changes when some elements in the system fail. The following example discusses one of such problems.

Example 1.4 (A network flow problem). Consider the network flow problem in Figure 1.3, where the arrows refer to the positive values of the flow variables.

Equations (1.16) and (1.17) become

$$\begin{pmatrix} 1 & 1 & 1 & 0 & 0 \\ -1 & 0 & 0 & 1 & 0 \\ 0 & -1 & 0 & 0 & 1 \\ 0 & 0 & -1 & -1 & -1 \end{pmatrix} \begin{pmatrix} x_{12} \\ x_{13} \\ x_{14} \\ x_{24} \\ x_{34} \end{pmatrix} = \begin{pmatrix} f_1 \\ f_2 \\ f_3 \\ f_4 \end{pmatrix}$$

$$\begin{aligned} x_{ij} &\leq m_{ij}; \ \forall i < j \\ -x_{ij} &\leq m_{ij}; \ \forall i < j \end{aligned} \tag{1.19}$$

where we assume that $m_{ij} = 4, \forall i < j$, and $(f_1, f_2, f_3, f_4) = (7, -4, -1, -2)$.

Assume also that $c_{ij} = 1; \forall i, j$. The optimization problem is to minimize

$$Z = x_{12} + x_{13} + x_{14} + x_{24} + x_{34}$$

subject to (1.19).

Using a standard LPP, the following solution can be obtained:

$$z = 5 \text{ at the point } \begin{pmatrix} x_{12} \\ x_{13} \\ x_{14} \\ x_{24} \\ x_{34} \end{pmatrix} = \lambda \begin{pmatrix} 0 \\ 3 \\ 4 \\ -4 \\ 2 \end{pmatrix} + (1-\lambda) \begin{pmatrix} 4 \\ -1 \\ 4 \\ 0 \\ -2 \end{pmatrix}; \quad 0 \le \lambda \le 1$$

This means that an infinite set of solutions can be obtained, all giving the same objective function value, $Z = 5$. In Chapter 5, Section A.6.5, we shall explain how all these optima can be obtained.

■

1.6 The Portfolio Problem

An investor owns shares of several different stocks. More precisely, it owns b_i shares of stock A_i, $i = 1, 2, ..., m$. The current values of these stocks are v_i. Assume that we can predict the dividends to be paid at the end of the year that starts now, and the values of the different stocks: A_i will pay d_i and will have a new price of w_i.

Our aim is to adjust our portfolio, namely, the number of shares of each stock, so as to maximize the dividends. Our unknowns are x_i, the change in the number of shares of stock A_i with respect to the number of shares we already have. Since we originally had b_i, after the adjustment we will own $b_i + x_i$, respectively.

The main elements in the portfolio problem are

1. **Data**

   m: the number of stocks

   b_i: the current number of shares of stock i

   v_i: the current value of stock i

   d_i: the dividend to be paid at the end of the year for stock i

   w_i: the new value of stock i

   r: minimum fraction r of the current value of the whole portfolio that cannot be underpassed after the adjustment

   s: minimum fraction of the total current value that cannot be underpassed by the total future value of our holdings, to keep up with inflation

2. **Variables**

   x_i: the change in the number of shares of stock i

3. **Constraints.** Even though we have not indicated so in the problem statement, we must care about certain conditions that a well-equilibrated portfolio should verify, as the following set of constraints. The holding of each stock must be nonnegative:

$$x_i \geq -b_i \tag{1.20}$$

The portfolio must be balanced in the sense that we would tend to avoid excessive reliance on a single stock; this condition can be enforced by requiring that the value of every stock after the adjustment represent at least a certain fraction r of the current value of the whole portfolio. This means:

$$r(\sum_i v_i(b_i + x_i)) \leq v_j(b_j + x_j); \ \ \forall j \tag{1.21}$$

The current value of our portfolio must not change in the adjustment since we assume there is no additional money for investment:

$$\sum_i v_i x_i = 0 \tag{1.22}$$

To keep up with inflation, the total future value of our holdings must be at least a certain fraction s greater that the total current value:

$$\sum_i w_i(b_i + x_i) \geq (1 + s)\sum_i v_i b_i \tag{1.23}$$

4. **Function to be optimized.** Our objective is to maximize dividends:

$$Z = \sum_i d_i(b_i + x_i) \tag{1.24}$$

Our task reduces to finding the maximum value of dividends subject to all the above constraints.

Example 1.5 (The portfolio problem). Consider the particular case in which we have shares of three stocks, 75 of A, 100 of B, and 35 of C, with values \$20, \$20 and \$100, respectively. In addition we have the information: A will pay no dividends with a new value of \$18, B will pay \$3 per share and the new value will be \$23, and C will pay \$5 per share with a new value of \$102. If we take the percentage r to be 0.25 and s, 0.03, all the constraints above become

$$\begin{array}{rcl} x_A & \geq & -75 \\ x_B & \geq & -100 \\ x_C & \geq & -35 \\ 0.25\,[20(75 + x_A) + 20(100 + x_B) + 100(35 + x_C)] & \leq & 20(75 + x_A) \\ 0.25\,[20(75 + x_A) + 20(100 + x_B) + 100(35 + x_C)] & \leq & 20(100 + x_B) \\ 0.25\,[20(75 + x_A) + 20(100 + x_B) + 100(35 + x_C)] & \leq & 100(35 + x_C) \\ 20x_A + 20x_B + 100x_C & = & 0 \\ 18(75 + x_A) + 23(100 + x_B) + 102(35 + x_C) & \geq & 1.03(7000) \end{array} \tag{1.25}$$

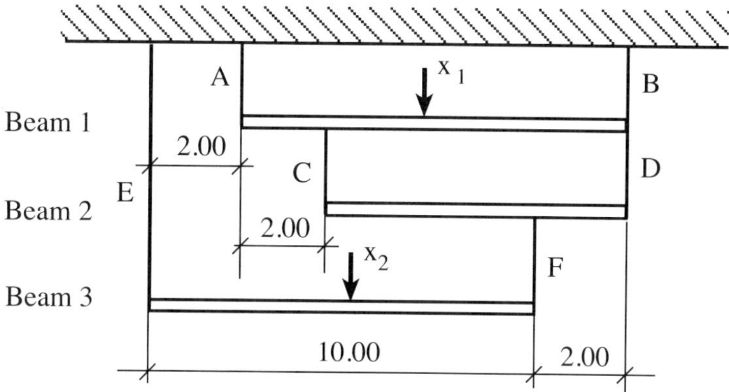

Figure 1.4: System with ropes and beams.

After several simplifications, they can be written as

$$
\begin{array}{rcl}
x_A & \geq & -75 \\
x_B & \geq & -100 \\
x_C & \geq & -35 \\
15x_A - 5x_B - 25x_C & \geq & 250 \\
-5x_A + 15x_B - 25x_C & \geq & -250 \\
-5x_A - 5x_B + 75x_C & \geq & -1750 \\
20x_A + 20x_B + 100x_C & = & 0 \\
18x_A + 23x_B + 102x_C & \geq & 270
\end{array}
\quad (1.26)
$$

The GAMS solution of this problem is

$$Z = \$612.5 \text{ and } x_A = 12.5, \quad x_B = 75, \quad x_C = -17.5$$

■

1.7 Scaffolding System

A scaffolding system consists of several ropes and beams connected in a definite way. Several external loads act downward in the mid point of some selected beams. The problem consists in determining the permissible total load that may withstand the system without collapsing, under equilibrium of forces and moments if we assume that the weight of ropes and beams is negligible.

The main elements of this situation are

1. **Data**

   I: set of loads

   S: set of ropes

   B: set of beams

T_s: maximum admissible load on rope $s \in S$

Ω_b: set of loads applied in the mid point of beam b Notice that $\Omega_b \subset I$ and it consists of a single load or else it is empty

Ψ_b: set of ropes supporting beam b; Ψ_b is a subset of S and it typically consists of two elements

Θ_b: set of ropes supported in beam b

dl_i: distance of load i to the left endpoint of the beam where it acts

dr_s: distance of supporting rope s to the left endpoint of the beam b it supports. $s \in \Psi_b$

2. **Variables.** The variables involved in this problem are the following:

   x_i: the load i

   t_s: tension generated on rope s under the action of the set of loads x_i, $i \in I$

3. **Constraints.** The condition of equilibrium of forces on each beam leads to the set of equations

$$\sum_{s\in\Psi_b} t_s = \sum_{i\in\Omega_b} x_i + \sum_{x\in\Theta_b} t_s$$

   for each $b \in B$, and the condition of equilibrium of moments (taken with respect to left endpoints of each beam) yields

$$\sum_{s\in\Psi_b} dr_s t_s = \sum_{i\in\Omega_b} dl_i x_i + \sum_{x\in\Theta_b} dr_s t_s$$

   for each beam $b \in B$. We must also respect the maximum permissible tension on each rope

$$0 \leq t_s \leq T_s$$

   for each $s \in S$, and the nonnegativity of each load i

$$x_i \geq 0$$

4. **Function to be optimized.** The problem consists in maximizing the total load:

$$\sum_i x_i$$

Example 1.6 (Scaffolding system). Consider, as a specific example, the scaffolding system of Figure 1.4, where loads x_1 and x_2 are applied in the middle point of beams 2 and 3, respectively. Ropes A and B may stand a maximum load of 300; C and D, 200; and E and F, 100.

The condition of equilibrium of forces on each beam leads to the equations

$$\begin{aligned} t_E + t_F &= x_2 \\ t_C + t_D &= t_F \\ t_A + t_B &= x_1 + t_C + t_D \end{aligned} \tag{1.27}$$

and the moment (taken with respect to E, C, and A, respectively) equilibrium conditions give

$$\begin{aligned} 10t_F &= 5x_2 \\ 8t_D &= 6t_F \\ 10t_B &= 5x_1 + 2t_C + 10t_D \end{aligned} \tag{1.28}$$

Solving successively these equations for the tensions, we arrive at

$$\begin{aligned} t_F &= \frac{x_2}{2} \\ t_E &= \frac{x_2}{2} \\ t_D &= \frac{3x_2}{8} \\ t_C &= \frac{x_2}{8} \\ t_B &= \frac{2x_2}{5} + \frac{x_1}{2} \\ t_A &= \frac{x_2}{10} + \frac{x_1}{2} \end{aligned} \tag{1.29}$$

Finally, considering that loads act downward ($x_1, x_2 \geq 0$), the constraints become:

$$\begin{aligned} \frac{x_2}{2} &\leq 100 \\ \frac{3x_2}{8} &\leq 200 \\ \frac{x_2}{8} &\leq 200, \\ \frac{x_1}{2} + \frac{2x_2}{5} &\leq 300 \\ \frac{x_1}{2} + \frac{x_2}{10} &\leq 300 \\ x_1 &\geq 0 \\ x_2 &\geq 0 \end{aligned} \tag{1.30}$$

In fact, among all these constraints there are some which are more restrictive than others. Thus, they can be reduced to the set of constraints:

$$\begin{aligned} 0 \leq x_2 &\leq 200 \\ 4x_2 + 5x_1 &\leq 3000 \\ x_1 &\geq 0 \end{aligned} \tag{1.31}$$

The problem consists in maximizing the total load:

$$x_1 + x_2$$

The GAMS solution, which will be obtained in Section 11.2.6, is

$$Z = 640 \text{ attained at the point } x_1 = 440; \;\; x_2 = 200$$

The corresponding rope tensions are

$$t_A = 240; \;\; t_B = 300; \;\; t_C = 25; \;\; t_D = 75; \;\; t_E = 100; \;\; t_F = 100$$

■

1.8 Electric Power Economic Dispatch

Power generators as well as power demands are located in different busses on an electric power network. The objective of the electric power economic dispatch problem is to calculate, for a single period of time, the output power of every generator so that all demands are satisfied at minimum cost, while satisfying different technical constraints of the network and the generators.

Each transmission line of a power network transmits power from its sending bus to its receiving bus. The amount of power transmitted is proportional to the difference of the angles of these buses (similarly, the flow of water through a pipeline connecting two water tanks is proportional to the difference of height of the water surface in the two tanks). The constant of proportionality has a funny name, electrical engineers call it *susceptance*. The power transmitted from bus i to bus j through line $i-j$ is therefore

$$B_{ij}(\delta_i - \delta_j) \tag{1.32}$$

where B_{ij} is the susceptance of line $i-j$, and δ_i and δ_j the angles of buses i and j, respectively.[1]

For physical reasons, the amount of power transmitted through a power line has a limit. This limit is related either to thermal or stability considerations. Therefore, a power line should be operated so that its transmission limit is not violated. This can be formulated as

$$-P_{ij}^{max} \leq B_{ij}(\delta_i - \delta_j) \leq P_{ij}^{max} \tag{1.33}$$

where P_{ij}^{max} is the maximum transmission capacity of line $i-j$.

It should be noted that power transmitted is proportional to a difference of angles and not to a given angle. Therefore, the value of an arbitrary angle can be fixed to 0 and taken as the origin:

$$\delta_k = 0 \tag{1.34}$$

where k is an arbitrary bus.

[1]More precise models are available, but the one presented here is good enough for many technical and economic studies.

One consequence of the possibility of arbitrarily selecting the origin is that angles are variables not limited in sign.

The power produced by a generator is a positive magnitude limited below and above. The lower limit is due to stability conditions (similarly, a car cannot move at a speed below a limit). The upper bound is due to thermal limits (similarly, a car cannot move at a speed above a given maximum). The preceding constraints can be expressed as

$$P_i^{min} \leq p_i \leq P_i^{max} \tag{1.35}$$

where p_i is the actual power produced by generator i and P_i^{min} and P_i^{max} are positive constants representing respectively the minimum and maximum output powers of generator i.

In every bus, the power reaching the bus should be equal to the power leaving the bus (energy conservation law), which can be written as

$$\sum_{j \in \Omega_i} B_{ij}(\delta_i - \delta_j) + p_i = D_i, \quad \forall i \tag{1.36}$$

where Ω_i is the set of buses connected through lines to bus i and D_i the demand at bus i.

As previously stated, the power transmitted through every line is limited, so

$$-P_{ij}^{max} \leq B_{ij}(\delta_i - \delta_j) \leq P_{ij}^{max}, \quad \forall j \in \Omega_i, \ \forall i \tag{1.37}$$

The main elements of this problem are

1. **Data**

   n: the number of generators

   P_i^{min}: the minimum output power of generator i

   P_i^{max}: the maximum output power of generator i

   B_{ij}: the susceptance of line $i - j$

   P_{ij}^{max}: the maximum transmission capacity of line $i - j$

   C_i: the cost of producing power from generator i

   Ω_i: the set of buses connected through lines to bus i

   D_i: the demand at bus i

2. **Variables**

   p_i: the actual power produced by generator i

   δ_i: the angle of bus i

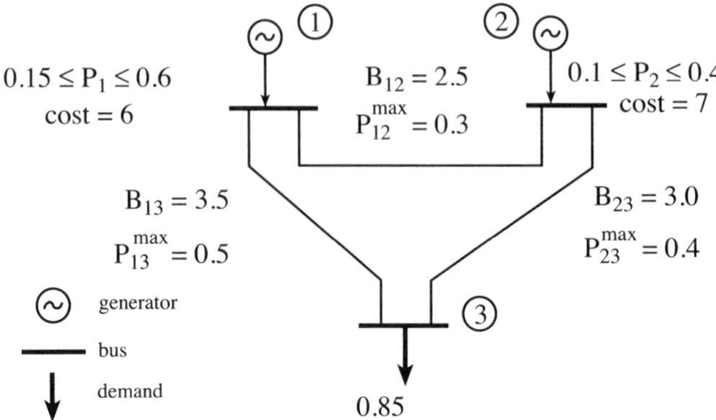

Figure 1.5: Illustration of the economic dispatch problem.

3. **Constraints.** The constraints in this problem are

$$
\begin{aligned}
\delta_k &= 0 \\
\sum_{j\in\Omega_i} B_{ij}(\delta_i - \delta_j) + p_i &= D_i \quad i = 1,2,\ldots,n \\
-P_{ij}^{max} \le B_{ij}(\delta_i - \delta_j) &\le P_{ij}^{max};\ \forall j \in \Omega_i,\ i = 1,2,\ldots,n \\
P_i^{min} \le p_i &\le P_i^{max};\ i = 1,2,\ldots,n
\end{aligned}
\tag{1.38}
$$

4. **Function to be minimized.** The objective of the economic dispatch problem is to minimize total production costs, which can be stated as

$$
z = \sum_{i=1}^{n} c_i\, p_i \tag{1.39}
$$

where c_i is the cost of producing power from generator i, and n the number of generators.

Example 1.7 (Economic dispatch). Consider a 3 bus 3 line system (see Figure 1.5). The generator of bus 1 produces at cost 6 and its lower and upper limits are respectively 0.15 and 0.6. The production cost of the generator of bus 2 is 7 and its power limits are respectively 0.1 and 0.4. Line 1–2 has a susceptance 2.5 and a maximum transmission limit of 0.3; line 1–3 a susceptance of 3.5 and a transmission limit of 0.5; and finally line 2–3 a susceptance of 3.0 and a transmission limit of 0.4. This system has a single demand located in bus 3 with a value of 0.85. A one hour time period is considered. The origin is taken in bus 3.

The economic dispatch problem for this example has the following form. Minimize

$$
6p_1 + 7p_2 \tag{1.40}
$$

subject to

$$\begin{array}{rcl} \delta_3 & = & 0 \\ 3.5(\delta_3 - \delta_1) + 2.5(\delta_2 - \delta_1) + p_1 & = & 0 \\ 3.0(\delta_3 - \delta_2) + 2.5(\delta_1 - \delta_2) + p_2 & = & 0 \\ 3.5(\delta_1 - \delta_3) + 3.0(\delta_2 - \delta_3) & = & 0.85 \\ 0.15 \ \le \ p_1 & \le & 0.6 \\ 0.10 \ \le \ p_2 & \le & 0.4 \\ -0.3 \ \le \ 2.5(\delta_1 - \delta_2) & \le & 0.3 \\ -0.4 \ \le \ 3.0(\delta_2 - \delta_3) & \le & 0.4 \\ -0.5 \ \le \ 3.5(\delta_1 - \delta_3) & \le & 0.5 \end{array} \tag{1.41}$$

The optimization variables are p_1, p_2, δ_1 and δ_2.

The optimal solution for the economic dispatch problem, that will be solved in Section 11.2.7, is:

$$Z = 5.385, \qquad \mathbf{p} = (0.565, 0.285)^T, \qquad \boldsymbol{\delta} = (-0.143, -0.117, 0)^T$$

This requires generator 1 to produce 0.565 and generator 2 to produce 0.285. ■

Exercises

1.1 Peter Pérez builds electrical cable of high quality using two types of metallic alloys, A and B. Alloy A contains 80% of copper and 20% of aluminum. Alloy B contains 68% of copper and 32% of aluminum. The alloy A market price is 80 Euros per ton, and the alloy B market price, 60 Euros per ton. Which are the quantities of alloys A and B, respectively that Peter has to use to produce a ton of cable that contains at least 20% of aluminum and whose production cost is the cheapest possible?

1.2 Three employees are supposed to work on six different tasks. The ith employee can do a_{ij} parts of the task j in one hour and is paid c_{ij} per hour in the jth task. The total number of work hours for the ith employee is b_{1i}, and the number of unities required to finish task j is b_{2j}. The minimum cost

$$C = \sum_{i=1}^{3} \sum_{j=1}^{6} c_{ij} x_{ij}$$

is to be found where x_{ij} represents the number of hours spent on task j by employee i. Write down this problem as a linear programming problem.

1.3 A furniture manufacturing company wishes to determine how many tables, chairs, desks, and bookcases it should make to optimize the use of its available resources. These products utilize two different types of lumber, and the company has on hand 1500 board-feet of the first type and 1000 board-feet of the second type. There are 800 labor-hours available for the job. Sales forecasts plus back orders require the company to make at least

Table 1.3: Distribution problem

Plants	Market_1	Market_2	Market_3
Plant_1	1	3	5
Plant_2	2	5	4

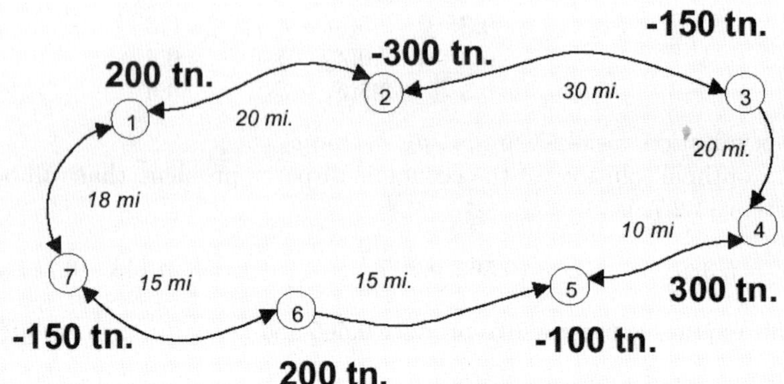

Figure 1.6: Illustration of the road and transportation data.

40 tables, 130 chairs, 30 desks, and no more than 10 bookcases. Each table, chair, desk, and bookcase requires $5, 1, 9,$ and 12 board-feet, respectively, of the first type of lumber and $2, 3, 4,$ and 1 board-feet of the second type. A table requires 3 labor-hours to make; a chair, 2 hours; a desk, 5 hours; and a bookcase, 10 hours. The company makes a total of \$12 profit on a table, \$5 on a chair, \$15 on a desk, and \$10 on a bookcase. Write out the linear programming model of this problem in terms of maximizing profit. Modify the problem to impose that four chairs be made for every table.

1.4 A firm producing a certain good P has two plants. Each plant produces 90 tons of P monthly, and the good is distributed in three different markets. Table 1.3 shows the unit costs of shipping 1 ton of good P from a given plant to a given market. The firm wants to send the same number of tons to each market and minimize the total cost. Formulate the corresponding linear programming problem.

1.5 The road of Figure 1.6 is being constructed. There are earthworks between points on the road. Figure 1.6 shows the amount of tones per each node that must be transported. If this amount is positive, the node is a source; otherwise it is a sink. The objective is to minimize the total amount of tones that must be carried away. Formulate the corresponding linear programming problem.

1.6 An electricity producer should plan its hourly energy production so that it maximizes its profits from selling energy during a planning horizon re-

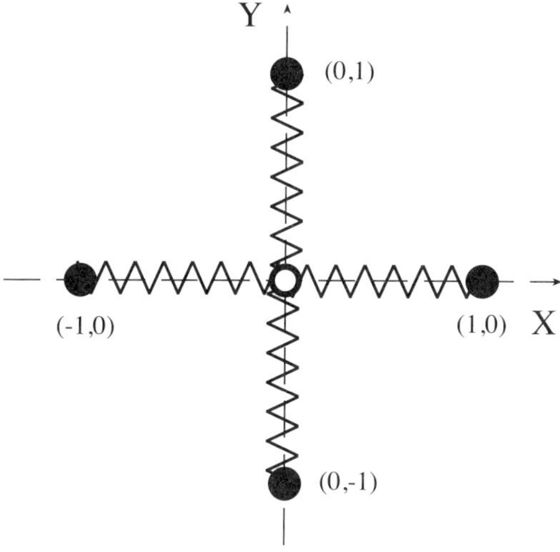

Figure 1.7: System of springs.

quiring a given number of hours. The producer does not produce before the planning horizon. Hourly energy prices can be forecasted and are considered known. The minimum energy production of the producer in each hour is zero and the maximum one a given quantity. Energy productions in two consecutive hours cannot differ in more that a pre-specified amount. The producer production cost is linear. Formulate a linear programming problem to maximize the profits of the electricity producer.

1.7 In the system of springs in Figure 1.7, black dots are fixed joints while white dots correspond to free points. Springs can rotate freely around each joint. Each spring is determined by a positive constant k_i in the appropriate units. The equilibrium position for the central free joint is found by solving the system

$$\sum_{i=1}^{4} k_i(x - x_i) = 0$$

if x_i is the position vector for joint i and x is the position of the free joint. We would like to find the best combination of springs that minimize the work done on the free joint by a known constant force F, if the constants for the four springs are restricted by the linear constraint

$$\sum_i k_i = c$$

where c is a positive, fixed constant. Formulate the problem as a LPP.

Chapter 2

Mixed-Integer Linear Programming

2.1 Introduction

In Chapter 1 we dealt with linear programming problems where the variables involved were real numbers. However, in many cases of real life, some variables are not real but integers, or they are even more restricted, as binary variables, that take values 0 or 1 only. We shall see in Chapter 7 that using integer variables adds more difficulties to the linear programming problem, because of the lack of continuity.

In this chapter we give some real examples of integer linear programming problems (ILPP), in some of which we use binary variables.

2.2 The 0–1 Knapsack Problem

An important class of integer programming problems are those where the variables of the problem can take only two values. This situation can be formulated using the *0–1 variables*. Each value is associated with one of the possibilities of a binary choice:

$$x = \begin{cases} 1 & \text{if the event occurs} \\ 0 & \text{otherwise} \end{cases}$$

A classical problem involving this type of variable is the *0–1 knapsack problem*. Consider a person who must pack a hike. Assume that there exist a set of items that have a utility for this person, and there exist a limitation on the number of items the person can carry. The problem consists of choosing a subset of items to maximize the sum of the utilities while not exceeding the carrying capacity of the hiker.

The problem has the following elements:

1. **Data**

   n: the number of objects
   a_j: the weight of the object j
   c_j: the utility of object j
   b: the capacity of the knapsack (hiker)

2. **Variables**

$$x_j = \begin{cases} 1 & \text{if the object } j \text{ is put in the knapsack} \\ 0 & \text{otherwise} \end{cases} \tag{2.1}$$

3. **Constraints.** The capacity is not exceeded:

$$\sum_{j=1}^{n} a_j x_j \leq b$$

4. **Function to be maximized.** The objective of this problem is to maximize the utility, which can be stated as

$$Z = \sum_{j=1}^{n} c_j x_j$$

Example 2.1 (The ship owner). A ship owner has a freighter with a capacity of 700 tons. The firm transports containers of different weights for a specific route. On the current trip the ship owner could ship some of the following containers:

Container	c_1	c_2	c_3	c_4	c_5	c_6	c_7	c_8	c_9	c_{10}
Weight	100	155	50	112	70	80	60	118	110	55

The decisionmaker's firm would determine the freight such that it maximizes the transported load.

This problem could be formulated as a 0–1 knapsack problem. The variables are:

$$x_j = \begin{cases} 1 & \text{if container } j \text{ is shipped} \\ 0 & \text{otherwise} \end{cases}$$

The objective is to maximize the freight that will be transported by the freighter:

$$\begin{aligned} Z &= 100x_1 + 155x_2 + 50x_3 + 112x_4 + 70x_5 \\ &\quad +80x_6 + 60x_7 + 118x_8 + 110x_9 + 55x_{10} \end{aligned}$$

and the constraint is that the freight cannot exceed the capacity of the ship:

$$\begin{aligned} &100x_1 + 155x_2 + 50x_3 + 112x_4 + 70x_5 + 80x_6 \\ &\qquad +60x_7 + 118x_8 + 110x_9 + 55x_{10} \leq 700 \end{aligned}$$

Note that here $a_i = c_i; \forall i$, because the utility coincides with the weight.

The optimal freight consists of using the containers: $c_1, c_3, c_4, c_5, c_6, c_7, c_8, c_9$. The optimal value is 700, which means that the ship is full.

■

2.3 Identifying Relevant Symptoms

Let $\mathcal{D} = \{D_1, D_2, \dots, D_n\}$ be a given set of possible diseases, and assume that physicians, when identifying the diseases associated with a set of patients, usually base their decisions on a set of symptoms $\mathcal{S} = \{S_1, S_2, \dots, S_m\}$. Assume that we want to identify a minimal subset of symptoms $\mathcal{S}_a \subset \mathcal{S}$, such that all diseases can be perfectly distinguished from each other according to the levels of symptoms in $\mathcal{S}_a$. Finding the minimum set of symptoms is important because it implies minimizing the cost of the diagnosis process.

The problem has the following elements:

1. **Data**

   $\mathcal{D}$: the set of diseases

   $\mathcal{S}$: the set of symptoms

   n: the number of diseases (cardinal of $\mathcal{D}$)

   m: the number of symptoms (cardinal of $\mathcal{S}$)

   c_{ij}: the level of symptom j associated with disease i

   d_{ikj}: discrepancy between diseases i and k due to symptom j

   a: the minimum required discrepancy level (to be explained below)

2. **Variables**

$$x_j = \begin{cases} 1 & \text{if the symptom } j \text{ belongs to } \mathcal{S}_a \\ 0 & \text{otherwise} \end{cases} \tag{2.2}$$

3. **Constraints.** The subset $\mathcal{S}_a$ must be sufficient for a clear distinction of all diseases:

$$\sum_{j=1}^{m} x_j d_{ikj} \geq a; \;\; \forall i, k \in \{1, 2, \dots, n\}, \; i \neq k \tag{2.3}$$

   where

$$d_{ikj} = \begin{cases} 1 & \text{if} \;\; c_{ij} \neq c_{kj} \\ 0 & \text{if} \;\; c_{ij} = c_{kj} \end{cases} \tag{2.4}$$

   measures the discrepancy between diseases D_i and D_k in terms of the symptoms in $\mathcal{S}_a$, and $a > 0$ is the discrepancy level we desire. Note that the larger the value of a, the larger the number of required symptoms (cardinal of $\mathcal{S}_a$). In this case

$$\sum_{j=1}^{m} x_j d_{ikj}$$

coincides with the number of symptoms in $\mathcal{S}_0$ that take different levels for diseases D_i and D_k, and a is the corresponding minimum number, for any pair (D_i, D_k) of diseases, that are required to have an acceptable subset $\mathcal{S}_a$. This means that $a - 1$ symptoms can be missing and we still can differentiate any pair of diseases (D_i, D_k).

4. **Function to be minimized.** The objective of this problem is to minimize the number of selected symptoms, the cardinal of the set $\mathcal{S}_0$:

$$Z = \sum_{j=1}^{m} x_j.$$

The problem as stated above allows us to determine a minimal subset $\mathcal{S}_0$, associated with $a = 0$, of symptoms of the set $\mathcal{S}$ which allows identification of the diseases in the set $\mathcal{D}$. However, if the diseases are to be identified with some missing information, the set $\mathcal{S}_0$ can become useless. So, we normally use $a > 0$.

Once we have selected the relevant symptoms to identify all diseases, we can determine the relevant symptoms associated with disease i. This can be done by minimizing

$$Z = \sum_{j=1}^{m} x_j$$

subject to

$$\sum_{j=1}^{m} x_j d_{ikj} > a; \quad k \in \{1, 2, \ldots, n\}, \ i \neq k \tag{2.5}$$

In other words, we find the minimal subset of $\mathcal{S}_{ai} \subseteq \mathcal{S}$ such that disease i has different symptoms when compared with all other diseases. This subset is called the *set of relevant symptoms* for disease i.

Example 2.2 (Identifying relevant symptoms). Assume that we have the set of diseases $\mathcal{D} = \{D_1, D_2, D_3, D_4, D_5\}$ and the set of symptoms $\mathcal{S} = \{S_1, S_2, \ldots, S_8\}$. Assume also that the symptoms associated with the different diseases are those listed in Table 2.1.

Then, minimizing the sum $Z = \sum_{j=1}^{m} x_j$ subject to (2.3), and two values of a, we conclude that the set of symptoms $\{2, 5\}$ is a minimal sufficient set of symptoms able to distinguish the 5 diseases. However, if we use a discrepancy level of $a = 3$, the required set is $\{1, 2, 4, 5, 7\}$. Note that in this case we can have two missing symptoms and the diagnostic would be still correct.

Table 2.1: Symptoms associated with all diseases in Example 2.2

	Symptoms							
Disease	S_1	S_2	S_3	S_4	S_5	S_6	S_7	S_8
D_1	2	3	1	1	1	2	1	2
D_2	1	1	1	1	3	1	2	1
D_3	3	4	2	3	2	2	3	2
D_4	2	2	2	2	2	1	2	3
D_5	1	1	1	2	1	1	1	2

Table 2.2: Relevant symptoms for all diseases in Example 2.2 for $a = 1$

Disease	Relevant symptoms
D_1	$\{2\}$
D_2	$\{5\}$
D_3	$\{2\}$
D_4	$\{2\}$
D_5	$\{2, 5\}$

Finally, Table 2.2 shows the required set of relevant symptoms for each disease and $a = 1$. Note in Table 2.1 that symtom 2 is sufficient to identify diseases D_1, D_3, and D_4, and that symptom 5 is sufficient to identify disease D_2. However, we need symptoms 2 and 5 to identify disease D_5.

■

2.4 The Academy Problem

The Academy of Engineering has m members and is involved in the process of selecting r new members among a set of J candidates. To this end, each actual member is allowed to support from a minimum of 0 to a maximum of r candidates. The r candidates with the largest number of supports are incorporated to the academy.

Before the final selection process, a previous test is performed to know the degree of support of each candidate. In this process each actual member can assign the scores in the list $\mathbf{p}$ to a maximum of S candidates, but need not to assign all scores.

Only the sum of scores of each candidate is known. The problem consists of knowing the minimum and maximum number of final supports of each candidate based on the results of the test, assuming that assigning a score to one candidate is equivalent to supporting such candidate by the actual member assigning the score.

The problem has the following elements:

1. **Data**

   I: the actual number of members in the Academy of Engineering

   J: the number of candidates

   S: the number of different scores that can be assigned

   p_s: the s-th score

   C_j: the total score associated with candidate j

2. **Variables**

   x_{ijs}: a binary variable that takes value 1 if member i assigns score p_s to candidate j; otherwise, it takes value 0

3. **Constraints**

   - Each member can assign at the most one score to each candidate:

   $$\sum_{s=1}^{S} x_{ijs} \leq 1; \ \forall i \in \{1, 2, \ldots, I\}, \ j \in \{1, 2, \ldots, J\}$$

   - Each member can assign score p_s to at most one candidate:

   $$\sum_{j=1}^{J} x_{ijs} \leq 1; \ \forall i \in \{1, 2, \ldots, I\}, \ s \in \{1, 2, \ldots, S\}$$

   - The total score obtained by each candidate must be the given value:

   $$\sum_{i=1}^{I} \sum_{s=1}^{S} p_s x_{ijs} = C_j; \ \forall j \in \{1, 2, \ldots, J\}$$

4. **Function to be optimized.** The objective of this problem consists of minimizing and maximizing this function for each candidate:

$$Z_j = \sum_{i=1}^{I} \sum_{s=1}^{S} x_{ijs}, \ \ j \in \{1, 2, \ldots, J\} \tag{2.6}$$

Example 2.3 (Academy problem). Assume that the Academy of Engineering has 20 members and that $r = 4$ new members are to be selected among $J = 8$ candidates, and that $\mathbf{p} \equiv \{10, 8, 3, 1\}$, which implies $S = 4$.

The available information consists of the last row in Table 2.3, namely, the total scores received by each candidate in the first round, and we look for the number of supports of each candidate (see the second last row in Table 2.4).

Table 2.3: Total scores received by the 8 candidates in Example 2.3

Candidate	1	2	3	4	5	6	7	8
Received score	71	14	139	13	137	18	24	8

Table 2.4: Scores received by the 8 candidates in Example 2.3

	Candidate							
Actual member	1	2	3	4	5	6	7	8
1	3	–	10	–	8	1	–	–
2	1	–	10	–	8	3	–	–
3	–	1	–	3	10	–	8	–
4	–	3	10	–	8	1	–	–
5	3	–	8	–	10	–	1	–
6	1	–	10	–	8	–	3	–
7	10	–	8	–	3	1	–	–
8	3	–	10	1	8	–	–	–
9	8	–	3	–	10	1	–	–
10	–	3	10	–	1	–	8	–
11	8	–	1	–	10	–	3	–
12	–	–	–	–	10	–	–	–
13	–	–	10	–	8	–	–	–
14	10	–	–	1	3	–	–	8
15	3	–	10	–	8	–	1	–
16	10	–	1	–	8	–	3	–
17	1	3	10	8	–	–	–	–
18	1	3	8	–	10	–	–	–
19	1	–	10	–	3	8	–	–
20	8	1	10	–	3	–	–	–
Number of supports	15	6	17	4	19	7	6	1
Total score	71	14	139	13	137	18	24	8

The actual scores received by each candidate from each actual member are those given in Table 2.4 (note that this information is not available, but has been given only for illustration).

If we minimize and maximize (2.6) for all candidates, we get the results shown in Table 2.5. The following conclusions can be drawn from this table:

1. Only candidates 3 and 5 have at least 15 guaranteed supports. Note that the next one, candidate 1, has only 8 guaranteed supports.

2. It is not clear from Table 2.5 that candidates 3 and 5 enter the academy, since candidates 6, 1 and 7 have a maximum of 18, 20, and 20 guaranteed supports, and they can get only 15, 16, or 17.

Table 2.5: Actual and bounds for the number of supports for the 8 candidates in Example 2.3

	Candidate							
Supports	1	2	3	4	5	6	7	8
Minimum	8	3	15	2	15	2	3	1
Maximum	20	14	20	13	20	18	20	8
Actual	15	6	17	4	19	7	6	1
Scores	71	14	139	13	137	18	24	8

3. To know, before the final election, whether candidate 3 enters the academy, it is necessary to add new constraints to the problem. For example, adding that the total number of supports of candidates $1, 5, 6$, and 7 are larger than the total number of supports of candidate 3:

$$\begin{array}{rcl}\sum_{i=1}^{I}\sum_{s=1}^{S} x_{i1s} & \geq & \sum_{i=1}^{I}\sum_{s=1}^{S} x_{i3s}\\ \sum_{i=1}^{I}\sum_{s=1}^{S} x_{i5s} & \geq & \sum_{i=1}^{I}\sum_{s=1}^{S} x_{i3s}\\ \sum_{i=1}^{I}\sum_{s=1}^{S} x_{i6s} & \geq & \sum_{i=1}^{I}\sum_{s=1}^{S} x_{i3s}\\ \sum_{i=1}^{I}\sum_{s=1}^{S} x_{i7s} & \geq & \sum_{i=1}^{I}\sum_{s=1}^{S} x_{i3s}\end{array}$$

Since this leads to an unfeasible problem, then we can guarantee that candidate 3 enters the Academy of Engineering.

■

2.5 School Timetable Problem

This example is a simple instance of the "school timetable problem". It consists of allocating classrooms and teaching hours for the subjects of one academic program divided in blocks.

We assume that n_c classrooms and n_h teaching hours are available, respectively, to teach n_s subjects. These subjects are grouped by (1) academic blocks and (2) instructors. Binary variable $v(s, c, h)$ is equal to 1 if subject s is taught in classroom c at hour h, and 0 otherwise.

We denote the set of all subjects, by Ω, the set of the n_i subjects taught by instructor i, by Ω_i, and the set of the n_b subjects grouped in academic block b, by Δ_b. Indices s, c, h, i, and b indicate respectively subject, classroom, hour, instructor and block.

The problem has the following elements:

1. **Data**

n_c: the number of classrooms

n_h: the number of available teaching hours

n_s: the number of subjects

n_i: the number of subjects taught by instructor i

n_b: the number of academic blocks

Ω: the set of all subjects to be taught

Ω_i: the set of subjects taught by instructor i

Δ_b: the set of subjects belonging to academic block b

2. **Variables**

$v(s,c,h)$: a binary variable that takes value 1 if subject s is taught in classroom c at hour h, and 0 otherwise

3. **Constraints**

(a) Every instructor teaches all his/her subjects:

$$\sum_{s \in \Omega_i} \sum_{c=1}^{n_c} \sum_{h=1}^{n_h} v(s,c,h) = n_i, \qquad \forall i \tag{2.7}$$

(b) Every instructor teaches at most 1 subject every hour:

$$\sum_{s \in \Omega_i} \sum_{c=1}^{n_c} v(s,c,h) \leq 1, \qquad \forall h, \quad \forall i \tag{2.8}$$

(c) Every subject is taught once:

$$\sum_{c=1}^{n_c} \sum_{h=1}^{n_h} v(s,c,h) = 1, \qquad \forall s \tag{2.9}$$

(d) In every classroom–hour combination at most 1 subject is taught:

$$\sum_{s \in \Omega} v(s,c,h) \leq 1, \qquad \forall c, \quad \forall h \tag{2.10}$$

(e) At every hour, at most 1 subject of any academic block is taught:

$$\sum_{s \in \Delta_b} \sum_{c=1}^{n_c} v(s,c,h) \leq 1, \qquad \forall h, \quad \forall b \tag{2.11}$$

4. **Function to be optimized.** Formulating an appropriate objective function to be minimized is not an easy task. However, in this example we consider a very simple objective function. The target is to produce a compact timetable. It can be formulated by minimizing

$$\sum_{s\in\Omega}\sum_{c=1}^{n_c}\sum_{h=1}^{n_h}(c+h)\ v(s,c,h)$$

subject to constraints (2.7)–(2.10).

This optimization function has been chosen because it penalizes the $v(s,c,h)$ variables taking on value 1 for high values of c and h. Thus, it tries to compact the teaching classrooms and hours. The smallest the classroom number and the hour the better.

Example 2.4 (School timetable problem). Consider 3 classrooms, 5 teaching hours, 8 subjects, 2 instructors, and 2 course blocks. The set of all subjects is $\Omega = \{s_1, s_2, \ldots, s_8\}$, the set of subjects of instructor 1 is $\Omega_1 = \{s_1, s_2, s_8\}$, the set of subjects of instructor 2 is $\Omega_2 = \{s_3, s_4, s_5, s_6, s_7\}$, the set of subjects of academic block 1 is $\Delta_1 = \{s_1, s_2, s_3, s_4\}$, and the subjects of academic block 2 is $\Delta_2 = \{s_5, s_6, s_7, s_8\}$. Note that $\Omega_1 \cup \Omega_2 = \Omega$ and $\Omega_1 \cap \Omega_2 = \emptyset$, and $\Delta_1 \cup \Delta_2 = \Omega$ and $\Delta_1 \cap \Delta_2 = \emptyset$.

The solution is provided in the tables below:

	$h=1$	$h=2$	$h=3$	$h=4$	$h=5$
$c=1$	s_7	s_6	s_3	s_4	s_5
$c=2$	s_2	s_1	s_8	–	–
$c=3$	–	–	–	–	–

The schedule for instructor 1 is

	$h=1$	$h=2$	$h=3$	$h=4$	$h=5$
$c=1$	–	–	–	–	–
$c=2$	s_2	s_1	s_8	–	–
$c=3$	–	–	–	–	–

The schedule for instructor 2 is

	$h=1$	$h=2$	$h=3$	$h=4$	$h=5$
$c=1$	s_7	s_6	s_3	s_4	s_5
$c=2$	–	–	–	–	–
$c=3$	–	–	–	–	–

The schedule for academic block 1 is

	$h=1$	$h=2$	$h=3$	$h=4$	$h=5$
$c=1$	–	–	s_3	s_4	–
$c=2$	s_2	s_1	–	–	–
$c=3$	–	–	–	–	–

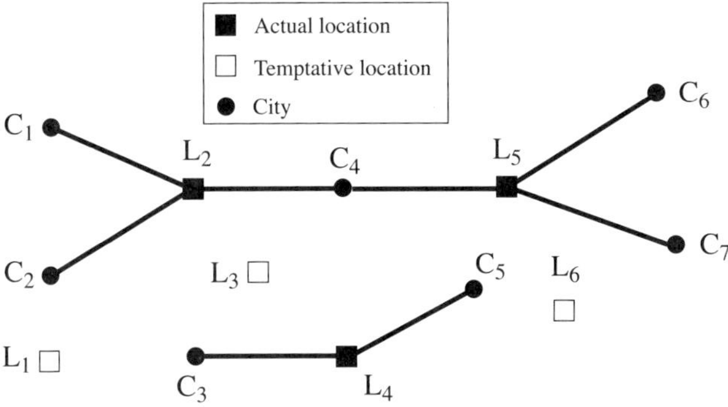

Figure 2.1: Solution of the example of the capacitated facility location problem.

The schedule for academic block 2 is

	$h=1$	$h=2$	$h=3$	$h=4$	$h=5$
$c=1$	s_7	s_6	–	–	s_5
$c=2$	–	–	s_8	–	–
$c=3$	–	–	–	–	–

■

2.6 Models of Discrete Location

In this example we describe one of the *discrete location models* or, more precisely, the *capacitated facility location problem*. This deals with deciding where to locate facilities within a finite set of sites, taking into account the needs of the clients to be served, and optimizing certain economic criteria. Usually, setting up a facility involves significant costs that do not depend on the production level of the facility.

The problem is motivated by a number of potential applications. For example, several plants are to be set up at some points of a transportation system to maximize the benefit by means of minimizing the production and shipment costs. Figure 2.1 shows a solution to the problem of locating plants that provide service to a set of clients.

Thus, the main elements in this problem are

1. **Data**

   I: a set $\{1, \ldots, n\}$ of n clients

   J: a set $\{1, \ldots, m\}$ of m sites where facilities can be located

   f_j: the fixed cost of the opening facility placed at j for $j \in J$

   c_{ij}: the profit per unit of sale of the goods originated at facility j to client i. Usually, the c_{ij} depend on the production costs at facility j, the demand and selling price for client i, and the transportation costs between client i and facility j

u_j: the capacity of the facility located at j

b_i: the demand of the ith client

2. **Variables.** The variables involved in this problem are the following:

y_j: a binary variable to model the choice of "opening" a facility at the site j. This is defined as follows:

$$y_j = \begin{cases} 1 & \text{if facility } j \text{ is open} \\ 0 & \text{otherwise} \end{cases} \tag{2.12}$$

x_{ij}: the quantity of commodities sent from facility j to client i.

3. **Constraints.** The constraints in this problem are as follows. Each client's demand must be satisfied:

$$\sum_{j \in J} x_{ij} = b_i, \quad \forall i \in I \tag{2.13}$$

Since a client i cannot be served from j unless a facility is placed at j, we have the following constraints:

$$\sum_{i \in I} x_{ij} \leq u_j y_j, \quad \forall j \in J \tag{2.14}$$

These linear inequalities take into account that the client i can be served from j only if a facility is located at node j, since $y_j = 0$ implies that $x_{ij} = 0$, $\forall i$ and $y_j = 1$ yields the constraint $\sum_{i \in I} x_{ij} \leq u_j$, which means that the production level of the facility j cannot exceed its capacity. In addition, the variable constraints are

$$y_j \in \{0, 1\}, \quad \forall j \in J \tag{2.15}$$

$$x_{ij} \geq 0 \; \forall i \in I, \quad \forall j \in J \tag{2.16}$$

4. **Function to be optimized.** In the so-called strong formulation of the uncapacitated facility location problem we maximize

$$Z = \sum_{i \in I} \sum_{j \in J} c_{ij} x_{ij} - \sum_{j \in J} f_j y_j \tag{2.17}$$

In this model, the allocation problem in the case of unlimited capacity of the sites is solved easily. In reality, in the presence of a feasible set of locations, the allocation problem is solved by means of assigning each client to the most profitable open facility. However, it may be unrealistic to assume that a facility can supply any number of clients. Thus, limited capacities must be dealt with.

Table 2.6: City demands

City	Demand
C_1	1.5
C_2	2.0
C_3	3.0
C_4	4.0
C_5	2.5
C_6	1.0
C_7	2.0

Table 2.7: Benefits according to different locations

	Cities (C_i)						
Locations (L_j)	C_1	C_2	C_3	C_4	C_5	C_6	C_7
L_1	4.0	4.5	2.5	0.5	1.0	0.5	−3.5
L_2	4.0	4.5	2.5	4.2	3.5	1.5	−0.5
L_3	3.5	5.0	4.0	3.5	4.5	1.5	0.0
L_4	1.3	3.0	5.0	3.3	5.5	1.8	1.3
L_5	0.5	1.0	1.5	5.0	4.0	5.5	3.0
L_6	−1.0	0.0	1.5	3.3	4.0	4.5	2.0

Example 2.5 (Location of industrial plants). A company wishes to build various industrial plants to supply 7 cities with a certain product. The demand of these cities based on demographic factors and social characteristics is estimated. These values are shown in Table 2.6.

A study has indicated 6 possible sites for these industrial plants. It is supposed that all the plants have the same characteristics. The maximum production capacity per plant is 6 units. The cost of investment recovering has been calculated in 10 monetary units for the period of study.

Table 2.7 shows the profit achieved by selling, to city i, one unit manufactured in one plant located at site j.

The decisionmaker needs to determine the amount of plants and their locations, so that the cities demand is satisfied, and locations are such that the demands are satisfied, and a maximum financial benefit is obtained. The optimization problem associated with this decision process consists of maximizing the total benefit including amortization costs, subject to the constraints. So, the problem can be stated as follows, by maximizing

$$Z = \sum_{i=1}^{7}\sum_{j=1}^{6} c_{ij}x_{ij} - \sum_{j=1}^{6} 10y_j$$

Table 2.8: Amount of production of each operating plant to be provided to each city

	Cities						
Locations	C_1	C_2	C_3	C_4	C_5	C_6	C_7
L_2	1.5	2.0		1.0			
L_4			3.0		2.5		
L_5				3.0		1.0	2.0

subject to

$$\begin{aligned} \sum_{j=1}^{6} x_{1j} &= 1.5; & \sum_{j=1}^{6} x_{2j} &= 2.0 \\ \sum_{j=1}^{6} x_{3j} &= 3.0; & \sum_{j=1}^{6} x_{4j} &= 4.0 \\ \sum_{j=1}^{6} x_{5j} &= 2.5; & \sum_{j=1}^{6} x_{6j} &= 1.0 \\ \sum_{j=1}^{6} x_{7j} &= 2.0 \end{aligned} \tag{2.18}$$

and

$$\sum_{i=1}^{6} x_{ij} \leq 6y_j; \; j = 1, \ldots, 7 \tag{2.19}$$

$$\begin{aligned} y_j &\in \{0,1\}; \; j = 1, \ldots, 6 \\ x_{ij} &\geq 0; \; i = 1, \ldots, 7; \; j = 1, \ldots, 6 \end{aligned} \tag{2.20}$$

where (2.18) and (2.19) are the demand and the production capacity constraints, respectively.

The solution of this problem is plotted in Figure 2.1 consists of placing 3 industrial plants in locations L_2, L_4, and L_3, and the production distribution by cities is the Table 2.8.

■

2.7 Unit Commitment of Thermal Power Units

The cost of starting up an electric power thermal unit after being offline for a couple of days is approximately half the cost of buying a 100 m^2 apartment in a distinguished neighborhood. Therefore, the planning of the startups and shutdowns of any thermal unit should be done carefully. The electric power thermal unit commitment problem consists of determining, for a planning horizon, the startup and shutdown schedule of every unit so that the electric demand is served and total operating costs are minimized, while satisfying different technical and security constraints.

A typical planning horizon is one day divided in hours. If time intervals are denoted by k, the planning horizon consists of the periods

$$k = 1, 2, \ldots, K \tag{2.21}$$

where K is typically equal to 24.

The startup cost is an exponential function of the time the unit has been offline, but it will be considered constant (this is a reasonable simplification in most cases). Every time a unit is started up, its startup cost is incurred, and this can be expressed as

$$C_j y_{jk} \tag{2.22}$$

where C_j is the startup cost of unit j and y_{jk} is a binary variable that is equal to 1 if unit j is started up at the beginning of period k and 0, otherwise.

The shutdown cost can be expressed in a similar fashion as the startup cost; thus

$$E_j z_{jk} \tag{2.23}$$

where E_j is the shutdown cost of unit j and z_{jk} a binary variable that is equal to 1 if unit j is shut down at the beginning of period k, and 0 otherwise.

The running costs consist of a fixed cost and a variable cost. The fixed cost can be expressed as

$$A_j \; v_{jk}, \tag{2.24}$$

where A_j is the fixed cost of unit j and v_{jk} is a binary variable that is equal to 1 if unit j is online during period k and 0, otherwise.

The variable cost can be considered proportional to the unit output power:[1]

$$B_j \; p_{jk} \tag{2.25}$$

where B_j is the variable cost of unit j and p_{jk} the output power of unit j during period k.

Thermal units cannot operate below a minimum output power and above a maximum output power. These technical constraints can be expressed as

$$\underline{P}_j \; v_{jk} \leq p_{jk} \leq \overline{P}_j \; v_{jk} \tag{2.26}$$

where $\underline{P}_j$ and $\overline{P}_j$ are respectively the minimum and maximum output powers of unit j.

The left-hand side of the preceding constraint expresses that if unit j is online during period k ($v_{jk} = 1$), its output power should be above the minimum output power. Analogously, the right-hand-side of the constraint above expresses

[1] A more precise modeling requires the variable cost to be a quadratic or cubic function of the power output.

that if unit j is online during period k ($v_{jk} = 1$), its output power should be below the maximum output power. If $v_{jk} = 0$, the preceding constraint forces $p_{jk} = 0$.

From one time period to the next one, any power unit cannot increase its output power above a maximum power increment, called the rampup limit. This can be written as

$$p_{jk+1} - p_{jk} \leq S_j \tag{2.27}$$

where S_j is the maximum rampup power increment of unit j.

For the first period of the planning horizon the above constraint becomes

$$p_{j1} - P_j^0 \leq S_j \tag{2.28}$$

where P_j^0 is the output power of unit j just before the first period of the planning horizon.

Similarly, any power unit cannot decrease its output power above a maximum power decrement, which is called the rampdown power limit. Therefore

$$p_{jk} - p_{jk+1} \leq T_j \tag{2.29}$$

where T_j is the maximum rampdown power decrement of unit j.

For the first period of the time horizon, the constraint above becomes

$$P_j^0 - p_{j1} \leq T_j \tag{2.30}$$

Any unit that is online can be shut down but not started up, and analogously, any unit that is offline can be started up but not shut down. This can be expressed as

$$y_{jk} - z_{jk} = v_{jk} - v_{jk-1} \tag{2.31}$$

For the first period the above constraint becomes

$$y_{j1} - z_{j1} = v_{j1} - V_j^0 \tag{2.32}$$

where V_j^0 is a binary constant that is equal to 1 if unit j is online the period preceding the first period of the planning horizon, and 0 otherwise. The reader is encouraged to verify these two conditions using examples.

In every period the power demand should be satisfied, so

$$\sum_{j=1}^{J} p_{jk} = D_k \tag{2.33}$$

where J is the number of power units and D_k the demand in period k.

For security reasons, the total output power available online should be larger than the actual demand by a specified amount. This is formulated as

$$\sum_{j=1}^{J} \overline{P}_j v_{jk} \geq D_k + R_k \tag{2.34}$$

where R_k is the amount of required reserve (over the demand) in period k.

The main elements in this problem are:

1. **Data**

   K: the number of time intervals

   C_j: the startup cost of unit j

   E_j: the shutdown cost of unit j

   A_j: the fixed cost of unit j

   B_j: the variable cost of unit j

   $\underline{P}_j$: the minimum output power of unit j

   $\overline{P}_j$: the maximum output power of unit j

   S_j: the maximum rampup power increment of unit j

   P_j^0: the output power of unit j just before the first period of the planning horizon

   T_j: the maximum rampdown power decrement of unit j

   V_j^0: a binary constant that is equal to 1; if unit j is online the period preceding the first period of the planning horizon, and 0, otherwise

   J: the number of power units

   D_k: the demand in period k

   R_k: the amount of required reserve (over the demand) in period k

2. **Variables.** The variables involved in this problem are the following:

   y_{jk}: a binary variable that is equal to 1, if unit j is started up at the beginning of period k and 0, otherwise

   z_{jk}: a binary variable that is equal to 1, if unit j is shut down at the beginning of period k, and 0, otherwise

   v_{jk}: a binary variable that is equal to 1, if unit j is online during period k and 0, otherwise

   p_{jk}: the output power of unit j during period k

3. **Constraints.** The constraints in this problem are as follows. Any unit at any time should operate above its minimum output power and below its maximum output power, then

$$\underline{P}_j v_{jk} \leq p_{jk} \leq \overline{P}_j v_{jk} \quad \forall j, k \tag{2.35}$$

Rampup constraints should be satisfied:

$$p_{jk+1} - p_{jk} \leq S_j, \;\; \forall j, k = 0, \dots, K-1 \tag{2.36}$$

where

$$p_{j0} = P_j^0$$

Rampdown constraints should also be satisfied:

$$p_{jk} - p_{jk+1} \leq T_j, \;\; \forall j, k = 0, \dots, K-1 \tag{2.37}$$

The logic of status changes (from online to offline and vice versa) should be preserved; therefore

$$y_{jk} - z_{jk} = v_{jk} - v_{jk-1}, \;\; \forall j, k = 1, \dots, K \tag{2.38}$$

where

$$v_{j0} = V_j^0, \;\; \forall j$$

The demand should be satisfied in every period; thus

$$\sum_{j=1}^{J} p_{jk} = D_k, \;\; \forall k \tag{2.39}$$

Finally, security constraints should be satisfied in all periods of the planning horizon; then

$$\sum_{j=1}^{J} \overline{P}_j \, v_{jk} \geq D_k + R_k, \;\; \forall k. \tag{2.40}$$

4. **Function to be minimized.** The objective of the unit commitment problem is to minimize total costs; the objective is therefore to minimize

$$Z = \sum_{k=1}^{K} \sum_{j=1}^{J} [A_j \, v_{jk} + B_j \, p_{jk} + C_j \, y_{jk} + E_j \, z_{jk}] \tag{2.41}$$

The problem illustrated in (2.35)–(2.41) is a simplified version of the electric power thermal unit commitment problem. It should be noted that it is a binary mixed-integer linear programming problem.

Example 2.6 (Unit commitment). A 3-hour planning horizon is considered. The demands in these hours are respectively 150, 500, and 400. Reserves are respectively 15, 50, and 40. Three power units are considered. Data for these units are given below:

Power unit number	1	2	3
Maximum output power	350	200	140
Minimum output power	50	80	40
Rampup limit	200	100	100
Rampdown limit	300	150	100
Fixed cost	5	7	6
Startup cost	20	18	5
Shutdown cost	0.5	0.3	1.0
Variable cost	0.100	0.125	0.150

All units are offline before the planning horizon.
The unit output powers for the optimal solution are

	Hour		
Unit	1	2	3
1	150	350	320
2	—	100	080
3	—	050	—
Total	150	500	400

The minimum cost is 191. Unit 1 is started up a the beginning of hour 1 and remains online for the 3 hours. Unit 2 is started up at the beginning of hour 2 and remains on line during hours 2 and 3. Unit 3 is started up at the beginning of hour 2 and shut down at the beginning of hour 3.

■

Exercises

2.1 Walter builds two types of transformers and has available 6 tons of ferromagnetic material and 28 hours of working time. Transformer 1 requires 2 tons of ferromagnetic material and 7 hours of work, and transformer 2 requires 1 unit of ferromagnetic material and 8 hours of work. Selling prices of transformers 1 and 2 are respectively 120 and 80 thousand Eurodollars. How many transformers or each type should manufacture Walter to maximize his benefits? Solve the problem graphically and analytically.

2.2 Consider a transportation network where several cities are connected by roads. The problem becomes one of finding the shortest route between two cities. We assume that the distance between two directly connected cities is known. Formulate this problem using integer linear programming (*The Shortest Path Problem*).

2.3 Consider a salesperson who wants to find the minimum cost tour that visits each of n given cities exactly once and returns to the originating city. Formulate this problem as an integer linear programming problem (*The Traveling Salesperson Problem*).

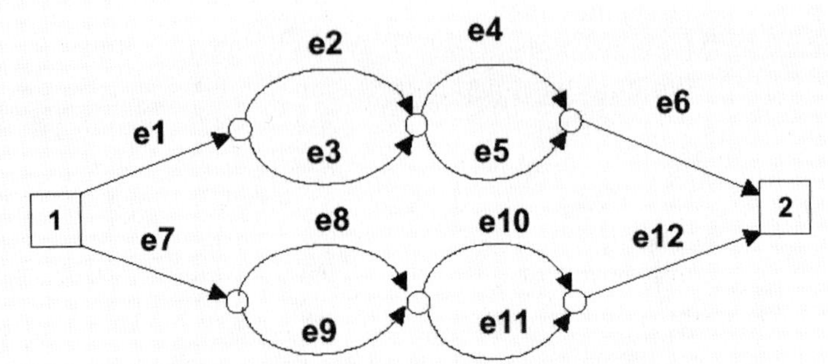

Figure 2.2: Network topology.

2.4 Consider the problem of finding the maximum number of paths in a communication network from the origin p to destination q such that they are link-disjoint. That means that two different paths do not have common links. Formulate this problem as an integer linear programming problem.

2.5 Given a connected graph $\mathcal{G} = (\mathcal{N}, \mathcal{A})$, where N is the set of nodes, A is the set of links, and a special set of "terminal vertices" $\mathcal{T} = \{t_1, t_2, \dots t_k\} \subset \mathcal{N}$, the network reliability problem consists of evaluating the probability of communication between the elements of $\mathcal{T}$ under a random degradation of the network. Let $p(e)$ be the probability of link e to be *operative*. The network reliability is the probability of all pairs of terminals being connected at least by an operative path. Consider the problem that one has to design the least-cost network that satisfies the requirement of a given reliability. This design consists of the selection of a subset of links of the original network. More specifically, consider as the original network the one shown in Figure 2.2 with terminal nodes $T = \{1, 2\}$ and assume that the desired network reliability is 0.90. The reliability and cost of the links are given in Table 2.9. Formulate the corresponding problem as an integer linear programming problem.

2.6 An electricity producer should plan its hourly energy production to maximize its profits from selling energy during a planning horizon of a given number of hours. Formulate a mixed-integer linear programming problem taking into account that:

(a) The producer does not produce before the planning horizon.

(b) Hourly energy prices can be forecasted and are considered known.

(c) If running, the minimum and maximum energy productions of the producer are known quantities, and the minimum is greater than zero.

(d) Energy productions in two consecutive hours cannot differ in more that a prespecified amount.

Table 2.9: Reliability and cost of the components of the network of Figure 2.2

Link	Reliability ($p(e_i)$)	Cost
e_1	0.90	1.90
e_2	0.85	1.85
e_3	0.95	1.95
e_4	0.70	1.70
e_5	0.80	1.80
e_6	0.90	1.90
e_7	0.90	1.90
e_8	0.50	1.35
e_9	0.60	1.45
e_{10}	0.60	1.20
e_{11}	0.30	1.30
e_{12}	0.90	1.90

Table 2.10: Times required for different processes

	Process				
Part	L	S	D	M	G
A	0.6	0.4	0.1	0.5	0.2
B	0.9	0.1	0.2	0.3	0.3
Availability	10	3	4	6	5

(e) Producer production cost is linear.

2.7 The manufacture of the two parts, A and B, of a certain machine requires the processes L, S, D, M, and G. The time of each process to operate on each part and the number of available processes are given in Table 2.10 (hours per unity). Each one can be used during 8 hours, 30 days per month.

(a) Determine the optimal production strategy to maximize the total number of parts A and B manufactured in a month.

(b) If the number of parts A must be equal to the number of parts B, what is the optimal strategy?

2.8 A hospital manager should plan the working timetable for the hospital staff. Determine the minimum weekly cost associated with the staff of this hospital if

(a) The daily working time is structured in 3 shifts.

(b) In every shift there should be at least 1 physician, 2 (male) nurses, and 3 assistants.

(c) The maximum total number of employees needed on every shift is 10.

(d) The salaries are: $50/shift for a physician, $20/shift for a nurse, and $10/shift for an assistant.

(e) The total number of employees is: 15 physicians, 36 nurses, and 49 assistants.

(f) Each employee should rest during at least two consecutive shifts.

Chapter 3

Nonlinear Programming

3.1 Introduction

Chapters 1 and 2 were devoted to linear programming problems, where the constraints and the function to be optimized were linear.

Even though linear programming problems are very common and cover a wide range of problems, in real life we are faced in many occasions with other types of problems that are not linear. When either the set of constraints, the function to be optimized, or both are nonlinear, we say that we are faced with a *nonlinear programming problem* (NLPP).

In this chapter we introduce some nonlinear programming problems. Some of them coincide with the problems discussed in previous chapters, but with different assumptions.

3.2 Some Geometrically Motivated Examples

In this section, we present some nonlinear programming problems that have a geometric motivation and can be solved analytically.

3.2.1 The Postal Package Example

A postal package is assumed to be a box, of dimensions x, y, and z (see Figure 3.1), that satisfies the following requirements to be accepted by the postal office. The height plus the base perimeter cannot exceed 108 cm

$$z + 2x + 2y \leq 108; \quad x, y, z \geq 0$$

because negative lengths are not admissible.

We look for the dimensions maximizing the volume

$$V(x, y, z) = xyz$$

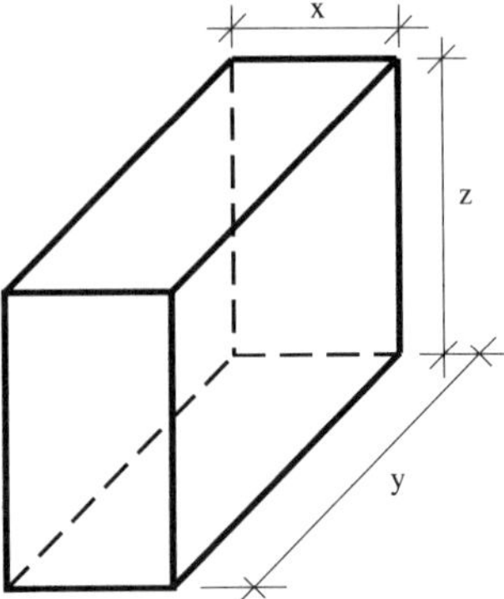

Figure 3.1: Illustration of the postal package dimensions.

3.2.2 The Tent Example

A tent consists of a square base of sides $2a$, four vertical walls of height b, and a roof that is a pyramid of height h (see Figure 3.2). If the total volume V and the total height H of the tent are given, find the optimal values for a, b, and h so that the resulting tent has minimum surface (a minimum of material is required to build the tent).

In this case the aim is to minimize the surface of the tent. Since the total surface is the sum of the surface of the four walls plus the roof surface, we have to minimize

$$S(a,b,h) = 4(2ab + a\sqrt{h^2 + a^2})$$

The constraints we must satisfy concern the total tent volume and the total height; thus, we must preserve the quantities

$$\begin{aligned} V &= 4a^2\left(b + \tfrac{h}{3}\right) \\ H &= b + h \end{aligned}$$

Moreover, all variables must be non-negative, i.e., $a \geq 0$, $b \geq 0$, $h \geq 0$.

3.2.3 The Lightbulb Example

A lightbulb is placed right above the center of a circle of radius $r = 30$ cm. Assuming that the intensity of light at any point of the circle is proportional to the square root of the sine of the angle with which the ray hits such a point, and to the inverse of the distance to the bulb, d, determine the optimal height of the bulb so that the intensity at the boundary of the circle is as large as possible.

We ought to maximize the intensity of light at the points of the circumference of radius $r = 30$ cm. Let us call I to the intensity, measured in appropriate units, which depends on the sine of the angle formed by the ray of light and the

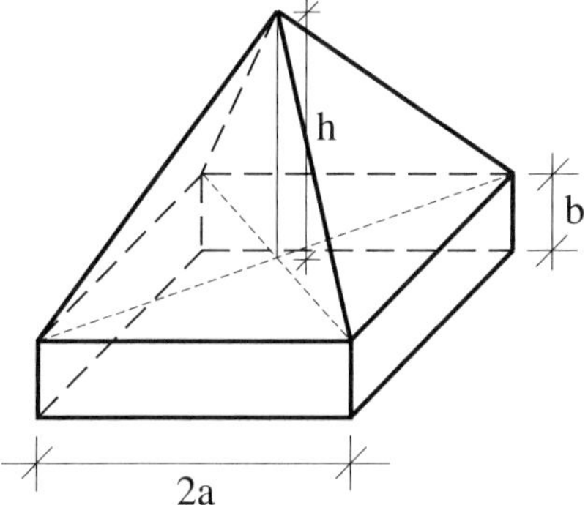

Figure 3.2: Illustration of the tent example.

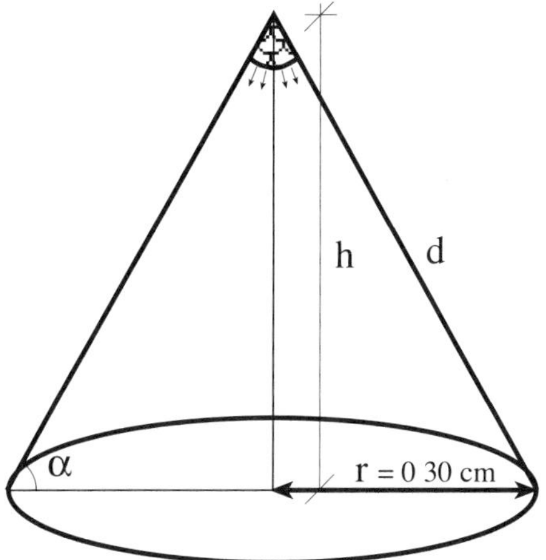

Figure 3.3: Illustration of the lightbulb example.

horizon. In the triangle of Figure 3.3 we see that

$$\sin \alpha = \frac{h}{\sqrt{h^2 + r^2}}; \quad d = \sqrt{h^2 + r^2}$$

Therefore we can write

$$I(h) = k \frac{h^{1/2}}{(h^2 + r^2)^{3/4}}$$

where $k > 0$ is a proportionality constant. This expression must be maximized with respect to h, keeping in mind the only constraint $h \geq 0$.

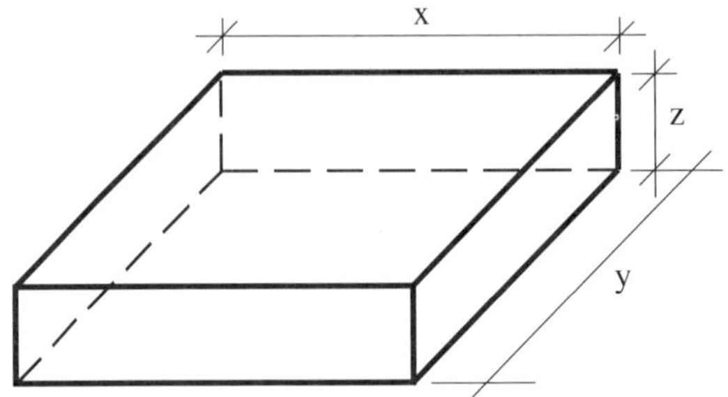

Figure 3.4: Illustration of the moving sand problem.

3.2.4 The Surface Example

To find the points on the surface in the three-dimensional space determined by the equation $xyz = 1$ that are closest to the origin, we must minimize the distance function to the origin

$$D(x, y, z) = \sqrt{x^2 + y^2 + z^2}$$

subject to the constraint that the points are on the surface:

$$xyz = 1$$

3.2.5 The Moving Sand Example

The cost of moving sand from one place to another in a box of dimensions x, y and z is \$2 per round trip (see Figure 3.4). Assuming that the cost of the construction material for the top and bottom and sides of the box are 3 times and twice, respectively, the cost of the ends, find the minimum cost of transporting $c = 50$ cubic units of sand.

If z is the height of the box, then the cost of making the box will be

$$k(3xy + 2(2xz + 2yz) + xy)$$

where $k = 4 > 0$ is a proportionality constant. We have to add the transportation cost, which is given by

$$2\frac{50}{xyz}$$

The total cost, we are supposed to minimize, is then

$$C(x, y, z) = k(3xy + 2(2xz + 2yz) + xy) + 2\frac{50}{xyz}$$

subject to the constraints $x, y, z \geq 0$.

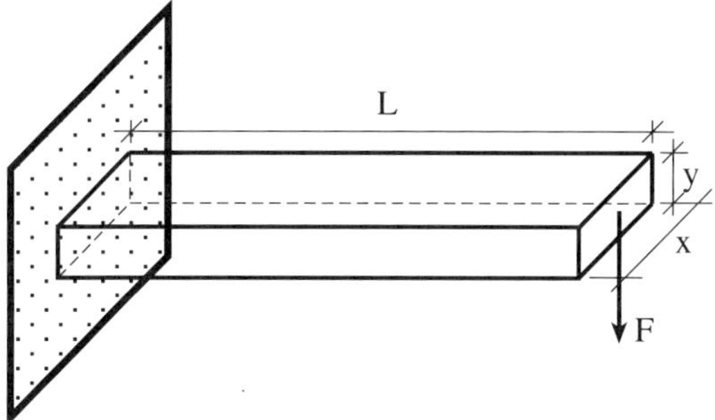

Figure 3.5: Cantilever beam.

3.3 Some Mechanically Motivated Examples

Another category of important examples is related to mechanical situations. We include several typical situations that can be found under different formats and varying conditions in many places. Some of them will be solved later.

3.3.1 The Cantilever Beam Example

We would like to design a cantilever beam of rectangular cross section and prescribed length so as to have minimum weight, and a limited deflection under the action of a given vertical load acting on its tip. The material we plan to use has a known unit weight.

Let x and y be the width and height (see Figure 3.5) we look for. Let L, F, S, and γ be the length, the load at the tip, the maximum allowable deflection, and unit weight, respectively, that is, the data of our problem. The objective is to minimize the weight

$$W(x, y) = \gamma Lxy.$$

Since the width x must be larger than 0.5 and, according to the *strength of materials* theory, the deflection at the tip is given by $FL^3/3EI$, where E is the Young modulus of the material the beam is made of, and $I = xy^3/12$ is the area moment of inertia of its rectangular cross section, we must minimize $W(x, y)$ subject to the constraints

$$\begin{array}{rcl} \dfrac{4FL^3}{Exy^3} & \leq & S \\ x & \geq & 0.5 \\ x, y & \geq & 0 \end{array}$$

3.3.2 The Two-Bar Truss Example

The two-bar truss in Figure 3.6 is to be designed according to three different desired objectives: a minimum weight, the maximum stresses not to exceed a

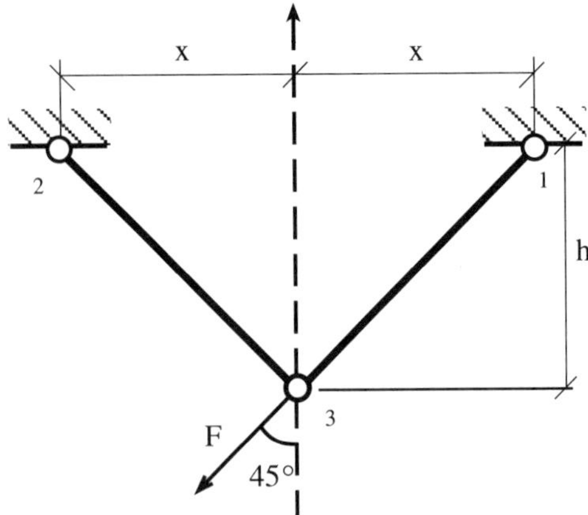

Figure 3.6: Two-bar truss used in the example

maximum, the known value, and the displacement of joint 3 not to exceed a certain value. The data set of the problem is as follows:

γ: the unit weight of the bar material

E: Young's modulus of the material that the truss is made of

F: the applied load at the fixed joint, making a 45° angle to the left of the Y axis

S_0: the maximum admissible stress

D_0: the maximum admissible displacement of joint 3

h: the trust height

The two design variables we need to determine for optimal design are

x: the distance from the fixed joints to the Y axis

z: the area of the cross section of the truss arms

D: the displacement of the joint 3

S^1: the stress at joint 1

S^2: the stress at joint 2

W: the total weight of the bar truss

Therefore we must minimize

$$W(x,z) = 2\gamma z\sqrt{x^2+h^2} \tag{3.1}$$

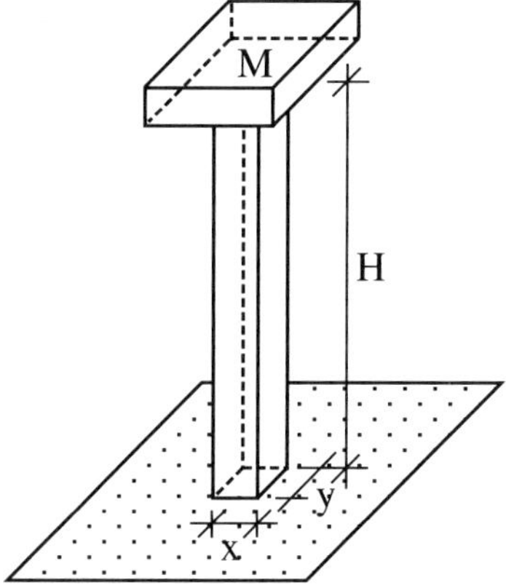

Figure 3.7: The buckling column.

subject to

$$\begin{aligned} D(x,z) &= \frac{F}{Eh^2 2\sqrt{2}} \frac{(h^2+x^2)^{3/2}(h^4+x^4)^{1/2}}{x^2 z} &\leq D_0 \\ S^1(x,z) &= \frac{F}{2\sqrt{2}h} \frac{(x+h)\sqrt{x^2+h^2}}{xz} &\leq S_0 \\ S^2(x,z) &= \frac{F}{2\sqrt{2}h} \frac{(h-x)\sqrt{x^2+h^2}}{xz} &\leq S_0 \\ x,z &\geq 0 \end{aligned}$$

3.3.3 The Column Example

A uniform, rectangular cross section column of known height is to be designed to support a given mass on top of it. On one hand, we would like to minimize the amount of material to be used, but on the other hand we are interested in maximizing the natural frequency of transversal vibration. Find the optimal dimensions of such a column avoiding failure due to compression and buckling (stability failure) (see Figure 3.7).

The data set of the problem is

M: the mass to be supported by the column

H: the height of the column

D: unit density of the material to be used

E: the Young's modulus of the material to be used

S: the maximum admissible stress (force per unit surface)

The design variables are the two dimensions, x and y, of the column cross section.

The first objective relates to minimizing the total mass of the column

$$W(x, y) = DHxy$$

The second objective is to maximize the frequency of the transversal vibration which, as given by elementary mechanics, is

$$\left(\frac{Exy^3}{4H^3(M + \frac{33}{140}DHxy)}\right)^{1/2}$$

Notice that maximizing this quantity is equivalent to minimizing its reciprocal.

The constraints we ought to respect are concerned with stresses. On one hand, the stress must not be greater than the maximum admissible stress S. On the other hand, it should be less than the buckling stress. The compression stress is given by Mg/xy where g is the acceleration of gravity, while the buckling stress is $\pi^2 Ey^2/48H^2$. Altogether we must enforce the constraints

$$\begin{array}{rcl} \frac{Mg}{xy} & \leq & S \\ \frac{Mg}{xy} & \leq & \frac{\pi^2 Ey^2}{48H^2} \\ x, y & \geq & 0 \end{array}$$

Assuming that both objective are equally desired, we would have to minimize

$$Z = DHxy - \left(\frac{Exy^3}{4H^3\left(M + \frac{33}{140}DHxy\right)}\right)^{1/2}$$

under the above mentioned constraints.

3.3.4 Scaffolding System

In this example we consider a modified version of the model given in Section 1.7. Essentially, we allow the data dl_i, the distance of the point where load i is applied on beam b ($i \in \Omega_b$), to become a variable instead of a data. To emphasize this fact we change the notation and use xl_i instead of dl_i. Moreover we need to keep in mind the need of an additional data, l_b, the total length of beam $b \in B$. The constraints for this new situation would be

$$\begin{array}{rcl} \sum\limits_{s \in \Psi_b} t_s & = & \sum\limits_{i \in \Omega_b} x_i + \sum\limits_{x \in \Theta_b} t_s, b \in B \\ \sum\limits_{s \in \Psi_b} dr_s t_s & = & \sum\limits_{i \in \Omega_b} xl_i x_i + \sum\limits_{x \in \Theta_b} dr_s t_s, b \in B \\ 0 \leq t_s & \leq & T_s, s \in S \\ 0 \leq xl_i & \leq & l_b, i \in \Omega_b \\ 0 & \leq & x_i \end{array} \tag{3.2}$$

and the function to be maximized is

$$\sum_i x_i$$

In particular, if in the specific example of Figure 1.4 we allow loads x_1 and x_2 to be applied on points a distance x_3 and x_4 from the left endpoints of beams 1 and 3, respectively, the balance equations change to

$$\begin{array}{rcl} t_E + t_F & = & x_2 \\ 10t_F & = & x_4 x_2 \\ t_C + t_D & = & t_F \\ 8t_D & = & 6t_F \\ t_A + t_B & = & x_1 + t_C + t_D \\ 10t_B & = & 2t_C + 10t_D + x_1 x_3 \end{array} \quad (3.3)$$

If we express the tensions on the ropes in terms of the independent variables of our problem, we conclude that we must satisfy the constraints

$$\begin{array}{rcccl} t_F & = & \dfrac{x_2 x_4}{10} & \leq & 100 \\ t_E & = & x_2 - \dfrac{x_2 x_4}{10} & \leq & 100 \\ t_D & = & \dfrac{3x_2 x_4}{40} & \leq & 200 \\ t_C & = & \dfrac{x_2 x_4}{40} & \leq & 200 \\ t_B & = & \dfrac{x_1 x_3}{10} + \dfrac{2x_2 x_4}{25} & \leq & 300 \\ t_A & = & x_1 - \dfrac{x_1 x_3}{10} + \dfrac{x_2 x_4}{50} & \leq & 300 \\ 0 & \leq & x_3 & \leq & 10 \\ 0 & \leq & x_4 & \leq & 10 \\ 0 & \leq & x_1 & & \\ 0 & \leq & x_2 & & \end{array} \quad (3.4)$$

which are nonlinear. Thus, we are in front of a nonlinear programming problem.

The solution obtained with the GAMS package is (see Section 11.4.9)

$$Z = 700 \text{ attained at point } x_1 = 500; \quad x_2 = 200; \quad x_3 = 4.4; \quad x_4 = 5.0$$

The corresponding rope tensions are

$$t_A = 300; \quad t_B = 300; \quad t_C = 25; \quad t_D = 75; \quad t_E = 100; \quad t_F = 100$$

3.4 Some Electrically Motivated Examples

This section includes some examples of electrical origin.

3.4.1 Power Circuit State Estimation

Electric power circuits span over countries and continents to deliver electric energy to every industry, business, and home.

Voltmeters measure voltage magnitudes with a given level of precision. A voltage magnitude measurement is available almost at every bus i, and is denoted by $\hat{v}_i$; its level of quality is given by the parameter σ_i^v. The smaller the value of σ_i^v, the larger the precision of the measurement.

Active power measurements are usually available at both endbuses of any power line. The measurement of *active power* leaving bus k toward bus l of line $k-l$ is denoted by $\hat{p}_{kl}$ and has a degree of precision denoted by σ_{kl}^p. Analogously, *reactive power* measurements are available at both ends of any line and denoted by $\hat{q}_{kl}$ with a degree of precision given by parameter σ_{kl}^q.

In addition to active power, reactive power also travels through power lines. The reactive power is a variable related with voltage magnitudes. If enough reactive power is available, the voltage magnitude profile is fine, but if not, the voltage magnitude profile goes down. Finally, if more than enough reactive power is available, the voltage magnitude profile goes up.

The state of the power circuit is determined by the bus voltages. Every bus voltage is a complex variable, usually expressed in polar form. The magnitude of this complex variable provide the actual voltage magnitude, and its angle is a measure of the relative "height" of this bus. The larger the height of a bus, the larger the flow of active power from this bus to other buses connected to it.

To know the state of the power network, it is necessary to know the state variables. A number of measurements larger than the minimum required to know the state of the network is usually available. But available measurements are not exact. A rational way to proceed is to use all measurements to generate the best estimate of the state of the circuit.

Voltages (state variables) across the network are denoted by $v_i \angle \delta_i \quad i \in \Omega$, where v_i is the voltage magnitude; δ_i the voltage angle; and Ω, the set of buses of the network. It should be noted that the angle of any bus can be taken as the origin; therefore $\delta_m = 0$, where m is an arbitrary bus.

Electrical engineers have found out that the active power going from bus k to bus l can be expressed as

$$p_{kl}(v_k, v_l, \delta_k, \delta_l) = \frac{v_k^2}{z_{kl}} \cos\theta_{kl} - \frac{v_k v_l}{z_{kl}} \cos(\theta_{kl} + \delta_k - \delta_l) \tag{3.5}$$

where p_{kl} is the active power going from bus k to bus l, and $z_{kl}\angle\theta_{kl}$ is a complex constant related to the physical parameters of line $k-l$.

Electrical engineers have also found out that the reactive power going from bus k to bus l through line $k-l$ can be expressed as

$$q_{kl}(v_k, v_l, \delta_k, \delta_l) = \frac{v_k^2}{z_{kl}} \sin\theta_{kl} - \frac{v_k v_l}{z_{kl}} \sin(\theta_{kl} + \delta_k - \delta_l). \tag{3.6}$$

The main elements involved in this problem are

1. **Data.** The data consist of

   $\hat{v}_i$: the measured voltage magnitude at bus i

   σ_i^v: the voltage measurement quality level

   $\hat{p}_{kl}$: the measurement of the active power leaving bus k toward bus l of line $k-l$

   $\hat{q}_{kl}$: the measurement of the reactive power leaving bus k toward bus l of line $k-l$

   σ_{kl}^p: the degree of precision of $\hat{p}_{kl}$

   σ_{kl}^q: the degree of precision of $\hat{q}_{kl}$

   Ω: the set of buses of the network

   Ω_k: the set of buses connected to bus k

   $z_{k\ell}$: the impedance magnitude associated with line $k-l$

   $\theta_{k\ell}$: the impedance angle associated with line $k-l$

2. **Variables**

   v_i: the voltage magnitude at bus i

   δ_i: the voltage angle at bus i

3. **Constraints:** In this case there are no constraints; that is, the problem above is an unconstrained nonlinear programming problem.

4. **Function to be minimized.** To estimate the state of the power network (i.e., to determine the value of the state variables) minimizing the quadratic error of every measurement with respect to its actual estimate, the problem below has to be solved. Minimize

$$\sum_{i\in\Omega} \frac{1}{\sigma_i^v}(v_i - \hat{v}_i)^2 + \sum_{k\in\Omega, l\in\Omega_k} \frac{1}{\sigma_{kl}^p}(p_{kl}(v_k, v_l, \delta_k, \delta_l) - \hat{p}_{kl})^2$$

$$+ \sum_{k\in\Omega, l\in\Omega_k} \frac{1}{\sigma_{kl}^q}(q_{kl}(v_k, v_l, \delta_k, \delta_l) - \hat{q}_{kl})^2$$

   where Ω_k is the set of buses connected to bus k, and $p_{kl}(v_k, v_l, \delta_k, \delta_l)$ and $q_{kl}(v_k, v_l, \delta_k, \delta_l)$ are given in (3.5) and (3.6). Note that the factors $1/\sigma_i^v$, $1/\sigma_{kl}^p$, and $1/\sigma_{kl}^q$ weight the errors according to the qualities of the measurements.

Example 3.1 (Power circuit state estimation). Consider a power circuit of two buses and a single power line connecting these two buses. The line is characterized through the constant $z_{12}\angle\theta_{12} = 0.15\angle 90°$. The voltage magnitude measurements in buses 1 and 2 are respectively 1.07 and 1.01. The active power measurements at both ends of the line are respectively 0.83 and 0.81, and the reactive power measurements at both ends of the line are respectively 0.73 and 0.58. The origin for angles is taken at bus 2. All meters have identical precision.

The power circuit state estimation problem is formulated by minimizing

$$
\begin{aligned}
Z = \; & (v_1 - 1.07)^2 + (v_2 - 1.01)^2 + (\tfrac{1}{0.15} v_1 v_2 \sin\delta_1 - 0.83)^2 \\
& + (-\tfrac{1}{0.15} v_1 v_2 \sin\delta_1 - 0.81)^2 + (\tfrac{1}{0.15} v_1^2 - \tfrac{1}{0.15} v_1 v_2 \cos\delta_1 - 0.73)^2 \\
& + (\tfrac{1}{0.15} v_2^2 - \tfrac{1}{0.15} v_1 v_2 \cos\delta_1 - 0.58)^2
\end{aligned}
$$

subject to no constraint.

The solution of this problem is

$$
\begin{aligned}
v_1 &= 1.045 \\
v_2 &= 1.033 \\
\delta_1 &= 0.002
\end{aligned}
$$

■

3.4.2 Optimal Power Flow

The purpose of a power transmission network is to transmit electric power from power generators to loads. The objective of the optimal power flow problem is to determine the power production of every generator so that all loads are supplied at minimum cost, while satisfying network constraints. In addition to supplying the load, voltage magnitudes throughout the network should be at acceptable levels. Reactive power has to be transmitted throughout the network, and reactive power demands should be satisfied.

The net active (generation minus demand) power reaching a bus can be expressed as a function of all voltage magnitudes and angles in the network

$$
p_{Gi} - P_{Di} = v_i \sum_{k=1}^{n} y_{ik} v_k \cos(\delta_i - \delta_k - \theta_{ik})
$$

where p_{Gi} is the active power generation in bus i P_{Di}, the active power demand in bus i; v_i, the voltage magnitude; δ_i, the angle of bus i; y_{ik}, the modulus; and θ_{ik}, the argument of a complex constant which depends on the topology and physical structure of the network; and n, the number of buses of the network.

Analogously, the net reactive power reaching a bus can be expressed as a function of all voltage magnitudes and angles in the network;

$$
q_{Gi} - Q_{Di} = v_i \sum_{k=1}^{n} y_{ik} v_k \sin(\delta_i - \delta_k - \theta_{ik})
$$

where q_{Gi} is the reactive power generation in bus i; and Q_{Di} is the reactive power demand in bus i.

The voltage magnitude of any bus is limited above and below

$$
\underline{V}_i \leq v_i \leq \overline{V}_i
$$

where $\underline{V}_i$ is the lower bound of voltage magnitude in bus i, and $\overline{V}_i$ is the upper bound of the voltage magnitude at bus i.

Generators can produce active power above a lower bound and below an upper bound

$$\underline{P}_{Gi} \le p_{Gi} \le \overline{P}_{Gi}$$

where $\underline{P}_i$ is the minimum active output power of generator i and $\overline{P}_i$ is the maximum one.

Similarly, generators can produce reactive power between a lower and an upper bound

$$\underline{Q}_{Gi} \le q_{Gi} \le \overline{Q}_{Gi}$$

where $\underline{Q}_i$ is the minimum reactive output power of generator i and $\overline{Q}_i$ is the maximum one.

The four main elements in this problem are

1. **Data**

   n: the number of buses in the network

   (y_{ik}, θ_{ik}): a complex constant given as its magnitude y_{ik}, and argument θ_{ik} which depends on the topology and physical structure of the network

   P_{Di}: the active power demand in bus i

   Q_{Di}: the reactive power demand in bus i

   $\underline{V}_i$: the lower bound of the voltage magnitude in bus i

   $\overline{V}_i$: the upper bound of the voltage magnitude at bus i

   $\underline{P}_{Gi}$: the minimum active output power of generator i

   $\overline{P}_{Gi}$: the maximum active output power of generator i

   $\underline{Q}_{Gi}$: the minimum reactive output power of generator i

   $\overline{Q}_{Gi}$: the maximum reactive output power of generator i

   C_i: the cost of producing one unit of active power using generator i

2. **Variables**

   v_i: the voltage at bus i

   δ_i: the angle at bus i

   p_{Gi}: the active power generation in bus i

   q_{Gi}: the reactive power generation in bus i

3. **Constraints.** There are several types of constraints.

   (a) Active power balance constraints:

$$p_{Gi} - P_{Di} = v_i \sum_{k=1}^{n} y_{ik} v_k \cos(\delta_i - \delta_k - \theta_{ik}), \quad \forall i = 1, 2, \ldots, n$$

(b) Reactive power balance constraints:

$$q_{Gi} - Q_{Di} = v_i \sum_{k=1}^{n} y_{ik} v_k \sin(\delta_i - \delta_k - \theta_{ik}), \;\; \forall i = 1, 2, \ldots, n$$

(c) Variable bounds:

$$\begin{array}{ccccll} \underline{V}_i & \leq & v_i & \leq & \overline{V}_i, & \forall i = 1, 2, \ldots, n \\ \underline{P}_{Gi} & \leq & p_{Gi} & \leq & \overline{P}_{Gi}, & \forall i = 1, 2, \ldots, n \\ \underline{Q}_{Gi} & \leq & q_{Gi} & \leq & \overline{Q}_{Gi}, & \forall i = 1, 2, \ldots, n \end{array}$$

(d) Bounds on angles:

$$-\pi \leq \delta_i \leq \pi, \;\; \forall i = 1, 2, \ldots, n$$

The origin for angles can be arbitrarily selected in any fixed bus (e.g., bus k, $\delta_k = 0$).

4. **Function to be optimized.** Given C_i, the cost of producing one unit of active power using generator i, the optimal power flow problem becomes minimization of

$$Z = \sum_{i=1}^{n} C_i p_{Gi}$$

subject to all the preceding constraints.

Example 3.2 (Optimal power flow). Consider a power network including 3 buses and 3 lines. Network parameters are as follows

$$\begin{array}{ll} y_{11} = 22.97, & \theta_{11} = -1.338 \\ y_{22} = 21.93, & \theta_{22} = -1.347 \\ y_{33} = 20.65, & \theta_{33} = -1.362 \\ y_{12} = y_{21} = 12.13, & \theta_{12} = \theta_{21} = 1.816 \\ y_{13} = y_{31} = 10.85, & \theta_{13} = \theta_{31} = 1.789 \\ y_{23} = y_{32} = 9.81, & \theta_{23} = \theta_{32} = 1.768 \end{array}$$

Bus 1 is a generating bus with lower and upper bounds on active power generation of 0 and 3, respectively, and of reactive power generation of -1 and 2, respectively. The active power generation lower and upper limits of generator 2 are, respectively, 0 and 3, and reactive power limits -1 and 2, respectively. Bus 3 is a load bus with a demand of 4.5 and 1.5. The voltage magnitude lower and upper limits for buses 2 and 3 are, respectively, 0.95 and 1.10, and for bus 1 are 0.95 and 1.13, respectively. The production cost from generator of bus 1 is 6 and from generator of bus 2 is 7. Consider a period of 1 hour and assume that the voltage angle reference in bus 3. This problem has the following structure. Minimize

$$6p_{G1} + 7p_{G2}$$

subject to

$$\begin{aligned}
0 &= p_{G1} - 22.97v_1^2 \cos(1.338) - 12.13v_1v_2 \cos(\delta_1 - \delta_2 - 12.127) \\
&\quad -10.85v_1v_3 \cos(\delta_1 - \delta_3 - 10.846) \\
0 &= p_{G2} - 21.93v_2^2 \cos(1.347) - 12.13v_2v_1 \cos(\delta_2 - \delta_1 - 12.127) \\
&\quad -9.81v_2v_3 \cos(\delta_2 - \delta_3 - 9.806) \\
0 &= -4.5 - 20.65v_3^2 \cos(1.362) - 10.85v_3v_1 \cos(\delta_3 - \delta_1 - 10.846) \\
&\quad -9.81v_3v_2 \cos(\delta_3 - \delta_2 - 9.806) \\
0 &= q_{G1} - 22.97v_1^2 \sin(1.338) - 12.13v_1v_2 \sin(\delta_1 - \delta_2 - 12.127) \\
&\quad -10.85v_1v_3 \sin(\delta_1 - \delta_3 - 10.846) \\
0 &= q_{G2} - 21.93v_2^2 \sin(1.347) - 12.13v_2v_1 \sin(\delta_2 - \delta_1 - 12.127) \\
&\quad -9.81v_2v_3 \sin(\delta_2 - \delta_3 - 9.806) \\
0 &= -1.5 - 20.65v_3^2 \sin(1.362) - 10.85v_3v_1 \sin(\delta_3 - \delta_1 - 10.846) \\
&\quad -9.81v_3v_2 \sin(\delta_3 - \delta_2 - 9.806)
\end{aligned}$$

$$\begin{array}{rcccr}
0.95 & \leq & v_1 & \leq & 1.13 \\
0.95 & \leq & v_2 & \leq & 1.10 \\
0.95 & \leq & v_3 & \leq & 1.10 \\
0 & \leq & p_{G1} & \leq & 3 \\
0 & \leq & p_{G2} & \leq & 3 \\
-1 & \leq & q_{G1} & \leq & 2 \\
-1 & \leq & q_{G2} & \leq & 2 \\
-\pi & \leq & \delta_1 & \leq & \pi \\
-\pi & \leq & \delta_2 & \leq & \pi \\
 & & \delta_3 & = & 0
\end{array}$$

The local optimal solution, that has been obtained using the GAMS package (see Section 11.4.11), is

$$\begin{array}{rcl}
Z & = & 30.312 \\
p_{G1} & = & 3.000 \\
p_{G2} & = & 1.759 \\
q_{G1} & = & 1.860 \\
q_{G2} & = & 0.746 \\
v_1 & = & 1.130 \\
v_2 & = & 1.100 \\
v_3 & = & 0.979 \\
\delta_1 & = & 0.190 \\
\delta_2 & = & 0.174
\end{array}$$

■

Table 3.1: Example of a contingency table

	t_{ij}				
	B_1	B_2	...	B_n	r_i
A_1	t_{11}	t_{12}	...	t_{1n}	r_1
A_2	t_{21}	t_{22}	...	t_{2n}	r_2
$\vdots$	$\vdots$	$\vdots$	$\vdots$	$\vdots$	$\vdots$
A_m	t_{m1}	t_{m2}	...	t_{mn}	r_m
c_j	c_1	c_2	...	c_n	

3.5 The Matrix Balancing Problem

Engineers, statisticians, sociologists, and other people use experimental information that is structured in contingency tables (see Table 3.1).

These tables use two criteria to classify the items. The first criterion, named A, has m categories and the second, named B, has n categories. Thus, an item can be classified in one of the $m \times n$ cells. The content of these cells can vary in time. The matrix balancing problem deals with updating these contingency tables assuming that the row and column sums of the updated tables are known.

For example, suppose that we are interested in knowing the people involved in domestic migration, and how this migration changes with time. Each migration can be classified according to its origin and destination. Our aim consists of estimating the number of people migrating from any pair of cities. To this end, we can only work with a census that, due to cost reasons, is made every 10 years. In addition, we have information about the net yearly migration in each city, which must be used to update the yearly contingency tables.

Consider an out-of-date $m \times n$ contingency table $\{t_{ij}\}$, and denote the updated one as $\mathbf{T} = (T_{ij})$, where T_{ij} is the cell $i - j$ content. Assume that the sum of columns an rows of the new matrix $\mathbf{T}$ are known, i.e., that we have the constraints

$$\sum_{j=1}^{n} T_{ij} = r_i, i = 1, \ldots, m \tag{3.7}$$

$$\sum_{i=1}^{m} T_{ij} = c_j, j = 1, \ldots, n \tag{3.8}$$

where the parameters $r_i; i = 1, \ldots, m$ and $c_j; j = 1, \ldots, n$ are known.

We need to introduce additional constraints to reflect that (in the majority of the applications) the cell represents counts of items (are nonnegative):

$$T_{ij} \geq 0, \ i = 1, \ldots, m; \ \ j = 1, \ldots, n$$

Our objective consists of obtaining a matrix with the same structure as the old contingency table but satisfying the constraints on the sum of rows and columns. Thus, the selected objective function is a distance between $\{t_{ij}\}$ and $\{T_{ij}\}$:

$$Z = F(\mathbf{T}, \mathbf{t}) \tag{3.9}$$

where F is a gauge between $\mathbf{T}$ and $\mathbf{t}$.

One possibility consists of using the weighted least-squares function

$$Z = \sum_{i=1}^{m} \sum_{j=1}^{n} \omega_{ij} (T_{ij} - t_{ij})^2 \tag{3.10}$$

where ω_{ij} (the weights) are given positive scalars. An alternative, is the (minus) entropy function

$$Z = \sum_{i=1}^{m} \sum_{j=1}^{n} \frac{T_{ij} \log T_{ij}}{t_{ij}} - T_{ij} + t_{ij} \tag{3.11}$$

Note that $Z = 0$ when $T_{ij} = t_{ij}; \forall i, j$, and that Z increases with the discrepancy between T_{ij} and t_{ij}.

To summarize, the main elements involved in this problem are

1. **Data**

   r_i: the sum of all elements in row i

   c_j: the sum of all elements in column j

   t_{ij}: the observed cell $i - j$ content

2. **Variables**

   T_{ij}: the estimated cell $i - j$ content

3. **Constraints**

$$\begin{aligned} \sum_{j=1}^{n} T_{ij} &= r_i, i = 1, \ldots, m & (3.12) \\ \sum_{i=1}^{m} T_{ij} &= c_j, j = 1, \ldots, n & (3.13) \\ T_{ij} &\geq 0, \; i = 1, \ldots, m; \; j = 1, \ldots, n & (3.14) \end{aligned}$$

4. **Function to be minimized.** Function (3.10) or (3.11).

Example 3.3 (Trip distribution). One example of a practical problem that can be addressed using this model is the problem of predicting the distribution matrix T of trips on a city. In this example we discuss this application. Consider a small area that has been divided into four zones and assume that a limited survey has resulted in the trip matrix in Table 3.2.

Table 3.2: Origin-destination matrix obtained after a survey

	t_{ij}				
	1	2	3	4	r_i
1	–	60	275	571	906
2	50	–	410	443	903
3	123	61	–	47	231
4	205	265	75	–	545
c_j	378	386	760	1061	2585

Table 3.3: Estimated future interzonal movements and the estimates for future total trip ends for each zone

Zones	Estimated future origins (r_i)	Estimated future destinations (c_j)
1	12,000	6,750
2	10,500	7,300
3	3,800	10,000
4	7,700	9,950

The two sets of constraints (3.7)–(3.8) reflect our knowledge about trip generation and attraction in the zones of the area under study. We are interested only in matrix entries that can be interpreted as trips. The estimated future inter-zonal movements and the estimates for future total trip ends for each zone are as given in Table 3.3.

In this problem the intrazonal variables are dropped, that is, the variables T_{ii} $i = 1, \ldots, 4$ have been dropped from the formulation.

We have used two gauges. The first one is based on (3.11), and the solution is given in Table 3.4. The second estimates are obtained using (3.9), with $\omega_i = 1/T_{ij}$. The solution is given in Table 3.5.

■

3.6 The Traffic Assignment Problem

Traffic planning and transportation analysis problems have motivated a large amount of mathematical models. The analysis of these models of traffic planning problems aids planners in predicting what effect changes in the traffic network topology will have on the network's performance. The typical transportation model used in the applications has four stages:

Table 3.4: Solution based on the entropy function

	T_{ij}				
	1	2	3	4	r_i
1	–	1997.893	4176.466	5825.642	12,000
2	1364.381	–	5293.379	3842.240	10,500
3	2322.858	1195.024	–	282.118	3,800
4	3062.761	4107.084	530.156	–	7,700
c_j	6750	7300	10,000	9950	34,000

Table 3.5: Solution based on the weighted least squares method

	T_{ij}				
	1	2	3	4	r_i
1	–	1700.324	4253.616	6046.060	12,000
2	1250.923	–	5484.550	3764.527	10,500
3	2396.886	1263.701	–	139.413	3,800
4	3102.191	4335.975	261.835	–	7,700
c_j	6750	7300	10,000	9950	34,000

1. **Trip generation phase.** This approach starts by considering a zoning and network system, and database of each zone of the study. These data, which include information about economic activity, social distribution, educational and recreational facilities, and shopping space, are used to estimate the model of the total number of trips generated and attracted by each zone of the study area.

2. **Distribution phase.** The following stage is the allocation of these trips to particular destinations: their distribution over space, forming a trip matrix.

3. **Modal split.** Next is the split modal step, which produces the allocation of trips to different transportation modes. In this phase the origin–destination matrices are obtained for each transportation mode. Their elements are the total number of trips by a transport mode for the origin–destination O–D pair ω.

4. **Assignment.** Finally, the last stage requires the *assignment* of these trips to the transportation network. This example deals with the assignment of private cars in the road network.

We must introduce a principle that governs the users' behavior in their choice of the route in the transport network. Wardrop [102]) was the first to formally enunciate this principle: "Under equilibrium conditions traffic arranges itself in congested networks in such a way that no individual trip maker can reduce his paths cost by switching routes."

This principle has been used as a framework to build equilibrium assignment models. A corollary of this principle is that if all tripmakers perceive costs in the same way, under the equilibrium conditions traffic arranges itself in congested networks such that all used routes between an O-D pair have equal and minimum costs while all unused routes have greater or equal costs.

Beckman et al. [10] formulated the following optimization problem to express the equilibrium condition derived from Wardrop's first principle. This model predicts the utilization level of the different arcs in the network. Thus, it can be used to answer questions such as what would happen in the network service level if a new road is built or the capacity of an existing road is modified.

The main elements of this problem are as follows:

1. **Data**

$(\mathcal{N}, \mathcal{A})$: a directed graph $(\mathcal{N}, \mathcal{A})$, which is viewed as a model of a road network with a set of nodes $\mathcal{N}$, and a set of arcs $\mathcal{A}$ that represent transportation links such as roads, highways, and streets. The set of nodes $\mathcal{N}$ of the graph represent junction points or the so-called centroids, which represent the zones of the study (origins and destinations).

W: the set of origin–destination pairs.

d_ω: input representing the number of trips by car from the origin i to destination j, for any origin–destination pair $\omega = (i, j)$. The origin–destination matrix $\{d_\omega\}_{\omega \in W}$ is obtained in the modal allocation phase.

$C_a(f_a)$: a cost function that produces the delay on arc $a \in \mathcal{A}$, for each arc $(i, j) \in \mathcal{A}$, as a function of the total flow f_a carried by the arc a. We assume that the vehicles use any road segment. Therefore the road become increasingly congested and so the delay on the road increases.

$\mathcal{R}_\omega$: the set of simple routes from $\omega = (i, j)$.

2. **Variables**

h_r: the flow on route r

f_a: the flow on link a

3. **Constraints.** The number of users of an origin–destination pair ω is the sum of the total amount of users in different paths that satisfy such a pair:

$$\sum_{r \in \mathcal{R}_\omega} h_r = d_\omega, \ \forall \omega \in W \tag{3.15}$$

Moreover, the path flow must be nonnegative:

$$h_r \geq 0, \ \forall r \in \mathcal{R}_\omega, \ \forall \omega \in W \tag{3.16}$$

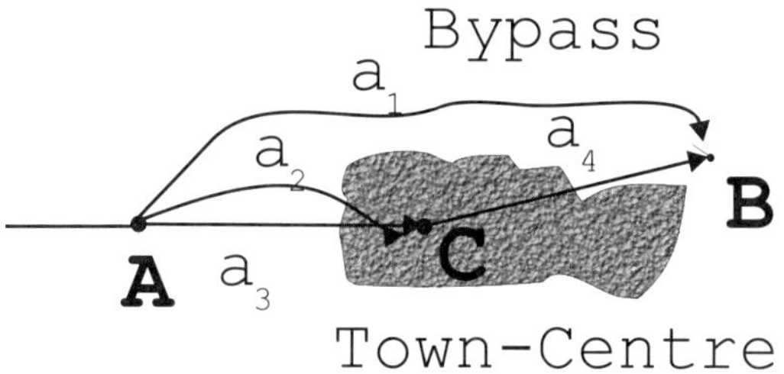

Figure 3.8: Diagram showing the road network.

The flow on each link a is the sum of the flow in all paths that use it:

$$\sum_{w \in W} \sum_{r \in \mathcal{R}_\omega} \delta_{a,r} h_r = f_a \quad \forall a \in \mathcal{A}$$

where $\delta_{a,r} = 1$ if $r \in \mathcal{R}_\omega$ contains arc a, and 0 otherwise.

4. **Function to be optimized.** The traffic assignment problem minimizes the following function:

$$Z = \sum_{a \in \mathcal{A}} \int_0^{f_a} C_a(x) dx \tag{3.17}$$

As a result of growing traffic volumes, the speed on a link tends to decrease. The function C_a, that is the travel time on the link a, takes into account this behavior. These functions are usually modeled as positive, nonlinear, and strictly increasing functions in the analysis of traffic systems. The basic parameters of a link performance function, relating travel time, C_a, on link a to the flow f_a, on the link, is the *free-flow travel time*, c_a^0, which is measure of the travel time at zero flow, and the practical capacity of the link, k_a, which is a measure of the flow from which the travel time will increase very rapidly if the flow is further increased. The most common expression for $C_a(f_a)$, called the BPR function, is

$$C_a(f_a) = c_a^0 + b_a \left(\frac{f_a}{k_a} \right)^{n_a} \tag{3.18}$$

Example 3.4 (Traffic assignment problem). Consider the problem of a town served by a bypass and several town-center routes as is illustrated in Figure 3.8. Assume that there are 4000 trips from A to B, and 2500 trips from A to C. The routes available to satisfy the pair of demand $\omega_1 = (A, B)$ are $r_1 = a_1$, $r_2 = a_2 - a_4$, and $r_3 = a_3 - a_4$, and the routes for the pair $\omega_2 = (A, C)$ are $r_4 = a_2$ and $r_5 = a_3$. In this example $W = \{w_1, w_2\}$, and $\mathcal{R}_{\omega_1} = \{r_1, r_2, r_3\}$ and $\mathcal{R}_{\omega_2} = \{r_4, r_5\}$. The path flow variables are $h_1, \ldots, h_5$, and the line flow variables are $f_1, \ldots, f_4$.

Table 3.6: .Parameters for BPR functions

Link a	k_a	c_a^0	b_a	n_a
1	500	5	1	4
2	400	7	1	4
3	400	10	1	4
4	500	2	1	4

Table 3.7: Estimated trips

Flow on links		Flow on routes	
a_1	3845.913	r_1	3845.913
a_2	2354.419	r_2	154.087
a_3	299.667	r_3	0.000
a_4	154.087	r_4	2200.333
		r_5	299.667

Since

$$\begin{aligned}\int_0^{f_a} C_a(x)dx &= \int_0^{f_a} c_a^0 + b_a \left(\frac{x}{k_a}\right)^{n_a} dx \\ &= c_a^0 f_a + \frac{b_a}{n_a+1}\left(\frac{f_a}{k_a}\right)^{n_a+1}\end{aligned}$$

the full formulation is as follows. Minimize

$$Z = \sum_{i=1}^{4} \int_0^{f_{a_i}} C_{a_i}(x)dx = \sum_{a\in\mathcal{A}} c_{a_i}^0 f_{a_i} + \frac{b_{a_i}}{n_{a_i}+1}\left(\frac{f_{a_i}}{k_{a_i}}\right)^{n_{a_i}+1} \tag{3.19}$$

subject to

$$\begin{aligned}h_1 + h_2 + h_3 &= 4000 &(3.20)\\ h_4 + h_5 &= 2500 &(3.21)\\ h_1 &= f_1 &(3.22)\\ h_2 + h_4 &= f_2 &(3.23)\\ h_3 + h_5 &= f_3 &(3.24)\\ h_2 + h_3 &= f_4 &(3.25)\\ h_1, \ldots, h_5 &\geq 0 &(3.26)\end{aligned}$$

In this example we have utilized the functions in (3.18), and Table 3.6 shows the used parameters. The corresponding solution is given in Table 3.7.

■

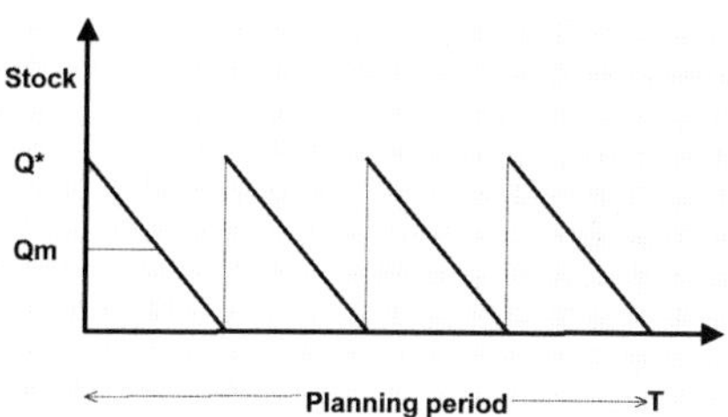

Figure 3.9: Stock evolution with time.

Exercises

3.1 A table is to be pushed through the right angle of a corridor of A units width to another one of B units width. What are the sizes of the table (a units width and b units long) that allow such movement?

3.2 A hospital is to be built between two cities a distant D apart. A decision is made about the exact location being the least contaminated place in the line joining the two towns. Knowing that contamination is proportional to industry and to the inverse of distance plus 1, and that the second city generates 3 times more industrial activity than the first, determine the location for the hospital.

3.3 Consider the problem of finding the minimum length of a ladder that must lean against a wall if a box of dimensions a and b is placed right at the corner of that same wall. Formulate this problem as a nonlinear programming one.

3.4 Consider the problem of finding the maximum length of a beam that can be pushed through the door of a room assuming that the height of the door is h and is located a distance d from the facing wall. The height of the room can be assumed infinite. Formulate it as a nonlinear programming problem.

3.5 Consider the following problem of management of inventories. The demand of good is constant along a planning period of 365 days. The warehouseperson takes into account a storage cost of $C_s = \$1$ per day and tons of the good, and a fixed cost to place a request (independently of the size of the lot) of $C_r = \$18$. The demand is 365 tons for the planning period. Figure 3.9 shows the evolution of the stock assuming that the depot is empty at the beginning of the considered period, and an order is placed at the initial time. The mean stock, Q_m, is computed as $Q/2$, where Q is the amount of request. Formulate a nonlinear problem to decide the size lot of the request in order to minimize the inventory cost.

3.6 Consider problem 3.5 but adding a new item. The parameters for each item are shown in Table 3.8. Assume that the depot has a maximum

Table 3.8: Storage and order cost of the two items

Good	C_s	C_r	Demand
A	\$1	\$18	365 tons
B	\$2	\$25	365 tons

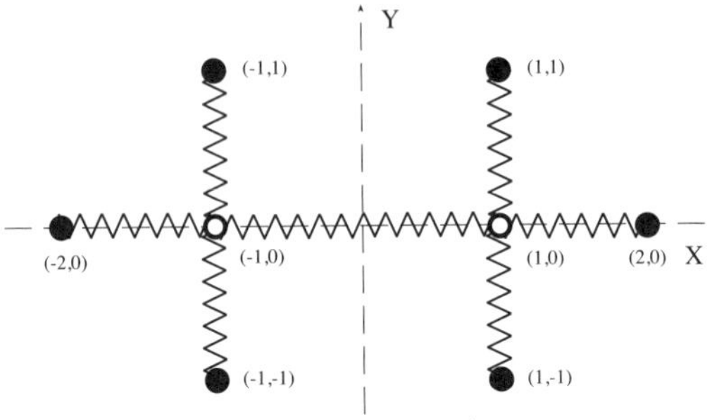

Figure 3.10: System of springs.

storage capacity of 60 tons. Formulate the corresponding problem as a nonlinear programming problem.

3.7 An electricity producer should plan its hourly energy production to maximize its profits from selling energy during a planning horizon of a given number of hours. Formulate a nonlinear programming problem taking into account that:

(a) The producer does not produce before the planning horizon.

(b) Hourly energy prices decrease linearly with the energy production of the producer in the corresponding hour.

(c) The minimum energy production of the producer in each hour is zero and the maximum one a given quantity.

(d) Energy productions in two consecutive hours cannot differ by more than a prespecified amount.

(e) Producer production cost is linear.

3.8 Consider the system of springs in Figure 3.10. The black dots are fixed joints; white dots are movable points. All joints are free to rotate so that the springs are always placed along the line joining its two endpoints. Each spring is characterized by a positive constant k_i, $i = 1, \ldots, 7$ in suitable units. We postulate that the equilibrium configuration of the white dots is the one minimizing the global "energy" of all the springs where the energy of one spring is proportional (with proportionality constant k_i) to the square of the distance between the endpoints. Formulate the problem as a NLPP.

Part II

Methods

Chapter 4

An Introduction to Linear Programming

4.1 Introduction

Linear programming has many interesting applications, as has been illustrated with the collection of examples in Chapter 1, where the main elements appearing in LPP were introduced. This chapter presents a more formal description of the linear programming problem. In particular, the most important concepts used in LPP are defined, and the different types of solutions are described.

More precisely, in Section 4.2 we introduce the linear programming problem and illustrate the cases of unique and multiple solutions, and the cases of unbounded and nonfeasible solutions. Section 4.3 describes the standard form of a linear programming problem and shows how a LPP can be transformed into standard form. Section 4.4 deals with basic solutions, which are the candidate solutions for the optimal values of the objective function. In Section 4.5 the problem of sensitivities with respect to the parameters used in the constraints is analyzed. Finally, in Section 4.6, we show how every linear programming problem has a dual one, and we give some examples with their corresponding interpretations of the primal and dual approaches.

4.2 Problem Statement and Basic Definitions

Linear programming deals with the problem of optimizing (minimizing or maximizing) a linear function of n variables subject to equality and/or inequality linear constraints. In other words, a linear programming problem consists of finding the optimum (maximum or minimum) of a linear function in a set that can be written as the intersection of a finite number of hyperplanes and halfspaces in $\mathbb{R}^n$.

We have the following definitions.

Definition 4.1 (Linear programming problem). *The most general form of a linear programming problem (LPP) is to minimize or maximize*

$$Z = f(\mathbf{x}) = \sum_{j=1}^{n} c_j x_j \tag{4.1}$$

subject to

$$\begin{array}{rcll} \sum_{j=1}^{n} a_{ij}x_j & = & b_i, & i = 1, 2, \ldots p-1 \\ \sum_{j=1}^{n} a_{ij}x_j & \geq & b_i, & i = p, \ldots, q-1 \\ \sum_{j=1}^{n} a_{ij}x_j & \leq & b_i, & i = q, \ldots, m \end{array} \tag{4.2}$$

where p, q, and m are positive integers such that

$$1 \leq p \leq q \leq m$$

■

What distinguishes a LPP from any other optimization problem is that all functions involved are linear. Just one nonlinear function in the statement of the problem is sufficient to refuse it as a LPP. In this book we always assume that we have a finite number of constraints.

The (linear) function in (4.1) is known as the objective or the cost function, and is the function to be optimized. Notice that we can have all different possibilities in (4.2) for the operators relating the two sides of the (linear) constraints, depending on the values of p and q. As special cases we can have only equalities, only inequalities of one type, inequalities of both types, equalities and inequalities, and so on.

In this book, unless otherwise indicated, we assume that we deal with minimization problems.

Definition 4.2 (Feasible solution). *A point $\mathbf{x} = (x_1, x_2, \ldots, x_n)$ satisfying all the constraints in (4.2) is called a feasible solution. The collection of all such solutions is the feasible region.* ■

Definition 4.3 (Optimal solution). *A feasible point $\tilde{\mathbf{x}}$ such that $f(\mathbf{x}) \geq f(\tilde{\mathbf{x}})$ for any other feasible point $\mathbf{x}$ is called an optimal solution of the problem.* ■

When one deals with an optimization problem, the main interest consists of finding a global optimum. Unfortunately, standard mathematical conditions for optima only guarantee local optima, if they exist (see Chapter 8). However, linear problems have the following interesting properties that facilitate the search for global optima:

- If the feasible region is bounded, the problem always has a solution (it is a sufficient but not necessary condition for existence of a solution).

- The optimum of a linear programming problem is always a global optimum.
- If $\mathbf{x}$ and $\mathbf{y}$ are optima of a linear programming problem, then any convex linear combination of them is also an optimum. Note that linear convex combinations of points with the same cost function values do not change the value of the cost function.
- The optimum is always attained, at least, at an extreme point of the feasible region. This is shown in Theorem 4.2, Section 4.2.

The following examples illustrate the cases of unique, multiple, unbounded, and nonfeasible solution.

Example 4.1 (Unique solution). Consider the following linear programming problem. Maximize

$$Z = 3x_1 + x_2$$

subject to

$$\begin{array}{rrrr} -x_1 & +x_2 & \leq & 2 \\ x_1 & +x_2 & \leq & 6 \\ x_1 & & \leq & 3 \\ 2x_1 & -x_2 & \leq & 4 \\ & -x_2 & \leq & 0 \\ -x_1 & -x_2 & \leq & -1 \\ -x_1 & & \leq & 0 \end{array} \tag{4.3}$$

Figure 4.1 shows the feasible region (shadowed region), the contour plots (dashed lines) of the objective function, and its gradient vector indicating the increasing direction of the objective function. The solution is reached at point P, because it is on the last contour plot with one feasible point in the indicated direction. P is therefore the intersection of the lines

$$\begin{array}{rrrr} x_1 & +x_2 & = & 6 \\ x_1 & & = & 3 \end{array}$$

So the maximum, $Z = 12$, is attained at the point

$$P = (3, 3)$$

■

Example 4.2 (Multiple solutions). If the objective function of Example 4.1 is replaced by

$$Z = x_1 + x_2$$

the resulting contour plots are parallel to one of the constraint lines (the second). In this case, the optimum is reached at the points of a polytope's edge, as shown in Figure 4.2.

Any point on the line segment between the points $(2, 4)^T$ and $(3, 3)^T$ leads to a global maximum of the problem ($Z = 6$). ■

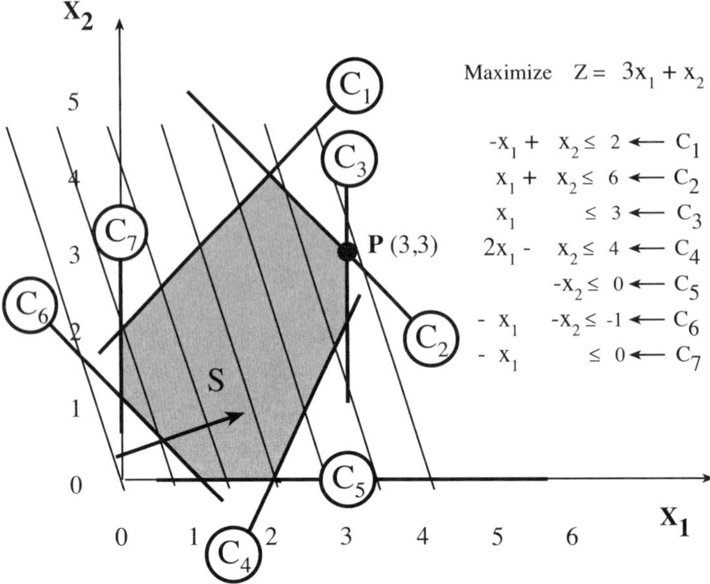

Figure 4.1: Graphical illustration of a linear programming problem with a unique solution.

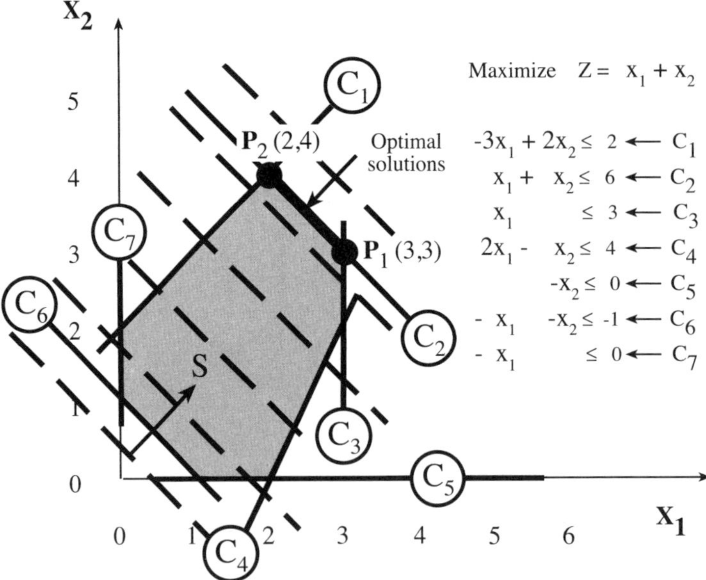

Figure 4.2: Graphical illustration of a linear programming problem with multiple solutions.

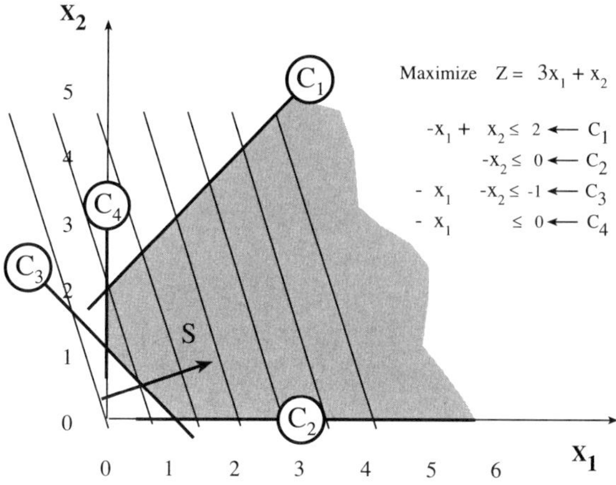

Figure 4.3: Graphical illustration of a linear programming problem with unbounded feasible region.

Example 4.3 (Unbounded example). Consider the following LPP, where we maximize

$$Z = 3x_1 + x_2 \tag{4.4}$$

subject to

$$\begin{array}{rrrr} -x_1 & +x_2 & \leq & 2 \\ & -x_2 & \leq & 0 \\ -x_1 & -x_2 & \leq & -1 \\ -x_1 & & \leq & 0 \end{array} \tag{4.5}$$

which has unbounded solution, because, as shown in Figure 4.3, the feasible region is unbounded in the increasing direction of the objective function.

■

Example 4.4 (Nonfeasible solution). Consider the following LPP. Maximize

$$Z = 3x_1 + x_2 \tag{4.6}$$

subject to

$$\begin{array}{rrrr} -x_1 & +x_2 & \leq & 2 \\ x_1 & +x_2 & \leq & 6 \\ x_1 & & \leq & 3 \\ 2x_1 & -x_2 & \leq & 4 \\ & -x_2 & \leq & 0 \\ -x_1 & -x_2 & \leq & -1 \\ -x_1 & & \leq & 0 \\ x_1 & +x_2 & \leq & 0 \end{array} \tag{4.7}$$

This problem is nonfeasible because the new constraint $x_1 + x_2 \leq 0$ is not compatible with the previous ones. ■

The vertices of the polytopes (polygons) in Figures 4.1–4.3 are called extreme points.

4.3 Linear Programming Problem in Standard Form

Since LPP can come in a variety of forms, to unify its understanding and treatment, it is convenient to transform it into what is usually called its *standard form.* Sometimes this process must be carried out before it can be solved and an optimal solution found.

The description of a LPP in standard form requires three main ingredients:

1. A vector $\mathbf{c} \in \mathbb{R}^n$
2. A nonnegative vector $\mathbf{b} \in \mathbb{R}^m$
3. An $m \times n$ matrix, $\mathbf{A}$

With these elements, the associated LPP in standard form is as follows. Minimize

$$Z = \mathbf{c}^T \mathbf{x} \tag{4.8}$$

subject to

$$\begin{array}{rcl} \mathbf{Ax} & = & \mathbf{b} \\ \mathbf{x} & \geq & \mathbf{0} \end{array} \tag{4.9}$$

Here, $\mathbf{c}^T\mathbf{x}$ denotes the dot product of the vectors $\mathbf{c}$ and $\mathbf{x}$, $\mathbf{Ax}$ is the product of the matrix $\mathbf{A}$ and the vector $\mathbf{x}$, and $\mathbf{x} \geq \mathbf{0}$ requires all components of feasible vectors to be nonnegative. It is in this form that a LPP is typically studied. Typically, n is much greater than m.

In summary, a problem is said to be in standard form iff (if and only if)

1. It is a minimization problem.
2. It contains only equality constraints.
3. The $\mathbf{b}$ vector is nonnegative.
4. The variables $\mathbf{x}$ are nonnegative.

Before starting to analyze such a problem, it is important to convince our readers that any problem in the form (4.1)–(4.2) can be suitably transformed into standard form.

4.3.1 Transformation to Standard Form

A linear programming problem can always be formulated in a standard form by performing some manipulations:

1. Unrestricted variables can be replaced by differences between two nonnegative variables. If some (or all) of the variables are not restricted to be nonnegative we can force them to be nonnegative by introducing the positive and negative parts of a variable x_i, defined by $x_i^+ = \max\{0, x_i\}$ and $x_i^- = \max\{0, -x_i\}$, respectively. It is elementary to check that $x = x_i^+ - x_i^-$, $|x| = x_i^+ + x_i^-$ and both x_i^+ and x_i^- are nonnegative. If the number of unrestricted variables is r, using this method implies adding r new variables. An alternative and *better solution* is given in Chapter 13, Section 13.2.1.

2. Inequality constraints can be converted to equality constraints by adding new variables known as *slack variables*:

   - If
   $$a_{i1}x_1 + a_{i2}x_2 + \cdots + a_{in}x_n \leq b_i$$
   then, there exists a variable $x_{n+1} \geq 0$ such that
   $$a_{i1}x_1 + a_{i2}x_2 + \cdots + a_{in}x_n + x_{n+1} = b_i$$

   - If
   $$a_{i1}x_1 + a_{i2}x_2 + \cdots + a_{in}x_n \geq b_i$$
   then, there exists a variable $x_{n+1} \geq 0$ such that
   $$a_{i1}x_1 + a_{i2}x_2 + \cdots + a_{in}x_n - x_{n+1} = b_i$$

3. A maximization problem is equivalent to a minimization one by a change of sign in the objective function. Specifically, what we mean is that maximizing
$$Z_{max} = \mathbf{c}^T\mathbf{x}$$
is equivalent to minimizing
$$Z_{min} = -\mathbf{c}^T\mathbf{x}$$
when both problems are subject to the same constraints. Note that the optimal values are attained at the same points, but $Z_{max} = -Z_{min}$.

4. The sign of an equation with the corresponding independent term b nonpositive can be changed to get an equivalent equation with its corresponding independent term nonnegative.

To clarify the transformation to standard form, we give several examples below.

Example 4.5 (Standard form). The standard form of the linear programming problem, maximizing

$$Z = 2x_1 - 3x_2 + 5x_3$$

subject to

$$\begin{array}{rrrcr} x_1 & +x_2 & & \leq & 2 \\ 3x_1 & +x_2 & -x_3 & \geq & 3 \\ & & x_1,\ x_2 & \geq & 0 \end{array} \tag{4.10}$$

is minimizing

$$Z = -2x_1 + 3x_2 - 5(x_6 - x_7)$$

subject to

$$\begin{array}{rrrrrcr} x_1 & +x_2 & & +x_4 & & = & 2 \\ 3x_1 & +x_2 & -(x_6 - x_7) & & -x_5 & = & 3 \\ & & & & x_1,\ x_2,\ x_4,\ x_5,\ x_6,\ x_7 & \geq & 0 \end{array} \tag{4.11}$$

■

Example 4.6 (Nonnegative variables: standard form). Assume that we are given the following LPP, where we maximize

$$Z = 3x_1 - x_3$$

subject to

$$\begin{array}{rrrcr} x_1 & +x_2 & +x_3 & = & 1 \\ x_1 & -x_2 & -x_3 & \leq & 1 \\ x_1 & & +x_3 & \geq & -1 \\ x_1 & & & \geq & 0 \end{array} \tag{4.12}$$

First, we change sign of the third constraint; second, we must have the nonnegativeness of all variables involved. To this end we write

$$\begin{array}{rcl} x_2 & = & y_2 - z_2 \\ x_3 & = & y_3 - z_3 \\ y_2, y_3, z_2, z_3 & \geq & 0 \end{array} \tag{4.13}$$

where $y_2 = x_2^+$, $y_3 = x_3^+$, and $z_2 = x_2^-$, $z_3 = x_3^-$. Thus, the initial problem transforms to maximization of

$$Z = 3x_1 - y_3 + z_3$$

subject to

$$\begin{array}{rrrrrcr} x_1 & +y_2 & -z_2 & +y_3 & -z_3 & = & 1 \\ x_1 & -y_2 & +z_2 & -y_3 & +z_3 & \leq & 1 \\ -x_1 & & & -y_3 & +z_3 & \leq & 1 \\ x_1, & y_2, & z_2, & y_3, & z_3 & \geq & 0 \end{array} \tag{4.14}$$

Second, by means of the slack variables, u_1 and u_2, inequalities can be transformed into equalities

$$\begin{array}{lllllllll} x_1 & -y_2 & +z_2 & -y_3 & +z_3 & +u_1 & & = & 1 \\ x_1 & & & +y_3 & -z_3 & & -u_2 & = & -1 \end{array} \tag{4.15}$$

Finally, we change the maximum to a minimum by a change of sign of the objective function and minimize

$$Z = -3x_1 + y_3 - z_3$$

subject to

$$\begin{array}{rrrrrrrcl} x_1 & +y_2 & -z_2 & +y_3 & -z_3 & & & = & 1 \\ x_1 & -y_2 & +z_2 & -y_3 & +z_3 & +u_1 & & = & 1 \\ -x_1 & & & -y_3 & +z_3 & & +u_2 & = & 1 \\ x_1, & y_2, & z_2, & y_3, & z_3 & u_1 & u_2 & \geq & 0 \end{array} \tag{4.16}$$

Once the solution of this problem is found, the relationship between the variables x_1, x_2, x_3 and y_2, y_3, z_2, z_3 will provide the optimal solution for the original problem. Namely, the triplet $(x_1, y_2 - z_2, y_3 - z_3)$ will be an optimal vector for the initial problem if $(x_1, y_2, y_3, z_2, z_3, u_1, u_2)$ is an optimal point for the problem in standard form. Notice that the slack variables are removed from the final solution because they are auxiliary variables.

Finally, the optimum value of the objective function in the original problem is the negative of the optimum value of the objective function in the problem in standard form. ∎

4.4 Basic Solutions

Consider a linear programming problem in its standard matrix form (4.8)–(4.9), where we shall assume, without loss of generality, that the rank of the $m \times n$ matrix $\mathbf{A}$ is m (remember $m \leq n$), and that the linear system $\mathbf{Ax} = \mathbf{b}$ is solvable. Otherwise, the associated optimization problem is equivalent to another one with less constraints, or there are no feasible solutions, respectively.

Definition 4.4 (Basic matrix). *A nonsingular $m \times m$ submatrix $\mathbf{B}$ of $\mathbf{A}$ is called a basic matrix or the basis. $\mathbf{B}$ is also called a feasible basic matrix if and only if $\mathbf{B}^{-1}\mathbf{b} \geq 0$.* ∎

Each basic matrix has an associated vector called *basic solution.* The process for calculating this solution is the following.

Let $\mathbf{x}_B$ be the vector of variables associated with the columns of $\mathbf{A}$ used to build $\mathbf{B}$. Variables in $\mathbf{x}_B$ are known as *basic variables* and the remaining ones are called *nonbasic variables.* Giving zero values to nonbasic variables

$$(\mathbf{B} \quad \mathbf{N}) \begin{pmatrix} \mathbf{x}_B \\ \mathbf{0} \end{pmatrix} = \mathbf{b}, \;\; \mathbf{B}\mathbf{x}_B = \mathbf{b}, \;\; \mathbf{x}_B = \mathbf{B}^{-1}\mathbf{b} \tag{4.17}$$

where $\mathbf{N}$ is such that $\mathbf{A} = (\mathbf{B} \quad \mathbf{N})$. Therefore $\mathbf{B}^{-1}\mathbf{b}$ allows us to obtain the basic solution associated with $\mathbf{B}$. If $\mathbf{B}$ is a feasible basic matrix, its basic solution is said to be feasible.

The number of feasible basic solutions of a bounded linear programming problem with a finite number of constraints is always finite, and each of them corresponds to an extreme point of its feasible region. More precisely, the following theorem establishes this relationship between basic feasible solutions (BFSs) and extreme points.

Theorem 4.1 (Characterization of extreme points). *Let* $S = \{\mathbf{x} : \mathbf{A}\mathbf{x} = \mathbf{b}, \mathbf{x} \geq \mathbf{0}\}$, *where* $\mathbf{A}$ *is an* $m \times n$ *matrix of rank* m, *and* $\mathbf{b}$ *is an* m *vector. A point* $\mathbf{x}$ *is an extreme point of* S *if and only if* $\mathbf{A}$ *can be decomposed into* $(\mathbf{B}, \mathbf{N})$ *such that*

$$\mathbf{x} = \begin{pmatrix} \mathbf{x}_B \\ \mathbf{x}_N \end{pmatrix} = \begin{pmatrix} \mathbf{B}^{-1}\mathbf{b} \\ \mathbf{0} \end{pmatrix}$$

where $\mathbf{B}$ *is an* $m \times m$ *invertible matrix satisfying* $\mathbf{B}^{-1}\mathbf{b} \geq \mathbf{0}$. ∎

For the proof of this theorem the reader is referred to Bazaraa et al. [9]. The following theorem explains why the basic solutions are important.

Theorem 4.2 (Main LPP theorem). *If a linear programming problem has an optimal solution, it also has an optimal basic feasible solution.* ∎

Proof. A polyhedral region can always be written as

$$\mathbf{x} = \sum_i \rho_i \mathbf{v}_i + \sum_j \pi_j \mathbf{w}_j + \sum_k \lambda_k \mathbf{q}_k$$

where $\rho_i \in \mathbb{R}$, $\pi_j \in \mathbb{R}^+$, and $0 \leq \lambda_k \leq 1; \sum_k \lambda_k = 1$.

The feasible region of a standard LPP does not contain a linear space, because of the constraint $\mathbf{x} \geq \mathbf{0}$. Thus, in this special case the polyhedral set can be written as the sum of a polytope plus a cone, the minimal set of generators of the polytope is the set of its extreme points, and the minimal set of generators of the cone is the set of its extreme directions.

Then, the value of the objective function $Z = \mathbf{c}^T\mathbf{x}$ to be minimized becomes

$$Z = \mathbf{c}^T\mathbf{x} = \sum_j \pi_j \mathbf{c}^T \mathbf{w}_j + \sum_k \lambda_k \mathbf{c}^T \mathbf{q}_k$$

For the problem to have a bounded solution, we should have

$$\mathbf{c}^T\mathbf{w}_j \geq 0, \ \forall j$$

Otherwise, the value of the function can be arbitrarily small by selecting the adequate values of $\pi_j = 0$ and/or λ_k. Then, the value of the objective function becomes

$$Z = \mathbf{c}^T\mathbf{x} = \sum_k \lambda_k \mathbf{c}^T \mathbf{q}_k$$

which attains the minimum for

$$\lambda_s = 1; \text{ for any } s \text{ such that } \mathbf{c}^T\mathbf{q}_s = \min(\mathbf{c}^T\mathbf{q}_1, \ldots, \mathbf{c}^T\mathbf{q}_r)$$

where r is the number of extreme points.

This implies that the minimum is attained at one of the extreme points $\{\mathbf{q}_1, \ldots, \mathbf{q}_r\}$ and, using Theorem 4.1, this point is a basic feasible solution. ■

4.5 Sensitivities

From the optimal solution of an LPP it is possible to derive very relevant sensitivity information. This is shown below.

Let B^* be the optimal basis; then

$$\begin{array}{rcl} \mathbf{x}_B^* & = & (\mathbf{B}^*)^{-1}\mathbf{b} \\ z^* & = & \mathbf{c}_B^T\mathbf{x}_B^* \end{array} \tag{4.18}$$

Let us consider a marginal change (i.e., a change so that the basis does not change) in the RHS vector $\mathbf{b}$:

$$\mathbf{b}^* \to \mathbf{b}^* + \Delta\mathbf{b}$$

This change in RHS vector originates changes in the minimizer and the optimal objective function values:

$$\begin{array}{c} \mathbf{x}_B^* \to \mathbf{x}_B^* + \Delta\mathbf{x}_B \\ z^* \to z^* + \Delta z \end{array}$$

Because equations (4.18) are linear, we can write

$$\begin{array}{c} \Delta\mathbf{x}_B = (\mathbf{B}^*)^{-1}\Delta\mathbf{b} \\ \Delta z = \mathbf{c}_B^T\Delta\mathbf{x}_B \end{array}$$

Combining these equations, we get

$$\Delta z = \mathbf{c}_B^T\Delta\mathbf{x}_B = \mathbf{c}_B^T(\mathbf{B}^*)^{-1}\Delta\mathbf{b}$$

Defining

$$\boldsymbol{\lambda}^{*T} = \mathbf{c}_B^T(\mathbf{B}^*)^{-1}$$

the previous equation becomes

$$\Delta z = \boldsymbol{\lambda}^{*T}\Delta\mathbf{b}$$

and for a given coordinate j the preceding expression is

$$\lambda_j^* = \frac{\Delta z}{\Delta b_j}$$

which means that λ_j^* provides the change in the objective function optimal value as a result of a marginal change in the j component of the RHS vector $\mathbf{b}$.

These sensitivity parameters play a fundamental role in engineering and science applications.

As will be seen in the sections below, sensitivity parameters are in fact dual variables.

4.6 Duality

This section deals with the concept and significance of duality in linear programming. The dual problem of a given linear programming problem is formulated, and then the mathematical relationship between these two problems is studied. We provide simple examples that lead to a better understanding of this important concept.

Given a linear programming problem, called a *primal problem*, there exists another linear programming problem, called a *dual problem*, closely associated with it. Both problems are said to be dual of each other. Under certain assumptions, the primal and dual problems have the same optimal objective function value; hence it is possible to solve the primal problem indirectly by solving the corresponding dual problem. This can lead to important computational advantages.

Definition 4.5 (Dual problem). *Given the linear programming problem minimizing*

$$Z = \mathbf{c}^T\mathbf{x}$$

subject to

$$\begin{array}{rcl} \mathbf{A}\mathbf{x} & \geq & \mathbf{b} \\ \mathbf{x} & \geq & \mathbf{0} \end{array} \tag{4.19}$$

its dual problem is maximization of

$$Z = \mathbf{b}^T\mathbf{y}$$

subject to

$$\begin{array}{rcl} \mathbf{A}^T\mathbf{y} & \leq & \mathbf{c} \\ \mathbf{y} & \geq & \mathbf{0} \end{array} \tag{4.20}$$

where $\mathbf{y} = (y_1, \ldots, y_m)^T$ *are called dual variables.* ∎

We shall identify the first problem as the primal problem, and the second one, as its dual counterpart. Notice how the same elements (the matrix **A**, and the vectors **b** and **c**) determine both problems. Although the primal problem is not in its standard form, this format enables us to see more clearly the symmetry between the primal and the corresponding dual problem, and in particular it shows us that the dual of the dual is the primal.

Remark 4.1 *Duality is a symmetric relationship, that is, if problem D is the dual of problem P, then P is also the dual of D.* ∎

To see this, we rewrite the dual problem above as a minimization problem with constraints of the following form $\geq$. Minimize

$$Z = -\mathbf{b}^T\mathbf{y}$$

subject to

$$\begin{array}{rcl} -\mathbf{A}^T\mathbf{y} & \geq & -\mathbf{c} \\ \mathbf{y} & \geq & \mathbf{0} \end{array} \tag{4.21}$$

Then, its dual is maximization of

$$Z = -\mathbf{c}^T\mathbf{x}$$

subject to

$$\begin{array}{rcl} -\mathbf{A}\mathbf{x} & \leq & -\mathbf{b} \\ \mathbf{x} & \geq & \mathbf{0} \end{array} \tag{4.22}$$

which is equivalent to the primal above.

Remark 4.2 *As can be observed, every constraint of the primal problem has a dual variable associated with it, the coefficients of the objective function of the primal problem are the RHS terms of the dual constraints and vice versa, and the coefficient matrix of the dual constraints is the transpose of the primal one. The primal is a minimization problem but its dual is a maximization problem.* ∎

4.6.1 Obtaining the Dual from a Primal in Standard Form

In this section we deal with the problem of finding the dual problem when the primal is given in its standard form. To answer this question all we have to do is to use the duality relations previously defined in Section 4.6. Assume, then, that we have the typical LPP, where we minimize

$$Z = \mathbf{c}^T\mathbf{x}$$

subject to

$$\begin{array}{rcl} \mathbf{A}\mathbf{x} & = & \mathbf{b} \\ \mathbf{x} & \geq & \mathbf{0} \end{array}$$

The equality $\mathbf{A}\mathbf{x} = \mathbf{b}$ can be replaced by the two equivalent inequalities $\mathbf{A}\mathbf{x} \geq \mathbf{b}$ and $-\mathbf{A}\mathbf{x} \geq -\mathbf{b}$. Then, we can write the preceding problem by minimizing

$$Z = \mathbf{c}^T\mathbf{x}$$

subject to

$$\begin{array}{rcl} \begin{pmatrix} \mathbf{A} \\ -\mathbf{A} \end{pmatrix}\mathbf{x} & \geq & \begin{pmatrix} \mathbf{b} \\ -\mathbf{b} \end{pmatrix} \\ \mathbf{x} & \geq & \mathbf{0} \end{array}$$

The dual of this problem is to maximize

$$Z = \mathbf{b}^T\mathbf{y}^{(1)} - \mathbf{b}^T\mathbf{y}^{(2)} = \mathbf{b}^T\mathbf{y}$$

where $\mathbf{y} = \mathbf{y}^{(1)} - \mathbf{y}^{(2)}$ is not restricted in sign, subject to

$$\begin{pmatrix} \mathbf{A}^T & -\mathbf{A}^T \end{pmatrix} \begin{pmatrix} \mathbf{y}^{(1)} \\ \mathbf{y}^{(2)} \end{pmatrix} = \mathbf{A}^T\mathbf{y} \leq \mathbf{c}$$

This is the form of the dual problem when the primal is given in its standard form.

4.6.2 Obtaining the Dual Problem

A linear programming problem whose formulation is not in the form (4.19) also has a dual problem. It can be constructed using the following rules:

Rule 1. An equality constraint in the primal (dual) implies that the corresponding dual (primal) variable is unrestricted.

Rule 2. A $\geq$ ($\leq$) inequality constraint in the primal (dual) implies that its dual (primal) variable is nonnegative.

Rule 3. A $\leq$ ($\geq$) inequality constraint in the primal (dual) implies that its dual (primal) variable is nonpositive.

Rule 4. A primal (dual) nonnegative variable has a $\leq$ ($\geq$) inequality constraint in the dual (primal) problem.

Rule 5. A primal (dual) nonpositive variable has a $\geq$ ($\leq$) inequality constraint in the dual (primal) problem.

Rule 6. A primal (dual) unrestricted variable has an equality constraint in the dual (primal) problem.

Example 4.7 (Dual problem). The dual of the linear programming problem minimization of

$$Z = x_1 + x_2 - x_3$$

subject to

$$\begin{array}{rrrcr} 2x_1 & +x_2 & & \geq & 3 \\ x_1 & & -x_3 & = & 2 \\ & & x_3 & \geq & 0 \end{array} \qquad (4.23)$$

is maximization of

$$Z = 3y_1 + 2y_2$$

subject to

$$\begin{array}{rrcr} 2y_1 & +y_2 & = & 1 \\ y_1 & & = & 1 \\ & -y_2 & \leq & -1 \\ & y_1 & \geq & 0 \end{array} \qquad (4.24)$$

To see this, we apply the rules above, as follows:

Rule 1. Since the second constraint in the primal is an equality constraint, the second dual variable y_2 is unrestricted.

Rule 2. Since the first constraint in the primal is $\geq$, the first dual variable y_1 is nonnegative.

Rule 3. Since the third primal variable x_3 is restricted, the third dual constraint is $\leq$.

Rule 4. Since the first and second primal variables x_1 and x_2 are unrestricted, the first and second dual constraints are equality constraints.

Using the same rules to the dual problem, we can recover the primal as follows:

Rule 1. Since the first and second constraints in the dual are equality constraints, the first and second primal variables x_1 and x_2 are unrestricted.

Rule 2. Since the third constraint in the dual is $\leq$, the third primal variable x_3 is nonnegative.

Rule 3. Since the first dual variable y_1 is restricted, the first primal constraint is $\geq$.

Rule 4. Since the second dual variable y_2 is unrestricted, the second primal constraint is an equality constraint.

■

4.6.3 Duality Theorems

The importance of the dual problem is established in the following theorems.

Lemma 4.1 (Weak duality lemma) *Let P be a LPP, and D its dual. Let $\mathbf{x}$ be a feasible solution of P and $\mathbf{y}$ a feasible solution of D. Then*

$$\mathbf{b}^T\mathbf{y} \leq \mathbf{c}^T\mathbf{x}$$

Proof. The proof is elementary. If $\mathbf{x}$ and $\mathbf{y}$ are feasible for P and D, respectively, then

$$\mathbf{A}\mathbf{x} = \mathbf{b}, \quad \mathbf{x} \geq \mathbf{0}, \quad \mathbf{A}^T\mathbf{y} \leq \mathbf{c}$$

Notice, by the nonnegativity of $\mathbf{x}$, that

$$\mathbf{b}^T\mathbf{y} = \mathbf{x}^T\mathbf{A}^T\mathbf{y} \leq \mathbf{x}^T\mathbf{c} = \mathbf{c}^T\mathbf{x}$$

■

Corollary 4.1 *If $\mathbf{b}^T\tilde{\mathbf{y}} = \mathbf{c}^T\tilde{\mathbf{x}}$ for some particular vectors $\tilde{\mathbf{x}}$ and $\tilde{\mathbf{y}}$, feasible for P and D, respectively, then*

$$\mathbf{c}^T\tilde{\mathbf{x}} = \mathbf{b}^T\tilde{\mathbf{y}} \leq \max_{\mathbf{y}}\{\mathbf{b}^T\mathbf{y}|\mathbf{A}^T\mathbf{y} \leq \mathbf{c}\} \leq \min_{\mathbf{x}}\{\mathbf{c}^T\mathbf{x}|\mathbf{A}\mathbf{x} = \mathbf{b}, \ \mathbf{x} \geq \mathbf{0}\} \leq \mathbf{c}^T\tilde{\mathbf{x}} = \mathbf{b}^T\tilde{\mathbf{y}}$$

Table 4.1: Data of the planing mill problem

	Table type		Available machine-hours
	1	2	per day
Cutting	4	3	40
Finishing	4	7	56
Profit	\$70	\$90	

Proof. This is also straightforward if we realize that we have just checked

$$\max_{\mathbf{y}}\{\mathbf{b}^T\mathbf{y}|\mathbf{A}^T\mathbf{y} \leq \mathbf{c}\} \leq \min_{\mathbf{x}}\{\mathbf{c}^T\mathbf{x}|\mathbf{A}\mathbf{x} = \mathbf{b},\ \ \mathbf{x} \geq \mathbf{0}\}$$

■

Hence, all inequalities are in reality equalities and $\tilde{\mathbf{x}}$ and $\tilde{\mathbf{y}}$ must be optimal solutions of P and D respectively, as claimed.

The strong duality theorem asserts that both problems, P and D, admit optimal solutions simultaneously.

Theorem 4.3 (Duality theorem). *If $\tilde{\mathbf{x}}$ is an optimal solution for P, then there is an optimal solution $\tilde{\mathbf{y}}$ for D, and the minimum in P and the maximum in D coincide with the common value $\mathbf{b}^T\tilde{\mathbf{y}} = \mathbf{c}^T\tilde{\mathbf{x}}$. Conversely, if $\tilde{\mathbf{y}}$ is an optimal solution for D, there exists an optimal solution for P, $\tilde{\mathbf{x}}$, and again the minimum and maximum in P and D coincide with $\mathbf{b}^T\tilde{\mathbf{y}} = \mathbf{c}^T\tilde{\mathbf{x}}$. Otherwise, one or both of the two sets of feasible vectors is empty.* ■

In a summary, if P is a LPP and D is its dual, then one of the following statements is true:

1. Both problems have optimal solution and their optimal values coincide.

2. One problem has unbounded optimum and the other one has an empty feasible region.

Example 4.8 (Primal and dual of a mill problem). A small mill manufactures two types of wood tables. Each table of type 1 requires 4 machine-hours for cutting from the wood stock to size and 4 machine-hours finishing time (assembly, painting, etc.). Similarly, each table of type 2 requires 3 machine-hours of cutting time and 7 machine-hours of finishing time. The cutting and finishing equipment capacities available per day are 40 and 56 machine–hours, respectively. Finally, each type 1 table provides \$70 profit, while each type 2 table provides \$90 profit. These data are summarized in Table 4.1.

We wish to know the quantity of tables produced per day of each type, in order to maximize the daily total profit. The linear programming model of the problem is maximizing

$$Z = 70x_1 + 90x_2$$

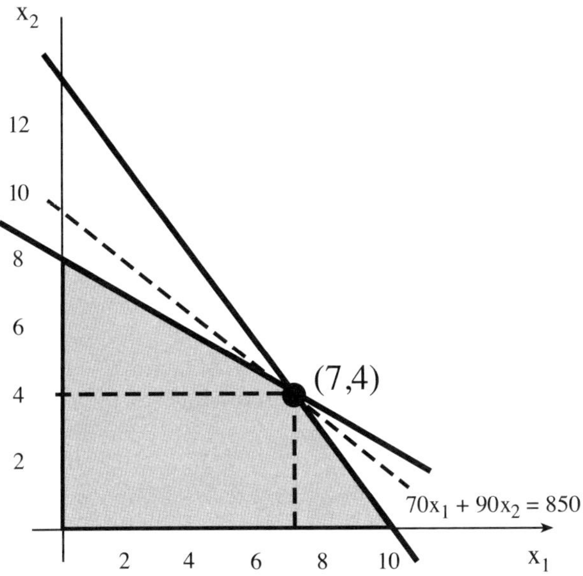

Figure 4.4: Graphical analysis of the planing mill problem.

subject to

$$\begin{array}{rrcr} 4x_1 & +3x_2 & \leq & 40 \\ 4x_1 & +7x_2 & \leq & 56 \\ & x_1,\, x_2 & \geq & 0 \end{array} \tag{4.25}$$

where x_1 and x_2 are the quantities of table types 1 and 2, respectively, to be produced per day.

The optimal solution of this model, as can be seen in Figure 4.4, indicates a daily production of 7 type 1 tables and 4 type 2 tables, and a daily maximum profit of \$850. This result indicates that both cutting and finishing resources are fully utilized, because both constraints are binding.

Now suppose that we wish to obtain higher daily profits. Then an expansion of the available capacities is necessary. Suppose that the available finishing capacity can be expanded from 56 machine-hours to 72 machine-hours per day. How does this expansion affect the total daily profit? The answer is given by the solution of the following linear programming problem, where we maximize

$$Z = 70x_1 + 90x_2$$

subject to

$$\begin{array}{rrcr} 4x_1 & +3x_2 & \leq & 40 \\ 4x_1 & +7x_2 & \leq & 72 \\ & x_1,\, x_2 & \geq & 0 \end{array} \tag{4.26}$$

In this case the optimal solution would be $x_1 = 4$ and $x_2 = 8$ with a maximum daily profit of $1000. This solution indicates that we can increase the daily profit by $150 if we increase the daily finishing capacity by $72 - 56 = 16$ machine–hours. The rate $1000-850/16=$150/16=$75/8, at which the objective function value will increase as the finishing capacity is increased by 1 hour, is called the *sensitivity or shadow price* (also *dual price*) of the finishing resource (see Section 4.5).

In general, the shadow price of a constraint is the change in the optimal value of the objective function, per unit change in the RHS value of the constraint, assuming that all other parameter values of the problem remain unchanged. For many linear programming applications the shadow prices are at least as important as the solution of the problem itself; they indicate whether certain changes in the use of productive capacities affect the objective function value. Shadow prices can be determined by formulating and solving the *dual problem.*

The dual of the planing mill problem (4.25) is stated as follows. Minimize

$$Z = 40y_1 + 56y_2$$

subject to

$$\begin{array}{rrcr} 4y_1 & +4y_2 & \geq & 70 \\ 3y_1 & +7y_2 & \geq & 90 \\ & y_1, y_2 & \geq & 0 \end{array} \tag{4.27}$$

The optimal solution of this problem is $y_1 = 65/8$, $y_2 = 75/8$, and the optimal value is 850. Note that y_1 and y_2 are the shadow prices of cutting and finishing resources, respectively, and optimal values of problems (4.25) and (4.27) coincide.

The dual problem (4.27) can be interpreted as follows. Assume that we are willing to sell machine time (rather than produce and sell tables), provided that we can obtain at least the same level of total profit. In this case, producing tables and selling machine time would be two equivalent alternatives; the y_1 and y_2 variables would represent the sale prices of one hour on the cutting and finishing machine, respectively. In order to maintain a competitive business we should set the total daily profit as low as possible, that is, minimize the function $40y_1 + 56y_2$, where the 40 and 56 coefficients represent the machine-hours available per day for sale of cutting and finishing machines. The constraints in (4.27) state that the total prices of the cutting and finishing hours required for producing one unit of each type of table must not be less than the profit expected from the sale of one unit of each type of table, and that prices must be positive quantities. ∎

Example 4.9 (Communication net). We are interested in sending a message through a communication net from one node A to some other node B with minimal cost. The network consists in four nodes, A, B, C, and D, and there are channels connecting A and C, A and D, D and C, C and B, and D and

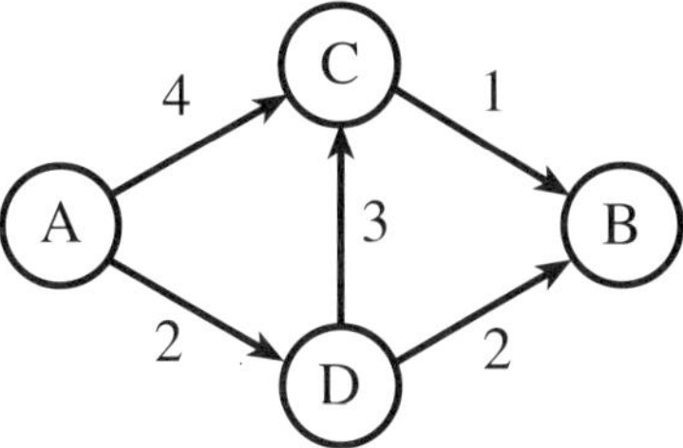

Figure 4.5: Communication net in Example 4.9.

B, with associated costs per unit of message 4, 2, 3, 1, and 2, respectively (see Figure 4.5).

Primal

The primal approach to this problem consists of minimizing

$$Z = 4x_{AC} + 2x_{AD} + 3x_{DC} + x_{CB} + 2x_{DB}$$

where each variable x_{PQ} denotes the fraction of the message going through the channel connecting any node P with any other node Q. This minimization means that we minimize the total cost of sending the messages.

The following constraints must be satisfied:

1. No fraction of the total message is lost in node C:

$$x_{AC} + x_{DC} = x_{CB} \tag{4.28}$$

2. No part of the message is lost in node D:

$$x_{AD} = x_{DC} + x_{DB} \tag{4.29}$$

3. The complete message arrives to B:

$$1 = x_{CB} + x_{DB} \tag{4.30}$$

If we write them in matrix form, we get

$$\begin{pmatrix} 1 & 0 & 1 & -1 & 0 \\ 0 & 1 & -1 & 0 & -1 \\ 0 & 0 & 0 & 1 & 1 \end{pmatrix} \begin{pmatrix} x_{AC} \\ x_{AD} \\ x_{DC} \\ x_{CB} \\ x_{DB} \end{pmatrix} = \begin{pmatrix} 0 \\ 0 \\ 1 \end{pmatrix} \tag{4.31}$$

Note that if we add the last two constraints and subtract the first, we get $x_{AC} + x_{AD} = 1$. This means that the whole message departs from A. Since it can be obtained as a linear combination of the other three, it cannot be included to avoid redundancy.

Dual

Alternatively, we can assume that we can buy the information of the messages at each node and sell them at a different node. Thus, the dual approach can be interpreted as finding the prices y_A, y_B, y_C, y_D of the fractions of the total message in each node, so that we assume that fractions of messages are bought in one node and sold in another. The net benefit obtained in such a transaction would be the difference of prices between the arriving and departing nodes. If we normalize by using the condition $y_A = 0$, such benefits will be

$$\begin{array}{l} y_C - y_A = y_C \\ y_D - y_A = y_D \\ y_C - y_D \\ y_B - y_C \\ y_B - y_D \end{array}$$

We must ensure that these benefits are not greater than the prices we already know:

$$\begin{array}{rcl} y_C & \leq & 4 \\ y_D & \leq & 2 \\ y_C - y_D & \leq & 3 \\ y_B - y_C & \leq & 1 \\ y_B - y_D & \leq & 2 \end{array}$$

These constraints can be written as

$$\begin{pmatrix} 1 & 0 & 0 \\ 0 & 1 & 0 \\ 1 & -1 & 0 \\ -1 & 0 & 1 \\ 0 & -1 & 1 \end{pmatrix} \begin{pmatrix} y_C \\ y_D \\ y_B \end{pmatrix} \leq \begin{pmatrix} 4 \\ 2 \\ 3 \\ 1 \\ 2 \end{pmatrix} \quad (4.32)$$

The objective is now to maximize the income when taking the total message from node A to node B, that is, to maximize y_B (in fact, $y_B - y_A$), subject to the above mentioned constraints.

Both formulations correspond to the primal and dual problems. ■

Exercises

4.1 Determine the feasible region and its vertices of the following linear programming problem. Minimize

$$Z = -4x_1 + 7x_2$$

subject to

$$\begin{array}{rrcl} x_1 & +x_2 & \geq & 3 \\ -x_1 & +x_2 & \leq & 3 \\ 2x_1 & +x_2 & \leq & 8 \\ & x_1, x_2 & \geq & 0 \end{array}$$

Obtain the optimal solution graphically.

4.2 Find the optimal solution of the following problems graphically

(a) Maximize

$$Z = 2x_1 + 6x_2$$

subject to

$$\begin{aligned} -x_1 + x_2 &\leq 1 \\ 2x_1 + x_2 &\leq 2 \\ x_1 &\geq 0 \\ x_2 &\geq 0 \end{aligned}$$

(b) Minimize

$$Z = -3x_1 + 2x_2$$

subject to

$$\begin{gathered} x_1 + x_2 \leq 5 \\ 0 \leq x_1 \leq 4 \\ 1 \leq x_2 \leq 6 \end{gathered}$$

4.3 Using the graphical method, solve the following LPP. Minimize

$$Z = 2x_1 + x_2$$

subject to

$$\begin{array}{rrcl} 3x_1 & +x_2 & \geq & 3 \\ 4x_1 & +3x_2 & \geq & 6 \\ x_1 & +2x_2 & \geq & 2 \\ & x_1,\, x_2 & \geq & 0 \end{array}$$

4.4 Write and solve the dual of the problem. Maximize

$$Z = 3x_1 - 4x_2 + 9x_3 + x_4$$

subject to

$$\begin{array}{rrrcr} x_1 & -5x_2 & +x_3 & \geq & 0 \\ 3x_1 & -5x_2 & +x_3 & \geq & 10 \\ & & x_2,\, x_3 & \geq & 0 \\ & & x_4 & \leq & 0 \end{array}$$

Table 4.2: Distribution problem

Plants	Markets		
	Market_1	Market_2	Market_3
Plant_1	1	3	5
Plant_2	2	5	4

4.5 Consider the following problem. Maximize

$$Z = 8x_1 - 4x_2$$

subject to

$$\begin{array}{rrcr} 4x_1 & -5x_2 & \leq & 2 \\ 2x_1 & +4x_2 & \geq & 6 \\ x_1 & -6x_2 & \leq & -14 \end{array}$$

(a) Draw the feasible region and identify the extreme points of the feasible region.

(b) Find a point in the feasible region at which the objective function attains the maximum.

4.6 Prove that the following problem does not have a finite optimum. Minimize

$$Z = -6x_1 + 2x_2$$

subject to

$$\begin{array}{rrcr} -3x_1 & +x_2 & \leq & 6 \\ 3x_1 & +5x_2 & \geq & 15 \\ & x_1,\ x_2 & \geq & 0 \end{array}$$

4.7 A refrigerator manufacturer has three factories that supply four retail stores. Consider the transportation problem shown in Figure 4.6; that is, one is interested in shipping 10 units from F_1, 15 units from F_2, and 17 units from F_3, and receiving 10 units in S_1, 9 units in S_2, 11 units in S_3, and 12 units in S_4. Let x_{ij} be the unknown number of refrigerators to be shipped from factory i to store j. State the associated linear programming problem.

4.8 A firm producing a product P has two plants. Each plant produces 90 tons of P monthly, and the product is distributed in three different markets. Table 4.2 shows the unit costs of shipping 1 ton of product P from a given plant to a given market. The firm wants to send the same number of tons to each market and minimize the total cost. Formulate the corresponding linear programming problem.

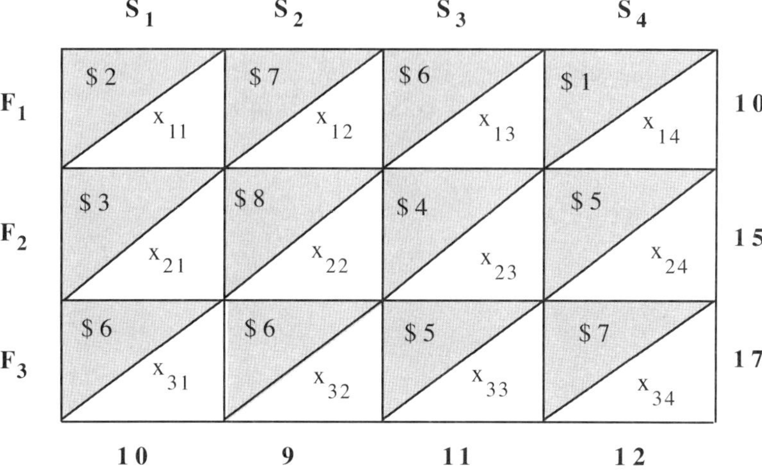

Figure 4.6: Transportation problem showing the origins and destinations.

4.9 A firm can produce 120 items of type A and 360 items of type B on a daily basis. The items are subject to a quality control process that can control 200 items per day. Items of type A are sold in the market at a price four times that of type B. Make a decision about the production in order to maximize profits.

4.10 Consider the following problem. Minimize

$$Z = \frac{\mathbf{p}^T\mathbf{x} + \alpha}{\mathbf{q}^t\mathbf{x} + \beta}$$

subject to

$$\begin{aligned} \mathbf{A}\mathbf{x} &= \mathbf{b} \\ \mathbf{x} &\geq \mathbf{0} \end{aligned}$$

where $\mathbf{p}$ and $\mathbf{q}$ are n vectors, $\mathbf{b}$ is an m vector, $\mathbf{A}$ is an $m \times n$ matrix, and α and β are scalars. Suppose that the feasible set is bounded, and suppose that $\mathbf{q}^T\mathbf{x} + \beta > 0$ for each feasible point $\mathbf{x}$. This problem is called a *linear fractional programming problem* (LFPP).

(a) Show that the the tranformation

$$z = \frac{1}{\mathbf{q}^T\mathbf{x} + \beta}; \quad \mathbf{y} = z\mathbf{x}$$

leads to an equivalent linear problem.

(b) Assume that the optimal value of a LFPP is Z^*. Prove that the following linear programming problem has the same solutions. Minimize

$$Z = \mathbf{p}^T\mathbf{x} + \alpha - Z^*(\mathbf{q}^t\mathbf{x} + \beta)$$

subject to

$$\begin{aligned} \mathbf{A}\mathbf{x} &= \mathbf{b} \\ \mathbf{x} &\geq \mathbf{0} \end{aligned}$$

Chapter 5

Understanding the Set of All Feasible Solutions

5.1 Introduction and Motivation

As indicated in Chapter 1, a mathematical programming problem has the following four elements: data, variables, constraints, and a function to be optimized. In this chapter we deal only with the set of constraints.

A LPP is said to be *well posed* if it has a bounded solution. If the problem has no solution, it is because it is overconstrained. If the problem is unbounded it is because it is underconstrained. So, to have a well-posed problem it is crucial to make a good selection of the adequate constraints.

Constraints are conditions that all candidate solutions must satisfy. They arise from mathematical, physical, or engineering considerations that are raised up during the problem statement. The set of constraints defines the *feasible set*, that is, the set of solutions that satisfy all the constraints. Thus, the knowledge of this set is important in itself, because it is the only set of solutions the problem designer is willing to accept and, consequently, the set of candidate solutions we are dealing with.

In previous chapters we have dealt with obtaining the optimal solution of a given function subject to a set of constraints. This means that we have selected, among all vectors satisfying the set of constraints, one that maximizes or minimizes a given function. It must be clear that the methods explained in other chapters allow us to obtain only a single solution of the problem, selected using the optimization function as the selection criterion. Since the optimization function is completely independent of the set of constraints, we ignore the optimization function in this chapter and deal only with the set of constraints.

To understand mathematical programming problems, we need to have a through knowledge of the structure of the different feasible regions we are working with. The constraints imposed by a set of linear equations and inequalities lead to feasible solutions that have the structure of linear spaces, cones, poly-

Table 5.1: Different structures arising as solutions of linear systems of equations and inequalities, and their representations

Algebraic structure	Definition in terms of constraints	Inner representation
Linear space	$\mathbf{Hx} = \mathbf{0}$	$\mathbf{x} = \sum_i \rho_i \mathbf{v}_i;\ \rho_i \in \mathbb{R}$
Affine space	$\mathbf{Hx} = \mathbf{a}$	$\mathbf{x} = \mathbf{q} + \sum_i \rho_i \mathbf{v}_i;\ \rho_i \in \mathbb{R}$
Cone	$\mathbf{Hx} \le \mathbf{0}$	$\mathbf{x} = \sum_j \pi_j \mathbf{w}_j;\ \pi_j \ge 0$
Polytope	$\mathbf{Hx} \le \mathbf{a}$	$\mathbf{x} = \sum_k \lambda_k \mathbf{q}_k;\ \lambda_k \ge 0;$ $\sum_k \lambda_k = 1$
Polyhedron	$\mathbf{Hx} \le \mathbf{a}$	$\mathbf{x} = \sum_i \rho_i \mathbf{v}_i + \sum_j \pi_j \mathbf{w}_j + \sum_k \lambda_k \mathbf{q}_k;$ $\rho_i \in \mathbb{R};\ \pi_j \ge 0;\ \lambda_k \ge 0;\ \sum_k \lambda_k = 1$

topes, and polyhedra (see Table 5.1). Thus, the knowledge of these structures is essential to understand the problems associated with these types of systems.

As shown in Table 5.1, the feasible set of solutions can be written in two forms: (1) by the set of constraints and (2) by its inner representation. The definition in terms of constraints is the natural form in which we are given the problem. However, it has the shortcoming that finding feasible solutions is difficult. In fact, it not easy to find a vector $\mathbf{x}$ satisfying the constraints. Alternatively, the inner representation allows us to generate all possible solutions of our problem by selecting values for the coefficients ρ_i, π_j, and λ_k.

To clarify this and to motivate this chapter, we use the simple case of a cube. Initially, if we consider an unconstrained problem in three dimensions, the set of feasible solutions is the linear space

$$\begin{pmatrix} x \\ y \\ z \end{pmatrix} = \rho_1 \begin{pmatrix} 1 \\ 0 \\ 0 \end{pmatrix} + \rho_2 \begin{pmatrix} 0 \\ 1 \\ 0 \end{pmatrix} + \rho_3 \begin{pmatrix} 0 \\ 0 \\ 1 \end{pmatrix}$$

Consider the set of constraints

$$\begin{array}{c} 0 \le x \le 1 \\ 0 \le y \le 1 \\ 0 \le z \le 1 \end{array} \tag{5.1}$$

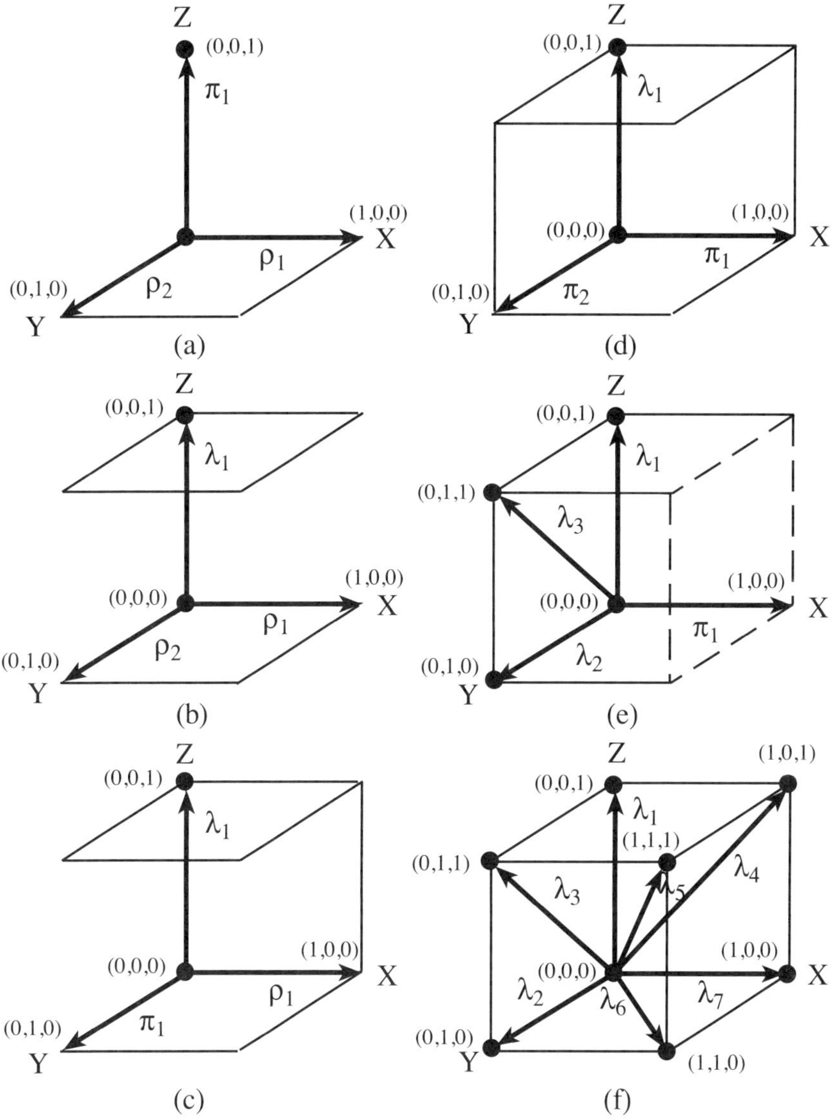

Figure 5.1: Illustration of how the solution evolves when adding constraints.

which together define a cubic region [see Figure 5.1(f)].

If starting from the empty set of constraints, we incorporate, one by one, the constraints in (5.1) to this set, the resulting solutions are illustrated in Figure 5.1 and Table 5.2, which show the corresponding systems of constraints and their solutions. The coefficients ρ, π, and λ refer to unrestricted, nonnegative, and nonnegative adding up to one, real numbers, respectively. They allow us to write linear combinations, linear nonnegative, and linear convex combinations

Table 5.2: Set of constraints and corresponding solutions[1]

System	Solution	Figure 5.1
$z \geq 0$	$\begin{pmatrix} x \\ y \\ z \end{pmatrix} = \pi_1 \begin{pmatrix} 0 \\ 0 \\ 1 \end{pmatrix} + \rho_1 \begin{pmatrix} 1 \\ 0 \\ 0 \end{pmatrix} + \rho_2 \begin{pmatrix} 0 \\ 1 \\ 0 \end{pmatrix}$	(a)
$z \geq 0$ $z \leq 1$	$\begin{pmatrix} x \\ y \\ z \end{pmatrix} = \lambda_1 \begin{pmatrix} 0 \\ 0 \\ 1 \end{pmatrix} + \rho_1 \begin{pmatrix} 1 \\ 0 \\ 0 \end{pmatrix} + \rho_2 \begin{pmatrix} 0 \\ 1 \\ 0 \end{pmatrix}$	(b)
$z \geq 0$ $z \leq 1$ $y \geq 0$	$\begin{pmatrix} x \\ y \\ z \end{pmatrix} = \lambda_1 \begin{pmatrix} 0 \\ 0 \\ 1 \end{pmatrix} + \rho_1 \begin{pmatrix} 1 \\ 0 \\ 0 \end{pmatrix} + \pi_1 \begin{pmatrix} 0 \\ 1 \\ 0 \end{pmatrix}$	(c)
$z \geq 0$ $z \leq 1$ $y \geq 0$ $x \geq 0$	$\begin{pmatrix} x \\ y \\ z \end{pmatrix} = \lambda_1 \begin{pmatrix} 0 \\ 0 \\ 1 \end{pmatrix} + \pi_1 \begin{pmatrix} 1 \\ 0 \\ 0 \end{pmatrix} + \pi_2 \begin{pmatrix} 0 \\ 1 \\ 0 \end{pmatrix}$	(d)
$z \geq 0$ $z \leq 1$ $y \geq 0$ $x \geq 0$ $y \leq 1$	$\begin{pmatrix} x \\ y \\ z \end{pmatrix} = \lambda_1 \begin{pmatrix} 0 \\ 0 \\ 1 \end{pmatrix} + \pi_1 \begin{pmatrix} 1 \\ 0 \\ 0 \end{pmatrix} + \lambda_2 \begin{pmatrix} 0 \\ 1 \\ 0 \end{pmatrix} + \lambda_3 \begin{pmatrix} 0 \\ 1 \\ 1 \end{pmatrix}$	(e)
$z \geq 0$ $z \leq 1$ $y \geq 0$ $x \geq 0$ $y \leq 1$ $x \leq 1$	$\begin{pmatrix} x \\ y \\ z \end{pmatrix} = \lambda_1 \begin{pmatrix} 0 \\ 0 \\ 1 \end{pmatrix} + \lambda_2 \begin{pmatrix} 0 \\ 1 \\ 0 \end{pmatrix} + \lambda_3 \begin{pmatrix} 0 \\ 1 \\ 1 \end{pmatrix} + \lambda_4 \begin{pmatrix} 1 \\ 0 \\ 1 \end{pmatrix}$ $+ \lambda_5 \begin{pmatrix} 1 \\ 1 \\ 1 \end{pmatrix} + \lambda_6 \begin{pmatrix} 1 \\ 1 \\ 0 \end{pmatrix} + \lambda_7 \begin{pmatrix} 1 \\ 0 \\ 0 \end{pmatrix}$	(f)

[1] The coefficients ρ, π, and λ refer to unrestricted, nonnegative, and nonnegative adding up to one, real numbers, respectively. The vectors with ρ, π, and λ coefficients define the linear space, the cone, and the polytope parts, respectively, of the associated solutions.

of vectors, which generate linear spaces, cones, and polytopes, respectively. For example, when we introduce the first constraint, $z \geq 0$, we restrict all z values to be nonnegative. Consequently, instead of a ρ_1 (real) coefficient, we use a π_1 (nonnegative) coefficient [see Figure 5.1(a) and Table 5.2, first row]. Next, we introduce the constraint $z \leq 1$, which limits the values of the π_1 coefficients to be between zero and one, this coefficient transforms to λ_1, which is allowed to range from zero to one [see Figure 5.1(b) and Table 5.2, second row]. When we introduce the constraint $y \geq 0$, the ρ_2 coefficient transforms to a π_1 coefficient [see Figure 5.1(c) and Table 5.2, third row]. The reader is encouraged to guess the changes produced in the coefficients when we introduce the other three constraints $x \geq 0$, $y \leq 1$, and $x \leq 1$. The final result is that thc regions in Figure 5.1(a)–(f) are a linear space plus a cone, a linear space plus a polytope, a linear space plus a cone plus a polytope, a cone plus a polytope, a cone plus a polytope, and a polytope, respectively.

Although obtaining the feasible set for the cube example is a simple task, the problem of obtaining the set of all feasible solutions associated with a set of linear constraints, the only one we deal with in this chapter, is not trivial. Castillo et al. [21] introduced the Γ algorithm, which is given in Appendix A, to solve this problem.

This chapter is structured as follows. In Section 5.2 we introduce convex sets and the important concepts of extreme points and directions, and show how a convex set can be written in terms of its extreme points and directions. Next we deal with some especial cases of convex sets, as linear spaces, polyhedral convex cones, polytopes, and polyhedra, in Sections 5.4–5.7, respectively, because they arise in a natural way when dealing with linear constraints.

5.2 Convex Sets

In this section we focus on convex sets. We start with the definition of a convex set and its properties, and then we define extreme points and extreme directions, two key concepts in convex analysis.

Convex sets are the natural and simplest type of sets arising in mathematical programming. A set S is convex if the line segment joining every two points of S lies entirely in S. Figure 5.2 shows a convex set and a nonconvex set in $\mathbb{R}^2$. A more precise and mathematical definition of convex set follows.

Definition 5.1 (Convex set). *A set S in $\mathbb{R}^n$ is said to be convex if and only if*

$$\lambda\mathbf{x} + (1-\lambda)\mathbf{y} \in S \tag{5.2}$$

for every $\lambda \in [0,1]$ and $\mathbf{x}, \mathbf{y} \in S$. ■

The following are some examples of convex sets:

1. (**Hyperplanes.**) The hyperplanes are sets defined by

$$S = \{\mathbf{x} \in \mathbb{R}^n \mid \mathbf{p}^T\mathbf{x} = \alpha\} \tag{5.3}$$

where $\mathbf{p}$ is a nonzero vector in $\mathbb{R}^n$ and α is a scalar. Using the definition of a convex set, it is very easy to prove that S is a convex set. Letting

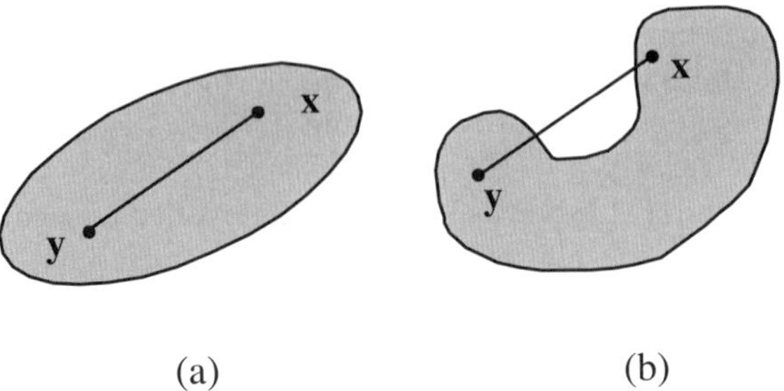

Figure 5.2: Illustration of convex (a) and nonconvex (b) sets in $\mathbb{R}^2$.

$\mathbf{x}, \mathbf{y} \in S$, by definition of S, we obtain that $\mathbf{p}^T\mathbf{x} = \alpha$ and $\mathbf{p}^T\mathbf{y} = \alpha$. For any $\lambda \in [0,1]$, we have

$$\mathbf{p}^T[\lambda\mathbf{x} + (1-\lambda)\mathbf{y}] = \lambda\mathbf{p}^T\mathbf{x} + (1-\lambda)\mathbf{p}^T\mathbf{y} = \lambda\alpha + (1-\lambda)\alpha = \alpha$$

Thus, any linear convex combination of vectors in S belongs to S.

2. (**Halfspaces.**) Figure 5.3 shows that the hyperplane

$$S = \{\mathbf{x} \in \mathbb{R}^n \mid x_1 + x_2 = 1\}$$

separates the region $\mathbb{R}^2$ into two parts. In one of them all the points $\mathbf{x} = (x_1, x_2)$ satisfy $x_1 + x_2 > 1$, and in the other, $x_1 + x_2 < 1$ holds. The sets $S^+ = \left\{\mathbf{x} \in \mathbb{R}^2 \mid x_1 + x_2 \geq 1\right\}$ and $S^- = \left\{\mathbf{x} \in \mathbb{R}^2 \mid x_1 + x_2 \leq 1\right\}$ are referred to as *halfspaces.* In general, if S is the hyperplane determined by (5.3), we define the halfspaces associated with S, as the sets

$$S^+ = \left\{\mathbf{x} \in \mathbb{R}^n \mid \mathbf{q}^T\mathbf{x} \geq \alpha\right\} \quad \text{and } S^- = \left\{\mathbf{x} \in \mathbb{R}^n \mid \mathbf{q}^T\mathbf{x} \leq \alpha\right\}$$

3. (**Convex Cones.**) A nonempty set C is called a *cone* if $\mathbf{x} \in C$ implies that $\lambda\mathbf{x} \in C$ for all $\lambda > 0$. If, in addition, C is convex, then C is called a *convex cone.* Figure 5.4 shows examples of convex and nonconvex cones.

The reader is encouraged to prove convexity in cases 2 and 3.

The following properties are immediate consequences of the definition of convexity:

- If $\{S_i\}_{i \in I}$ is a family of convex sets, then the set

$$\cap_{i \in I} S_i \equiv \{\mathbf{x} \mid \mathbf{x} \in S_i \text{ for all } i \in I\}$$

is also a convex set (see Figure 5.5).

- If S_1 and S_2 are convex sets, then the set

$$S_1 + S_2 \equiv \{\mathbf{y} = \mathbf{x}_1 + \mathbf{x}_2 \mid \mathbf{x}_1 \in S_1 \text{ and } \mathbf{x}_2 \in S_2\}$$

is also a convex set.

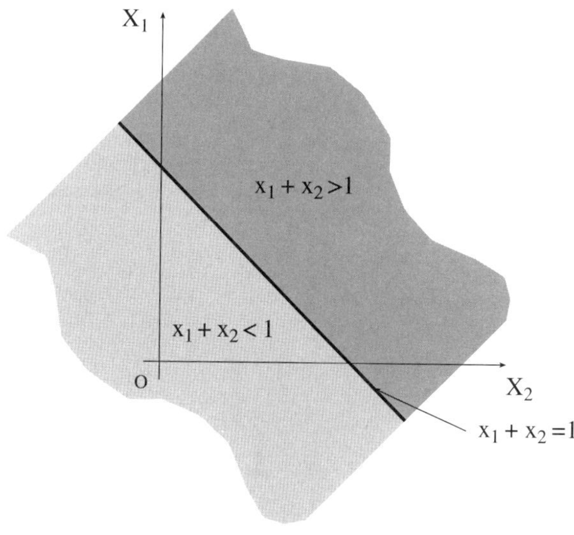

Figure 5.3: Two regions separated by a hyperplane.

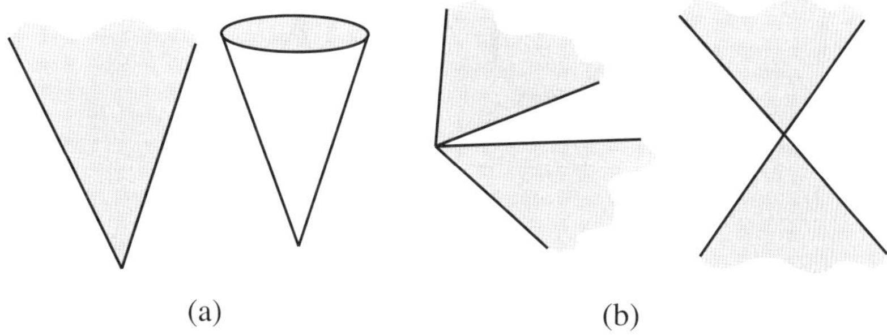

Figure 5.4: Illustration of convex (a) and nonconvex (b) cones.

These properties can be used to define new convex sets based on halfspaces. We can consider the intersection, finite or infinite, of halfspaces, and the resulting set is convex.

Next, we introduce the concepts of extreme point and extreme direction for convex sets.

Definition 5.2 (Extreme point). *Let S be a nonempty convex set in $\mathbb{R}^n$. A vector $\mathbf{x} \in S$ is called an extreme point of S if $\mathbf{x} = \lambda \mathbf{x}_1 + (1 - \lambda)\mathbf{x}_2$ with $\mathbf{x}_1, \mathbf{x}_2 \in S$ and $\lambda \in (0, 1)$ implies that $\mathbf{x} = \mathbf{x}_1 = \mathbf{x}_2$.* ∎

In a circle the number of extreme points is infinite but in a triangle and a square this number is finite (see Figure 5.6). The following are the sets E of extreme points of some convex sets S:

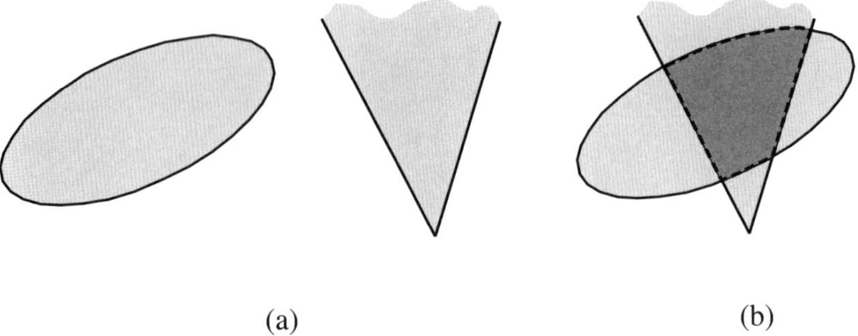

Figure 5.5: Convex sets (a) as the intersection (b) of convex sets.

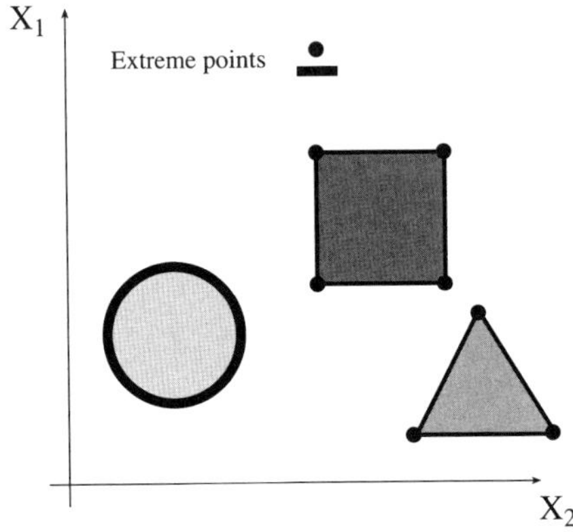

Figure 5.6: Extreme points in a circle, a square, and a triangle.

1. A circle

$$S = \{(x_1, x_2) \in \mathbb{R}^2 \mid x_1^2 + x_2^2 \leq 1\}$$
$$E = \{(x_1, x_2) \in \mathbb{R}^2 \mid x_1^2 + x_2^2 = 1\}$$

2. A triangle

$$S = \{(x_1, x_2) \in \mathbb{R}^2 \mid x_1 \geq 0,\ x_1 - x_2 \geq 0,\ x_1 + x_2 \leq 1\}$$
$$E = \{(0,0),\ (\tfrac{1}{2}, \tfrac{1}{2}),\ (0,1)\}$$

3. A square

$$S = \{(x_1, x_2) \in \mathbb{R}^2 \mid x_1 \geq 0,\ x_2 \geq 0,\ x_1 \leq 1,\ x_2 \leq 1\}$$
$$E = \{(0,0),\ (0,1),\ (1,1),\ (1,0)\}$$

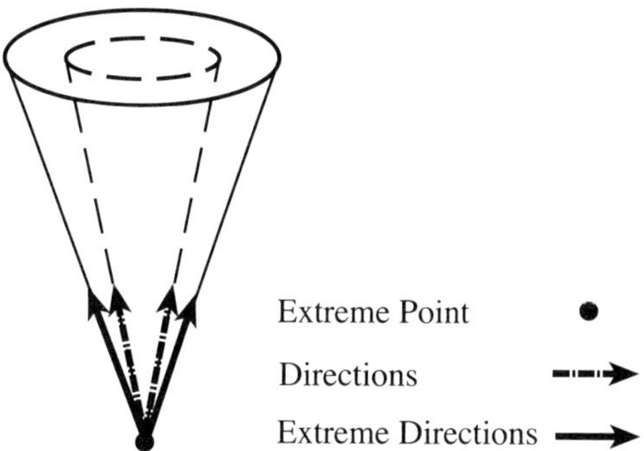

Figure 5.7: A cone, two of its infinite directions and extreme directions, and its unique extreme point.

Theorem 5.1 (Representation theorem for finite convex sets). *If a convex set is bounded and closed, any of its points can be written as a linear convex combination of its extreme points.* ■

Theorem 5.1 turns out to be false for a unbounded convex set. For example let $C = \{\mathbf{x} \in \mathbb{R}^3 \mid x_1^2 + x_2^2 \leq x_3^2,\ x_3 \geq 0\}$. This set is shown in Figure 5.7. Note that C is convex and closed. However, S contains only the extreme point, $(0, 0, 0)$, namely, the origin, and obviously C is not equal to the collection of all convex combinations of its extreme points. The difficulty appears because C is not bounded. To deal with unbounded sets, the notion of extreme directions is needed.

Definition 5.3 (Direction). *Let S be a nonempty, closed convex set in $\mathbb{R}^n$. We say that a unit vector $\mathbf{d}$ is a direction of S, if for each $\mathbf{x} \in S$, $\mathbf{x} + \pi\mathbf{d} \in S$ for all $\pi \geq 0$.* ■

Definition 5.4 (Extreme direction). *A direction $\mathbf{d}$ is called an extreme direction if it cannot be written as a positive linear combination of two distinct directions, that is, if $\mathbf{d} = \pi_1\mathbf{d}_1 + \pi_2\mathbf{d}_2$ for $\pi_1,\ \pi_2 > 0$, then $\mathbf{d} = \mathbf{d}_1 = \mathbf{d}_2$.* ■

To illustrate these definitions, consider the set C shown in Figure 5.7. The set of directions is $D = \{\mathbf{d} \in \mathbb{R}^3 \mid d_1^2 + d_2^2 \leq d_3^2,\ d_1^2 + d_2^2 + d_3^2 = 1\}$, and the extreme directions $D_E = \{\mathbf{d} \in \mathbb{R}^3 \mid d_1^2 + d_2^2 = d_3^2 = 1/2\}$.

In the following sections we describe especial cases of convex sets.

5.3 Linear Spaces

Linear spaces are the best known class of convex sets. They arise as linear combinations of sets of vectors.

Definition 5.5 (Linear combination). *Consider a matrix* $\mathbf{A}$, *and the set of its column vectors*

$$\mathbf{A} \equiv \{\mathbf{a}_1, \ldots, \mathbf{a}_k\}$$

A vector $\mathbf{x}$ *is said to be a linear combination of the vectors in the set* A *if it can be written as*

$$\mathbf{x} = \rho_1 \mathbf{a}_1 + \cdots + \rho_k \mathbf{a}_k \tag{5.4}$$

where $\rho_i \in \mathbb{R}, \forall i$. ■

By simplicity of notation, in the following we shall identify the set $\mathbf{A}$ with the matrix $\mathbf{A}$. We denote the set of all linear combinations of the vectors in $\mathbf{A}$ by $\mathbf{A}_\rho$.

Definition 5.6 (Linear space). *The set*

$$\mathbf{A}_\rho \equiv \{\mathbf{x} \in \mathbb{R}^n \mid \mathbf{x} = \rho_1 \mathbf{a}_1 + \ldots + \rho_k \mathbf{a}_k \quad \textit{with} \quad \rho_i \in \mathbb{R}; i = 1, \ldots, k\}$$

of all vectors, of the Euclidean space $\mathbb{R}^n$, *that can be generated by linear combinations of vectors in this set* $\mathbf{A}$ *is called a linear space (the linear space generated by* $\mathbf{A}$*), and the set* $\mathbf{A}$ *is called the set of its generators.* ■

It is important to know when a subset of a linear space is a linear subspace of it. To this end, we give, without proof, the following theorem.

Theorem 5.2 (Identifying subspaces). *Let* $\mathcal{V}$ *be a linear space and* $\mathcal{W}$ *a subset of it. Then,* $\mathcal{W}$ *is a linear subspace of* $\mathcal{V}$ *if and only if*

1. $\mathbf{a}, \mathbf{b} \in \mathcal{W} \Rightarrow \mathbf{a} + \mathbf{b} \in \mathcal{W}$

2. $\mathbf{a} \in \mathcal{W}, \alpha \in \mathbb{R} \Rightarrow \alpha \mathbf{a} \in \mathcal{W}$

■

The proof of this can be seen in Castillo et al. [21]. Condition 2 shows that a linear subspace is a cone.

The following theorem shows that the set of all feasible solutions of an homogeneous linear system of equations is a linear space, and that any linear space can be written as an homogeneous system of linear equations.

Theorem 5.3 (Representation of a linear space). *Let* S *be a subset of* $\mathbb{R}^n$. *The set* S *is a linear subspace of* $\mathbb{R}^n$ *if and only if there exists a matrix* $m \times n$, $\mathbf{H}$, *such that*

$$S = \{\mathbf{x} \in \mathbb{R}^n | \mathbf{H}\mathbf{x} = \mathbf{0}\} \tag{5.5}$$

■

Example 5.1 (Feasible set of solutions of an homogeneous system of equations). The general solution of the system of linear equations:

$$\begin{array}{ccccccc} x_1 & -x_2 & & -x_4 & = & 0 \\ x_1 & & -x_3 & +x_4 & = & 0 \\ & x_2 & -x_3 & +2x_4 & = & 0 \end{array} \tag{5.6}$$

is the linear subspace

$$\begin{pmatrix} x_1 \\ x_2 \\ x_3 \\ x_4 \end{pmatrix} = \rho_1 \begin{pmatrix} 1 \\ 1 \\ 1 \\ 0 \end{pmatrix} + \rho_2 \begin{pmatrix} -1 \\ -2 \\ 0 \\ 1 \end{pmatrix}; \quad \rho_1, \rho_2 \in \mathbb{R} \tag{5.7}$$

which can be obtained using standard packages, as Mathematica [103], for example. Note that we have 2 degrees of freedom (ρ_1 and ρ_2) to generate solutions of (5.6). ■

5.4 Polyhedral Convex Cones

In this section we define polyhedral convex cones, which are characterized by means of nonnegative linear combinations of vectors.

Definition 5.7 (Nonnegative linear combination). *We say that a vector* $\mathbf{x}$ *is a nonnegative linear combination of vectors of* $\mathbf{A} = \{\mathbf{a}_1, \ldots, \mathbf{a}_k\}$ *if and only if*

$$\mathbf{x} = \pi_1 \mathbf{a}_1 + \cdots + \pi_k \mathbf{a}_k \tag{5.8}$$

where $\pi_i \in \mathbb{R}^+, \forall i$, $\mathbb{R}^+$ *is the set of nonnegative real numbers.* ■

We denote the set of all nonnegative linear combinations of the vectors in $\mathbf{A}$ by $\mathbf{A}_\pi$.

Definition 5.8 (Polyhedral convex cone). *Let* $\mathbf{A} = \{\mathbf{a}_1, \ldots, \mathbf{a}_k\}$. *The set*

$$\mathbf{A}_\pi \equiv \{\mathbf{x} \in \mathbb{R}^n \mid \mathbf{x} = \pi_1 \mathbf{a}_1 + \ldots + \pi_k \mathbf{a}_k \quad \textit{with} \quad \pi_j \geq 0; j = 1, \ldots, k\}$$

of all nonnegative linear combinations of vectors of $\mathbf{A}$ *is known as a polyhedral convex cone. The vectors* $\mathbf{a}_1, \ldots, \mathbf{a}_k$ *are the cone generators.* ■

In the following, for simplicity, we shall refer to polyhedral convex cones as cones. So the set C is a cone if and only if there exist a set of vectors $\mathbf{A}$ such that $C = \mathbf{A}_\pi$.

Remark 5.1 *Note that convex cones and polyhedral convex cones are different concepts.* ■

Remark 5.2 *Note that the cone generators are directions of the cone.* ■

A cone C that is not a linear subspace arises when we deal with restricted (nonnegative or nonpositive) variables.

Definition 5.9 (General form of a polyhedral convex cone). *If we have a cone $\mathbf{A}_\pi$, we can classify its generators $\mathbf{A} = (\mathbf{a}_1, \ldots, \mathbf{a}_m)$ into two groups:*

1. *The generators whose opposite vectors belong to the cone:*

$$\mathbf{B} \equiv \{\mathbf{a}_i \mid -\mathbf{a}_i \in \mathbf{A}_\pi\}$$

2. *The generators whose opposite vectors do not belong to the cone:*

$$\mathbf{C} \equiv \{\mathbf{a}_i \mid -\mathbf{a}_i \notin \mathbf{A}_\pi\}$$

Thus, we can express the cone in the following form:

$$\mathbf{A}_\pi \equiv (\mathbf{B} : -\mathbf{B} : \mathbf{C})_\pi \equiv \mathbf{B}_\rho + \mathbf{C}_\pi \tag{5.9}$$

which is known as the general form of a cone. ■

The following theorem shows that the set of all feasible solutions of an homogeneous linear system of inequalities is a cone, and that any cone can be written as an homogeneous system of linear inequalities.

Theorem 5.4 (Representation of a cone). *Let S be a subset of $\mathbb{R}^n$. The set S is a cone with origin at $\mathbf{0}$ if and only if there exists a matrix $m \times n$, $\mathbf{H}$, such that*

$$S = \{\mathbf{x} \in \mathbb{R}^n | \mathbf{H}\mathbf{x} \leq \mathbf{0}\} \tag{5.10}$$

■

Example 5.2 (Feasible set of solutions of a homogeneous system of inequalities). The general solution of the system of linear inequalities:

$$\begin{array}{ccccc} x_1 & -x_2 & & -x_4 & \leq 0 \\ x_1 & & -x_3 & +x_4 & \leq 0 \\ & x_2 & -x_3 & +2x_4 & \leq 0 \end{array} \tag{5.11}$$

is the cone

$$\begin{pmatrix} x_1 \\ x_2 \\ x_3 \\ x_4 \end{pmatrix} = \rho_1 \begin{pmatrix} 1 \\ 1 \\ 1 \\ 0 \end{pmatrix} + \rho_2 \begin{pmatrix} -1 \\ -2 \\ 0 \\ 1 \end{pmatrix} + \pi_1 \begin{pmatrix} -1 \\ -1 \\ 0 \\ 0 \end{pmatrix} + \pi_2 \begin{pmatrix} -1 \\ 0 \\ 0 \\ 0 \end{pmatrix} \tag{5.12}$$

$$\rho_1, \rho_2 \in \mathbb{R}; \pi_1, \pi_2 \in \mathbb{R}^+$$

Obtaining the cone generators is not a trivial task that can be done using the concept of a dual cone (see the Appendix A).

As we can see we have 4 degrees of freedom (ρ_1, ρ_2, π_1 and π_2) to generate solutions of (5.11); however, two of the coefficients (π_1 and π_2) are restricted to be nonnegative. Note that the cone is written as the sum of a linear space plus a proper cone (a cone such that its linear space component does not exist), i.e., in its general form. ■

Remark 5.3 *A linear space* $\mathbf{A}_\rho$ *is a particular case of a cone:*

$$\mathbf{A}_\rho \equiv (\mathbf{A} : -\mathbf{A})_\pi, \tag{5.13}$$

where $-\mathbf{A}$ *is the negative of* $\mathbf{A}$. *In other words, a linear space is the cone generated by its generators and their opposite vectors.*

Expression (5.13) shows that the cone concept is broader than the linear space concept and, more importantly, allows us to replace the usual methods of classical linear algebra by the specific methods of cones, which are more efficient, because cone is a wider concept than linear space. ■

5.5 Polytopes

Definition 5.10 (Linear convex combination). *We say that a vector* $\mathbf{x}$ *is a linear convex combination of the vectors of* $\mathbf{A} = \{\mathbf{a}_1, \ldots, \mathbf{a}_k\}$ *if and only if*

$$\mathbf{x} = \lambda_1 \mathbf{a}_1 + \cdots + \lambda_k \mathbf{a}_k, \tag{5.14}$$

where $\lambda_i \in \mathbb{R}^+, \forall i$ *and* $\sum_{i=1}^{k} \lambda_i = 1$. ■

We denote the set of all linear convex combinations of vectors of $\mathbf{A}$ by $\mathbf{A}_\lambda$.

Definition 5.11 (Polytope). *Let* $\mathbf{A} = \{\mathbf{a}_1, \ldots, \mathbf{a}_k\}$. *The set*

$$S = \mathbf{A}_\lambda \equiv \left\{ \mathbf{x} \in \mathbb{R}^n \mid \mathbf{x} = \lambda_1 \mathbf{p}_1 + \cdots + \lambda_k \mathbf{p}_k \quad \textit{with} \quad \lambda_i \geq 0\,; \sum_{i=1}^{k} \lambda_i = 1 \right\},$$

of all linear convex combinations of vectors of $\mathbf{A}$ *is known as a polytope or the convex hull generated by* $\{\mathbf{a}_1, \ldots, \mathbf{a}_k\}$. *If the set of vectors* $\{\mathbf{a}_1, \ldots, \mathbf{a}_k\}$ *is a minimal set, then this is the set of the* k *extreme points of* S. ■

Figure 5.8 shows two examples of convex polytopes. Case (a) is a tetrahedron, because the points associated with the generators $\mathbf{a}_1, \mathbf{a}_2, \mathbf{a}_3$, and $\mathbf{a}_4$ do not belong to the same plane. However, in case (b) these points are coplanar and the polytope degenerates into a simple quadrilateral.

The following theorem shows that any polytope can be written as a complete system of linear inequalities.

Theorem 5.5 (Representation of a polytope). *A polytope S can be written as the intersection of a finite set of halfspaces:*

$$S = \{\mathbf{x} \in \mathbb{R}^n | \mathbf{H}\mathbf{x} \leq \mathbf{b}\} \tag{5.15}$$

■

Example 5.3 (Representation of a polytope). Consider the polytope defined by the set of inequalities

$$\begin{array}{rrcl} -x_1 & -x_2 & \leq & 0 \\ x_1 & -x_2 & \leq & 1 \\ -x_1 & +x_2 & \leq & 0 \\ x_1 & +x_2 & \leq & 1 \end{array} \tag{5.16}$$

illustrated in Figure 5.9. The set of its extreme points is the set

$$\mathbf{P} = \left\{ (0,0)^T, \left(\frac{1}{2}, \frac{1}{2}\right)^T, (1,0)^T, \left(\frac{1}{2}, -\frac{1}{2}\right)^T \right\}.$$

Then, S can be represented by

$$\begin{pmatrix} x_1 \\ x_2 \end{pmatrix} = \lambda_1 \begin{pmatrix} 0 \\ 0 \end{pmatrix} + \lambda_2 \begin{pmatrix} \frac{1}{2} \\ \frac{1}{2} \end{pmatrix} + \lambda_3 \begin{pmatrix} 1 \\ 0 \end{pmatrix} + \lambda_4 \begin{pmatrix} \frac{1}{2} \\ -\frac{1}{2} \end{pmatrix} \tag{5.17}$$

where

$$\begin{array}{ccccccc} \lambda_1 & +\lambda_2 & +\lambda_3 & +\lambda_4 & = & 1 \\ \lambda_1 & & & & \geq & 0 \\ & \lambda_2 & & & \geq & 0 \\ & & \lambda_3 & & \geq & 0 \\ & & & \lambda_4 & \geq & 0 \end{array} \tag{5.18}$$

■

5.6 Polyhedra

Definition 5.12 (Polyhedron). *A polyhedron is the intersection of a finite set of halfspaces:*

$$S = \{\mathbf{x} \in \mathbb{R}^n | \mathbf{H}\mathbf{x} \leq \mathbf{b}\} \tag{5.19}$$

When S is bounded, then S is a polytope.

■

Expression (5.19) shows that the set of all feasible solutions of a linear system of inequalities is a polyhedron.

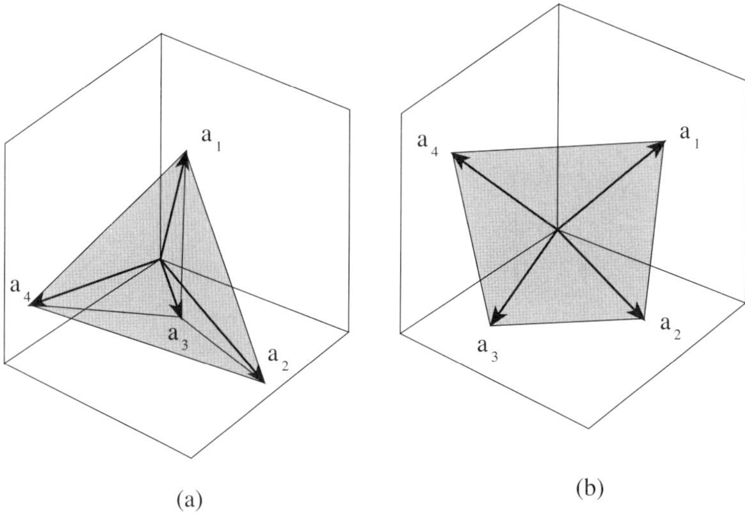

Figure 5.8: Two examples of convex polytopes.

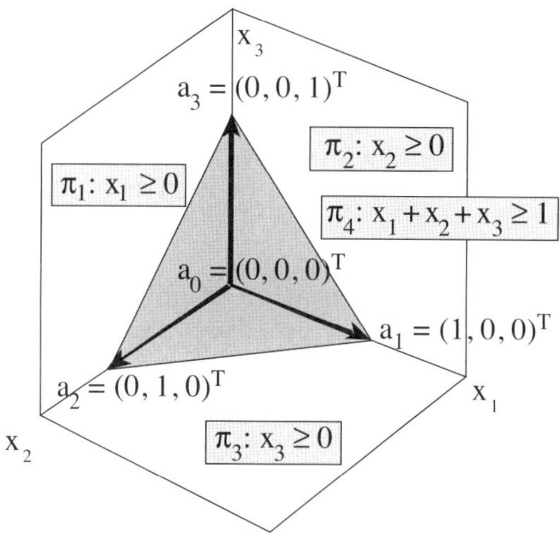

Figure 5.9: Two different definitions of a polytope: by its extreme points, and by a set of inequalities.

Example 5.4 (Representation of polyhedral sets). The solutions of the system of linear inequalities

$$\begin{array}{cccccc} x_1 & -x_2 & & -x_4 & \leq & 3 \\ x_1 & & -x_3 & +x_4 & \leq & 1 \\ & x_2 & -x_3 & +2x_4 & \leq & 0 \end{array} \tag{5.20}$$

can be obtained using the Γ algorithm (see Section A.1), and is the polyhedron

$$\begin{pmatrix} x_1 \\ x_2 \\ x_3 \\ x_4 \end{pmatrix} = \lambda_1 \begin{pmatrix} 1 \\ 0 \\ 0 \\ 0 \end{pmatrix} + \lambda_2 \begin{pmatrix} 1 \\ -2 \\ 0 \\ 0 \end{pmatrix} + \pi_1 \begin{pmatrix} -1 \\ -1 \\ 0 \\ 0 \end{pmatrix} + \pi_2 \begin{pmatrix} -1 \\ 0 \\ 0 \\ 0 \end{pmatrix} + \pi_3 \begin{pmatrix} -1 \\ -2 \\ 0 \\ 1 \end{pmatrix} + \pi_4 \begin{pmatrix} 1 \\ 2 \\ 0 \\ -1 \end{pmatrix} \tag{5.21}$$

where

$$\begin{array}{rcl} \lambda_1 + \lambda_2 & = & 1 \\ \lambda_1,\ \lambda_2 & \geq & 0 \\ \pi_1,\ \pi_2,\ \pi_3,\ \pi_4 & \geq & 0 \end{array} \tag{5.22}$$

This representation has 6 degrees of freedom (λ_1, λ_2, π_1, π_2, π_3, and π_4) to generate solutions of (5.20). Note that the polyhedron is written as the sum of a polytope plus a cone. ■

5.6.1 General Representation of Polyhedra

The set of constraints in a LPP defines a polyhedron. Particular cases of polyhedra are linear spaces, cones, and polytopes (see Figure 5.10). The following theorem shows that any polyhedron can be written as the sum of these three basic structures.

Theorem 5.6 (General representation of polyhedra). *Any polyhedron S can be written as the sum of a linear space plus a cone plus a polytope:*

$$S = \mathbf{V}_\rho + \mathbf{W}_\pi + \mathbf{Q}_\lambda.$$

This means that $\mathbf{x} \in S$ if and only if $\mathbf{x}$ can be written as

$$\mathbf{x} = \sum_i \rho_i \mathbf{v}_i + \sum_j \pi_j \mathbf{w}_j + \sum_k \lambda_k \mathbf{q}_k;\ \rho_i \in \mathbb{R};\ \pi_j \geq 0;\ \lambda_k \geq 0;\ \sum_k \lambda_k = 1 \tag{5.23}$$

Some of this summands can be empty. ■

Corollary 5.1 *If the linear space is empty, a minimal set of generators of the cone and the polytope are the sets of extreme directions and extreme points, respectively. In this case the set of extreme directions is* $\mathbf{W}$, *and the set of extreme points is* $\mathbf{Q}$.

All these results are summarized in Table 5.1 and Figure 5.10.

5.7 Bounded and Unbounded LPP

We end this chapter showing the close relation of the feasible region structure and the bounded or unbounded character of the LPP.

Consider the following problem. Minimize

$$Z = \mathbf{c}^T \mathbf{x}$$

subject to one of the constraints in Table 5.1. Then, according to this table, we have the following cases:

Linear space. Then, the value of the cost function becomes

$$\mathbf{c}^T \mathbf{x} = \sum_i \rho_i \mathbf{c}^T \mathbf{v}_i$$

which leads to a bounded problem only if $\mathbf{c}^T$ is orthogonal to all $\mathbf{v}_i$. Otherwise, it is unbounded, because the values of ρ_i can be chosen as large as possible.

Cone. Then, the value of the cost function becomes

$$\mathbf{c}^T \mathbf{x} = \sum_j \pi_j \mathbf{c}^T \mathbf{w}_i; \ \pi_j \geq 0$$

which leads to a bounded problem only if $\mathbf{c}^T$ is orthogonal to all $\mathbf{w}_j$, it is a bounded from below problem only if $\mathbf{c}^T \mathbf{w}_j \geq 0; \ \forall j$, and from above, if $\mathbf{c}^T \mathbf{w}_j \leq 0; \ \forall j$. Otherwise, it is unbounded.

Polytope. Then, the value of the cost function becomes

$$\mathbf{c}^T \mathbf{x} = \sum_k \lambda_k \mathbf{c}^T \mathbf{q}_k; \ \lambda_k \geq 0$$

which always leads to a bounded problem.

Polyhedron. Then, the value of the cost function becomes

$$\mathbf{c}^T \mathbf{x} = \sum_i \rho_i \mathbf{v}_i + \sum_j \pi_j \mathbf{w}_j + \sum_k \lambda_k \mathbf{c}^T \mathbf{q}_k$$

and we have the following cases:

1. A bounded problem only if $\mathbf{c}^T$ is orthogonal to all $\mathbf{v}_i$ and $\mathbf{w}_j$
2. A bounded from below problem only if $\mathbf{c}^T$ is orthogonal to all $\mathbf{v}_i$ and $\mathbf{c}^T \mathbf{w}_j \geq 0; \ \forall j$
3. A bounded from above problem only if $\mathbf{c}^T$ is orthogonal to all $\mathbf{v}_i$ and $\mathbf{c}^T \mathbf{w}_j \leq 0; \ \forall j$
4. An unbounded problem, otherwise

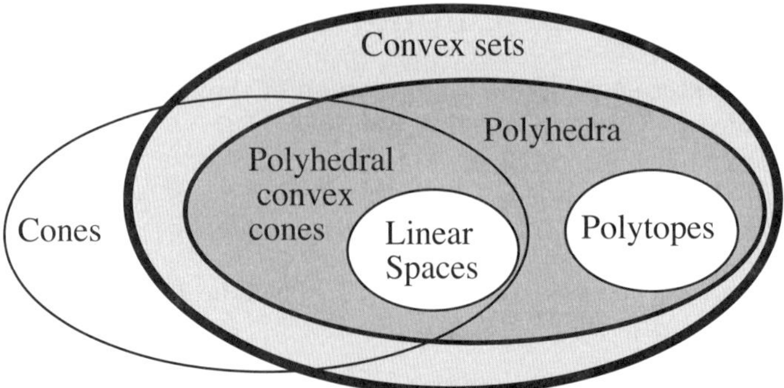

Figure 5.10: The different structures arising as solutions of linear systems of equations and inequalities.

Exercises

5.1 Show that the set $S = \{\mathbf{x} \in \mathbb{R}^3 \mid x_1^2 + x_2^2 \leq x_3^2\}$ is a cone.

5.2 Given the following polyhedral sets

$$\begin{aligned}
S_1 &= \left\{\mathbf{x} \in \mathbb{R}^2 \mid x_1 + x_2 \leq 1,\ x_1 \geq 0,\ x_2 \geq 0\right\} \\
S_2 &= \left\{\mathbf{x} \in \mathbb{R}^2 \mid x_1 - x_2 \leq 0,\ x_1 \geq 0,\ x_2 \geq 0\right\} \\
S_3 &= \left\{\mathbf{x} \in \mathbb{R}^2 \mid x_1 - x_2 \leq 0,\ x_1 + x_2 \leq 1,\ x_1 \geq 0,\ x_2 \geq 0\right\} \\
S_4 &= \left\{\mathbf{x} \in \mathbb{R}^2 \mid x_2 - x_1 \leq 0,\ x_1 - x_2 \leq 0,\ x_1 + x_2 \leq 1,\ x_1 \geq 0,\ x_2 \geq 0\right\} \\
S_5 &= \left\{\mathbf{x} \in \mathbb{R}^2 \mid x_1 + x_2 \leq 1,\ 2x_1 + 2x_2 \geq 3\right\}
\end{aligned}$$

(a) Represent these sets by means of extreme points and directions.

(b) Draw these sets, and locate the extreme points and directions on the figures.

5.3 Why is it not possible to represent the set $S = \{\mathbf{x} \in \mathbb{R}^3 \mid x_1 \geq 0\}$ using only extreme points and directions?

5.4 Let $\Phi(\mathbf{x}) = \mathbf{A}\mathbf{x}$ be a linear mapping, where $\mathbf{A}$ is an $m \times n$ matrix. Show that if S is a (bounded) polyhedral set in $\mathbb{R}^n$, then $\Phi(S)$ is a (bounded) polyhedral set in $\mathbb{R}^m$.

5.5 If S_1 and S_2 are polyhedral sets in $\mathbb{R}^n$, then $S_1 + S_2$ is also a polyhedral set. (*Hint*: Use the previous result.)

5.6 Consider the following sets of constraints:

(a)

$$-2x_1 \quad + x_2 \quad = \quad 0$$

(b)

$$\begin{array}{rrcr} -2x_1 & + x_2 & \leq & 0 \\ & - x_2 & \leq & 0 \end{array}$$

(c)

$$\begin{array}{rrcr} -2x_1 & + x_2 & \leq & 0 \\ & - x_2 & \leq & 0 \\ x_1 & - x_2 & \leq & 2 \end{array}$$

(d)

$$\begin{array}{rrcr} -2x_1 & + x_2 & \leq & 0 \\ & - x_2 & \leq & 0 \\ x_1 & - x_2 & \leq & 2 \\ x_1 & + x_2 & \leq & 6 \end{array}$$

(i) Draw these sets, and locate the extreme points and directions on the figure.

(ii) Give a representation of these sets by means of its extreme points and directions.

(iii) Identify them as linear spaces, cones, polytopes, or polyhedra.

(iv) Obtain graphically the minimum and maximum of the objective function $Z = x_1 + x_2$.

(v) Give an optimization problem for which there exist multiple solutions.

5.7 Consider the following sets of constraints:

(a)

$$-x_1 \quad + x_3 = 0$$

(b)

$$\begin{array}{rrrcr} -x_1 & & + x_3 & \leq & 0 \\ & & - x_3 & \leq & 0 \\ & -x_2 & + x_3 & \leq & 0 \end{array}$$

(c)

$$\begin{array}{rrrcr} -x_1 & & + x_3 & \leq & 0 \\ & & - x_3 & \leq & 0 \\ & -x_2 & + x_3 & \leq & 0 \\ x_1 & +x_2 & & \leq & 2 \end{array}$$

(i) Draw these sets, and locate the extreme points and directions on the figure.

(ii) Give a representation of these sets by means of its extreme points and directions.

(iii) Identify them as linear spaces, cones, polytopes, or polyhedra.

(iv) Obtain graphically the minimum and maximum of the objective function $Z = 3x_1 + x_2 - x_3$.

(v) Give an optimization problem for which there exist multiple solutions.

Chapter 6

Solving the Linear Programming Problem

6.1 Introduction

In Chapter 5 we developed a method for obtaining the set of all feasible solutions defined by the set of constraints based on obtaining the set of all extreme points and directions using the Γ algorithm. However, this method is computationally intensive and is not an adequate method for LPP. In this chapter we introduce methods specially developed for this type of problems.

Economists of the former Soviet Union were the first in applying linear programming techniques in the organization and planning of production. However, it was during World War II that linear programming gained prominence. The U.S. Air Force created the SCOOP project (Scientific Computation of Optima Programs) directed by G. B. Dantzig [30]. The best known method for solving linear programming problems, *the simplex method*, is due to Dantzig, who presented the method in 1947. Fortunately, the growth of large and fast computing facilities has aided in the use of the developed techniques.

During the last few decades very intensive work and research has been devoted to the linear programming problem (LPP). While implementation of the simplex method algorithm has undergone major modification, the underlying concept has withstood the test of time. A description of the computer implementation of this method can be seen, for example, in the *IBM Optimization Subroutine Library* (OSL) [56] or in Kuenzi et al. [64].

For many linear programming problems, the simplex method is still the best method. However, several improvements have been introduced, such as the revised simplex, the dual, or the primal–dual methods. The primal–dual methods work simultaneously with both the primal problem and its dual. This can have advantages because we can exploit the well–known relationships between primal and dual problems (see Forrest and Tomlin [40]). However, as indicated, for example, by Press et al. [89]: "There seems to be no clearcut evidence that these methods are superior to the usual method by any factor substantially larger than the 'tender-loving-care factor' which reflects the programming effort of the proponents." For most cases the primal is the preferred method. However, in some special cases the other methods give superior performance.

The interior point method (IPM) was introduced by Karmarkar [57] as an alternative to the simplex method (SM) (see, Gill et al. [48]). The primal–dual method has been superseded by the predictor–corrector method, which is a modified version of the primal–dual originated by Mehrotra [75]. This method usually requires less iterations and is the interior point method of choice.

Unlike the SM, which has exponential computational complexity, the IPM has polynomial computational complexity. As a result, for large problems (more than 2000 constraints and variables) interior point methods outperform the SM. However, the SM attains a solution within a finite number of steps, whereas the IPM does not necessarily, because it converges asymptotically to the optimal solution. If rubustness is at stake, the difference described above can be a determinant.

In this chapter, we focus on computational methods to solve linear programming problems. We assume that the analysis on data, variables, constraints, and objective function has been carried out previously, and we have the complete and coherent formulation of the problem at our disposal.

In what follows we first describe, in a insightful fashion, the simplex method (SM), then the exterior point method (EPM), which is a particular implementation of the dual simplex. The interior point method (IPM) will be described in Chapter 9.

6.2 The Simplex Method

Before starting with the description of the simplex method (SM), we illustrate its main basic idea with one example.

6.2.1 Motivating Example

Consider the following LPP in standard form. Minimize

$$Z = -11 + x_2 + 6x_3 + 2x_4 - x_9 \tag{6.1}$$

subject to

$$\begin{array}{ccccccccccc}
x_1 & & -x_3 & +x_4 & & & & & & = & 6 \\
 & x_2 & +3x_3 & & & & & & -x_9 & = & 4 \\
 & & & & x_5 & & & & +x_9 & = & 3 \\
 & & 2x_3 & -x_4 & & +x_6 & & & & = & 1 \\
 & & & -2x_4 & & & +x_7 & & -x_9 & = & 2 \\
 & & -x_3 & +x_4 & & & & +x_8 & -x_9 & = & 5
\end{array}$$

and

$$x_1, x_2, \ldots, x_9 \geq 0$$

Since we have 6 equality constraints, we can obtain 6 variables, say, x_1, x_2, x_5, x_6, x_7, and x_8, as functions of the remaining variables:

$$\begin{array}{rcrrrr} x_1 & = & 6 & +x_3 & -x_4 & \\ x_2 & = & 4 & -3x_3 & & +x_9 \\ x_5 & = & 3 & & & -x_9 \\ x_6 & = & 1 & -2x_3 & +x_4 & \\ x_7 & = & 2 & & +2x_4 & +x_9 \\ x_8 & = & 5 & +x_3 & -x_4 & x_9 \end{array} \tag{6.2}$$

This implies partitioning the set of variables $\{x_1, x_2, x_3, x_4, x_5, x_6, x_7, x_8, x_9\}$ into two sets $\{x_1, x_2, x_5, x_6, x_7, x_8\}$ and $\{x_3, x_4, x_9\}$, which are known as the set of *basic* and *nonbasic* variables, respectively.

Next, we replace the basic variables (6.2) in the objective function (6.1), in terms of nonbasic variables, to get

$$Z = -7 + x_9 + 3x_3 + 2x_4.$$

Thus, the initial LPP is equivalent to the LPP, where we minimize

$$Z = -7 + x_9 + 3x_3 + 2x_4$$

subject to

$$\begin{array}{rcrrrr} x_1 & = & 6 & +x_3 & -x_4 & \\ x_2 & = & 4 & -3x_3 & & +x_9 \\ x_5 & = & 3 & & & -x_9 \\ x_6 & = & 1 & -2x_3 & +x_4 & \\ x_7 & = & 2 & & +2x_4 & +x_9 \\ x_8 & = & 5 & +x_3 & -x_4 & +x_9 \end{array}$$

and

$$x_1, x_2, \ldots, x_9 \geq 0$$

The solution of this problem is $z = -7$ and is attained at the point

$$(x_1, x_2, x_3, x_4, x_5, x_6, x_7, x_8, x_9) = (6, 4, 0, 0, 3, 1, 2, 5, 0)$$

because

1. The nonnegative coefficients in the function z to be minimized guarantee that no value smaller than -7 can be obtained because the variables x_9, x_3 and x_4 are nonnegative. This minimum value is attained when $x_9 = x_3 = x_4 = 0$.

2. The values of the basic variables, which can be calculated using the constraints and replacing $x_9 = x_3 = x_4 = 0$, are nonnegative; thus, they also satisfy the nonnegativity constraints. This occurs because the independent terms in the constraints are nonnegative.

Consequently, if we have a function to be minimized with all its coefficients being nonnegative, then the minimum value is attained for all the variables appearing in it taking the value zero. This solution is feasible if and only if the independent terms appearing in the constraints are nonnegative.

The simplex method starts with a function to be minimized and a set of constraints that normally do not satisfy these conditions. In the so-called regulating stage we transform the set of constraints to another equivalent set with nonnegative independent terms, and in the so-called standard iterations we try to get nonnegative coefficients in the transformed function to be optimized while keeping nonnegative independent terms. If this is possible we obtain the optimal solution. Otherwise the problem is unbounded or infeasible.

6.2.2 General Description

There are many variants of the simplex method (SM), although all of them rely on the same central idea. In this chapter we present one of those existing versions.

The simplex method is applied to a LPP in the following standard form. Minimize

$$f(\mathbf{x}) = \mathbf{c}^T\mathbf{x} \tag{6.3}$$

subject to

$$\mathbf{A}\mathbf{x} = \mathbf{b}; \quad \mathbf{b} \geq \mathbf{0} \tag{6.4}$$

and

$$\mathbf{x} \geq \mathbf{0} \tag{6.5}$$

where $\mathbf{c} = (c_1, c_2, \ldots, c_n)^T$ is the column matrix of the cost coefficients, $\mathbf{x} = (x_1, x_2, \ldots, x_n)^T$ is the column vector of the initial variables, and $\mathbf{A}$ is an $m \times n$ matrix containing the constraints coefficients.

Theorems 4.1 and 4.2, respectively, guarantee that the optimal solution of a LPP, if it exists, is attained at a basic feasible solution (an extreme point). The SM generates a sequence of basic feasible solutions that sequentially improve the objective function value. The SM is a procedure to organize this process. The simplex method operates in two stages:

1. An initialization stage, in which
   (a) The initial set of equality constraints is transformed into another equivalent set of equality constraints, associated with a basic solution.
   (b) The values of the basic variables are transformed into nonnegative values (a basic feasible solution is obtained). This will be called the *regulating stage*.
2. A standard iteration stage, in which the coefficients of the cost function are transformed into nonnegative values and the value of the cost function is iteratively improved until the optimum solution is obtained, no feasible solution is detected, or unbounded solution is found to exist. In this

iterative process different basic feasible solutions are obtained. To this end the elemental pivoting operation is used.

6.2.3 Initialization Stage

One essential peculiarity of the SM consists of incorporating a new variable Z, equal to the objective function of the problem, and the associated constraint

$$Z = c_1 x_1 + \cdots + c_n x_n \tag{6.6}$$

Using standard simplex notation, the constraint equations are written as

$$\begin{pmatrix} \mathbf{B} & \mathbf{N} \end{pmatrix} \begin{pmatrix} \mathbf{x_B} \\ - \\ \mathbf{x_N} \end{pmatrix} = \mathbf{b} \tag{6.7}$$

and the objective function as

$$Z = \begin{pmatrix} \mathbf{c}_\mathbf{B}^T & \mathbf{c}_\mathbf{N}^T \end{pmatrix} \begin{pmatrix} \mathbf{x_B} \\ - \\ \mathbf{x_N} \end{pmatrix} \tag{6.8}$$

or

$$Z = \begin{pmatrix} 0 & \mathbf{c}_\mathbf{B}^T & \mathbf{c}_\mathbf{N}^T \end{pmatrix} \begin{pmatrix} 1 \\ - \\ \mathbf{x_B} \\ - \\ \mathbf{x_N} \end{pmatrix} \tag{6.9}$$

where $(\mathbf{B} \quad \mathbf{N})$ is a partition of matrix $\mathbf{A}$ and $\mathbf{x_B}$ and $\mathbf{x_N}$ define another partition of $\mathbf{x}$, in the basic and nonbasic variables, respectively

Using (6.7), one gets

$$\mathbf{B}\mathbf{x_B} + \mathbf{N}\mathbf{x_N} = \mathbf{b} \tag{6.10}$$

and

$$\mathbf{x_B} = \mathbf{B}^{-1}\mathbf{b} - \mathbf{B}^{-1}\mathbf{N}\mathbf{x_N} = \mathbf{v} + \mathbf{U}\mathbf{x_N} \tag{6.11}$$

where

$$\mathbf{v} = \mathbf{B}^{-1}\mathbf{b} \tag{6.12}$$

and

$$\mathbf{U} = -\mathbf{B}^{-1}\mathbf{N} \tag{6.13}$$

On the other hand, from (6.8) and (6.11), one gets

$$Z = \mathbf{c}_\mathbf{B}^T(\mathbf{v} + \mathbf{U}\mathbf{x_N}) + \mathbf{c}_\mathbf{N}^T\mathbf{x_N} \tag{6.14}$$

and

$$Z = \mathbf{c}_\mathbf{B}^T\mathbf{v} + (\mathbf{c}_\mathbf{B}^T\mathbf{U} + \mathbf{c}_\mathbf{N}^T)\mathbf{x_N} \tag{6.15}$$

and

$$Z = u_0 + \mathbf{w}\mathbf{x_N} \tag{6.16}$$

where

$$u_0 = \mathbf{c}_\mathbf{B}^T\mathbf{v} \tag{6.17}$$

and

$$\mathbf{w} = \mathbf{c}_\mathbf{B}^T\mathbf{U} + \mathbf{c}_\mathbf{N}^T \tag{6.18}$$

Equations (6.11) and (6.16) together are

$$\begin{cases} Z = u_0 + \mathbf{w}\mathbf{x_N} \\ \mathbf{x_B} = \mathbf{v} + \mathbf{U}\mathbf{x_N} \end{cases} \tag{6.19}$$

and finally

$$\begin{pmatrix} Z \\ \mathbf{x_B} \end{pmatrix} = \begin{pmatrix} u_0 & \mathbf{w} \\ \mathbf{v} & \mathbf{U} \end{pmatrix} \begin{pmatrix} 1 \\ \mathbf{x_N} \end{pmatrix} \tag{6.20}$$

The SM starts with the set of constraints (6.4) and (6.6) written as

$$\begin{pmatrix} Z \\ -- \\ \mathbf{x}_B \end{pmatrix} = \begin{pmatrix} u_0^{(0)} & | & \mathbf{w}^{(0)T} \\ - & + & - \\ \mathbf{v}^{(0)} & | & \mathbf{U}^{(0)} \end{pmatrix} \begin{pmatrix} 1 \\ -- \\ \mathbf{x}_N \end{pmatrix} = \mathbf{Z}^{(0)} \begin{pmatrix} 1 \\ -- \\ \mathbf{x}_N \end{pmatrix} \tag{6.21}$$

where $\mathbf{x}_B \cup \mathbf{x}_N$ is a partition of the set of variables $\{x_1, x_2, \dots, x_n\}$, the matrices $\mathbf{v}^{(0)}$ and $\mathbf{U}^{(0)}$ are obtained by solving (6.4) in $\mathbf{x}_B$, as indicated in (4.17), and replacing it in (6.3):

$$u^{(0)} = \mathbf{c}_B^T\mathbf{v}^{(0)}, \quad \mathbf{w}^{(0)} = \mathbf{c}_B^T\mathbf{U}^{(0)} + \mathbf{c}_N^T \tag{6.22}$$

where $\mathbf{c}_B^T$ and $\mathbf{c}_N^T$ are the cost coefficients associated with $\mathbf{x}_B$ and $\mathbf{x}_N$, respectively.

Then, the SM replaces (6.21), in each iteration, by a new equivalent set of constraints with the same structure

$$\begin{pmatrix} Z^{(t)} \\ -- \\ \mathbf{x}_B^{(t)} \end{pmatrix} = \begin{pmatrix} u_0^{(t)} & | & \mathbf{w}^{(t)T} \\ - & + & - \\ \mathbf{v}^{(t)} & | & \mathbf{U}^{(t)} \end{pmatrix} \begin{pmatrix} 1 \\ -- \\ \mathbf{x}_N^{(t)} \end{pmatrix} = \mathbf{Z}^{(t)} \begin{pmatrix} 1 \\ -- \\ \mathbf{x}_N^{(t)} \end{pmatrix} \tag{6.23}$$

where t refers to the iteration number and $t = 0$ is used to refer to the initial iteration.

The most important transformation used in the SM is the elemental pivoting operation, which is described next.

6.2.4 Elemental Pivoting Operation

The SM reaches the solution of the LPP by means of elemental transformations, which act initially on $\mathbf{Z}^{(0)}$, to obtain $\mathbf{Z}^{(1)}$ and sequentially $\mathbf{Z}^{(2)}, \dots, \mathbf{Z}^{(t)}$, until the solution is obtained or no solution can be found.

The idea consists of replacing a basic variable by a nonbasic variable, and the theoretical basis of this transformation is given in the following theorem.

Theorem 6.1 (Invariance property of the pivoting transformation). *Let $z_{\alpha\beta}^{(t)}$ be a nonnull element of $\mathbf{Z}^{(t)}$, which is called a pivot. If we transform the system of constraints $\mathbf{y}_B^{(t)} = \mathbf{Z}^{(t)}\mathbf{y}_N^{(t)}$ into the system of constraints $\mathbf{y}_B^{(t+1)} = \mathbf{Z}^{(t+1)}\mathbf{y}_N^{(t+1)}$, by the following transformation*

$$z_{ij}^{(t+1)} = \left\{ \begin{array}{ll} \dfrac{z_{i\beta}^{(t)}}{z_{\alpha\beta}^{(t)}}, & \text{if} \quad i \neq \alpha, \ \ j = \beta \\ z_{ij}^{(t)} - \dfrac{z_{\alpha j}^{(t)}}{z_{\alpha\beta}^{(t)}} z_{i\beta}^{(t)} & \text{if} \quad i \neq \alpha, \ \ j \neq \beta \\ \dfrac{1}{z_{\alpha\beta}^{(t)}} & \text{if} \quad i = \alpha, \ \ j = \beta \\ -\dfrac{z_{\alpha j}^{(t)}}{z_{\alpha\beta}^{(t)}} & \text{if} \quad i = \alpha, \ \ j \neq \beta \end{array} \right\} \tag{6.24}$$

then both systems of constraints are equivalent, that is, they have the same solutions (same feasible set). ■

Proof. Consider the set $\mathbf{y}_B^{(t)} = \mathbf{Z}^{(t)}\mathbf{y}_N^{(t)}$. If we obtain y_β from the equation associated with the row α of $\mathbf{Z}^{(t)}$

$$y_\alpha = \mathbf{z}_\alpha^{(t)} \ \mathbf{y}_N^{(t)} = \sum_j z_{\alpha j}^{(t)} y_{N j}$$

then we find

$$y_\beta = \sum_{j \neq \beta} \left(\frac{-z_{\alpha j}^{(t)}}{z_{\alpha\beta}^{(t)}} y_{N j} \right) + \frac{1}{z_{\alpha\beta}^{(t)}} y_\alpha = \mathbf{z}_\beta^{(t+1)} \ \mathbf{y}_N^{(t+1)} \tag{6.25}$$

and, replacing y_β in the remaining constraints, we get

$$y_i = \sum_{j \neq \beta} \left(z_{ij}^{(t)} - \frac{z_{\alpha j}^{(t)}}{z_{\alpha\beta}^{(t)}} z_{i\beta}^{(t)} \right) y_{N j} + \frac{z_{i\beta}^{(t)}}{z_{\alpha\beta}^{(t)}} y_\alpha = \sum_{j \neq \beta} z_{ij}^{(t+1)} \ y_{N j} + z_{i\alpha}^{(t+1)} \ y_\alpha; \ \ i \neq \beta. \tag{6.26}$$

Thus, from (6.25) and (6.26) we get

$$\mathbf{y} = \mathbf{Z}^{(t+1)} \ \mathbf{y}_N^{(t+1)}$$

The transformation (6.24), that is, (6.25) and (6.26), can be written in matrix form as

$$\mathbf{Z}^{(t+1)} = \mathbf{Z}^{(t)}\mathbf{T}^{(t)} + \mathbf{S}^{(t)}$$

Table 6.1: Initial and transformed matrices

Initial matrix					
	x_{N1}	$\cdots$	x_β	$\cdots$	x_{Nm-n}
x_{B1}	$z_{11}^{(t)}$	$\cdots$	$z_{1\beta}^{(t)}$	$\cdots$	$z_{1m-n}^{(t)}$
$\vdots$	$\vdots$	$\cdots$	$\vdots$	$\cdots$	$\vdots$
x_α	$z_{\alpha 1}^{(t)}$	$\cdots$	$z_{\alpha\beta}^{(t)}$	$\cdots$	$z_{\alpha m-n}^{(t)}$
$\vdots$	$\vdots$	$\cdots$	$\vdots$	$\cdots$	$\vdots$
x_{Bn}	$z_{n1}^{(t)}$	$\cdots$	$z_{n\beta}^{(t)}$	$\cdots$	$z_{nm-n}^{(t)}$
Transformed matrix					
	x_{N1}	$\cdots$	x_α	$\cdots$	x_{Nm-n}
x_{B1}	$z_{11}^{(t)} - \frac{z_{\alpha 1}^{(t)}}{z_{\alpha\beta}^{(t)}} z_{1\beta}^{(t)}$	$\cdots$	$\frac{z_{1\beta}^{(t)}}{z_{\alpha\beta}^{(t)}}$	$\cdots$	$z_{1m-n}^{(t)} - \frac{z_{\alpha m-n}^{(t)}}{z_{\alpha\beta}^{(t)}} z_{1\beta}^{(t)}$
$\vdots$	$\vdots$	$\cdots$	$\vdots$	$\cdots$	$\vdots$
x_β	$-\frac{z_{\alpha 1}^{(t)}}{z_{\alpha\beta}^{(t)}}$	$\cdots$	$\frac{1}{z_{\alpha\beta}^{(t)}}$	$\cdots$	$-\frac{z_{\alpha m-n}^{(t)}}{z_{\alpha\beta}^{(t)}}$
$\vdots$	$\vdots$	$\cdots$	$\vdots$	$\cdots$	$\vdots$
x_{Bn}	$z_{n1}^{(t)} - \frac{z_{\alpha 1}^{(t)}}{z_{\alpha\beta}^{(t)}} z_{n\beta}^{(t)}$	$\cdots$	$\frac{z_{n\beta}^{(t)}}{z_{\alpha\beta}^{(t)}}$	$\cdots$	$z_{nm-n}^{(t)} - \frac{z_{\alpha m-n}^{(t)}}{z_{\alpha\beta}^{(t)}} z_{n\beta}^{(t)}$

where $\mathbf{T}^{(t)}$ is an identity matrix with its β row replaced by

$$\frac{1}{z_{\alpha\beta}} \left(-z_{\alpha 1} \quad \cdots \quad -z_{\alpha\beta-1} \quad 1 \quad -z_{\alpha\beta+1} \quad \cdots \quad -z_{\alpha n} \right) \tag{6.27}$$

and $\mathbf{S}^{(t)}$, a null matrix with its α row replaced by

$$\frac{1}{z_{\alpha\beta}} \left(-z_{\alpha 1} \quad \cdots \quad -z_{\alpha\beta-1} \quad 1 - z_{\alpha\beta} \quad -z_{\alpha\beta+1} \quad \cdots \quad -z_{\alpha n} \right) \tag{6.28}$$

Since all these transformations are linear and have associated matrices with nonnull determinant, they lead to an equivalent system of equations.

Table 6.1 shows the initial and the transformed matrices: the initial and transformed set of constraints.

Note that, after this transformation, the roles of variables x_α and x_β have been exchanged, specifically, x_α changes from basic to nonbasic, and x_β, from nonbasic to basic.

■

6.2.5 Identifying an Optimal Solution

When the LPP is written in the form (6.23) we can easily know when we have reached a solution. The following theorem answers this question.

Theorem 6.2 (When a solution has been reached). *If*

$$\mathbf{w}^{(t)} \geq \mathbf{0}, \quad \mathbf{v}^{(t)} \geq \mathbf{0} \tag{6.29}$$

then, an optimal solution of the LPP is

$$Z^{(t)} = u_0^{(t)}; \ \mathbf{x}_B^* = \mathbf{v}^{(t)}, \quad \mathbf{x}_N^{(t)} = \mathbf{0} \tag{6.30}$$

where the asterisk refers to the optimal solution. ∎

Proof. From (6.23) we have

$$Z = u_0^{(t)} + \mathbf{w}^{(t)^T} \mathbf{x}_N^{(t)} \tag{6.31}$$

Then, since all elements in $\mathbf{w}^{(t)}$ are nonnegative, the minimum of the sum on the right-hand side of (6.31) is $u_0^{(t)}$ and is attained at $\mathbf{x}_N^{(t)} = \mathbf{0}, \mathbf{x}_B^{(t)} = \mathbf{v}^{(t)}$. Since, in addition, $\mathbf{v}^{(t)} \geq \mathbf{0}$, $\{\mathbf{x}_N^{(t)} = \mathbf{0}; \mathbf{x}_B^{(t)} = \mathbf{v}^{(t)}\}$ is the optimum feasible solution. ∎

This implies that the solution appears in the first column of the matrix of the constraints coefficients, and the maximum value of the objective function is $u_0^{(t)}$. This is the result we have illustrated in the motivating example.

Conditions (6.29) justify the following two stages:

1. **Regulating iteration.** We use a regulating iteration to get $\mathbf{v}^{(t)} \geq \mathbf{0}$; that is, we look for nonnegative values of the basic variables.

2. **Standard iterations.** We perform elemental transformations such that

   (a) The condition
   $$\mathbf{v}^{(t)} \geq \mathbf{0} \tag{6.32}$$
   is satisfied, and

   (b) we improve in satisfying the condition
   $$\mathbf{w}^{(t)} \geq \mathbf{0} \tag{6.33}$$
   which implies a bounded solution, or we detect unboundedness or no solution.

This transformation consists of changing the partition of $\mathbf{x}$ in basic and nonbasic variables in such a way that one basic variable and one nonbasic variable are exchanged.

6.2.6 Regulating Iteration

In the regulating iteration the initial system of constraints is transformed into another one that satisfies

$$\mathbf{v}^{(t)} \geq \mathbf{0}$$

that is, we obtain a basic feasible solution.

There are two possible situations:

1. There exists a nonbasic variable x_β, such that one of the following conditions is satisfied
$$\left\{\begin{array}{lll} u_{i\beta}^{(0)} > 0 & \text{if} & v_i^{(0)} < 0 \\ u_{i\beta}^{(0)} \geq 0 & \text{if} & v_i^{(0)} \geq 0 \end{array}\right. \quad \forall i \tag{6.34}$$

2. There is no such constraint. In this case we introduce an additional variable $x_{n+1} \geq 0$, add the term Mx_{n+1} to the objective function, where M is a very large positive constant, and modify all the constraints adding the term x_{n+1} to all basic variables. This implies adding one more element, M, to matrix $\mathbf{w}$, and adding a column of ones to matrix $\mathbf{U}^{(0)}$, which satisfies (6.34), and choose $x_\beta = x_{n+1}$.

In both cases, we satisfy (6.34) after using the pivoting transformation (6.24) with the pivot $u_{\alpha\beta}^{(0)}$ given by

$$\frac{v_\alpha^{(0)}}{u_{\alpha\beta}^{(0)}} = \min_{v_i^{(0)}<0} \frac{v_i^{(0)}}{u_{i\beta}^{(0)}} \tag{6.35}$$

Remark 6.1 *In the situation 2 above, and with the convenient selection of M, if the LPP has feasible solution, then $x_{n+1} = 0$.* ∎

6.2.7 Detecting Unboundedness

In this subsection we explain how the unboundedness conditions can be detected.

Consider the following LPP. Minimize

$$Z = -x_2$$

subject to

$$\begin{array}{lcrrr} x_3 & = & & x_1 & \\ x_4 & = & & & x_2 \\ x_5 & = & -1 & +x_1 & +x_2 \end{array}$$

Note that this problem is unbounded because we can make the value of x_2 as large as desired, and null the remaining nonbasic variables, to get nonnegative values for the basic variables and an unbounded value of the objective function.

This occurs when there exists a column β with all nonnegative values and $w_\beta < 0$:

$$w_\beta < 0 \text{ and } u_{i\beta} \geq 0\ \forall i$$

In the general case and at iteration t we get a value of the objective function

$$Z = u_0^{(t)} + \mathbf{w}^{(t)^T} \mathbf{x}_N^{(t)} \tag{6.36}$$

that is as small as possible, but keeping the solution in the feasible region, because

$$\mathbf{x}_B^{(t)} = \mathbf{v}^{(t)} + \mathbf{U}^{(t)} \mathbf{x}_N^{(t)} \geq \mathbf{0}$$

6.2.8 Detecting Infeasibility

In this subsection we explain how the infeasibility condition can be detected.

This occurs when the solution condition holds with $x_{n+1} > 0$ (see Remark 6.1).

6.2.9 Standard Iterations Stage

With the standard iterations the cost function is transformed into a function with all nonnegative coefficients; that is, we transform $\mathbf{w}^{(0)}$ into $\mathbf{w}^{(1)} \geq \mathbf{0}$.

In this stage, we use transformation (6.24) to get $\mathbf{Z}^{(2)}, \ldots, \mathbf{Z}^{(t)}$, such that

1. An equivalent system of constraints is obtained:

$$\mathbf{x}_B^{(0)} = \mathbf{v}^{(0)} + \mathbf{U}^{(0)}\, \mathbf{x}_N^{(0)} \Leftrightarrow \mathbf{x}_B^{(t)} = \mathbf{v}^{(t)} + \mathbf{U}^{(t)}\, \mathbf{x}_N, \forall t \geq 1 \tag{6.37}$$

2. The condition

$$\mathbf{v}^{(t)} \geq \mathbf{0} \tag{6.38}$$

still holds.

3. The value of the objective function does not increase:

$$z^{(2)} \geq \cdots \geq z^{(t)} \tag{6.39}$$

4. The pivot elements are selected to maximize the decrease of $Z^{(t)} = z^{(t)}$.

Thus, in each iteration the LPP is equivalent to minimization of Z subject to

$$\mathbf{x}_B = \mathbf{v}^{(t)} + \mathbf{U}^{(t)}\, \mathbf{x}_N;\; \mathbf{x}_N \geq \mathbf{0}, \mathbf{x}_B \geq \mathbf{0} \tag{6.40}$$

In the standard iteration stage, the SM works in the following way:

1. Each iteration has an associated feasible point that is obtained by giving zero values to the set of nonbasic variables, $\mathbf{x}_N$. Thus, since $\mathbf{x}_N^{(t)} = \mathbf{0}$, $u_0^{(t)}$ always contains the value of the objective function, and $\mathbf{x}_B$, the values of the basic variables.

2. The essence of the SM consists of adequately changing the sets of basic and nonbasic variables in the successive steps of the method. In each iteration, a basic variable, x_α, and a nonbasic variable, x_β, interchange their roles. This exchange is done in such a way that the new set of constraints is equivalent to the old set, in the sense of having the same solutions. Thus we need to

   (a) Give a rule to choose the variable x_α entering the set $\mathbf{x}_N$ and leaving the set $\mathbf{x}_B$.

   (b) Give a rule to choose the variable x_β leaving the set $\mathbf{x}_N$ and entering the set $\mathbf{x}_B$.

3. The selection of the entering and leaving variables is done by decreasing the value of the objective function while trying to obtain a feasible point.

Once the entering and leaving variables have been selected, we transform the system of constraints to an equivalent system, using the pivoting transformation.

Selecting the Variable x_β Entering $\mathbf{x}_B$

According to the previous results, if the solution has not been attained when we are in state t, and the infeasibility and unboundedness conditions have not appeared yet, a certain i and j must exist such that $w_j^{(t)} < 0$ and $u_{ij}^{(t)} < 0$.

The variable x_β leaving $\mathbf{x}_N$ is the one satisfying

$$w_\beta^{(t)} = \min_{w_j^{(t)}<0} w_j^{(t)} \tag{6.41}$$

Note that, according to the pivoting transformation, the new value of $v_\alpha^{(t)}$ is

$$v_\alpha^{(t+1)} = -\frac{v_\alpha^{(t)}}{u_{\alpha\beta}} \geq 0 \tag{6.42}$$

In this way, according to the previous assumptions, we get

$$z^{(0)} \geq \cdots \geq z^{(t)} \Rightarrow f(\mathbf{x}^{(0)}) \geq \cdots \geq f(\mathbf{x}^{(t)})$$

a relation that, interpreted in a geometric sense, gives a sequence of points $\mathbf{x}^{(0)}, \ldots, \mathbf{x}^{(t)}$, which approach the hyperplanes $f(\mathbf{x}) = f(\mathbf{x}^{(t)})$ to the optimal point $\mathbf{x}^* \in X$.

Selecting the Variable x_α Leaving $\mathbf{x}_B$

Given the entering variable x_β, the pivot $u_{\alpha\beta}^{(t)}$ of the transformation (6.24) is selected for $\mathbf{v}^{(t)}$ to satisfy (6.38). Thus, the pivot row α and the pivot value $u_{\alpha\beta}^{(t)}$ are obtained by means of

$$\Lambda_\alpha^{(t)} = \frac{v_\alpha^{(t)}}{u_{\alpha\beta}^{(t)}} = \max_{u_{i\beta}^{(t)}<0} \frac{v_i^{(t)}}{u_{i\beta}^{(t)}} \leq 0 \tag{6.43}$$

where $\Lambda_\alpha^{(t)}$ is the *function-pivot ratio*. This guarantees that $\mathbf{v}^{t+1} \geq 0$. Thus the variable associated with row α leaves the set $\mathbf{x}_B$.

The following is a justification of (6.43). According to the pivoting transformation (6.24), the new values $v_i^{(t+1)}$ must be guaranteed to be nonnegative for all i, that is, we must have

$$
\begin{aligned}
v_\alpha^{(t+1)} &= -\frac{v_\alpha^{(t)}}{u_{\alpha\beta}^{(t)}} \geq 0 && (6.44)\\
v_i^{(t+1)} &= v_i^{(t)} - \Lambda_\alpha^{(t)} u_{i\beta}^{(t)} \geq 0; \;\; \forall i \neq \alpha && (6.45)
\end{aligned}
$$

Condition (6.44) holds because of (6.43). To check that (6.45) also holds, we have two possibilities:

1. If $u_{i\beta}^{(t)} > 0$, then (6.45) holds because $v_i^{(t)}$, $-\Lambda_\alpha^{(t)}$ and $u_{i\beta}^{(t)}$, on its right hand side, are non negative.

2. If $u_{i\beta}^{(t)} < 0$, because of the definition of max, we have

$$
\Lambda_\alpha^{(t)} \geq \frac{v_i^{(t)}}{u_{i\beta}^{(t)}} \Leftrightarrow u_{i\beta}^{(t)} \Lambda_\alpha^{(t)} \leq v_i^{(t)} \tag{6.46}
$$

   which implies (6.45).

6.2.10 The Revised Simplex Algorithm

All the preceding discussion allows us to build our particular version of the revised simplex algorithm, which works with the $\mathbf{U}^{(t)}$ matrix, until we obtain the optimal solution or detecting infeasibility or unboundedness.

Algorithm 6.1 (The simplex algorithm)

- **Input.** The starting tableau, $\{\mathbf{U}^{(0)}, \mathbf{v}^{(0)}, \mathbf{w}^{(0)}, u^{(0)}\}$ of the LPP, including (in an embedded fashion) the objective function, and the problem constraints.

- **Output.** The solution of the minimization problem or a message indicating infeasibility or unbounded solution.

Regulating Iteration

Step 1 (Initialization). If $\{\mathbf{v}^{(0)} \geq \mathbf{0}\}$, go to step 4. Otherwise, transform $\mathbf{v}^{(0)}$ to nonnegative coefficients. To this end, check if there exists a column β in $\mathbf{U}^{(0)}$ satisfying one of the two conditions

$$
\left\{
\begin{array}{lll}
u_{i\beta}^{(0)} > 0 & \text{if} & v_i^{(0)} < 0 \\
u_{i\beta}^{(0)} \geq 0 & \text{if} & v_i^{(0)} \geq 0
\end{array}
\right. \quad \forall i \tag{6.47}
$$

If the answer is positive, continue with step 2. Otherwise, introduce the artificial variable $x_{n+1} \geq 0$, modify the cost function adding the term Mx_{n+1}, where M is a large positive constant, add the term x_{n+1} to all constraints (basic variables), and choose $x_\beta = x_{n+1}$.

Step 2 (Finding the pivot). Find the pivot row α using

$$\frac{v_\alpha^{(0)}}{u_{\alpha\beta}^{(0)}} = \min_{v_i^{(0)}<0} \frac{v_i^{(0)}}{u_{i\beta}^{(0)}} \tag{6.48}$$

and go to step 3.

Standard Iterations

Step 3 (Pivoting). Perform the pivoting transformation

$$z_{ij}^{(t+1)} = \left\{ \begin{array}{ll} \dfrac{z_{i\beta}^{(t)}}{z_{\alpha\beta}^{(t)}}, & \text{if} \quad i \neq \alpha, \quad j = \beta \\ z_{ij}^{(t)} - \dfrac{z_{\alpha j}^{(t)}}{z_{\alpha\beta}^{(t)}} z_{i\beta}^{(t)} & \text{if} \quad i \neq \alpha, \quad j \neq \beta \\ \dfrac{1}{z_{\alpha\beta}^{(t)}} & \text{if} \quad i = \alpha, \quad j = \beta \\ -\dfrac{z_{\alpha j}^{(t)}}{z_{\alpha\beta}^{(t)}} & \text{if} \quad i = \alpha, \quad j \neq \beta \end{array} \right\} \tag{6.49}$$

Step 4 (Selecting the entering variable x_β). If the solution condition

$$\mathbf{w}^{(t)} \geq \mathbf{0}, \quad \mathbf{v}^{(t)} \geq \mathbf{0} \tag{6.50}$$

holds with the unbounded condition $x_{n+1} > 0$, print the infeasible solution message; if the solution condition holds with $x_{n+1} = 0$, print the solution

$$\mathbf{x}_B^* = \mathbf{v}^{(t)}; \ \mathbf{x}_N^* = \mathbf{0} \tag{6.51}$$

and quit. Otherwise, the variable x_β entering $\mathbf{x}_B$ is selected using

$$w_\beta^{(t)} = \min_{w_j^{(t)}<0} w_j^{(t)} \tag{6.52}$$

Step 5 (Selecting the leaving variable). If all $u_{i\beta} \geq 0$, the problem has an unbounded solution. Otherwise, the variable x_α leaving $\mathbf{x}_B$ is selected using

$$\frac{v_\alpha^{(t)}}{u_{\alpha\beta}^{(t)}} = \max_{u_{i\beta}^{(t)}<0} \frac{v_i^{(t)}}{u_{i\beta}^{(t)}} \tag{6.53}$$

Next, return to step 3. ∎

Remark 6.2 *If the set of constraints is given as inequalities, the initial tableau can be written directly from the optimization function and the inequalities matrix coefficients, as illustrated in the following example.* ■

6.2.11 Some Illustrative Examples

In this section we illustrate the SM by several examples.

Example 6.1 (First example). Consider again the following LPP. Minimize

$$Z = -3x_1 - x_2 \tag{6.54}$$

subject to

$$\begin{array}{rrcr} -3x_1 & +2x_2 & \leq & 2 \\ x_1 & +x_2 & \leq & 6 \\ x_1 & & \leq & 3 \\ 2x_1 & -x_2 & \leq & 4 \\ & -x_2 & \leq & 0 \\ -x_1 & -x_2 & \leq & -1 \end{array} \tag{6.55}$$

with

$$x_1, x_2 \geq 0. \tag{6.56}$$

First we need to transform this into a set of equalities using six slackness variables and get

$$\begin{array}{rrrrrrrrcr} -3x_1 & +2x_2 & +x_3 & & & & & & = & 2 \\ x_1 & +x_2 & & +x_4 & & & & & = & 6 \\ x_1 & & & & +x_5 & & & & = & 3 \\ 2x_1 & -x_2 & & & & +x_6 & & & = & 4 \\ & -x_2 & & & & & +x_7 & & = & 0 \\ -x_1 & -x_2 & & & & & & +x_8 & = & -1 \end{array} \tag{6.57}$$

with

$$x_1, x_2, x_3, x_4, x_5, x_6, x_7, x_8 \geq 0 \tag{6.58}$$

Constraints (6.57) can be written as

$$\begin{array}{rcrrr} x_3 & = & 2 & +3x_1 & -2x_2 \\ x_4 & = & 6 & -x_1 & -x_2 \\ x_5 & = & 3 & -x_1 & \\ x_6 & = & 4 & -2x_1 & +x_2 \\ x_7 & = & & & +x_2 \\ x_8 & = & -1 & +x_1 & +x_2 \end{array} \tag{6.59}$$

Expressions (6.54) and (6.59) lead to the initial $\mathbf{Z}^{(0)}$ in Table 6.2, where the pivot elements are boldfaced.

Table 6.2: Initial tableau with matrices $u^{(0)}, \mathbf{w}^{(0)}, \mathbf{v}^{(0)}$, and $\mathbf{U}^{(0)}$ and transformed tableau after adding the artificial variable x_9 (Example 6.1)

$\mathbf{Z}^{(0)}$				$\mathbf{Z}^{(1)}$				
	1	x_1	x_2		1	x_1	x_2	x_9
Z	0	–3	–1	Z	0	–3	–1	20
x_3	2	3	–2	x_3	2	3	–2	1
x_4	6	–1	–1	x_4	6	–1	–1	1
x_5	3	–1	0	x_5	3	–1	0	1
x_6	4	–2	1	x_6	4	–2	1	1
x_7	0	0	1	x_7	0	0	1	1
x_8	–1	1	1	x_8	–1	1	1	**1**

Using the SM algorithm the following results are obtained:

Step 1. Table 6.2 shows the initial tableau $\mathbf{Z}^{(0)}$ with matrices $\mathbf{u}^{(0)}, \mathbf{w}^{(0)}, \mathbf{v}^{(0)}$, and $\mathbf{U}^{(0)}$, where we can see that some of the elements in the first column are negative. Thus, we need to transform them to nonnegative values. Since no column β satisfies

$$\left\{ \begin{array}{lll} u_{i\beta}^{(0)} > 0 & \text{if} & v_i^{(0)} < 0 \\ u_{i\beta}^{(0)} \geq 0 & \text{if} & v_i^{(0)} \geq 0 \end{array} \right. \quad \forall i$$

we introduce the artificial variable $x_9 \geq 0$, modify the objective function adding the term $20x_9$, add the term x_9 to all constraints (basic variables), and make $x_\beta = x_9$, to get $\mathbf{Z}^{(1)}$.

Step 2. We find the pivot using

$$\frac{v_\alpha^{(0)}}{u_{\alpha\beta}^{(0)}} = \min_{v_i^{(0)}<0} \frac{v_i^{(0)}}{u_{i\beta}^{(0)}} = \min\{\frac{-1}{1}\} = -1 \tag{6.60}$$

which means $x_\alpha = x_8$.

Step 3. We perform the pivoting process and get $\mathbf{Z}^{(2)}$ in Table 6.3, where pivot elements are again boldfaced. Note that now all the $\mathbf{v}$ coefficients, in the first column, are nonnegative (they will remain nonnegative from now on).

Step 4. Since the solution condition is not satisfied, because $\mathbf{w} \geq \mathbf{0}$ does not hold, we proceed to find the entering variable using

$$w_\beta^{(t)} = \min_{w_j^{(t)}<0} w_j^{(t)} = \min\{-23, -21\} = -23 \tag{6.61}$$

which implies $x_\beta = x_1$.

Table 6.3: Tables associated with iterations 2 to 7 showing matrices $u^{(0)}, \mathbf{w}^{(0)}, \mathbf{v}^{(0)}$, and $\mathbf{U}^{(0)}$ (Example 6.1). Pivot elements are boldfaced.

$\mathbf{Z}^{(2)}$					$\mathbf{Z}^{(3)}$				
	1	x_1	x_2	x_8		1	x_7	x_2	x_8
Z	20	-23	-21	20	Z	-3	23	-21	-3
x_3	3	2	-3	1	x_3	5	-2	-3	3
x_4	7	-2	-2	1	x_4	5	2	-2	-1
x_5	4	-2	-1	1	x_5	2	2	-1	-1
x_6	5	-3	0	1	x_6	2	3	0	-2
x_7	1	**-1**	0	1	x_1	1	-1	0	1
x_9	1	-1	-1	1	x_9	0	1	**-1**	0
$\mathbf{Z}^{(4)}$					$\mathbf{Z}^{(5)}$				
	1	x_7	x_9	x_8		1	x_7	x_9	x_6
Z	-3	2	21	-3	Z	-6	-5/2	21	3/2
x_3	5	-5	3	3	x_3	8	-1/2	3	-3/2
x_4	5	0	2	-1	x_4	4	-3/2	2	1/2
x_5	2	1	1	-1	x_5	1	**-1/2**	1	1/2
x_6	2	3	0	**-2**	x_8	1	3/2	0	-1/2
x_1	1	-1	0	1	x_1	2	1/2	0	-1/2
x_2	0	1	-1	0	x_2	0	1	-1	0
$\mathbf{Z}^{(6)}$					$\mathbf{Z}^{(7)}$				
	1	x_5	x_9	x_6		1	x_5	x_9	x_4
Z	-11	5	16	-1	Z	-12	2	17	1
x_3	7	1	2	-2	x_3	5	-5	4	2
x_4	1	3	-1	**-1**	x_6	1	3	-1	-1
x_7	2	-2	2	1	x_7	3	1	1	-1
x_8	4	-3	3	1	x_8	5	0	2	-1
x_1	3	-1	1	0	x_1	3	-1	1	0
x_2	2	-2	1	1	x_2	3	1	0	-1

Step 5. We select the leaving variable using

$$\frac{v_\alpha^{(t)}}{u_{\alpha\beta}^{(t)}} = \max_{u_{i\beta}^{(t)}<0} \frac{v_i^{(t)}}{u_{i\beta}^{(t)}} = \max\left\{\frac{7}{-2}, \frac{4}{-2}, \frac{5}{-3}, \frac{1}{-1}, \frac{1}{-1}\right\} = -\frac{1}{1} \tag{6.62}$$

which leads to a tie between x_7 and x_9, which we resolve by arbitrarily selecting the first variable.

Step 3. We perform the pivoting process and get $\mathbf{Z}^{(3)}$ in Table 6.3.

Step 4. Since the solution condition is not satisfied, because $\mathbf{w} \geq \mathbf{0}$ does not hold, we proceed to find the entering variable using

$$w_\beta^{(t)} = \min_{w_j^{(t)}<0} w_j^{(t)} = \min\{-21, -3\} = -21 \tag{6.63}$$

which implies $x_\beta = x_2$.

Step 5. We select the leaving variable using

$$\frac{v_\alpha^{(t)}}{u_{\alpha\beta}^{(t)}} = \max_{u_{i\beta}^{(t)}<0} \frac{v_i^{(t)}}{u_{i\beta}^{(t)}} = \max\left\{\frac{5}{-3}, \frac{5}{-2}, \frac{2}{-1}, \frac{0}{-1}\right\} = 0 \tag{6.64}$$

which implies $x_\alpha = x_9$.

Step 3. We perform the pivoting process and get $\mathbf{Z}^{(4)}$ in Table 6.3.

Step 4. Since the solution condition is not satisfied, because $\mathbf{w} \geq \mathbf{0}$ does not hold, we proceed to find the entering variable using

$$w_\beta^{(t)} = \min_{w_j^{(t)}<0} w_j^{(t)} = \min\{-3\} = -3 \tag{6.65}$$

which implies $x_\beta = x_8$.

Step 5. We select the leaving variable using

$$\frac{v_\alpha^{(t)}}{u_{\alpha\beta}^{(t)}} = \max_{u_{i\beta}^{(t)}<0} \frac{v_i^{(t)}}{u_{i\beta}^{(t)}} = \max\left\{\frac{5}{-1}, \frac{2}{-1}, \frac{2}{-2}\right\} = -\frac{2}{2} \tag{6.66}$$

which implies $x_\alpha = x_6$.

Step 3. We perform the pivoting process and get $\mathbf{Z}^{(5)}$ in Table 6.3.

Step 4. Since the solution condition is not satisfied, because $\mathbf{w} \geq \mathbf{0}$ does not hold, we proceed to find the entering variable using

$$w_\beta^{(t)} = \min_{w_j^{(t)}<0} w_j^{(t)} = \min\left\{\frac{-5}{2}\right\} = -\frac{5}{2} \tag{6.67}$$

which implies $x_\beta = x_7$.

Step 5. We select the leaving variable using

$$\frac{v_\alpha^{(t)}}{u_{\alpha\beta}^{(t)}} = \max_{u_{i\beta}^{(t)}<0} \frac{v_i^{(t)}}{u_{i\beta}^{(t)}} = \max\left\{\frac{8}{-1/2}, \frac{4}{-3/2}, \frac{1}{-1/2}\right\} = -2 \qquad (6.68)$$

which implies $x_\alpha = x_5$.

Step 3. We perform the pivoting process and get $\mathbf{Z}^{(6)}$ in Table 6.3.

Step 4. Since the solution condition is not satisfied, because $\mathbf{w} \geq \mathbf{0}$ does not hold, we proceed to find the entering variable using

$$w_\beta^{(t)} = \min_{w_j^{(t)}<0} w_j^{(t)} = \min\{-1\} = -1 \qquad (6.69)$$

which implies $x_\beta = x_6$.

Step 5. We select the leaving variable using

$$\frac{v_\alpha^{(t)}}{u_{\alpha\beta}^{(t)}} = \max_{u_{i\beta}^{(t)}<0} \frac{v_i^{(t)}}{u_{i\beta}^{(t)}} = \max\left\{\frac{7}{-2}, \frac{1}{-1}\right\} = -1 \qquad (6.70)$$

which implies $x_\alpha = x_4$.

Step 3. We perform the pivoting process and get $\mathbf{Z}^{(7)}$ in Table 6.3.

Step 4. Since the solution condition is satisfied, because $\mathbf{w} \geq \mathbf{0}$, and $x_9 = 0$, we have found the following optimal solution to our problem:

$$\begin{array}{rcl} Z &=& -12; \\ (x_1, x_2, x_3, x_4, x_5, x_6, x_7, x_8, x_9) &=& (3, 3, 5, 0, 0, 1, 3, 5, 0) \end{array}$$

■

Example 6.2 (A more complex example). Consider the following LPP. Minimize

$$Z = -2x_1 - 4x_2 - 3x_3 + x_4 - 2x_5 - x_6 - 2x_7 - 2x_8 \qquad (6.71)$$

subject to

$$\begin{array}{rrrrrrrrcr} 5x_1 & +4x_2 & +3x_3 & +2x_4 & +x_5 & +2x_6 & +3x_7 & +4x_8 & \leq & 17 \\ 2x_1 & -x_2 & +4x_3 & -2x_4 & +3x_5 & +3x_6 & +x_7 & -x_8 & \leq & 11 \\ 3x_1 & +2x_2 & +3x_3 & -2x_4 & +x_5 & -x_6 & +2x_7 & -x_8 & \leq & 8 \\ x_1 & -x_2 & +2x_3 & +2x_4 & -x_5 & & +x_7 & +x_8 & \leq & 4 \\ 4x_1 & +3x_2 & +x_3 & -x_4 & -2x_5 & +x_6 & +4x_7 & +3x_8 & \leq & 14 \end{array} \qquad (6.72)$$

Table 6.4: Initial tableau with matrices $u^{(0)}, \mathbf{w}^{(0)}, \mathbf{v}^{(0)}$, and $\mathbf{U}^{(0)}$ (Example 6.2)

$\mathbf{Z}^{(0)}$									
	1	x_1	x_2	x_3	x_4	x_5	x_6	x_7	x_8
Z	0	–2	–4	–3	1	–2	–1	–2	–2
x_9	17	–5	–4	–3	–2	–1	–2	–3	–4
x_{10}	11	–2	1	–4	2	–3	–3	–1	1
x_{11}	8	–3	–2	–3	2	–1	1	–2	1
x_{12}	4	–1	1	–2	–2	1	0	–1	–1
x_{13}	14	–4	–3	–1	1	2	–1	–4	–3

with

$$x_1, \ldots, x_8 \geq 0 \tag{6.73}$$

Tables 6.4 and 6.5 (where pivot entries are again boldfaced), give the tableaux corresponding to all the iterations. We leave it to the reader to apply the SM algorithm and check that the final solution is as follows:

$$\begin{aligned} Z &= \frac{-772}{39} \\ (x_1, x_2, \ldots, x_{13}) &= \left(0, \frac{83}{39}, 0, 0, \frac{61}{13}, 0, 0, \frac{37}{39}, 0, 0, 0, \frac{385}{39}, \frac{184}{13}\right) \end{aligned}$$

■

Example 6.3 (Unbounded example). Consider again the following LPP Minimize

$$Z = -3x_1 - x_2 \tag{6.74}$$

subject to

$$\begin{array}{rrcr} -3x_1 & +2x_2 & \leq & 2 \\ & -x_2 & \leq & 0 \\ -x_1 & -x_2 & \leq & -1 \end{array} \tag{6.75}$$

with

$$x_1, x_2 \geq 0 \tag{6.76}$$

Step 1. Table 6.6 (where pivot elements are again boldfaced), shows the initial tableau $\mathbf{Z}^{(0)}$ with matrices $u^{(0)}, \mathbf{w}^{(0)}, \mathbf{v}^{(0)}$, and $\mathbf{U}^{(0)}$, where we can see that some of the elements in the first column are negative. Thus, we need to transform them to nonnegative values.

Since the column associated with $x_\beta = x_1$ satisfies

$$\left\{ \begin{array}{lll} u_{i\beta} > 0 & \text{if} & v_i < 0 \\ u_{i\beta} \geq 0 & \text{if} & v_i \geq 0 \end{array} \right. \quad \forall i \tag{6.77}$$

we proceed to step 2.

Table 6.5: Tables associated with iterations 1 to 4 showing matrices $u^{(0)}, \mathbf{w}^{(0)}, \mathbf{v}^{(0)}$, and $\mathbf{U}^{(0)}$ (Example 6.2)

$\mathbf{Z}^{(1)}$									
	1	x_1	x_2	x_3	x_4	x_5	x_6	x_7	x_8
Z	0	–2	–4	–3	1	–2	–1	–2	–2
x_9	17	–5	–4	–3	–2	–1	–2	–3	–4
x_{10}	11	–2	1	–4	2	–3	–3	–1	1
x_{11}	8	–3	**–2**	–3	2	–1	1	–2	1
x_{12}	4	–1	1	–2	–2	1	0	–1	–1
x_{13}	14	–4	–3	–1	1	2	–1	–4	–3
$\mathbf{Z}^{(2)}$									
	1	x_1	x_{11}	x_3	x_4	x_5	x_6	x_7	x_8
Z	–16	4	2	3	–3	0	–3	2	–4
x_9	1	1	2	3	–6	1	–4	1	**–6**
x_{10}	15	$\frac{-7}{2}$	$\frac{-1}{2}$	$\frac{-11}{2}$	3	$\frac{-7}{2}$	$\frac{-5}{2}$	-2	$\frac{3}{2}$
x_2	4	$\frac{-3}{2}$	$\frac{-1}{2}$	$\frac{-3}{2}$	1	$\frac{-1}{2}$	$\frac{1}{2}$	–1	$\frac{1}{2}$
x_{12}	8	$\frac{-5}{2}$	$\frac{-1}{2}$	$\frac{-7}{2}$	–1	$\frac{1}{2}$	$\frac{1}{2}$	–2	$\frac{-1}{2}$
x_{13}	2	$\frac{1}{2}$	$\frac{3}{2}$	$\frac{7}{2}$	–2	$\frac{7}{2}$	$\frac{-5}{2}$	–1	$\frac{-9}{2}$
$\mathbf{Z}^{(3)}$									
	1	x_1	x_{11}	x_3	x_4	x_5	x_6	x_7	x_9
Z	$\frac{-50}{3}$	$\frac{10}{3}$	$\frac{2}{3}$	1	1	$\frac{-2}{3}$	$\frac{-1}{3}$	$\frac{4}{3}$	$\frac{2}{3}$
x_8	$\frac{1}{6}$	$\frac{1}{6}$	$\frac{1}{3}$	$\frac{1}{2}$	–1	$\frac{1}{6}$	$\frac{-2}{3}$	$\frac{1}{6}$	$\frac{-1}{6}$
x_{10}	$\frac{61}{4}$	$\frac{-13}{4}$	0	$\frac{-19}{4}$	$\frac{3}{2}$	$\boldsymbol{\frac{-13}{4}}$	$\frac{-7}{2}$	$\frac{-7}{4}$	$\frac{-1}{4}$
x_2	$\frac{49}{12}$	$\frac{-17}{12}$	$\frac{-1}{3}$	$\frac{-5}{4}$	$\frac{1}{2}$	$\frac{-5}{12}$	$\frac{1}{6}$	$\frac{-11}{12}$	$\frac{-1}{12}$
x_{12}	$\frac{95}{12}$	$\frac{-31}{12}$	$\frac{-2}{3}$	$\frac{-15}{4}$	$\frac{-1}{2}$	$\frac{5}{12}$	$\frac{5}{6}$	$\frac{-25}{12}$	$\frac{1}{12}$
x_{13}	$\frac{5}{4}$	$\frac{-1}{4}$	0	$\frac{5}{4}$	$\frac{5}{2}$	$\frac{11}{4}$	$\frac{1}{2}$	$\frac{-7}{4}$	$\frac{3}{4}$
$\mathbf{Z}^{(4)}$									
	1	x_1	x_{11}	x_3	x_4	x_{10}	x_6	x_7	x_9
Z	$\frac{-772}{39}$	4	$\frac{2}{3}$	$\frac{77}{39}$	$\frac{9}{13}$	$\frac{8}{39}$	$\frac{5}{13}$	$\frac{22}{13}$	$\frac{28}{39}$
x_8	$\frac{37}{39}$	0	$\frac{1}{3}$	$\frac{10}{39}$	$\frac{-12}{13}$	$\frac{-2}{39}$	$\frac{-11}{13}$	$\frac{1}{13}$	$\frac{-7}{39}$
x_5	$\frac{61}{13}$	–1	0	$\frac{-19}{13}$	$\frac{6}{13}$	$\frac{-4}{13}$	$\frac{-14}{13}$	$\frac{-7}{13}$	$\frac{-1}{13}$
x_2	$\frac{83}{39}$	–1	$\frac{-1}{3}$	$\frac{-25}{39}$	$\frac{4}{13}$	$\frac{5}{39}$	$\frac{8}{13}$	$\frac{-9}{13}$	$\frac{-2}{39}$
x_{12}	$\frac{385}{39}$	–3	$\frac{-2}{3}$	$\frac{-170}{39}$	$\frac{-4}{13}$	$\frac{-5}{39}$	$\frac{5}{13}$	$\frac{-30}{13}$	$\frac{2}{39}$
x_{13}	$\frac{184}{13}$	–3	0	$\frac{-36}{13}$	$\frac{49}{13}$	$\frac{-11}{13}$	$\frac{-32}{13}$	$\frac{-42}{13}$	$\frac{7}{13}$

Table 6.6: Initial tableau with matrices $u^{(0)}, \mathbf{w}^{(0)}, \mathbf{v}^{(0)}$, and $\mathbf{U}^{(0)}$ and transformed tableau after pivoting (Example 6.3)

$\mathbf{Z}^{(0)}$				$\mathbf{Z}^{(1)}$			
	1	x_1	x_2		1	x_5	x_2
Z	0	–3	–1	Z	–3	–3	2
x_3	2	3	–2	x_3	5	3	–5
x_4	0	0	1	x_4	0	0	1
x_5	–1	**1**	1	x_1	1	1	–1

Step 2. We find the pivot using

$$\frac{v_\alpha^{(0)}}{u_{\alpha\beta}^{(0)}} = \min_{v_i^{(0)}<0} \frac{v_i^{(0)}}{u_{i\beta}^{(0)}} = \min\left\{\frac{-1}{1}\right\} = -1, \tag{6.78}$$

which means $x_\alpha = x_5$.

Step 3. We perform the pivoting process and get $\mathbf{Z}^{(1)}$ in Table 6.6. Note that now all the $\mathbf{v}$ coefficients, in the first column, are nonnegative.

Step 4. Since the solution condition is not satisfied, because $\mathbf{w} \geq \mathbf{0}$ does not hold, we proceed to find the leaving variable using

$$w_\beta^{(t)} = \min_{w_j^{(t)}<0} w_j^{(t)} = \min\{-3\} = -3 \tag{6.79}$$

which implies $x_\beta = x_1$.

Step 5. Since all u entries in the column associated with x_1 are nonnegative, the problem is unbounded. The descent direction is

$$d = (0, 0, 5, 0, 1)^T$$

■

Example 6.4 (Infeasible example). Consider the following LPP. Minimize

$$Z = -3x_1 - x_2 \tag{6.80}$$

subject to

$$\begin{array}{rrcr} -3x_1 & +2x_2 & \leq & 2 \\ x_1 & +x_2 & \leq & 6 \\ x_1 & & \leq & 3 \\ 2x_1 & -x_2 & \leq & 4 \\ & -x_2 & \leq & 0 \\ -x_1 & -x_2 & \leq & -1 \\ x_1 & +x_2 & \leq & 0 \end{array} \tag{6.81}$$

Table 6.7: Initial tableau with matrices $u^{(0)}, \mathbf{w}^{(0)}, \mathbf{v}^{(0)}$, and $\mathbf{U}^{(0)}$ and transformed tableau after adding the artificial variable x_{10} (Example 6.4)

$\mathbf{Z}^{(0)}$				$\mathbf{Z}^{(1)}$				
	1	x_1	x_2		1	x_1	x_2	x_{10}
Z	0	–3	–1	Z	0	–3	–1	20
x_3	2	3	–2	x_3	2	3	–2	1
x_4	6	–1	–1	x_4	6	–1	–1	1
x_5	3	–1	0	x_5	3	–1	0	1
x_6	4	–2	1	x_6	4	–2	1	1
x_7	0	0	1	x_7	0	0	1	1
x_8	–1	1	1	x_8	–1	1	1	**1**
x_9	0	–1	–1	x_9	0	–1	–1	1

with

$$x_1, x_2 \geq 0 \tag{6.82}$$

Using the SM algorithm the following results are obtained:

Step 1. Table 6.7 shows the initial tableau $\mathbf{Z}^{(0)}$ with matrices $u^{(0)}, \mathbf{w}^{(0)}, \mathbf{v}^{(0)}$, and $\mathbf{U}^{(0)}$, where we can see that some of the elements in the first column are negative. Thus, we need to transform them to nonnegative values.

Since no column β satisfies

$$\left\{ \begin{array}{lll} u_{i\beta}^{(0)} > 0 & \text{if} & v_i^{(0)} < 0 \\ u_{i\beta}^{(0)} \geq 0 & \text{if} & v_i^{(0)} \geq 0 \end{array} \right. \quad \forall i$$

we introduce the artificial variable $x_{10} > 0$, modify the objective function adding the term $20x_{10}$, add the term x_{10} to all constraints (basic variables), and make $x_\beta = x_{10}$.

Step 2. We find the pivot using

$$\frac{v_\alpha^{(0)}}{u_{\alpha\beta}^{(0)}} = \min_{v_i^{(0)}<0} \frac{v_i^{(0)}}{u_{i\beta}^{(0)}} = \min \left\{ \frac{-1}{1} \right\} = -1 \tag{6.83}$$

which means $x_\alpha = x_8$.

Step 3. We perform the pivoting process and get $\mathbf{Z}^{(2)}$ in Table 6.8 (where the pivot element is bolfaced). Note that now all the $\mathbf{v}$ coefficients, in the first column, are nonnegative.

Table 6.8: Tableaux associated with iterations 2 and 3 showing matrices $u^{(0)}, \mathbf{w}^{(0)}, \mathbf{v}^{(0)}$, and $\mathbf{U}^{(0)}$ (Example 6.4)

$\mathbf{Z}^{(2)}$					$\mathbf{Z}^{(3)}$				
	1	x_1	x_2	x_8		1	x_9	x_2	x_8
Z	20	–23	–21	20	Z	17/2	23/2	2	17/2
x_3	3	2	–3	1	x_3	4	–1	–5	2
x_4	7	–2	–2	1	x_4	6	1	0	0
x_5	4	–2	–1	1	x_5	3	1	1	0
x_6	5	–3	0	1	x_6	$\frac{7}{2}$	$\frac{3}{2}$	3	$\frac{-1}{2}$
x_7	1	–1	0	1	x_7	$\frac{1}{2}$	$\frac{1}{2}$	1	$\frac{1}{2}$
x_{10}	1	–1	–1	1	x_{10}	$\frac{1}{2}$	$\frac{1}{2}$	0	$\frac{1}{2}$
x_9	1	**–2**	–2	1	x_1	$\frac{1}{2}$	$\frac{-1}{2}$	–1	$\frac{1}{2}$

Step 4. Since the solution condition is not satisfied, because $\mathbf{w} \geq \mathbf{0}$ does not hold, we proceed to find the entering variable using

$$w_\beta^{(t)} = \min_{w_j^{(t)}<0} w_j^{(t)} = \min\{-23, -21\} = -23 \tag{6.84}$$

which implies $x_\beta = x_1$.

Step 5. We select the leaving variable using

$$\frac{v_\alpha^{(t)}}{u_{\alpha\beta}^{(t)}} = \max_{u_{i\beta}^{(t)}<0} \frac{v_i^{(t)}}{u_{i\beta}^{(t)}} = \max\left\{\frac{7}{-2}, \frac{4}{-2}, \frac{5}{-3}, \frac{1}{-1}, \frac{1}{-1}, \frac{1}{-2}\right\} = \frac{1}{-2} \tag{6.85}$$

which implies $x_\alpha = x_9$.

Step 3. We perform the pivoting process and get $\mathbf{Z}^{(3)}$ in Table 6.8.

Step 4. Since the solution condition is satisfied, but $x_{10} = \frac{1}{2} > 0$, the problem is infeasible.

■

6.3 The Exterior Point Method

It should be noted that the exterior point algorithm is equivalent to the dual simplex algorithm.

To understand how the exterior point method (EPM) works and its main differences from the previously described simplex method, we start this section

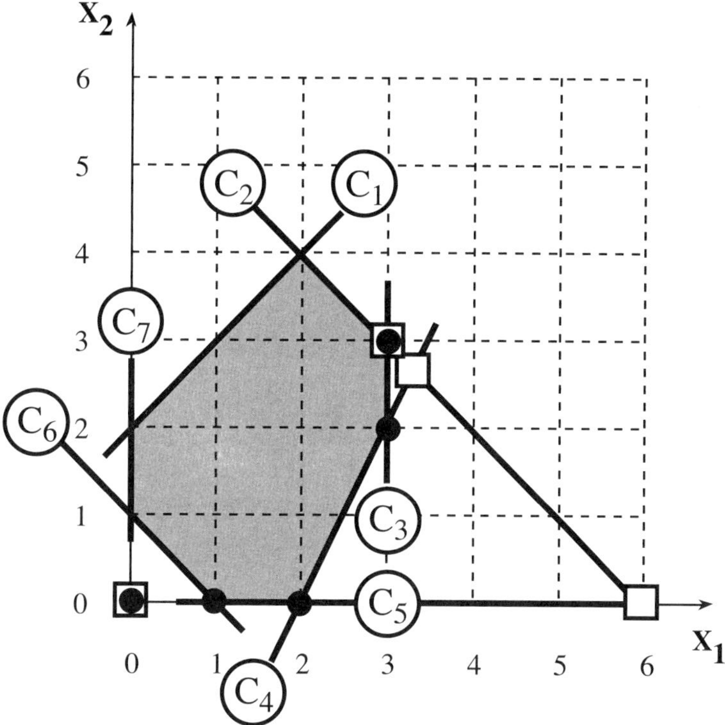

Figure 6.1: Illustration of the differences between the EPM (empty squares) and the SM (black bullets).

with a simple example. Consider the following LPP. Minimize

$$Z = -3x_1 - x_2 \tag{6.86}$$

subject to

$$\begin{array}{rrcr} -3x_1 & +2x_2 & \leq & 2 \\ x_1 & +x_2 & \leq & 6 \\ x_1 & & \leq & 3 \\ 2x_1 & -x_2 & \leq & 4 \\ & -x_2 & \leq & 0 \\ -x_1 & -x_2 & \leq & -1 \end{array} \tag{6.87}$$

with

$$x_1, x_2 \geq 0 \tag{6.88}$$

The set of points satisfying these constraints defines the feasible region; the remaining points form the infeasible region (see Figure 6.1).

In this case, the simplex method begins at the origin and moves, in the following iterations, to vertices $(1,0)^T, (2,0)^T$, and $(3,2)^T$ before reaching the optimum at vertex $(3,3)^T$. Observe that all points, but the initial (the origin), belong to the feasible region of the problem.

On the contrary, the EPM works outside the feasible region. It also starts at the origin but moves to the points $(6,0)^T$ and $(\frac{10}{3},\frac{8}{3})^T$ before reaching the optimum at the vertex $(3,3)^T$.

The geometric interpretation of both methods shown in Figure 6.1 helps in understanding the EPM, whose theoretical foundations are given below.

As in the simplex method, the EPM is applied to a LPP in the following form. Minimize

$$f(\mathbf{x}) = \mathbf{c}^T\mathbf{x} \tag{6.89}$$

subject to

$$\mathbf{A}\mathbf{x} = \mathbf{b} \tag{6.90}$$

and

$$\mathbf{x} \geq \mathbf{0} \tag{6.91}$$

where $\mathbf{c} = (c_1, c_2, \ldots, c_n)^T$ is the column matrix of the cost coefficients, $\mathbf{x} = (x_1, x_2, \ldots, x_n)^T$ is the column vector of the initial variables, and $\mathbf{A}$ is an $m \times n$ matrix containing the constraints coefficients.

The exterior point method operates in three stages:

1. An initial stage, in which the initial set of equality constraints is transformed into another equivalent set of equality constraints

2. A regulating stage, in which the cost coefficients are transformed to non-negative values

3. An iteration stage, in which the value of the cost function is iteratively improved until the optimal solution is obtained, no feasible solution is detected, or an unbounded solution is found to exist

6.3.1 Initial Stage

One of the essential peculiarities of the EPM consists of incorporating a new variable Z, equal to the objective function of the problem, and the associated constraint

$$Z = 0x_0 + c_1x_1 + \cdots + c_nx_n \tag{6.92}$$

where $x_0 = 1$.

The EPM starts with the set of constraints (6.90) and (6.92) written as

$$\begin{pmatrix} Z \\ -- \\ \mathbf{x}_B \end{pmatrix} = \begin{pmatrix} u_0^{(0)} & | & \mathbf{w}^{(0)^T} \\ -- & + & -- \\ \mathbf{v}^{(0)} & | & \mathbf{U}^{(0)} \end{pmatrix} \begin{pmatrix} 1 \\ -- \\ \mathbf{x}_N \end{pmatrix} = \mathbf{Z}^{(0)} \begin{pmatrix} 1 \\ -- \\ \mathbf{x}_N \end{pmatrix} \tag{6.93}$$

where $\mathbf{x}_B \cup \mathbf{x}_N$ is a partition of the set of variables $\{x_1, x_2, \ldots, x_n\}$, the matrices $\mathbf{v}^{(0)}$ and $\mathbf{U}^{(0)}$ are obtained by solving (6.90) for $\mathbf{x}_B$, and replacing it in (6.89):

$$u^{(0)} = \mathbf{c}_B^T\mathbf{v}^{(0)}, \quad \mathbf{w}^{(0)} = \mathbf{c}_B^T\mathbf{U}^{(0)} + \mathbf{c}_N^T \tag{6.94}$$

where $\mathbf{c}_B^T$ and $\mathbf{c}_N^T$ are the cost coefficients associated with $\mathbf{x}_B$ and $\mathbf{x}_N$, respectively.

Then, the EPM replaces (6.93), in each iteration, by a new equivalent set of constraints with the same structure

$$\begin{pmatrix} Z^{(t)} \\ -- \\ \mathbf{x}_B^{(t)} \end{pmatrix} = \begin{pmatrix} u_0^{(t)} & | & \mathbf{w}^{(t)^T} \\ -- & + & -- \\ \mathbf{v}^{(t)} & | & \mathbf{U}^{(t)} \end{pmatrix} \begin{pmatrix} 1 \\ -- \\ \mathbf{x}_N^{(t)} \end{pmatrix} = \mathbf{Z}^{(t)} \begin{pmatrix} 1 \\ -- \\ \mathbf{x}_N^{(t)} \end{pmatrix} \tag{6.95}$$

where t refers to the iteration number and $t = 0$ is used to refer to the initial iteration.

The most important transformation used in the EPM is the previously described elemental pivoting operation.

The optimal solution conditions

$$\mathbf{w}^{(t)} \geq \mathbf{0}, \ \ \mathbf{v}^{(t)} \geq \mathbf{0} \tag{6.96}$$

justify the following two stages:

1. **Regulating iteration**. We use a regulating iteration to get $\mathbf{w}^{(t)} \geq \mathbf{0}$; that is, we look for a cost function with all nonnegative cost coefficients.

2. **Standard iterations**. We perform elemental transformations such that we have
$$\mathbf{w}^{(t)} \geq \mathbf{0} \tag{6.97}$$
until we have
$$\mathbf{v}^{(t)} \geq \mathbf{0} \tag{6.98}$$
which is the bounded solution, or we detect unboundedness or no solution.

6.3.2 Regulating Stage

In the regulating iteration we transform the cost function into a function with all nonnegative coefficients; that is, we transform $\mathbf{w}^{(0)}$ into $\mathbf{w}^{(1)} \geq \mathbf{0}$.
We can have one of two situations:

1. There exists a constraint (row), associated with x_α, such that
$$\left.\begin{array}{l} u_{\alpha j}^{(0)} < 0 \text{ if } w_j^{(0)} \geq 0 \\ u_{\alpha j}^{(0)} \leq 0 \text{ if } w_j^{(0)} < 0 \end{array}\right\} \forall j \neq 0 \tag{6.99}$$

2. There is no such constraint. In this case we introduce the additional constraint
$$x_1 + \cdots + x_n \leq M \Rightarrow x_{m+1} = M - x_1 - \cdots - x_n = \mathbf{u}_{m+1}^{(0)}\, \mathbf{x}_N \tag{6.100}$$
where x_{m+1} is the new slack variable, M is a number much larger than any other in $\mathbf{U}^{(0)}$, which obviously satisfies (6.99), and choose $x_\alpha = x_{m+1}$.

In both cases, we satisfy (6.99) after using the pivoting transformation with the pivot $u_{\alpha\beta}^{(0)}$ given by

$$\frac{w_\beta^{(0)}}{u_{\alpha\beta}^{(0)}} = \max_{w_j^{(0)}<0} \frac{w_j^{(0)}}{u_{\alpha j}^{(0)}} \tag{6.101}$$

Remark 6.3 *With the convenient selection of M, if the LPP has bounded solution, then $x_{m+1} > 0$.* ■

6.3.3 Detecting Infeasibility and Unboundedness

In this subsection we explain how the infeasibility or the unboundedness conditions can be detected.

1. **Unbounded solution situation.** This occurs when we have $x_{m+1}^* = 0$ and $\mathbf{w}^{(t)} \geq \mathbf{0}$.

   Proof. This is obvious, since $\mathbf{x}^* = \mathbf{v}^{(t)}$ (the fictitious constraint is active) gives the solution of the LPP and because of Remark 6.3, the initial problem (without the fictitious constraint) is unbounded. ■

2. **Infeasible solution situation.** This occurs when, for any t, there exists a row i such that

$$v_i^{(t)} < 0, \;\; u_{ij}^{(t)} \leq 0, \forall j \tag{6.102}$$

   Proof. Since $x_{B_i}^{(t)} = v_i^{(t)} + \sum_j u_{ij}^{(t)} x_{N\,j}^{(t)}$ and $v_i^{(t)}$ and all elements of $u_{ij}^{(t)}$ are nonpositive, then, $x_{B_i}^{(t)}$ must be negative, for any feasible values of $\mathbf{x}_N^{(t)}$. Thus, no feasible solution exists. ■

6.3.4 Standard Iterations Stage

In this stage, transformation (6.24) is used in obtaining $\mathbf{Z}^{(1)}, \ldots, \mathbf{Z}^{(t)}$, such that

1. We obtain an equivalent system of constraints:

$$\mathbf{x}_B^{(0)} = \mathbf{v}^{(0)} + \mathbf{U}^{(0)}\,\mathbf{x}_N^{(0)} \Leftrightarrow \mathbf{x}_B^{(t)} = \mathbf{v}^{(t)} + \mathbf{U}^{(t)}\,\mathbf{x}_N^{(t)}, \forall t \geq 1 \tag{6.103}$$

2. The condition

$$\mathbf{w}^{(t)} \geq \mathbf{0} \tag{6.104}$$

   still holds.

3. The value of the objective function does not increase:

$$u_0^{(1)} \geq \cdots \geq u_0^{(t)} \tag{6.105}$$

4. The pivot elements are selected to maximize the decrease of $u_0^{(t)}$.

Thus, in each iteration the LPP is equivalent to minimizing Z subject to

$$\begin{array}{rcl} \mathbf{x}_B & = & \mathbf{v}^{(t)} + \mathbf{U}^{(t)}\mathbf{x}_N \\ \mathbf{x}_N & \geq & \mathbf{0} \\ \mathbf{x}_B & \geq & \mathbf{0} \end{array} \tag{6.106}$$

In the standard iteration stage, the EPM works in the following way:

1. Each iteration has an associated infeasible point that is obtained by giving zero values to the set of nonbasic variables, $\mathbf{x}_N$. Thus, since $\mathbf{x}_N^{(t)} = \mathbf{0}$, $u_0^{(t)}$ always contains the value of the objective function, and $\mathbf{x}_B$, the values of the basic variables.

2. The essence of the exterior point method consists of adequately changing the sets of basic and nonbasic variables in the successive steps of the method. In each iteration, a basic variable, x_α, and a nonbasic variable, x_β, interchange their roles. This exchange is done in such a way that the new set of constraints is equivalent to the old set, in the sense of having the same solutions. Thus we need to

   (a) Give a rule to choose the variable x_α leaving the set $\mathbf{x}_B$ and entering the set $\mathbf{x}_N$

   (b) Give a rule to choose the variable x_β entering the set $\mathbf{x}_B$ and leaving the set $\mathbf{x}_N$

3. The selection of the entering and leaving variables is done by decreasing the value of the cost function and trying to obtain a feasible point.

Once the entering and leaving variables have been selected, we transform the system of constraints to an equivalent system, using the pivoting transformation.

Selecting the Variable Leaving $\mathbf{x}_B$

According to the previous results, if when we are in state t, the solution has not been attained and the infeasibility and unboundedness conditions have not appeared yet, a certain i and j must exist such that $v_i^{(t)} < 0$ and $u_{ij}^{(t)} > 0$.

The variable x_α leaving $\mathbf{x}_B$ is the one satisfying

$$v_\alpha^{(t)} = \min_{v_i^{(t)} < 0} v_i^{(t)} \tag{6.107}$$

Note that, according to the pivoting transformation, the new value of $v_\alpha^{(t)}$ is

$$v_\alpha^{(t+1)} = -\frac{v_\alpha^{(t)}}{u_{\alpha\beta}} \geq 0 \tag{6.108}$$

In this way, according to the previous assumptions, we get

$$u_0^{(0)} \geq \cdots \geq u_0^{(t)} \Rightarrow Z^{(0)} \geq \cdots \geq Z^{(t)}$$

a relation that, interpreted in a geometric sense, gives a sequence of points $\mathbf{x}^{(0)}, \ldots, \mathbf{x}^{(t)}$, which approach the hyperplanes $f(\mathbf{x}) = f(\mathbf{x}^{(t)})$ to the contact point $\mathbf{x}^* \in X$. This justifies the use of the *exterior point* name, which has been selected for the proposed method.

Selecting the Variable Entering $\mathbf{x}_B$

Given the leaving variable x_α, the pivot $u_{\alpha\beta}^{(t)}$ of the transformation (6.24) is selected for $\mathbf{w}^{(t)}$ to satisfy (6.104). Thus, the pivot column β and the pivot value $u_{\alpha\beta}^{(t)}$ are obtained by means of

$$\Lambda_\beta^{(t)} = \frac{w_\beta^{(t)}}{u_{\alpha\beta}^{(t)}} = \min_{u_{\alpha j}^{(t)}>0} \frac{w_j^{(t)}}{u_{\alpha j}^{(t)}} \geq 0 \tag{6.109}$$

where $\Lambda_\beta^{(t)}$ is the *function-pivot ratio.* This guarantees that $\mathbf{w}^{(t+1)} \geq \mathbf{0}$. Thus, the variable associated with column β leaves the set $\mathbf{x}_N$.

A justification of (6.109) is the following. According to the pivoting transformation (6.24), the new values of $w_j^{(t+1)}$ must be guaranteed to be nonnegative for all j:

$$w_\beta^{(t+1)} = \frac{w_\beta^{(t)}}{u_{\alpha\beta}^{(t)}} \geq 0 \tag{6.110}$$

$$w_j^{(t+1)} = w_j^{(t)} - \Lambda_\beta^{(t)} u_{\alpha j}^{(t)} \geq 0; \;\; \forall j \neq \beta \tag{6.111}$$

Condition (6.110) holds because of (6.109). To check that (6.111) also holds, we have two possibilities:

1. If $u_{\alpha j}^{(t)} < 0$, then (6.111) holds because $w_j^{(t)}$ and $\Lambda_\beta^{(t)}$ on its RHS, are nonnegative.

2. If $u_{\alpha j}^{(t)} > 0$, then because of the definition of min, we have

$$\Lambda_\beta^{(t)} \leq \frac{w_j^{(t)}}{u_{\alpha j}^{(t)}} \Leftrightarrow \Lambda_\beta^{(t)} u_{\alpha j}^{(t)} \leq w_j^{(t)} \tag{6.112}$$

   which implies (6.111).

6.3.5 The EPM Algorithm

Next, we give the EPM (exterior point method) algorithm, which works with the $\mathbf{U}^{(t)}$ matrix.

Algorithm 6.2 (The EPM algorithm)

- **Input.** The starting tableau, $\{\mathbf{U}^{(0)}, \mathbf{v}^{(0)}, \mathbf{w}^{(0)}, u^{(0)}\}$ of the LPP, including the objective function, and the problem constraints.

- **Output.** The solution of the minimization problem or a message indicating infeasibility or unbounded solution.

Regulating Iteration

Step 1 (Initialization). If $\{\mathbf{w}^{(0)} \geq \mathbf{0}\}$, go to step 4. Otherwise, transform the objective function to negative coefficients. To this end, check if there exists a constraint α satisfying condition

$$\left.\begin{array}{l} u_{\alpha j}^{(0)} < 0 \text{ if } w_j^{(0)} \geq 0 \\ u_{\alpha j}^{(0)} \leq 0 \text{ if } w_j^{(0)} < 0 \end{array}\right\} \forall j \neq 0 \tag{6.113}$$

If the answer is "yes", continue with step 2. Otherwise, introduce the artificial constraint

$$x_{m+1} = M - x_1 - \cdots - x_n \tag{6.114}$$

and choose $x_\alpha = x_{m+1}$.

Step 2 (Finding the pivot). Find the pivot column β of row α using

$$\frac{w_\beta^{(0)}}{u_{\alpha\beta}^{(0)}} = \max_{w_j^{(0)}<0} \frac{w_j^{(0)}}{u_{\alpha j}^{(0)}} \tag{6.115}$$

and go to step 3.

Standard Iterations

Step 3 (Pivoting). Perform the pivoting transformation

$$z_{ij}^{(t+1)} = \left\{\begin{array}{ll} \dfrac{z_{i\beta}^{(t)}}{z_{\alpha\beta}^{(t)}}, & \text{if} \quad i \neq \alpha, \ j = \beta \\[2ex] z_{ij}^{(t)} - \dfrac{z_{\alpha j}^{(t)}}{z_{\alpha\beta}^{(t)}} z_{i\beta}^{(t)} & \text{if} \quad i \neq \alpha, \ j \neq \beta \\[2ex] \dfrac{1}{z_{\alpha\beta}^{(t)}} & \text{if} \quad i = \alpha, \ j = \beta \\[2ex] -\dfrac{z_{\alpha j}^{(t)}}{z_{\alpha\beta}^{(t)}} & \text{if} \quad i = \alpha, \ j \neq \beta \end{array}\right\} \tag{6.116}$$

Step 4 (Selecting the leaving variable x_α). If the solution condition

$$\mathbf{w}^{(t)} \geq \mathbf{0}, \ \ \mathbf{v}^{(t)} \geq 0 \tag{6.117}$$

and the unbounded condition $x^*_{m+1} = 0$ hold, print the unbounded condition message; If only the solution condition holds, print the solution

$$\mathbf{x}^*_B = \mathbf{v}^{(t)};\ \mathbf{x}^*_N = \mathbf{0};\ Z = u_0^{(t)}, \tag{6.118}$$

and quit. Otherwise, the variable x_α leaving $\mathbf{x}_B$ is selected using

$$v_\alpha^{(t)} = \min_{v_i^{(t)}<0} v_i^{(t)} \tag{6.119}$$

If none of them hold, continue with step 5.

Step 5 (Selecting the entering variable x_β). The variable x_β leaving $\mathbf{x}_N$ is selected using

$$\frac{w_\beta^{(t)}}{u_{\alpha\beta}^{(t)}} = \min_{u_{\alpha j}^{(t)}>0} \frac{w_j^{(t)}}{u_{\alpha j}^{(t)}} \tag{6.120}$$

If the infeasible solution conditions

$$v_i^{(t)} < 0, u_{ij}^{(t)} \le 0, \forall j \tag{6.121}$$

hold, send the corresponding message and quit. Otherwise, return to step 3. ■

6.3.6 Some Illustrative Examples

In this section we illustrate the EPM by several examples.

Example 6.5 (First example). Consider again the following LPP. Minimize

$$Z = -3x_1 - x_2 \tag{6.122}$$

subject to

$$\begin{array}{rrcr} -3x_1 & +2x_2 & \le & 2 \\ x_1 & +x_2 & \le & 6 \\ x_1 & & \le & 3 \\ 2x_1 & -x_2 & \le & 4 \\ & -x_2 & \le & 0 \\ -x_1 & -x_2 & \le & -1 \end{array} \tag{6.123}$$

with

$$x_1, x_2 \ge 0 \tag{6.124}$$

Using the EPM algorithm the following results are obtained:

Step 1. Table 6.9 (where the pivot element is boldfaced), shows the initial tableau with matrices $u^{(0)}, \mathbf{w}^{(0)}, \mathbf{v}^{(0)}$, and $\mathbf{U}^{(0)}$, where we can see that some of the elements in the first row are negative. Thus, we need to transform them to nonnegative values. Since the constraint associated with variable x_4 satisfies the regulating condition (6.113), we make $x_\alpha = x_4$ and continue with step 2.

Table 6.9: Initial tableau with matrices $u^{(0)}, \mathbf{w}^{(0)}, \mathbf{v}^{(0)}$, and $\mathbf{U}^{(0)}$

$\mathbf{Z}^{(0)}$			
	x_0	x_1	x_2
Z	0	–3	–1
x_3	2	3	–2
x_4	6	**–1**	–1
x_5	3	–1	0
x_6	4	–2	1
x_7	0	0	1
x_8	–1	1	1

Step 2. Since

$$\max\left(\frac{-3}{-1}, \frac{-1}{-1}\right) = 3$$

the pivot is -1, which corresponds to $x_\beta = x_1$.

Step 3. We perform the pivoting transformation (6.49) and get Table 6.10 [$\mathbf{Z}^{(1)}$] (where pivot elements are boldfaced).

Step 4. Since the solution condition does not hold and

$$\min_{v_i^{(t)}<0} v_i^{(t)} = \min(-3, -8) = -8$$

the variable entering is $x_\alpha = x_6$.

Step 5. Since

$$\min_{u_{6j}^{(t)}>0} \frac{w_j^{(t)}}{u_{6j}^{(t)}} = \min\left(\frac{3}{2}, \frac{2}{3}\right) = \frac{2}{3}$$

the pivot is 3 that corresponds to $x_\beta = x_2$.

Step 3. We perform the pivoting transformation and get Table 6.10 [$\mathbf{Z}^{(2)}$].

Step 4. Since the solution condition does not hold and

$$\min_{v_i^{(t)}<0} v_i^{(t)} = \min\left(\frac{-1}{3}\right) = -\frac{1}{3}$$

the variable entering $x_\alpha = x_5$.

Table 6.10: Tableau at the end of step 3 for iterations 1, 2, and 3

$\mathbf{Z}^{(1)}$				$\mathbf{Z}^{(2)}$				$\mathbf{Z}^{(3)}$			
	x_0	x_4	x_2		x_0	x_4	x_6		x_0	x_4	x_5
Z	–18	3	2	Z	$\frac{-38}{3}$	$\frac{5}{3}$	$\frac{2}{3}$	Z	–12	1	2
x_3	20	–3	–5	x_3	$\frac{20}{3}$	$\frac{1}{3}$	$\frac{-5}{3}$	x_3	5	2	–5
x_1	6	–1	–1	x_1	$\frac{10}{3}$	$\frac{-1}{3}$	$\frac{-1}{3}$	x_1	3	0	–1
x_5	–3	1	1	x_5	$\frac{-1}{3}$	$\frac{1}{3}$	$\mathbf{\frac{1}{3}}$	x_6	1	–1	3
x_6	–8	2	**3**	x_2	$\frac{8}{3}$	$\frac{-2}{3}$	$\frac{1}{3}$	x_2	3	–1	1
x_7	0	0	1	x_7	$\frac{8}{3}$	$\frac{-2}{3}$	$\frac{1}{3}$	x_7	3	–1	1
x_8	5	–1	0	x_8	5	–1	0	x_8	5	–1	0

Table 6.11: Tableau at the end of step 3 for iteration 3.[1]

$\mathbf{Z}^{(3)}$			
	x_0	x_4	x_5
Z	–12	1	2
x_3	5	2	–5
x_1	3	0	–1
x_6	1	–1	3
x_2	3	–1	1
x_7	3	–1	1
x_8	5	–1	0

[1] This table satisfies the solution conditions.

Step 5. Since

$$\min_{u_{5j}^{(t)}>0} \frac{w_j^{(t)}}{u_{5j}^{(t)}} = \min\left(\frac{5/3}{1/3}, \frac{2/3}{1/3}\right) = 2$$

the pivot is $\frac{1}{3}$ which corresponds to $x_\beta = x_6$.

Step 3. We perform the pivoting transformation and get Table 6.10 [$\mathbf{Z}^{(3)}$].

Step 4. Since all values in the 1 column are positive, the solution is found. The value of the objective function is -12, which is attained at

$$x_1 = 3, x_2 = 3$$

Figure 6.2 shows the vertices visited by the SM (black bullets) and the EPM

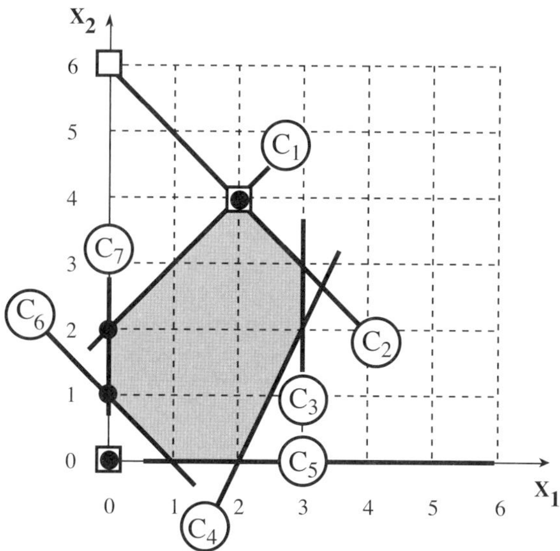

Figure 6.2: Illustration of the differences between the EPM (empty squares) and the SM (black bullets).

(empty squares).

■

Example 6.6 (A revisited example). Consider again the following LPP. Minimize

$$Z = -2x_1 - 4x_2 - 3x_3 + x_4 - 2x_5 - x_6 - 2x_7 - 2x_8 \tag{6.125}$$

subject to

$$\begin{array}{rrrrrrrrcr} 5x_1 & +4x_2 & +3x_3 & +2x_4 & +x_5 & +2x_6 & +3x_7 & +4x_8 & \leq & 17 \\ 2x_1 & -x_2 & +4x_3 & -2x_4 & +3x_5 & +3x_6 & +x_7 & -x_8 & \leq & 11 \\ 3x_1 & +2x_2 & +3x_3 & -2x_4 & +x_5 & -x_6 & +2x_7 & -x_8 & \leq & 8 \\ x_1 & -x_2 & +2x_3 & +2x_4 & -x_5 & & +x_7 & +x_8 & \leq & 4 \\ 4x_1 & +3x_2 & +x_3 & -x_4 & -2x_5 & +x_6 & +4x_7 & +3x_8 & \leq & 14 \end{array} \tag{6.126}$$

with

$$x_1, \ldots, x_8 \geq 0. \tag{6.127}$$

Using the EPM algorithm the following results are obtained:

Step 1. Table 6.12 (where the pivot element is boldfaced) shows the initial tableau. Since some of the elements in the first row are negative, we need to transform them in nonnegative values. Since the constraint associated with variable x_9, satisfies the regulating condition (6.113), we make $x_\alpha = x_9$.

Table 6.12: Initial tableau with matrices $u^{(0)}, \mathbf{w}^{(0)}, \mathbf{v}^{(0)}$, and $\mathbf{U}^{(0)}$

$\mathbf{Z}^{(0)}$									
	x_0	x_1	x_2	x_3	x_4	x_5	x_6	x_7	x_8
Z	0	–2	–4	–3	1	–2	–1	–2	–2
x_9	17	–5	–4	–3	–2	**–1**	–2	–3	–4
x_{10}	11	–2	1	–4	2	–3	–3	–1	1
x_{11}	8	–3	–2	–3	2	–1	1	–2	1
x_{12}	4	–1	1	–2	–2	1	0	–1	–1
x_{13}	14	–4	–3	–1	1	2	–1	–4	–3

Step 2. Since

$$\max\left(\frac{-2}{-5}, \frac{-4}{-4}, \frac{-3}{-3}, \frac{-2}{-1}, \frac{-1}{-2}, \frac{-2}{-3}, \frac{-2}{-4}\right) = 2$$

the pivot is -1, which corresponds to the column of x_5 and the row of x_9.

Step 3. We perform the pivoting transformation and get Table 6.13 (where the pivot element is again boldfaced).

Step 4. Since the solution condition does not hold and

$$\min_{v_i^{(t)}<0} v_i^{(t)} = \min(-40, -9) = -40$$

the variable entering is $x_\alpha = x_{10}$.

Step 5. Since

$$\min_{u_{10j}^{(t)}>0} \frac{w_j^{(t)}}{u_{10j}^{(t)}} = \min\left(\frac{8}{13}, \frac{4}{13}, \frac{3}{5}, \frac{5}{8}, \frac{2}{3}, \frac{3}{3}, \frac{4}{8}, \frac{6}{13}\right) = \frac{4}{13}$$

the pivot is 13, that corresponds to the column of x_2 and the row of x_{10}.

Step 3. We perform the pivoting transformation and get Table 6.14 (where the pivot element is boldfaced).

Step 4. Since the solution condition does not hold and

$$\min_{v_i^{(t)}<0} v_i^{(t)} = \min\left(\frac{-37}{13}\right) = -\frac{37}{13}$$

the variable entering is $x_\alpha = x_{11}$.

Table 6.13: Tableau at the end of step 3 for iteration 1

$\mathbf{Z}^{(1)}$									
	x_0	x_1	x_2	x_3	x_4	x_9	x_6	x_7	x_8
Z	–34	8	4	3	5	2	3	4	6
x_5	17	–5	–4	–3	–2	–1	–2	–3	–4
x_{10}	–40	13	**13**	5	8	3	3	8	13
x_{11}	–9	2	2	0	4	1	3	1	5
x_{12}	21	–6	–3	–5	–4	–1	–2	–4	–5
x_{13}	48	–14	–11	–7	–3	–2	–5	–10	–11

Table 6.14: Tableau at the end of step 3 for iteration 2

$\mathbf{Z}^{(2)}$									
	x_0	x_1	x_{10}	x_3	x_4	x_9	x_6	x_7	x_8
Z	$\frac{-282}{13}$	4	$\frac{4}{13}$	$\frac{19}{13}$	$\frac{33}{13}$	$\frac{14}{13}$	$\frac{27}{13}$	$\frac{20}{13}$	2
x_5	$\frac{61}{13}$	–1	$\frac{-4}{13}$	$\frac{-19}{13}$	$\frac{6}{13}$	$\frac{-1}{13}$	$\frac{-14}{13}$	$\frac{-7}{13}$	0
x_2	$\frac{40}{13}$	–1	$\frac{1}{13}$	$\frac{-5}{13}$	$\frac{-8}{13}$	$\frac{-3}{13}$	$\frac{-3}{13}$	$\frac{-8}{13}$	–1
x_{11}	$\frac{-37}{13}$	0	$\frac{2}{13}$	$\frac{-10}{13}$	$\frac{36}{13}$	$\frac{7}{13}$	$\frac{33}{13}$	$\frac{-3}{13}$	**3**
x_{12}	$\frac{153}{13}$	–3	$\frac{-3}{13}$	$\frac{-50}{13}$	$\frac{-28}{13}$	$\frac{-4}{13}$	$\frac{-17}{13}$	$\frac{-28}{13}$	–2
x_{13}	$\frac{184}{13}$	–3	$\frac{-11}{13}$	$\frac{-36}{13}$	$\frac{49}{13}$	$\frac{7}{13}$	$\frac{-32}{13}$	$\frac{-42}{13}$	0

Step 5. Since

$$\min_{u_{11j}^{(t)}>0} \frac{w_j^{(t)}}{u_{11j}^{(t)}} = \min\left(\frac{4/13}{2/13}, \frac{33/13}{36/13}, \frac{14/13}{7/13}, \frac{27/13}{33/13}, \frac{2}{3}\right) = \frac{2}{3}$$

the pivot is 3, which corresponds to the column of x_8 and the row of x_{11}.

Step 3. We perform the pivoting transformation and get Table 6.15.

Step 4. Since all values in the 1 column are positive, the solution is found. The value of the objective function is $-772/39$, which is attained at

$$x_1 = 0, x_2 = \frac{83}{39}, x_3 = 0, x_4 = 0, x_5 = \frac{61}{13}, x_6 = 0, x_7 = 0, x_8 = \frac{37}{39}.$$

■

Table 6.15: Tableau at the end of step 3 for iteration 3

$\mathbf{Z}^{(3)}$									
	x_0	x_1	x_{10}	x_3	x_4	x_9	x_6	x_7	x_{11}
Z	$\frac{-772}{39}$	4	$\frac{8}{39}$	$\frac{77}{39}$	$\frac{9}{13}$	$\frac{28}{39}$	$\frac{5}{13}$	$\frac{22}{13}$	$\frac{2}{3}$
x_5	$\frac{61}{13}$	-1	$\frac{-4}{13}$	$\frac{-19}{13}$	$\frac{6}{13}$	$\frac{-1}{13}$	$\frac{-14}{13}$	$\frac{-7}{13}$	0
x_2	$\frac{83}{39}$	-1	$\frac{5}{39}$	$\frac{-25}{39}$	$\frac{4}{13}$	$\frac{-2}{39}$	$\frac{8}{13}$	$\frac{-9}{13}$	$\frac{-1}{3}$
x_8	$\frac{37}{39}$	0	$\frac{-2}{39}$	$\frac{10}{39}$	$\frac{-12}{13}$	$\frac{-7}{39}$	$\frac{-11}{13}$	$\frac{1}{13}$	$\frac{1}{3}$
x_{12}	$\frac{385}{39}$	-3	$\frac{-5}{39}$	$\frac{-170}{39}$	$\frac{-4}{13}$	$\frac{2}{39}$	$\frac{5}{13}$	$\frac{-30}{13}$	$\frac{-2}{3}$
x_{13}	$\frac{184}{13}$	-3	$\frac{-11}{13}$	$\frac{-36}{13}$	$\frac{49}{13}$	$\frac{7}{13}$	$\frac{-32}{13}$	$\frac{-42}{13}$	0

Example 6.7 (Unbounded example). Consider again the following LPP. Minimize

$$Z = -3x_1 - x_2 \tag{6.128}$$

subject to

$$\begin{array}{rrcr} -3x_1 & +2x_2 & \leq & 2 \\ & -x_2 & \leq & 0 \\ -x_1 & -x_2 & \leq & -1 \end{array} \tag{6.129}$$

with

$$x_1, x_2 \geq 0 \tag{6.130}$$

Using the EPM algorithm the following results are obtained:

Step 1. Table 6.16 shows the initial tableau. Since some of the elements in the first row are negative, we need to transform them in nonnegative values. Since no constraint satisfies the regulating condition (6.113), we add an artificial constraint, associated with the basic variable $x_\alpha = x_6$ with $M = 20$.

Step 2. Since

$$\max\left(\frac{-3}{-1}, \frac{-1}{-1}\right) = 3$$

the pivot is -1, which corresponds to the column of x_1 and the row of x_6.

Step 3. We perform the pivoting transformation and get $\mathbf{Z}^{(1)}$ in Table 6.16.

Step 4. Since the solution conditions hold with $x_6 = 0$, the problem is unbounded.

■

Table 6.16: Initial tableau with matrices $u^{(0)}, \mathbf{w}^{(0)}, \mathbf{v}^{(0)}$, and $\mathbf{U}^{(0)}$

$\mathbf{Z}^{(0)}$				$\mathbf{Z}^{(0)}$			
	1	x_1	x_2		1	x_6	x_2
Z	0	-3	-1	Z	60	3	2
x_3	2	3	-2	x_3	62	-3	-5
x_4	0	0	1	x_4	0	0	1
x_5	-1	1	1	x_5	19	-1	0
x_6	20	**-1**	-1	x_1	20	-1	-1

Example 6.8 (Revised nonfeasible LPP). Consider again the following LPP. Minimize

$$Z = -3x_1 - x_2 \tag{6.131}$$

subject to

$$\begin{array}{rrcr} -3x_1 & +2x_2 & \leq & 2 \\ x_1 & +x_2 & \leq & 6 \\ x_1 & & \leq & 3 \\ 2x_1 & -x_2 & \leq & 4 \\ & -x_2 & \leq & 0 \\ -x_1 & -x_2 & \leq & -1 \\ x_1 & x_2 & \leq & 0 \end{array} \tag{6.132}$$

with

$$x_1 \geq 0, x_2 \geq 0 \tag{6.133}$$

Using the EPM algorithm the following results are obtained:

Step 1. Table 6.17 (where the pvot element is again boldfaced) shows the initial tableau. Since some of the elements in the first row are negative, we need to transform them into nonnegative values. Since the constraint associated with x_4 satisfies the regulating condition (6.113), we make $x_\alpha = x_4$.

Step 2. Since

$$\max\left(\frac{-3}{-1}, \frac{-1}{-1}\right) = 1$$

the pivot is -1 which corresponds to the column of x_1 and the row of x_4.

Step 3. We perform the pivoting transformation and get $\mathbf{Z}^{(1)}$ in Table 6.17.

Step 4. Since the solution condition does not hold and

$$\min_{v_i^{(t)}<0} v_i^{(t)} = \min(-3, -8, -6) = -8$$

Table 6.17: Initial tableau with matrices $u^{(0)}, \mathbf{w}^{(0)}, \mathbf{v}^{(0)}$, and $\mathbf{U}^{(0)}$

$\mathbf{Z}^{(0)}$				$\mathbf{Z}^{(1)}$			
	1	x_1	x_2		1	x_4	x_2
Z	0	–3	–1	Z	–18	3	2
x_3	2	3	–2	x_3	20	–3	–5
x_4	6	**–1**	–1	x_1	6	–1	–1
x_5	3	–1	0	x_5	–3	1	1
x_6	4	–2	1	x_6	–8	2	**3**
x_7	0	0	1	x_7	0	0	1
x_8	–1	1	1	x_8	5	–1	0
x_9	0	–1	–1	x_9	–6	1	0

the leaving variable is $x_\alpha = x_6$.

Step 5. Since

$$\min_{u_{6j}^{(t)}>0} \frac{w_j}{u_{6j}^{(t)}} = \min\left(\frac{3}{2}, \frac{2}{3}\right) = \frac{2}{3}$$

the pivot is 3, which corresponds to the column of x_2 and the row of x_6.

Step 3. We perform the pivoting transformation and get $\mathbf{Z}^{(2)}$ in Table 6.18.

Step 4. Since the solution condition does not hold and

$$\min_{v_i^{(t)}<0} v_i^{(t)} = \min\left(\frac{-1}{3}, -6\right) = -6,$$

the leaving variable is $x_\alpha = x_9$.

Step 5. Since

$$\min_{u_{9j}^{(t)}>0} \frac{w_j}{u_{9j}^{(t)}} = \min\left(\frac{5/3}{1}\right) = \frac{5}{3}$$

the pivot is 1, which corresponds to the column of x_4 and the row of x_9.

Step 3. We perform the pivoting transformation and get $\mathbf{Z}^{(3)}$ in Table 6.18.

Step 4. Since the solution condition does not hold and

$$\min_{v_i^{(t)}<0} v_i^{(t)} = \min\left(\frac{-4}{3}, \frac{-4}{3}\right) = -\frac{4}{3}$$

the leaving variable is $x_\alpha = x_2$.

Table 6.18: Initial tableau with matrices $u^{(0)}, \mathbf{w}^{(0)}, \mathbf{v}^{(0)}$, and $\mathbf{U}^{(0)}$

	$\mathbf{Z}^{(2)}$				$\mathbf{Z}^{(3)}$				$\mathbf{Z}^{(4)}$		
	1	x_4	x_6		1	x_9	x_6		1	x_9	x_2
Z	$\frac{-38}{3}$	$\frac{5}{3}$	$\frac{2}{3}$	Z	$\frac{-8}{3}$	$\frac{5}{3}$	$\frac{2}{3}$	Z	0	3	2
x_3	$\frac{20}{3}$	$\frac{1}{3}$	$\frac{-5}{3}$	x_3	$\frac{26}{3}$	$\frac{1}{3}$	$\frac{-5}{3}$	x_3	2	-3	-5
x_1	$\frac{10}{3}$	$\frac{-1}{3}$	$\frac{-1}{3}$	x_1	$\frac{4}{3}$	$\frac{-1}{3}$	$\frac{-1}{3}$	x_1	0	-1	-1
x_5	$\frac{-1}{3}$	$\frac{1}{3}$	$\frac{1}{3}$	x_5	$\frac{5}{3}$	$\frac{1}{3}$	$\frac{1}{3}$	x_5	3	1	1
x_2	$\frac{8}{3}$	$\frac{-2}{3}$	$\frac{1}{3}$	x_2	$\frac{-4}{3}$	$\frac{-2}{3}$	$\mathbf{\frac{1}{3}}$	x_6	4	2	3
x_7	$\frac{8}{3}$	$\frac{-2}{3}$	$\frac{1}{3}$	x_7	$\frac{-4}{3}$	$\frac{-2}{3}$	$\frac{1}{3}$	x_7	0	0	1
x_8	5	-1	0	x_8	-1	-1	0	x_8	-1	-1	0
x_9	-6	**1**	0	x_4	6	1	0	x_4	6	1	0

Step 5. Since

$$\min_{u_{2j}^{(t)}>0} \frac{w_j}{u_{2j}^{(t)}} = \min\left(\frac{2/3}{1/3}\right) = \frac{2/3}{1/3}$$

the pivot is $\frac{1}{3}$, which corresponds to the column of x_6 and the row of x_2.

Step 3. We perform the pivoting transformation and get $\mathbf{Z}^{(4)}$ in Table 6.18.

Step 4. Since the solution condition does not hold and

$$\min_{v_i^{(t)}<0} v_i^{(t)} = \min(-1) = -1$$

the leaving variable is $x_\alpha = x_8$.

Step 5. Since the infeasible conditions hold, because all the values u_{8j} are nonpositive, the problem has no feasible solution.

■

Exercises

6.1 Determine the feasible region and its vertices of the following linear programming problem. Minimize

$$Z = -4x_1 + 7x_2$$

subject to

$$\begin{array}{rrcl} x_1 & +x_2 & \geq & 3 \\ -x_1 & +x_2 & \leq & 3 \\ 2x_1 & +x_2 & \leq & 8 \\ & x_1,\, x_2 & \geq & 0 \end{array}$$

Obtain the optimal solution using the SM and the EPM methods.

6.2 Prove that the following problem does not have a finite optimum, using the SM and the EPM methods. Minimize

$$Z = -6x_1 + 2x_2$$

subject to

$$\begin{array}{rrcr} -3x_1 & +x_2 & \leq & 6 \\ 3x_1 & +5x_2 & \geq & 15 \\ & x_1,\, x_2 & \geq & 0 \end{array}$$

6.3 Prove, using the SM and the EPM methods, that the following problem, involving maximization of

$$Z = -6x_1 + 2x_2$$

subject to

$$\begin{array}{rrcr} -3x_1 & +x_2 & \leq & 6 \\ 3x_1 & +5x_2 & \geq & 15 \\ & x_1,\, x_2 & \geq & 0 \end{array}$$

has an infinite maxima.

6.4 Solve the dual of the problem. Minimize

$$Z = 5x_1 - 6x_2 + 7x_3 + x_4$$

subject to

$$\begin{array}{rrrrcr} x_1 & +2x_2 & -x_3 & -x_4 & = & -7 \\ 6x_1 & -3x_2 & +x_3 & +7x_4 & \geq & 14 \\ -2x_1 & -7x_2 & +4x_3 & +2x_4 & \leq & -3 \\ & & & x_1,\, x_2 & \geq & 0 \end{array}$$

6.5 Solve the following problem. Maximize

$$Z = 240x_1 + 104x_2 + 60x_3 + 19x_4$$

subject to:

$$\begin{array}{rcl} 20x_1 + 9x_2 + 6x_3 + x_4 & \leq & 20 \\ 10x_1 + 4x_2 + 2x_3 + x_4 & \leq & 10 \\ x_i & \geq & 0, \quad i = 1, 2, 3, 4 \end{array}$$

looking at its dual.

6.6 Determine, using the SM and the EPM methods, the maximum value of the cost function

$$Z = 18x_1 + 4x_2 + 6x_3$$

subject to the constraints

$$\begin{aligned} 3x_1 + x_2 &\leq -3 \\ 2x_1 + x_3 &\leq -5 \\ x_i &\leq 0, \text{ for } i = 1, 2, 3 \end{aligned}$$

6.7 An electricity producer should plan its hourly energy production to maximize its profits from selling energy during a planning horizon of 2 hours. Formulate and solve a linear programming problem to maximize the profits of the electricity producer if

(a) The producer produces 5 energy units before the planning horizon.

(b) Hourly energy prices are 6 and 2 monetary units per energy unit.

(c) The minimum energy production of the producer in each hour is 0 and the maximum one 10 energy units.

(d) Energy productions in two consecutive hours cannot differ by more than 4 energy units.

(e) Producer production cost is 3 monetary units per energy unit.

6.8 Solve by the SM, Exercise 1.7, where $c = 1$, and the four fixed points are

$$(1, 0), (0, 1), (-1, 0), (0, -1)$$

Chapter 7

Mixed-Integer Linear Programming

7.1 Introduction

A mixed-integer linear programming problem (MILPP) is a linear programming problem (LPP) in which some of the variables have to be integers. If all the integer variables have to be binary (0/1), the problem is denominated 0/1 MILPP. If, on the other hand, all variables have to be integer the problem is called integer linear programming problem (ILPP). In engineering practice, MILPP are the most frequent problems. Mixed-integer linear programming problems provide a flexible and efficient framework to formulate and solve many engineering problems. This chapter gives an introduction to MILPP and the solution techniques available to solve them. An example of the set of feasible solutions in this integer case is shown in Figure 7.1, where this set of feasible points is associated with dots.

A general MILPP is formulated in standard form as minimizing

$$Z = \sum_{j=1}^{n} c_j x_j$$

subject to

$$\begin{array}{rcl} \sum_{j=1}^{n} a_{ij} x_j & = & b_i; \ \ i = 1, 2, \ldots, m \\ x_j & \geq & 0; \ \ j = 1, 2, \ldots, n \\ x_j & \in & \mathbb{N}; \ \ \text{for all or some } j = 1, 2, \ldots, n \end{array}$$

where $\mathbb{N}$ will be used in this chapter to refer to the set $\{0, 1, 2, \ldots\}$.

Relevant references to MILPP are the classical manual of Garfinkel and Nemhauser [47], the excellent monography by Nemhauser and Wolsey [79], the

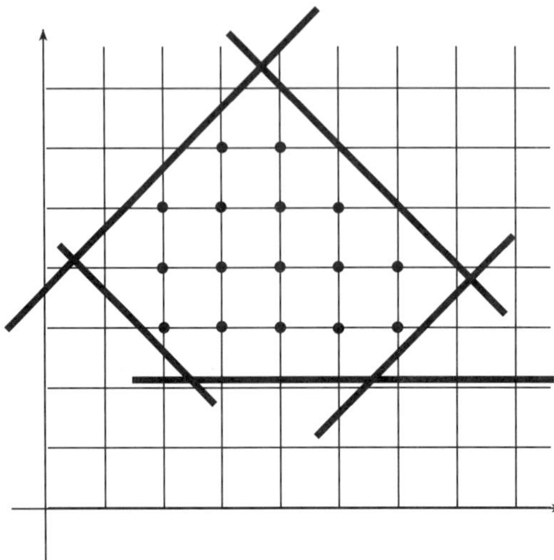

Figure 7.1: Illustration of the set of feasible solutions (set of points associated with dots) of an ILPP.

practical manual of Bradley et al. [15], and the introductory manual by Rao [92].

Two solution techniques for MILPP are explained in this chapter: the branch–bound (BB) and the Gomory cuts (GC) methods. The BB is the most used and usually the most computationally efficient technique. However, more recently, a hybrid highly efficient technique of both methods, denominated branch–cut (BC), is being successfully applied.

Binary variables constitute a powerful tool to model nonlinearities of diverse nature which quite often occur in engineering. Some examples are

1. Alternative sets of constraints.

2. Conditional constraints

3. Discontinuous functions

4. Piecewise non convex functions

The modeling of nonlinearities using binary variables is explained with the help of several examples in Chapter 13, Section 13.2.9.

7.2 The Branch–Bound Method

7.2.1 Introduction

The BB method solves MILPP by solving a sequence of LPP obtained by relaxing integrality conditions and including additional constraints. The number of additional constraints increases as the BB procedure progresses. These constraints separate the feasible region into complementary feasible subregions.

The BB method initially sets up lower and upper bounds for the optimal solution (objective function optimal value). The branching strategy iteratively decreases the upper bound and increases the lower bound. The difference between those bounds is a measure of the proximity of the current solution to the optimal solution if it exists.

When minimizing, a lower bound for the optimal solution can be found by relaxing the integrality constraints of the original MILPP and solving the resulting LPP. Analogously, the objective function value for any solution (satisfying integrality conditions) of the original MILPP is an upper bound of the optimal solution.

7.2.2 The BB Algorithm for MILPP

Input. A MILPP to be solved.

Output. Its solution or a message informing that it is unbounded or infeasible.

Step 1. (Initialization). Set up an upper bound (∞) and a lower bound ($-\infty$) of the optimal solution. Solve the initial MILPP relaxing integrality conditions. If the relaxed problem is infeasible, the original MILPP is also infeasible and there is no solution. If the attained solution satisfies integrality conditions it is optimal. Otherwise, the lower bound is updated with the value of the optimal solution attained.

Step 2. (Branching). Using an integer-to-be variable x_k that is not an integer, two branching problems from the original problem are generated as follows. If the value of the integer-to-be variable x_k is $a.b$, where a and b are its integer and fractional parts, respectively, the first branch problem is the relaxed original MILPP plus the constraint $x_k \leq a$ and the second branch problem the relaxed original MILPP plus the complementary constraint $x_k \geq a + 1$. These problems are placed in a processing list and are considered sequentially or in parallel. It should be noted that the proposed branching strategy completely covers the solution space.

Step 3. (Solution). Solve next problem in the processing list.

Step 4. (Bound updating). If the solution of the current problem satisfies integrality conditions and its corresponding objective function value is smaller than the current upper bound, the upper bound is updated to the current objective function value and the current minimizer is stored as the best candidate to minimizer. If, on the other hand, the solution does not satisfy integrality conditions and the objective function value is in between the upper and lower bounds, the lower bound is updated to the objective function value, the current problem is branched, and the generated problems are added to the processing list.

Step 5. (Cutting). If the solution provided by the current problem satisfies integrality conditions, no further branching is possible. The branch is therefore cut as a result of integrality. If the solution does not satisfy integrality conditions and the objective function value is greater than the current upper bound, no better solution can be obtained further along that branch. The branch is therefore cut due to bounds. If the current problem is infeasible, no further branching is possible through that branch. The branch is therefore cut due to infeasibility.

Step 6. (Optimality). If the processing list is not empty, we continue with step 3. Otherwise, the procedure concludes. If there is a current candidate to the minimizer, the minimizer is such a candidate. Otherwise, the problem is infeasible.

The BB algorithm returns the optimal solution or informs about infeasibility in either step 1 or step 6.

The branching procedure therefore can stop for three reasons:

1. The current problem is infeasible

2. The current solution satisfies integrality conditions

3. The lower bound obtained is greater than the current upper bound

A branch is therefore cut for infesibility, integrality or bounds.

7.2.3 Branching and Processing Strategies

Any of the integer-to-be variables that are not integer in the current solution is a candidate for branching. Which one to choose is a nontrivial question whose solution should be based on the knowledge of the problem under study.

Problems in the processing list can be processed *breadth-first* or *depth-first* or using a mixture of these two strategies. Figure 7.2 illustrates the two alternatives. Usually the engineering knowledge of the problem provides insight to select the most appropriate strategy.

A depth-first strategy quickly generates highly constrained problems that usually provide good upper and lower bounds. It also quickly provides infeasible problems and therefore desirable branch cutting. A breadth-first strategy processes very similar problems, and this can be computationally exploited. Parallel computation techniques are applicable to any branching strategy.

Example 7.1 (BB method: integer problem). Consider the following ILPP. Minimize

$$Z = -x_1 - x_2$$

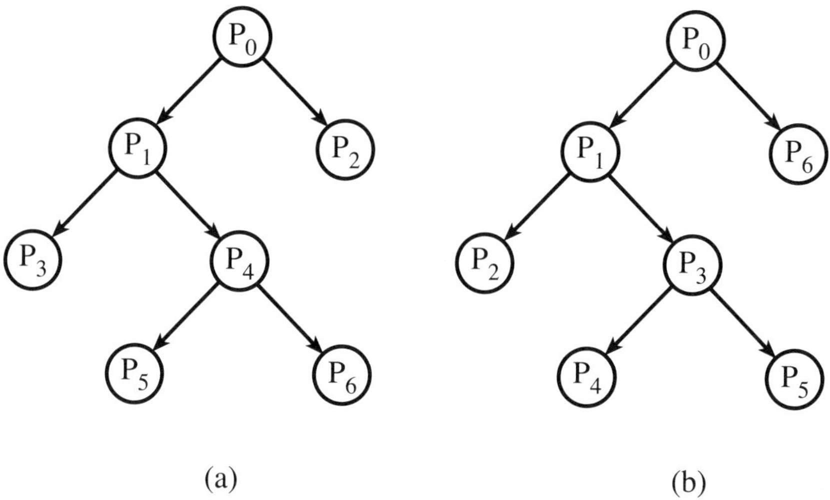

Figure 7.2: Illustration of (a) the 'breadth-first', and (b) the 'depth-first' strategies.

subject to

$$\begin{array}{rrcl} -x_1 & & \leq & 0 \\ 2x_1 & -2x_2 & \leq & 1 \\ & 2x_2 & \leq & 9 \\ & x_1, x_2 & \in & \mathbb{N} \end{array}$$

which is the dotted feasible region shown in Figure 7.3.

Step 1. (Initialization). The initial upper bound is $+\infty$ and the initial lower bound is $-\infty$. The relaxed problem, designated P_0, is as follows. Minimize

$$Z = -x_1 - x_2$$

subject to

$$\begin{array}{rrcl} -x_1 & & \leq & 0 \\ 2x_1 & -2x_2 & \leq & 1 \\ & 2x_2 & \leq & 9 \end{array}$$

Its solution is

$$x_1 = 5, \quad x_2 = 4.5; \quad Z = -9.5$$

which is the point P_1 in Figure 7.3. This solution does not satisfy integrality conditions ($x_2 \notin \mathbb{N}$). The objective function value is used to update the lower bound, from $-\infty$ to -9.5.

Step 2. (Branching). The integer-to-be variable x_2, through branching, generates the two problems presented below. Additional constraints are $x_2 \leq 4$ and

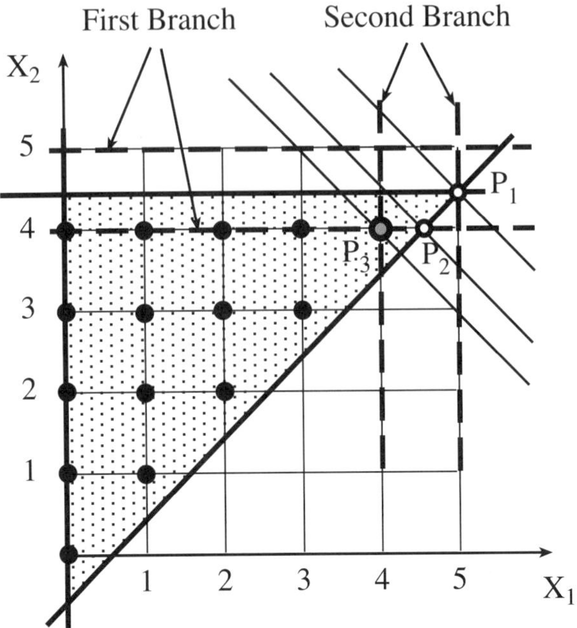

Figure 7.3: Graphical illustration of the BB method in Example 7.1.

$x_2 \geq 5$, which are shown in Figure 7.3 as horizontal dashed lines.

Problem P_1. Minimize

$$Z = -x_1 - x_2$$

subject to

$$\begin{array}{rrcl} -x_1 & & \leq & 0 \\ 2x_1 & -2x_2 & \leq & 1 \\ & 2x_2 & \leq & 9 \\ & x_2 & \leq & 4 \end{array}$$

Problem P_2. Minimize

$$Z = -x_1 - x_2$$

subject to

$$\begin{array}{rrcl} -x_1 & & \leq & 0 \\ 2x_1 & -2x_2 & \leq & 1 \\ & 2x_2 & \leq & 9 \\ & x_2 & \geq & 5 \end{array}$$

Step 3. (Solution). The solution of problem P_1 is

$$x_1 = 4.5; \quad x_2 = 4; \quad Z = -8.5$$

which is the point P_2 in Figure 7.3.

Step 4. (Bound updating). Since this solution does not satisfy integrality conditions ($x_1 \notin \mathbb{N}$), and the objective function value -8.5 is in between lower and upper bounds, the current lower bound value is updated from -9.5 to -8.5 (the optimal solution is therefore within -8.5 and ∞). Then, the problem is branched again. The integer-to-be variable x_1, through branching, generates the two problems below. Additional constrains are $x_1 \leq 4$ and $x_1 \geq 5$, which are shown in Figure 7.3 as vertical dashed lines.

Problem P_3. Minimize

$$Z = -x_1 - x_2$$

subject to

$$\begin{array}{rrcl} -x_1 & & \leq & 0 \\ 2x_1 & -2x_2 & \leq & 1 \\ & 2x_2 & \leq & 9 \\ & x_2 & \leq & 4 \\ x_1 & & \leq & 4 \end{array}$$

Problem P_4. Minimize

$$Z = -x_1 - x_2$$

subject to

$$\begin{array}{rrcl} -x_1 & & \leq & 0 \\ 2x_1 & -2x_2 & \leq & 1 \\ & 2x_2 & \leq & 9 \\ & x_2 & \leq & 4 \\ x_1 & & \geq & 5 \end{array}$$

Step 5. (Cutting). Nothing occurs at this step.

Step 6. Optimality). Since the processing list is not empty, we continue with problem P_2 and step 3.

Step 3. (Solution). The problem P_2 is infeasible; thus, nothing occurs at this step.

Step 4. (Bound updating). Nothing occurs at this step.

Step 5. (Cutting). Since the problem is infeasible, no further cut is possible through this branch and it is cut.

Step 6. (Optimality). Since the processing list is not empty, we continue with problem P_3 and step 3.

Step 3. (Solution). The solution of problem P_3 is

$$x_1 = 4; \quad x_2 = 4; \quad z = -8$$

which is point P_3 in Figure 7.3.

Step 4. (Bound updating). Since this solution satisfies integrality conditions ($x_1, x_2 \in \mathbb{N}$), and the objective function value -8 is in between lower and upper bounds, the current upper bound value is updated from $+\infty$ to -8, and the current minimizer is stored as the best candidate minimizer.

Step 5. (Cutting). Since the current solution satisfies integrality condition, no further branching is possible and the branch is cut.

Step 6. (Optimality). Since the processing list is not empty, we proceed with step 3.

Step 3. (Solution). The problem P_4 is infeasible; thus, nothing occurs at this step.

Step 4. (Bound updating). Nothing occurs at this step.

Step 5. (Cutting). Since the problem is infeasible, no further cut is possible through this branch and it is cut.

Step 6. (Optimality). Since the processing list is empty, and there is a candidate to minimizer, this is the solution of the initial problem:

$$x_1^* = 4; \quad x_2^* = 4; \quad Z^* = -8$$

■

Example 7.2 (BB method: Mixed-integer problem). Consider the following MILPP. Minimize

$$Z = 3x_1 + 2x_2$$

subject to

$$\begin{array}{rrrrcl}
x_1 & -2x_2 & +x_3 & & = & \frac{5}{2} \\
2x_1 & +x_2 & & +x_4 & = & \frac{3}{2} \\
 & & & x_1, x_2, x_3, x_4 & \geq & 0 \\
 & & & x_2, x_3 & \in & \mathbb{N}
\end{array}$$

Step 1. (Initialization). The initial upper bound is $+\infty$ and the initial lower bound is $-\infty$. The relaxed problem, denominated P_0, is to Minimize

$$Z = 3x_1 + 2x_2$$

subject to

$$\begin{array}{rrrrcl} x_1 & -2x_2 & +x_3 & & = & \frac{5}{2} \\ 2x_1 & +x_2 & & +x_4 & = & \frac{3}{2} \\ & & & x_1, x_2, x_3, x_4 & \geq & 0 \end{array}$$

Its solution is

$$x_1 = x_2 = 0; \quad x_3 = 2.5; \quad x_4 = 1.5; \quad Z = 0$$

This solution does not satisfy integrality conditions ($x_3 \notin \mathbb{N}$). The objective function value is used to update the lower bound, from $-\infty$ to 0.

Step 2. (Branching). The integer-to-be variable x_3, through branching, generates the two problems below. Additional constrains are $x_3 \leq 2$ and $x_3 \geq 3$.

Problem P_1. Minimize

$$Z = 3x_1 + 2x_2$$

subject to

$$\begin{array}{rrrrcl} x_1 & -2x_2 & +x_3 & & = & \frac{5}{2} \\ 2x_1 & +x_2 & & +x_4 & = & \frac{3}{2} \\ & & x_3 & & \leq & 2 \\ & & & x_1, x_2, x_3, x_4 & \geq & 0 \end{array}$$

Problem P_2. Minimize

$$Z = 3x_1 + 2x_2$$

subject to

$$\begin{array}{rrrrcl} x_1 & -2x_2 & +x_3 & & = & \frac{5}{2} \\ 2x_1 & +x_2 & & +x_4 & = & \frac{3}{2} \\ & & x_3 & & \geq & 3 \\ & & & x_1, x_2, x_3, x_4 & \geq & 0 \end{array}$$

Step 3. (Solution). The solution of problem P_1 is

$$x_1 = 0.5; \quad x_2 = 0; \quad x_3 = 2; \quad x_4 = 0.5; \quad Z = 1.5$$

Step 4. (Bound updating). Since this solution satisfies integrality conditions ($x_2, x_3 \in \mathbb{N}$), and the objective function value, 1.5, is smaller than the current upper bound value, $+\infty$, it is updated from $+\infty$ to 1.5 (the optimal solution is therefore within 0 and 1.5), and the solution is stored as the best candidate minimizer.

Step 5. (Cutting). Since this solution satisfies integrality conditions (in x_2 and x_3), no further branching is possible through this branch and it is cut. Next, we continue with step 3.

Step 3. (Solution). The solution to problem P_2 is

$$x_1 = 0; \quad x_2 = 0.25; \quad x_3 = 3; \quad x_4 = 1.25; \quad Z = 0.5$$

Step 4. (Bound updating). Since this solution does not satisfy integrality conditions ($x_2 \notin \mathbb{N}$) and the objective function value 0.5 is between the lower bound 0 and the upper bound 1.5, the lower bound is updated from 0 to 0.5. Therefore, the optimal solution is in the interval $(0.5, 1.5)$. Next, the current problem is branched. The branching variable x_2 originates the following two problems:

Problem P_3. Minimize

$$Z = 3x_1 + 2x_2$$

subject to

$$\begin{array}{rrrrcl}
x_1 & -2x_2 & +x_3 & & = & \frac{5}{2} \\
2x_1 & +x_2 & & +x_4 & = & \frac{3}{2} \\
 & & x_3 & & \geq & 3 \\
 & x_2 & & & \leq & 0 \\
 & & & x_1, x_2, x_3, x_4 & \geq & 0
\end{array}$$

Problem P_4. Minimize

$$Z = 3x_1 + 2x_2$$

subject to

$$
\begin{array}{rrrrrl}
x_1 & -2x_2 & +x_3 & & = & \frac{5}{2} \\
2x_1 & +x_2 & & +x_4 & = & \frac{3}{2} \\
 & & x_3 & & \geq & 3 \\
 & x_2 & & & \geq & 1 \\
 & & & x_1, x_2, x_3, x_4 & \geq & 0
\end{array}
$$

which are added to the processing list. Note that the additional constraints are respectively $x_2 \leq 0$ and $x_2 \geq 1$.

Step 5. (Cutting). Nothing is done in this step, and we continue with step 6.

Step 6. (Optimality). Since the list is not empty, we continue with step 3.

Step 3. (Solution). Problem P_3 is infeasible.

Step 4. (Bound updating). Nothing occurs in this step.

Step 5. (Cutting). Since the problem is infeasible, no further branching is possible, and we continue with step 6.

Step 6. (Optimality). Since the processing list is not empty, we continue with step 3.

Step 3. (Solution). Problem P_4 is feasible. Its solution is:

$$x_1 = 0; \quad x_2 = 1; \quad x_3 = 4.5; \quad x_4 = 0.5; \quad Z = 2$$

Step 4. (Bound updating). Nothing occurs in this step.

Step 5. (Cutting). Since this solution does not satisfy integrality conditions ($x_3 \notin \mathbb{N}$), and its objective function value is greater than the current upper bound, no better solution will be found by further branching. Next, we continue with step 6.

Step 6). (Optimality). Since the processing list is empty and there is a candidate to minimizer, this is the solution of our problem. The optimal solution is therefore

$$x_1^* = 0.5 \;\; x_2^* = 0 \;\; x_3^* = 2 \;\; x_4^* = 0.5 \;\; Z^* = 1.5$$

The entire process is illustrated in Figure 7.4.

It should be noted that in this simple example, a breadth-first strategy has been used.

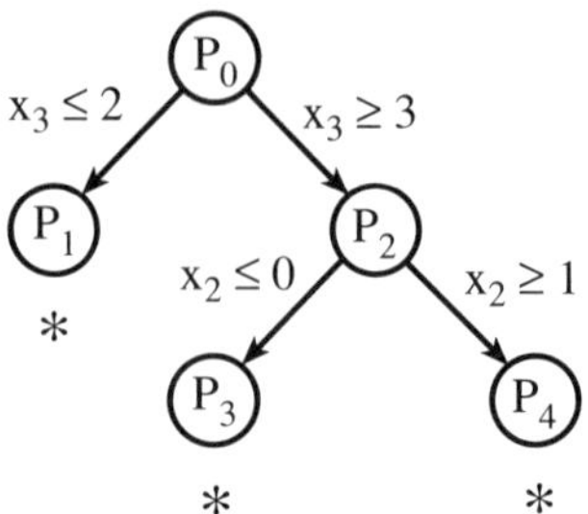

Figure 7.4: Illustration of the process in Example 7.2.

■

7.2.4 Other Mixed-Integer Linear Programming Problems

Two particular (further constrained) MILPPs are of interest: (i) integer linear programming problems (ILPPs) and (ii) 0/1 mixed-integer linear programming problems (0/1 MILPPs).

ILPP are solved using the BB technique in a similar way as with the MILPP. All variables are in principle candidates for branching. 0/1 MILPP are also solved using the BB technique. In these problems, branching is equivalent to fix binary variables either to 0 or 1.

7.3 The Gomory Cuts Method

7.3.1 Introduction

An alternative solution technique for MILPP is the Gomory cuts method. At each iteration of this technique the relaxed original problem is solved including additional constraints which effectively reduce the feasible region without excluding any solution satisfying the integrality conditions. At each iteration a new additional constraint is added, which is denominated a Gomory cut. This technique progressively generates a convex hull of the mixed-integer feasible region and in this way the solution is forced to satisfy integrality conditions. Additional information can be found in Rao [92], Bronson [17], and the original Gomory [50] paper.

The Gomory algorithm for ILPP is explained below. Its generalization to MILPP is proposed as an exercise for the reader.

7.3.2 Cut Generation

The feasible region of the relaxed original problem is $\mathbf{Ax} = \mathbf{b}$, which, using the standard simplex partition, can be written as

$$\begin{pmatrix} \mathbf{B} & \mathbf{N} \end{pmatrix} \begin{pmatrix} \mathbf{x}_B \\ \mathbf{x}_N \end{pmatrix} = \mathbf{b}$$

or

$$\mathbf{B}\mathbf{x}_B + \mathbf{N}\mathbf{x}_N = \mathbf{b}$$

Solving for $\mathbf{x}_B$ we get

$$\mathbf{x}_B - \mathbf{B}^{-1}\mathbf{N}\mathbf{x}_N = \mathbf{B}^{-1}\mathbf{b}$$

and in matrix form

$$\begin{pmatrix} \mathbf{I} & \mathbf{B}^{-1}\mathbf{N} \end{pmatrix} \begin{pmatrix} \mathbf{x}_B \\ \mathbf{x}_N \end{pmatrix} = \mathbf{B}^{-1}\mathbf{b},$$

which, using standard simplex notation, becomes

$$\begin{pmatrix} \mathbf{I} & \mathbf{U} \end{pmatrix} \begin{pmatrix} \mathbf{x}_B \\ \mathbf{x}_N \end{pmatrix} = \tilde{\mathbf{b}}$$

or

$$\mathbf{x}_B + \mathbf{U}\mathbf{x}_N = \tilde{\mathbf{b}}$$

Let x_{B_i} be a basic noninteger but an integer-to-be variable. Its corresponding row in the immediately preceding equation is

$$x_{B_i} + \sum_j u_{ij} x_{N_j} = \tilde{b}_i \tag{7.1}$$

Because the current solution is basic and feasible, variables x_{N_j}, which are nonbasic, are null; then, $\tilde{\mathbf{b}} = \mathbf{x}_B$, and because x_{B_i} is noninteger, $\tilde{b}_i$ has to be noninteger.

On the other hand, every element u_{ij} can be expressed as the sum of an integer (positive or negative) and a nonnegative fraction f_{ij} smaller than 1:

$$u_{ij} = i_{ij} + f_{ij}; \quad \forall j \tag{7.2}$$

Analogously, $\tilde{b}_i$ can be decomposed as

$$\tilde{b}_i = \tilde{i}_i + \tilde{f}_i \tag{7.3}$$

where $\tilde{i}_i$ is integer (positive or negative) and $\tilde{f}_i$ a nonnegative fraction smaller than 1. It should be noted that some f_{ij} values can be null, but $\tilde{f}_i$ is necessarily positive.

Substitution of (7.2) and (7.3) into (7.1) leads to

$$x_{B_i} + \sum_j (i_{ij} + f_{ij}) x_{N_j} = \tilde{i}_i + \tilde{f}_i$$

or

$$x_{B_i} + \sum_j i_{ij} x_{N_j} - \tilde{i}_i = \tilde{f}_i - \sum_j f_{ij} x_{N_j}$$

The left-hand side has to be an integer because all variables are integers, therefore, its RHS has to be an integer, too.

On the other hand, f_{ij} for all j, and x_{N_j} for all j are nonnegative, and therefore $\sum_j f_{ij} x_{N_j}$ is nonnegative.

The RHS of the preceding equation, $\tilde{f}_i - \sum_j f_{ij} x_{N_j}$, is simultaneously

- An integer, and
- Smaller than a positive fraction $\tilde{f}_i$ smaller than 1

and since the only integers satisfying this condition are $0, -1, -2, \ldots$, then $\tilde{f}_i - \sum_j f_{ij} x_{N_j}$ is a nonpositive integer. Then

$$\tilde{f}_i - \sum_j f_{ij} x_{N_j} \leq 0 \tag{7.4}$$

or

$$\sum_j f_{ij} x_{N_j} - \tilde{f}_i \geq 0 \tag{7.5}$$

This last inequality is denominated the Gomory cut associated with the basic noninteger but integer-to-be variable x_{B_i}.

7.3.3 The Gomory Cuts Algorithm for an ILPP

Input. An ILPP to be solved.

Output. Its optimal solution or a message informing that it is unbounded or infeasible.

Step 1. (Initialization). Solve the initial problem relaxing integrality conditions. If it is unbounded (infeasible), stop; the original problem is unbounded (infeasible).

Step 2. (Optimality check). If the current solution satisfies integrality conditions, stop, it is optimal. Otherwise, go to the next step.

Step 3. (Cut generation). Use an integer-to-be basic variable with a noninteger value to generate a Gomory cut.

Step 4. (Problem solution). Add the Gomory cut to the current problem, solve it, and continue with step 2.

It should be noted that the number of constraints grows with the number of iterations. As one additional constraint is added at every iteration, the EPM (dual simplex) should be used. This is so because the original problem becomes infeasible but its dual becomes non optimal but feasible, and we can use the optimal table to restart the new problem. The example below clarifies the statements presented above.

In the primal problem we minimize

$$Z = \begin{pmatrix} -3 & -5 \end{pmatrix} \begin{pmatrix} x_1 \\ x_2 \end{pmatrix}$$

subject to

$$\begin{pmatrix} -1 & 0 \\ 0 & -1 \\ -3 & -2 \end{pmatrix} \begin{pmatrix} x_1 \\ x_2 \end{pmatrix} \geq \begin{pmatrix} -4 \\ -6 \\ -18 \end{pmatrix}$$

$$\begin{pmatrix} x_1 \\ x_2 \end{pmatrix} \geq \begin{pmatrix} 0 \\ 0 \end{pmatrix}$$

Its minimizer is

$$\begin{pmatrix} x_1^* \\ x_2^* \end{pmatrix} = \begin{pmatrix} 2 \\ 6 \end{pmatrix}$$

and the optimal value of the objective function is $z^* = -36$.

In the dual problem of the previous primal one we maximize

$$Z = \begin{pmatrix} y_1 & y_2 & y_3 \end{pmatrix} \begin{pmatrix} -4 \\ -6 \\ -18 \end{pmatrix}$$

subject to

$$\begin{pmatrix} y_1 & y_2 & y_3 \end{pmatrix} \begin{pmatrix} -1 & 0 \\ 0 & -1 \\ -3 & -2 \end{pmatrix} \leq \begin{pmatrix} -3 & -5 \end{pmatrix}$$

$$\begin{pmatrix} y_1 & y_2 & y_3 \end{pmatrix} \geq \begin{pmatrix} 0 & 0 & 0 \end{pmatrix}$$

Its maximizer is

$$\begin{pmatrix} y_1^* \\ y_2^* \\ y_3^* \end{pmatrix} = \begin{pmatrix} 0 \\ 3 \\ 1 \end{pmatrix}$$

and the objective function optimal value is $z^* = -36$.

If the additional constraint $x_2 \leq 4$ is added to the primal problem, it becomes minimization of

$$\begin{pmatrix} -3 & -5 \end{pmatrix} \begin{pmatrix} x_1 \\ x_2 \end{pmatrix}$$

subject to

$$\begin{pmatrix} -1 & 0 \\ 0 & -1 \\ -3 & -2 \\ 0 & -1 \end{pmatrix} \begin{pmatrix} x_1 \\ x_2 \end{pmatrix} \geq \begin{pmatrix} -4 \\ -6 \\ -18 \\ -4 \end{pmatrix}$$

$$\begin{pmatrix} x_1 \\ x_2 \end{pmatrix} \geq \begin{pmatrix} 0 \\ 0 \end{pmatrix}$$

Note that the original primal problem solution

$$\begin{pmatrix} x_1^* \\ x_2^* \end{pmatrix} = \begin{pmatrix} 2 \\ 6 \end{pmatrix}$$

is infeasible for this primal problem.

The dual problem of the (further constrained) primal one above is to maximize

$$Z = \begin{pmatrix} y_1 & y_2 & y_3 & y_4 \end{pmatrix} \begin{pmatrix} -4 \\ -6 \\ -18 \\ -4 \end{pmatrix}$$

subject to

$$\begin{pmatrix} y_1 & y_2 & y_3 & y_4 \end{pmatrix} \begin{pmatrix} -1 & 0 \\ 0 & -1 \\ -3 & -2 \\ 0 & -1 \end{pmatrix} \leq \begin{pmatrix} -3 & -5 \end{pmatrix}$$

$$\begin{pmatrix} y_1 & y_2 & y_3 & y_4 \end{pmatrix} \geq \begin{pmatrix} 0 & 0 & 0 & 0 \end{pmatrix}$$

Note that the optimal solution of the original dual problem

$$\begin{pmatrix} y_1^* \\ y_2^* \\ y_3^* \end{pmatrix} = \begin{pmatrix} 0 \\ 3 \\ 1 \end{pmatrix}$$

with

$$y_4 = 0$$

is a feasible, but not optimal solution of the preceding dual problem.

Example 7.3 (Gomory cuts algorithm). The following example illustrates the Gomory cuts algorithm.

Consider the following ILPP. Maximize

$$Z = 120x_1 + 80x_2$$

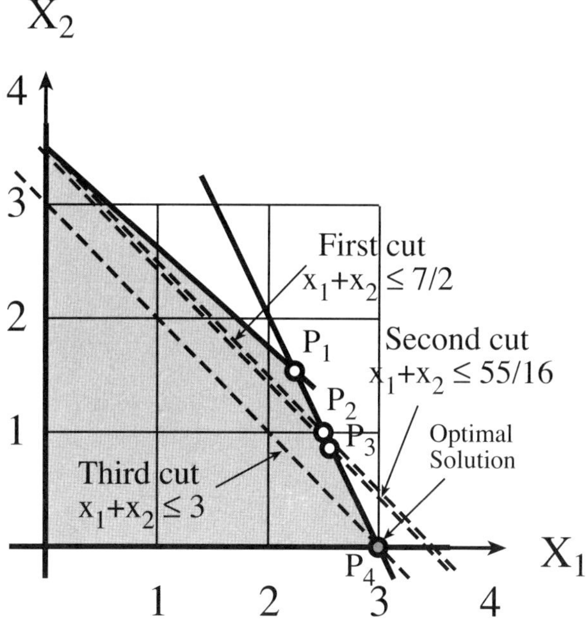

Figure 7.5: Graphical illustration of the Gomory cuts algorithm in Example 7.3.

subject to

$$
\begin{array}{rrcl}
2x_1 & +x_2 & \leq & 6 \\
7x_1 & +8x_2 & \leq & 28 \\
x_1 & & \geq & 0 \\
 & x_2 & \geq & 0 \\
x_1 & & \in & \mathbb{N}
\end{array}
$$

whose feasible region is shown, as a shadowed set, in Figure 7.5.

The above problem, relaxed, is to maximize

$$Z = 120x_1 + 80x_2$$

subject to

$$
\begin{array}{rrrrcl}
2x_1 & +x_2 & +x_3 & & = & 6 \\
7x_1 & +8x_2 & & +x_4 & = & 28 \\
 & & & x_1, x_2, x_3, x_4 & \geq & 0
\end{array}
$$

The solution to this problem is (see point P_1 in Figure 7.5)

$$z^* = 391.11$$

and we have

$$\tilde{\mathbf{b}} = \mathbf{x}_B^* = \begin{pmatrix} x_1^* \\ x_2^* \end{pmatrix} = \begin{pmatrix} \frac{20}{9} \\ \\ \frac{14}{9} \end{pmatrix}, \tag{7.6}$$

$$\mathbf{U}^* = \begin{pmatrix} \frac{8}{9} & -\frac{1}{9} \\ -\frac{7}{9} & \frac{2}{9} \end{pmatrix}. \tag{7.7}$$

If we use x_2 to generate a cut, Equation (7.1) in this case becomes $x_2 - \frac{7}{9}x_3 + \frac{2}{9}x_4 = \frac{14}{9}$. Thus, we obtain:

$$\begin{array}{rclcrclcrcl} u_{21} & = & i_{21} + f_{21} & \Leftrightarrow & -\frac{7}{9} & = & -1 + \frac{2}{9} & \Rightarrow & f_{21} & = & \frac{2}{9} \\ u_{22} & = & i_{22} + f_{22} & \Leftrightarrow & \frac{2}{9} & = & 0 + \frac{2}{9} & \Rightarrow & f_{22} & = & \frac{2}{9} \\ \tilde{b}_2 & = & \tilde{i}_2 + \tilde{f}_2 & \Leftrightarrow & \frac{14}{9} & = & 1 + \frac{5}{9} & \Rightarrow & \tilde{f}_2 & = & \frac{5}{9} \end{array}$$

and the (7.4) cut becomes $\frac{5}{9} - \begin{pmatrix} \frac{2}{9} & \frac{2}{9} \end{pmatrix} \begin{pmatrix} x_3 \\ x_4 \end{pmatrix} \leq 0$ or $x_3 + x_4 \geq \frac{5}{2}$.

It should be noted that the cut can be expressed as a function of the original variables x_1 and x_2, as follows. Using the equality constraints of the relaxed problem in standard form, x_3 and x_4 are expressed as a function of x_1 y x_2:

$$\begin{array}{rcl} x_3 & = & 6 - 2x_1 - x_2, \\ x_4 & = & 28 - 7x_1 - 8x_2 \end{array}$$

to get the first cut (see dashed line in Figure 7.5): $x_1 + x_2 \leq \frac{7}{2}$. If this constraint is represented in $\mathbb{R}^2$ over the feasible region of the original problem (expressed in canonical form), an effective reduction of the feasible region is observed. However, no integral solution is excluded.

The next problem to solve is to maximize

$$Z = 120x_1 + 80x_2$$

subject to

$$\begin{array}{rrrrrcl} 2x_1 & +x_2 & +x_3 & & & = & 6 \\ 7x_1 & +8x_2 & & +x_4 & & = & 28 \\ & & x_3 & +x_4 & -x_5 & = & \dfrac{5}{2} \\ & & \multicolumn{3}{r}{x_1, x_2, x_3, x_4, x_5} & \geq & 0 \end{array}$$

Note that the Gomory cut has been incorporated using an additional variable x_5.

Its solution is (see point P_2 in Figure 7.5)

$$Z^* = 380; \;\; \mathbf{x}_B^* = \begin{pmatrix} x_1^* \\ x_4^* \\ x_2^* \end{pmatrix} = \begin{pmatrix} \frac{5}{2} \\ \frac{5}{2} \\ 1 \end{pmatrix}; \;\; \mathbf{U}^* = \begin{pmatrix} 1 & -\frac{1}{9} \\ 1 & -1 \\ -1 & \frac{2}{9} \end{pmatrix}$$

Using x_1 to generate a new cut, since $\tilde{b}_1 = x_1^* = \frac{5}{2}$, Equation (7.1) becomes $x_1 + x_3 - \frac{1}{9}x_5 = \frac{5}{2}$.

Thus, we can write

$$\begin{array}{rclcrclcrcl} u_{11} & = & i_{11} + f_{11} & \Leftrightarrow & 1 & = & 1+0 & \Rightarrow & f_{11} & = & 0 \\ u_{12} & = & i_{12} + f_{12} & \Leftrightarrow & -\frac{1}{9} & = & -1+\frac{8}{9} & \Rightarrow & f_{21} & = & \frac{8}{9} \\ \tilde{b}_1 & = & \tilde{i}_1 + \tilde{f}_1 & \Leftrightarrow & \frac{5}{2} & = & 2+\frac{1}{2} & \Rightarrow & \tilde{f}_1 & = & \frac{1}{2} \end{array}$$

and the (7.4) cut becomes $\frac{1}{2} - \left(0 \quad \frac{8}{9}\right)\begin{pmatrix} x_3 \\ x_5 \end{pmatrix} \le 0$ or $x_5 \ge \frac{9}{16}$

This second cut as a function of the original variables x_1 y x_2 has the form (see dashed line in Figure 7.5): $x_1 + x_2 \le \frac{55}{16}$.

The next problem to solve is maximization of

$$Z = 120x_1 + 80x_2$$

subject to

$$\begin{array}{rrrrrrcl} 2x_1 & +x_2 & +x_3 & & & & = & 6 \\ 7x_1 & +8x_2 & & +x_4 & & & = & 28 \\ & & x_3 & +x_4 & -x_5 & & = & \dfrac{5}{2} \\ & & & & x_5 & -x_6 & = & \dfrac{9}{16} \\ & & & & & x_1, x_2, x_3, x_4, x_5, x_6 & \ge & 0 \end{array}$$

The solution is (see point P_3 in Figure 7.5)

$$Z^* = 377.5; \quad \mathbf{x}_B^* = \begin{pmatrix} x_1^* \\ x_4^* \\ x_5^* \\ x_2^* \end{pmatrix} = \begin{pmatrix} \frac{41}{16} \\ \frac{49}{16} \\ \frac{9}{16} \\ \frac{7}{8} \end{pmatrix}; \quad \mathbf{U}^* = \begin{pmatrix} 1 & -\frac{1}{9} \\ 1 & -1 \\ 0 & -1 \\ -1 & \frac{2}{9} \end{pmatrix}$$

Using x_2 to generate a new cut, since $\tilde{b}_2 = \frac{7}{8}$, Equation (7.1) becomes $x_2 - x_3 + \frac{2}{9}x_6 = \frac{7}{8}$.

Thus, we can write

$$\begin{array}{rclcrclcrcl} u_{41} & = & i_{41} + f_{41} & \Leftrightarrow & -1 & = & -1+0 & \Rightarrow & f_{41} & = & 0 \\ u_{42} & = & i_{42} + f_{42} & \Leftrightarrow & \frac{2}{9} & = & 0+\frac{2}{9} & \Rightarrow & f_{42} & = & \frac{2}{9} \\ \tilde{b}_4 & = & \tilde{i}_4 + \tilde{f}_4 & \Leftrightarrow & \frac{7}{8} & = & 0+\frac{7}{8} & \Rightarrow & \tilde{f}_4 & = & \frac{7}{8} \end{array}$$

and the (7.4) cut becomes $\frac{7}{8} - (0 \quad \frac{2}{9}) \begin{pmatrix} x_3 \\ x_6 \end{pmatrix} \leq 0$ or $x_6 \geq \frac{63}{16}$.
As a function of the original variables the third cut becomes

$$x_1 + x_2 \leq 3$$

Next problem to solve is maximization of

$$Z = 120x_1 + 80x_2$$

subject to

$$\begin{array}{ccccccccc} 2x_1 & +x_2 & +x_3 & & & & & = & 6 \\ 7x_1 & +8x_2 & & +x_4 & & & & = & 28 \\ & & x_3 & +x_4 & -x_5 & & & = & \frac{5}{2} \\ & & & & x_5 & -x_6 & & = & \frac{9}{16} \\ & & & & & x_6 & -x_7 & = & \frac{63}{16} \\ & & & & & & x_1, x_2, x_3, x_4, x_5, x_6, x_7 & \geq & 0 \end{array}$$

The solution is (see point P_4 in Figure 7.5):

$$z^* = 360; \quad \mathbf{x}_B^* = \begin{pmatrix} x_3^* \\ x_4^* \\ x_5^* \\ x_6^* \\ x_1^* \end{pmatrix} = \begin{pmatrix} 0 \\ 7 \\ \frac{9}{2} \\ \frac{63}{16} \\ 3 \end{pmatrix}.$$

This solution satisfies integrality conditions and is therefore the minimizer of the original problem. Thus, the solution of the original problem is

$$x_1^* = 3; \quad x_2^* = 0; \quad z^* = 360$$

■

Exercises

7.1 A manufacturer transforms raw material into a certain product in two possible forms and has available 6 units of raw material and 28 hours of working time. The manufacture of 1 unit of type I product requires 2 units of raw material and 7 hours of work, and of type II product 1 unit of raw material and 8 hours of work. Selling prices for type I and II products are respectively \$120 and \$80. How many products of each type should the manufacturer produce to maximize his benefits? Solve the problem using the Branch and Bound and the Gomory Cuts algorithms.

7.2 A businessperson owns two electric lamp warehouses containing respectively 1200 and 1000 lamps. The businessperson serves 3 markets whose demands are respectively 100, 700, and 500 lamps. Transportation costs are given in the table below.

	Market 1	Market 2	Market 3
Warehouse 1	14	13	11
Warehouse 2	13	13	12

How many lamps should the businessperson send from each warehouse to each market so that his benefit is maximum? Solve the problem using the Branch–bound and the Gomory cuts algorithms.

7.3 Preventive maintenance of a production plant requires to carry out 4 successive tasks. The maintenance company in charge has available 6 workers able to perform the maintenance tasks. Times in minutes required by every worker to carry out each task are given in the table below.

	Task 1	Task 2	Task 3	Task 4
Worker 1	65	73	63	57
Worker 2	67	70	65	58
Worker 3	68	72	69	55
Worker 4	67	75	70	59
Worker 5	71	69	75	57
Worker 6	69	71	66	59

Because one worker can perform only one maintenance task, formulate an ILPP to determine which worker should perform each task so that total maintenance time is minimized. This problem is an instance of the assignment problem.

7.4 A specialized worker has to fix a small business facility located in a remote mountainous location. It is convenient for the worker to carry 5 different repair equipments. However, the maximum weight the worker can carry is 60 weight units. The weight and a measure of the relative usefulness of each equipment is provided in the table below.

Equipment	1	2	3	4	5
Weight units	52	23	35	15	7
Usefulness measure (%)	100	60	70	15	15

Which equipment should the worker take along? (*Hint*: This problem is an instance of the knapsack problem.)

7.5 An electricity producer should plan its hourly energy production to maximize its profits from selling energy during a planning horizon comprising 2 hours. Formulate and solve a mixed-integer linear programming problem to maximize the profits of the electricity producer, if

(a) The producer produces 5 energy units before the planning horizon.

(b) Hourly energy prices are 6 and 2 monetary units per energy unit.

(c) If running, the minimum energy production of the producer in each hour is 2 and the maximum one 10 energy units.

(d) Energy productions in two consecutive hours cannot differ in more than 4 energy units.

(e) Producer production cost is 3 monetary units per energy unit.

7.6 An electric energy consumer may self-produce energy or buy it from the electricity market. During the next 2 hours its demands are 8 and 10 energy units, respectively. Energy prices in the market are 6 and 8 monetary units per unit of energy. Its self-production facility, if working, produces between 2 and 6 energy units, cannot change its production between 2 consecutive hours more than 4 energy units, and its linear running cost is 7 monetary units per unit of energy produced. It is actually producing 3 energy units before the considered 2 hours study horizon. Determine how much energy the consumer should self-produce (and how much should buy from the market) in each of the 2 hours considered.

Chapter 8

Optimality and Duality in Nonlinear Programming

Linear programming is a major contribution to the field of scientific decision-making. The versatility and adaptability of this model allows generating applications in almost every field of engineering and science. However, certain classes of applications require taking into account the nonlinear aspects of the problems. Thus, researchers started to develop theoretical, as well as computational, foundations for general nonlinear programming problems, which generated new mathematical models of complex systems. In this chapter we describe and deal with many generalizations of the linear model that take us into the field of nonlinear programming, or into the more inclusive field of *mathematical programming*. In Section 8.1 we describe the problem and introduce some important concepts. In Section 8.2 necessary optimality conditions are given, and the important Karush–Kuhn–Tucker optimality conditions are described and discussed. Section 8.3 is devoted to sufficient conditions, and the case of convex functions is analyzed. Section 8.4 deals with duality theory and the special case of convex programs. Finally, in Section 8.6 we discuss some constraint qualifications.

8.1 Introduction

The general problem of mathematical programming, also referred to as the *nonlinear programming problem* (NLPP), can be stated as follows. Minimize

$$Z = f(x_1, \ldots, x_n)$$

subject to

$$\begin{array}{rcl} h_1(x_1, \ldots, x_n) & = & 0 \\ \vdots & \vdots & \vdots \\ h_\ell(x_1, \ldots, x_n) & = & 0 \\ g_1(x_1, \ldots, x_n) & \leq & 0 \\ \vdots & \vdots & \vdots \\ g_m(x_1, \ldots, x_n) & \leq & 0 \end{array}$$

In compact form the previous model can be stated as follows. Minimize

$$Z = f(\mathbf{x})$$

subject to

$$\begin{array}{rcl} \mathbf{h}(\mathbf{x}) & = & \mathbf{0} \\ \mathbf{g}(\mathbf{x}) & \leq & \mathbf{0} \end{array}$$

where $\mathbf{x} = (x_1, \ldots, x_n)^T$ is the vector of the *decision variables*, $f : \mathbb{R}^n \to \mathbb{R}$ is the objective function, and $\mathbf{h} : \mathbb{R}^n \to \mathbb{R}^\ell$ and $\mathbf{g} : \mathbb{R}^n \to \mathbb{R}^m$, where $\mathbf{h}(\mathbf{x}) = (h_1(\mathbf{x}), \ldots, h_\ell(\mathbf{x}))^T$ and $\mathbf{g}(\mathbf{x}) = (g_1(\mathbf{x}), \ldots, g_m(\mathbf{x}))^T$, are the *equality* and the *inequality* constraints, respectively. For this problem to be nonlinear, at least one of the functions involved in its formulation must be nonlinear. Any vector $\mathbf{x} \in \mathbb{R}^n$ that satisfies the constraints is said to be a *feasible solution*, and the set of all feasible solutions is referred to as the *feasible region*.

We would like to point out that nonlinear programming problems are much more difficult to solve than their linear counterparts. These troubles arise even in the simplest case, when one deals with the minimization of a function of a single variable on $\mathbb{R}$, which can be formulated as minimization of

$$Z = f(x)$$

subject to

$$x \in \mathbb{R}.$$

The French mathematician Pierre de Fermat already dealt with this problem in his *Methodus ad disquirendam maximan et minimam* in the seventeenth century (see Boyer [14]). The problem was studied by means of the tangent lines. Figure 8.1(a) shows that the minimum of the problem is achieved in the set of points where the tangent line is horizontal, as it is studied in elementary Calculus. There are other types of relative minima that do not satisfy the previous condition [see Figure 8.1(b)].

However, if we would consider the problem of finding the minimum of the function (see Figure 8.2):

$$f(x) = (x+2)^{2/3} - (x-2)^{2/3}$$

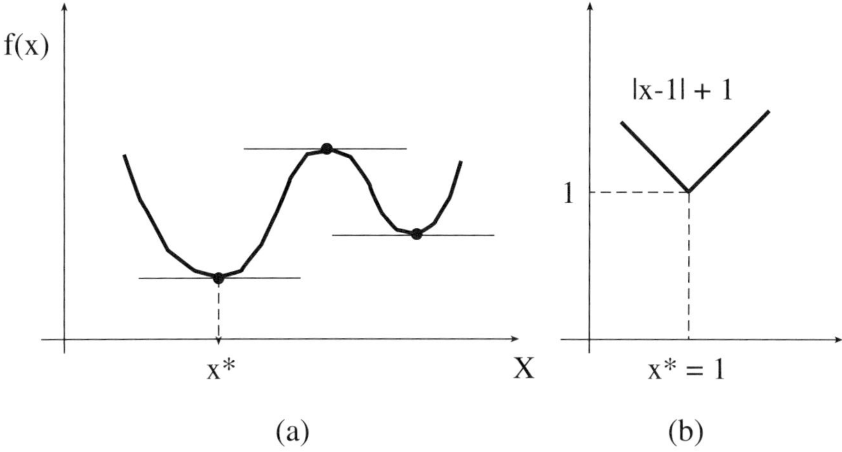

Figure 8.1: Global and local minima.

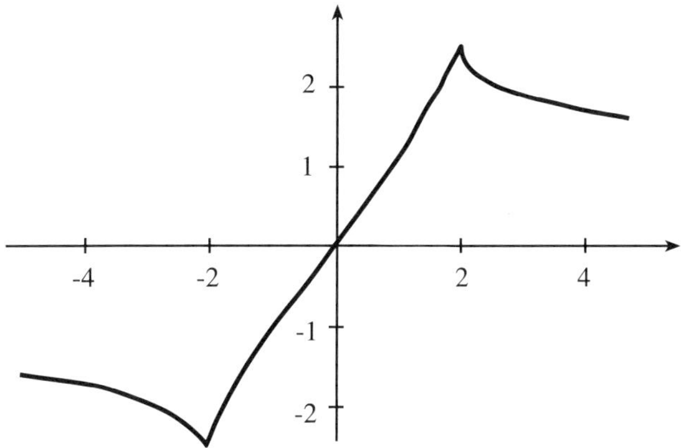

Figure 8.2: Graph of the function $f(x) = (x+2)^{2/3} - (x-2)^{2/3}$.

on $\mathbb{R}$, we would face a major difficulty. Applying the criterion of looking for points with vanishing derivative, we would find none. But, on the other hand, the fact that f tends to zero as x goes to $\pm\infty$ and that it takes on negative values indicates that f must attain its minimum somewhere. Where is the contradiction? A closer analysis of f reveals that the minimum is attained at $x = -2$, but f is not differentiable at this point. This simple example shows that extreme care must be taken when the involved functions are not differentiable.

Since this issue is beyond the scope of this book, we study only nonlinear programming problems in which all involved functions are assumed to be differentiable. We call the related optimization theory and computational methods *differentiable nonlinear optimization*. Generalizations of the concept of differentiability have been developed to deal with more general optimization problems.

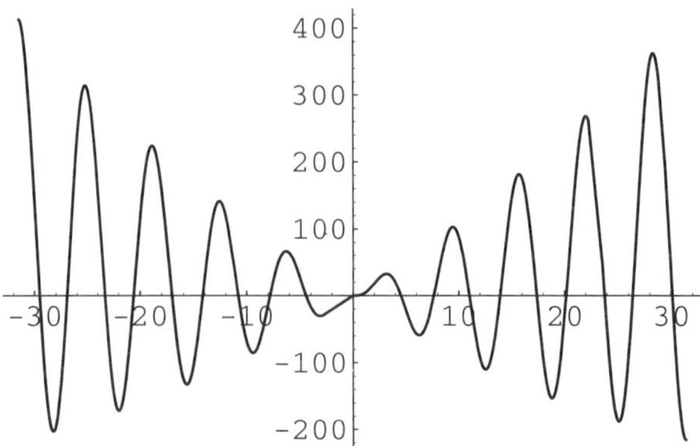

Figure 8.3: Graph of the function $f(x) = (1/10)x^2 + 10(\sin x - x \cos x)$.

At present, the most consistent theories are based on (subgradient) subdifferentials (see Rockafellar [95]) and the generalized gradient due to Clarke [26]. The interested readers can consult an introduction to these problems in Jahn [59] and Lemaréchal [65]. See also Shimizu ey al. [97], who deal with the theoretical and computational methods that make use of the subgradients and generalized gradients in the case of nondifferentiable programming problems.

There is a further, equally relevant, issue concerning differentiable nonlinear programming problems. To illustrate it, let us now look at the example (see Figure 8.3)

$$f(x) = \frac{1}{10}x^2 + 10(\sin x - x \cos x)$$

This function is everywhere differentiable, but it has an infinite set of points where the tangent is horizontal, that is, points where $f'(x) = 0$. These points are called stationary points and all of them except one are local optima, because if we restrict our attention to a small neighborhood of such points, they become either minima or maxima. The equation $f'(x) = 0$ cannot be solved in closed form, and then we must use numerical methods. The existence of an infinite set of candidates and the lack of a method to generate them explicitly leads to the impossibility of knowing, with full certainty, whether a given candidate is an absolute minimum.

One of the main behavioral differences between the linear and nonlinear programming problems is the existence of local minima. Feasible points where the objective function is optimized over a neighborhood of such points have to be considered in general nonlinear programs, although they may not optimize the function over the complete range of the feasible region. This difficulty motivates a new desirable constraint on the family of functions to be analyzed. We are interested in the class of functions such that their local minima are also global minima. In Figure 8.4 the graph of a function with a relative local minimum that is not an absolute minimum is shown. It can be observed that there are points in the interval $[\bar{x}, x^*]$ that are above the segment joining the relative and absolute

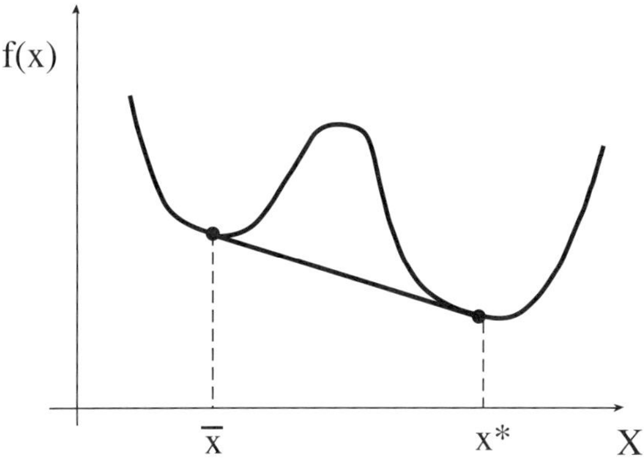

Figure 8.4: Illustration of the convexity property.

minima. Convexity appears as a sufficient condition to avoid this behavior. To be convex, the function is required to have its graph, in any interval, below the segment joining its extremes.

Finally, it is important to point out that in general a nonlinear programming problem (as can occur with its linear counterpart) could lack an optimum solution, essentially due to one of the following reasons:

1. The function is not bounded on the feasible region S. For example, the scalar function $f(x) = x$, where $x \in \mathbb{R}$ decreases without limit to $-\infty$ as x tends to $-\infty$. In this case we write

$$\text{Infimum}_{\ x \in S} \ f(x) = -\infty$$

2. The function is bounded on S, but the greatest lower bound is not achieved by the objective function on S. This value is denoted as $\text{infimum}_{x \in S} \ f(x)$. For example, the scalar function $f(x) = e^{-x}$ is bounded on $S = \mathbb{R}$ and the greatest lower bound is 0 but it is not achieved by $f(x)$.

In many cases it is important to know that there exists at least one solution of the optimization problem and we need sufficient conditions guaranteeing the existence of such a solution. An important condition is supplied by the Weierstrass theorem. The existence of at least one global minimum is guaranteed if f is a continuous function and S is a closed bounded set.

Theorem 8.1 (Weierstrass). *Let S be a closed bounded non empty set of $\mathbb{R}^n$ and $f : S \rightarrow \mathbb{R}$ a continuous function. The programming problem of minimizing*

$$Z = f(\mathbf{x})$$

subject to

$$\mathbf{x} \in S$$

admits at least one optimal solution. ∎

Often the set determined by the constraints in a mathematical programming problem is unbounded. In this case a corollary of the previous theorem can be applied to guarantee the existence of optimal solutions.

Corollary 8.1 (Existence of optimal solutions.) *Let S be a closed nonempty set (possibly unbounded) of $\mathbb{R}^n$ and $f : S \to \mathbb{R}$ a continuous function. If*

$$\lim_{\mathbf{x}\to\infty, \mathbf{x}\in S} f(\mathbf{x}) = +\infty$$

then the programming problem involving minimization of

$$Z = f(\mathbf{x})$$

subject to

$$\mathbf{x} \in S$$

admits at least one optimal solution.

These results can be even more explicit, as we shall see, when the objective function f is convex.

In the following section, we shall analyze the important necessary conditions for a constrained NLPP, known as the Karush–Kuhn–Tucker optimality conditions. These must be satisfied by all local minima of such programming problem. In particular, global minima should also be solutions of such optimality conditions. We shall try to motivate how these conditions arise, and defer to the end of the chapter a more formal treatment. Next, we shall be concerned with the assumptions under which these necessary conditions turn out to be sufficient conditions for global minima, and this will lead us to a discussion of convexity, as has been outlined above. Finally, duality will also be studied because of its great importance in numerical methods (Chapter 9).

8.2 Necessary Optimality Conditions

8.2.1 Differentiability

In the previous subsection we have mentioned the concepts of *differentiability* and *convexity* for unconstrained optimization problems. In this and the following sections we shall formalize them for the case of general optimization problems, not necessarily unconstrained.

Differentiability is a property that permits characterizing local extremes (either minima or maxima), thus providing necessary conditions for optimality. Since we shall focus on minima, for the sake of clarity we shall give here precise definitions for this type of point.

Definition 8.1 (Global minimum). *A function $f(\mathbf{x})$ has a global minimum (respectively, a strict global minimum) at the point $\mathbf{x}^*$, of a set of points S, if and only if $f(\mathbf{x}^*) \le f(\mathbf{x})$ [respectively, $f(\mathbf{x}^*) < f(\mathbf{x})$] for all $\mathbf{x}$ in S.* ∎

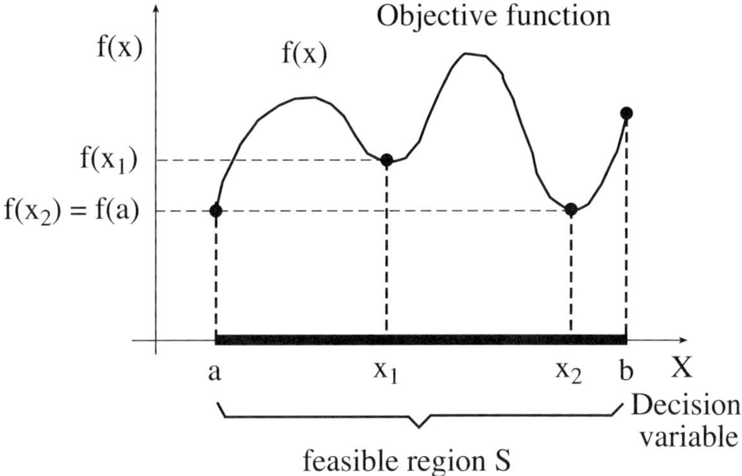

Figure 8.5: A function with three local minima and two global minima.

Definition 8.2 (Local minimum). *A function $f(\mathbf{x})$ has a local minimum (respectively, a strict local minimum) at the point $\bar{\mathbf{x}}$, of a set of points S, if and only if there exists a positive number ε such that $f(\bar{\mathbf{x}}) \le f(\mathbf{x})$ [respectively, $f(\bar{\mathbf{x}}) < f(\mathbf{x})$] for all $\mathbf{x}$ in S such that $0 < \|\bar{\mathbf{x}} - \mathbf{x}\| < \varepsilon$.* ■

From these definitions we conclude that a global minimum is also a local minimum.

In one dimension, it is easy to illustrate the concept of local and global minima for a function $f(\mathbf{x})$ (see Figure 8.5). In that figure, S is the segment $[a, b]$. $\bar{S} = \{a, x_1, x_2\}$ is the set of local minima, and $S^* = \{a, x_2\}$ is the set of global minima.

We also remind the reader of the concept of differentiability.

Definition 8.3 (Differentiability). *We say that $f : \mathbb{R}^n \to \mathbb{R}$ is differentiable at $\mathbf{x}$ if the partial derivatives $\partial f / \partial x_i$, $i = 1, \ldots, n$, exist, and*

$$\lim_{\mathbf{y} \to \mathbf{x}} \frac{f(\mathbf{y}) - f(\mathbf{x}) - \nabla f(\mathbf{x})^T (\mathbf{y} - \mathbf{x})}{\|\mathbf{y} - \mathbf{x}\|} = 0$$

■

Remember that the gradient of f at $\mathbf{x}$ is the (column) vector defined as

$$\nabla f(\mathbf{x}) = \left(\frac{\partial f(\mathbf{x})}{\partial x_1}, \ldots, \frac{\partial f(\mathbf{x})}{\partial x_n} \right)^T$$

It is also well known from elementary calculus that the gradient is a vector emanating from $\mathbf{x}$ that is orthogonal to the contour surface [the set of points such that $f(\mathbf{x}) = k$ for a constant k] passing through $\mathbf{x}$, and it points in the direction of maximum increase of f.

Definition 8.4 (Continuously differentiable function). *A function f is said to be continuously differentiable at $\bar{\mathbf{x}}$ if all partial derivatives are continuous at $\bar{\mathbf{x}}$. In this case the function is also differentiable.* ■

Throughout this chapter we consider the following NLPP. Minimize

$$Z = f(\mathbf{x}) \tag{8.1}$$

subject to

$$\begin{array}{rcl} \mathbf{h}(\mathbf{x}) & = & \mathbf{0} \\ \mathbf{g}(\mathbf{x}) & \leq & \mathbf{0} \end{array} \tag{8.2}$$

where $f : \mathbb{R}^n \to \mathbb{R}, \mathbf{h}: \mathbb{R}^n \to \mathbb{R}^\ell, \mathbf{g}: \mathbb{R}^n \to \mathbb{R}^m$ with $\mathbf{h}(\mathbf{x}) = (h_1(\mathbf{x}), \ldots, h_\ell(\mathbf{x}))^T$ and $\mathbf{g}(\mathbf{x}) = (g_1(\mathbf{x}), \ldots, g_m(\mathbf{x}))^T$ are continuously differentiable functions over the feasible region $S = \{\mathbf{x} | \mathbf{h}(\mathbf{x}) = \mathbf{0}, \mathbf{g}(\mathbf{x}) \leq \mathbf{0}\}$.

Problem (8.1) without constraints ($S = \mathbb{R}^n$) is referred to as an *unconstrained optimization problem.* When the constraints (8.2) are present, the problem is called a *constrained optimization problem.*

8.2.2 Karush–Kuhn–Tucker Optimality Conditions

The most important theoretical results in the field of nonlinear programming are the conditions of Karush, Kuhn, and Tucker. They must be satisfied at any constrained optimum, local or global, of any linear and most nonlinear programming problems. They form the basis for the development of many computational algorithms. In addition, the criteria for stopping many algorithms, specifically, for recognizing when a local constrained optimum has been achieved, are derived directly from them.

In unconstrained differentiable problems the gradient is equal to zero at the local minima. In constrained differentiable problems the gradient is not necessarily equal to zero, as is illustrated in Figure 8.6 for the point $\bar{\mathbf{x}} = \mathbf{a}$. This is due to the constraints of the problem. Karush–Kuhn–Tucker conditions generalize the necessary conditions for unconstrained problems to constrained problems.

Definition 8.5 (Karush–Kuhn–Tucker Conditions (KKTC)). *The vector $\bar{\mathbf{x}} \in \mathbb{R}^n$ satisfies the KKTCs for the NLPP (8.1)–(8.2) if there exists a pair of vectors $\boldsymbol{\mu} \in \mathbb{R}^m$ and $\boldsymbol{\lambda} \in \mathbb{R}^\ell$ such that*

$$\nabla f(\bar{\mathbf{x}}) + \sum_{k=1}^{\ell} \lambda_k \nabla h_k(\bar{\mathbf{x}}) + \sum_{j=1}^{m} \mu_j \nabla g_j(\bar{\mathbf{x}}) = \mathbf{0} \tag{8.3}$$

$$h_k(\bar{\mathbf{x}}) = 0, k = 1, \ldots, \ell \tag{8.4}$$

$$g_j(\bar{\mathbf{x}}) \leq 0, j = 1, \ldots, m \tag{8.5}$$

$$\mu_j g_j(\bar{\mathbf{x}}) = 0, j = 1, \ldots, m \tag{8.6}$$

$$\mu_j \geq 0, j = 1, \ldots, m \tag{8.7}$$

■

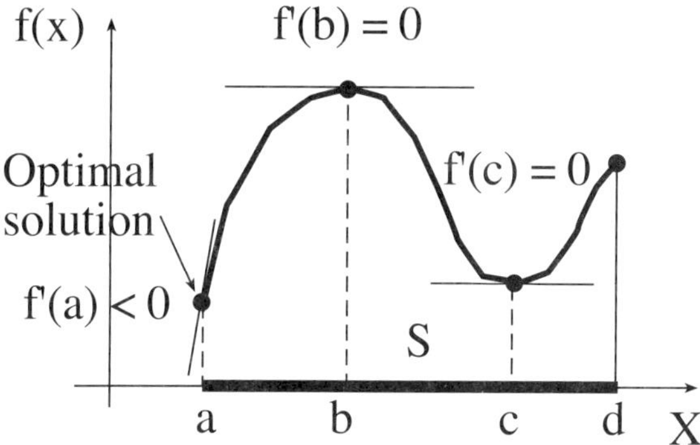

Figure 8.6: In constrained differentiable problems the gradient is not necessarily equal to zero at the optimal point.

The vectors $\boldsymbol{\mu}$ and $\boldsymbol{\lambda}$ are called the *Kuhn–Tucker multipliers*. Condition (8.6) is known as the *complementary slackness condition*, condition (8.6) requires the nonnegativity of the multipliers of the inequality constraints, and is referred to as the *dual feasibility conditions*, and (8.5)–(8.4) are called *the primal feasibility* conditions.

The genesis of these first-order optimality conditions (KKTC) can be motivated in the case of two independent variables for the case of one equality or inequality constraint and two inequality constraints, as shown in Figures 8.7, 8.8, and 8.9.

Consider the case of one equality constraint (see Figure 8.7). Satisfying the constraint is equivalent to moving along the curve that represents the actual constraint. When moving along this curve, the objective function contour curves must be cut in such a way that the objective function value decreases. The projection of the constraint gradient at a given point on the objective function gradient is negative. This fact indicates that the movement is advantageous when attempting to decrease the objective function value. As the movement progresses, the projection value increases until a maximum is reached. This maximum corresponds to a point where the objective function contour curve and the constraint curve are tangent, and hence, the gradients of the objective function and the constraint are parallel (linearly dependent). If the movement should progress further, the objective function values would increase, and this possibility is discarded from the perspective of minimizing the objective function. As a result, the (local) minimum coincides with a point where the gradients of the objective function and the constraint are linearly dependent. This is what the first order optimality conditions state (see Figure 8.7).

Now consider the case of an inequality constraint (see Figure 8.8), which separates the plane $\mathbb{R}^2$ into two regions. In one of them the constraint is satisfied, and in the other it is not. Feasible points are those in the feasible region including its border curve. If the minimum of the objective function

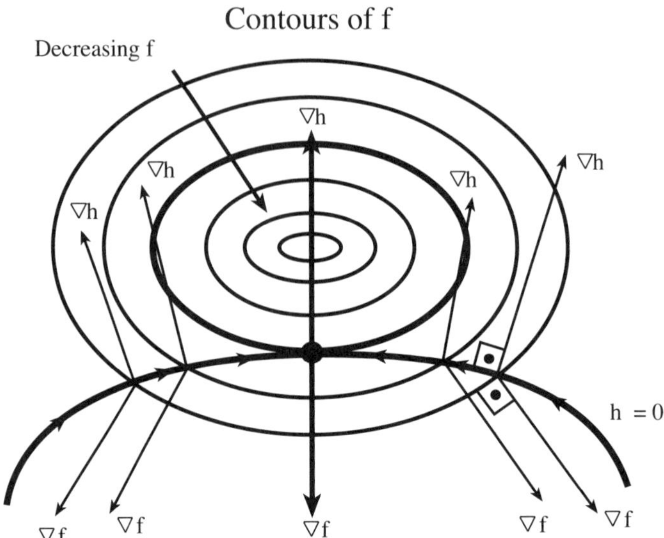

Figure 8.7: Illustration of Karush–Kuhn–Tucker conditions for the case of one equality constraint in the bidimensional case.

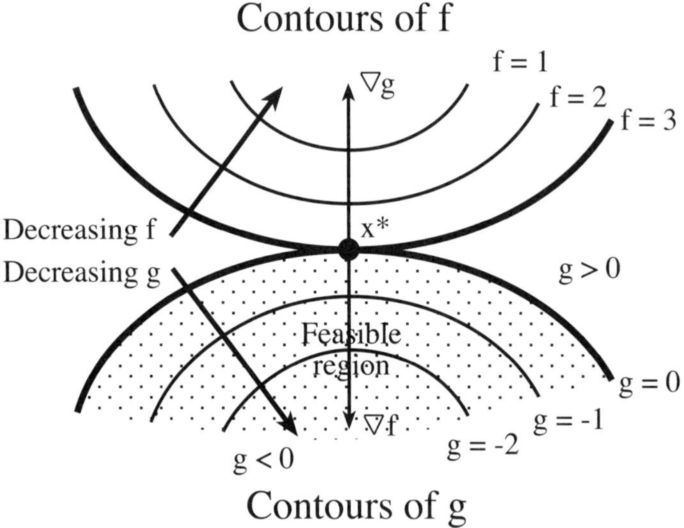

Figure 8.8: Illustration of Karush–Kuhn–Tucker conditions for the case of one inequality constraint in the bidimensional case.

is attained at the interior of the feasible region, the constraint is not binding and the associated multiplier vanishes (see Figure 8.9). If, on the contrary, the minimum is attained at the boundary, the constraint is binding. The problem

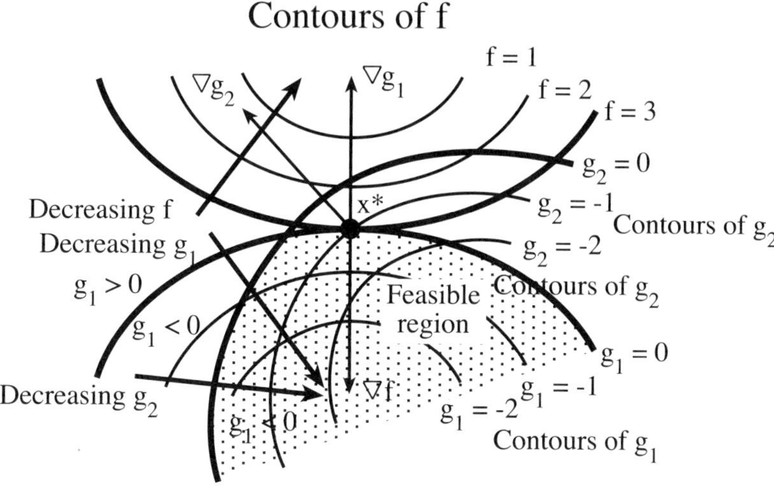

Figure 8.9: Illustration of Karush–Kuhn–Tucker conditions for the case of two inequality constraints in the bidimensional case.

is then equivalent to one with the corresponding equality constraint, and per the discussion in the previous paragraph, at the minimum the gradients of the objective function and the constraint must be parallel. They must be pointing in opposite directions because the objective function increases when we move toward the interior of the feasible region, while the constraint function becomes negative and thus it decreases. The multiplier is then either positive or zero. This is what optimality conditions say in this situation.

Remark 8.1 (Special cases.) *If a type of constraint is lacking in a NLPP then the multiplier associated with the "absent" constraint is equal to zero, and the constraint is dropped from the formulation of the KKTCs. The form of the KKTCs for these cases are:*

1. **(Unconstrained problems.)** *In this case we have only the condition*

$$\nabla f(\bar{\mathbf{x}}) = \mathbf{0}.$$

2. **(Problems with only equality constraints).** *The KKTCs are an extension of the classical principle of the multipliers method. This method appear when NLPP have only equality constraints, and the KKTCs has the form*

$$\begin{aligned} \nabla f(\bar{\mathbf{x}}) + \sum_{k=1}^{\ell} \lambda_k \nabla h_k(\bar{\mathbf{x}}) &= \mathbf{0} \\ h_k(\bar{\mathbf{x}}) &= 0, \ k = 1, \dots, \ell \end{aligned} \tag{8.8}$$

3. **(Problems with only inequality constraints)**. *KKTC conditions read*

$$\begin{aligned}
\nabla f(\bar{\mathbf{x}}) + \sum_{j=1}^{m} \mu_j \nabla g_j(\bar{\mathbf{x}}) &= \mathbf{0} \\
g_j(\bar{\mathbf{x}}) &\leq 0, \; j = 1, \ldots, m \\
\mu_j g_j(\bar{\mathbf{x}}) &= 0, \; j = 1, \ldots, m \\
\mu_j &\geq 0, \; j = 1, \ldots, m
\end{aligned} \tag{8.9}$$

■

If we define the Lagrangian by

$$\mathcal{L}(\mathbf{x}, \boldsymbol{\mu}, \boldsymbol{\lambda}) = f(\mathbf{x}) + \boldsymbol{\lambda}^T \mathbf{h}(\mathbf{x}) + \boldsymbol{\mu}^T \mathbf{g}(\mathbf{x})$$

we can write the KKTCs as

$$\begin{aligned}
\nabla_{\mathbf{x}} \mathcal{L}(\bar{\mathbf{x}}, \boldsymbol{\mu}, \boldsymbol{\lambda}) &= \mathbf{0} \\
\nabla_{\boldsymbol{\lambda}} \mathcal{L}(\bar{\mathbf{x}}, \boldsymbol{\mu}, \boldsymbol{\lambda}) &= \mathbf{0} \\
\nabla_{\boldsymbol{\mu}} \mathcal{L}(\bar{\mathbf{x}}, \boldsymbol{\mu}, \boldsymbol{\lambda}) &\leq \mathbf{0} \\
\boldsymbol{\mu}^T \nabla_{\boldsymbol{\mu}} \mathcal{L}(\bar{\mathbf{x}}, \boldsymbol{\mu}, \boldsymbol{\lambda}) &= 0 \\
\boldsymbol{\mu} &\geq \mathbf{0}
\end{aligned}$$

Note that $\boldsymbol{\mu}^T \nabla_{\boldsymbol{\mu}} \mathcal{L}(\bar{\mathbf{x}}, \boldsymbol{\mu}, \boldsymbol{\lambda}) = 0$ is equivalent to $\mu_j g_j(\bar{\mathbf{x}}) = 0, j = 1, \ldots, m$ only because $\nabla_{\boldsymbol{\mu}} \mathcal{L}(\bar{\mathbf{x}}, \boldsymbol{\mu}, \boldsymbol{\lambda}) \leq 0$ and $\mu \geq \mathbf{0}$.

The Special Case of Equality Constraints

Consider the following problem. Minimize

$$Z = f(\mathbf{x}) \tag{8.10}$$

subject to

$$\mathbf{h}(\mathbf{x}) = \mathbf{0} \tag{8.11}$$

The KKTCs for this problem with only equality constraints constitute the system of nonlinear equations

$$\begin{aligned}
\nabla_{\mathbf{x}} \mathcal{L}(\overline{\mathbf{x}}, \overline{\boldsymbol{\lambda}}) &= \mathbf{0} \\
\mathbf{h}(\overline{\mathbf{x}}) &= \mathbf{0}
\end{aligned} \tag{8.12}$$

where $\mathcal{L}(\mathbf{x}, \boldsymbol{\lambda}) = f(\mathbf{x}) + \boldsymbol{\lambda}^T \mathbf{h}(\mathbf{x})$.

The system of $n + \ell$ nonlinear equations presented above can be solved by the Newton algorithm. If $\mathbf{z}$ denotes $(\mathbf{x}, \boldsymbol{\lambda})$ and $\mathbf{F}(\mathbf{z})$ denotes system (8.12), the Taylor expansion of this system is

$$\mathbf{F}(\mathbf{z} + \boldsymbol{\Delta}\mathbf{z}) \approx \mathbf{F}(\mathbf{z}) + \nabla_{\mathbf{z}} \mathbf{F}(\mathbf{z}) \, \boldsymbol{\Delta}\mathbf{z}$$

for $||\mathbf{\Delta z}||$ sufficiently small.

To achieve $\mathbf{F}(\bar{\mathbf{z}}) = \mathbf{0}$, it is convenient to find a direction $\mathbf{\Delta z}$ so that $\mathbf{F}(\mathbf{z} + \mathbf{\Delta z}) = \mathbf{0}$. This direction is computed from

$$\nabla_{\mathbf{z}}\mathbf{F}(\mathbf{z})\mathbf{\Delta z} = -\mathbf{F}(\mathbf{z})$$

where $\nabla_{\mathbf{z}}\mathbf{F}(\mathbf{z})$ can be expressed as

$$\nabla_{\mathbf{z}}\mathbf{F}(\mathbf{z}) = \nabla_{(\mathbf{x},\boldsymbol{\lambda})}\mathbf{F}(\mathbf{x},\boldsymbol{\lambda}) = \begin{pmatrix} \nabla_{\mathbf{xx}}\mathcal{L}(\mathbf{x},\boldsymbol{\lambda}) & \boldsymbol{\nabla}_{\mathbf{x}}^T\mathbf{h}(\mathbf{x}) \\ \boldsymbol{\nabla}_{\mathbf{x}}\mathbf{h}(\mathbf{x}) & \mathbf{0} \end{pmatrix}$$

and where $\boldsymbol{\nabla}_{\mathbf{x}}\mathbf{h}(\mathbf{x})$ is the Jacobian of $\mathbf{h}(\mathbf{x})$.

The matrix above is denominated the KKT matrix of problem (8.10)–(8.11), and the system

$$\begin{pmatrix} \nabla_{\mathbf{xx}}\mathcal{L}(\mathbf{x},\boldsymbol{\lambda}) & \boldsymbol{\nabla}_{\mathbf{x}}^T\mathbf{h}(\mathbf{x}) \\ \boldsymbol{\nabla}_{\mathbf{x}}\mathbf{h}(\mathbf{x}) & \mathbf{0} \end{pmatrix}\begin{pmatrix} \mathbf{\Delta x} \\ \boldsymbol{\Delta\lambda} \end{pmatrix} = -\begin{pmatrix} \boldsymbol{\nabla}_{\mathbf{x}}\mathcal{L}(\mathbf{x},\boldsymbol{\lambda}) \\ \mathbf{h}(\mathbf{x}) \end{pmatrix}$$

is denominated the KKT system of problem (8.10)–(8.11). It constitutes a Newton iteration to solve system (8.12).

Example 8.1 (KKTCs for equality constraints). Consider the optimization problem involving minimization of

$$x^2 + y^2$$

subject to

$$\begin{array}{rcl} xy & \geq & 4 \\ x & \geq & 0 \end{array} \tag{8.13}$$

The Lagrangian function is

$$\mathcal{L}(x, y; \mu, \lambda) = x^2 + y^2 + \mu_1(4 - xy) - \mu_2 x$$

and the KKTCs

$$\frac{\partial \mathcal{L}}{\partial x} = 2x - \mu_1 y - \mu_2 = 0 \tag{8.14}$$

$$\frac{\partial \mathcal{L}}{\partial y} = 2y - \mu_1 x = 0 \tag{8.15}$$

$$xy \geq 4 \tag{8.16}$$

$$x \geq 0 \tag{8.17}$$

$$\mu_1(4 - xy) = 0 \tag{8.18}$$

$$\mu_2 x = 0 \tag{8.19}$$

$$\mu_1 \geq 0 \tag{8.20}$$

$$\mu_2 \geq 0. \tag{8.21}$$

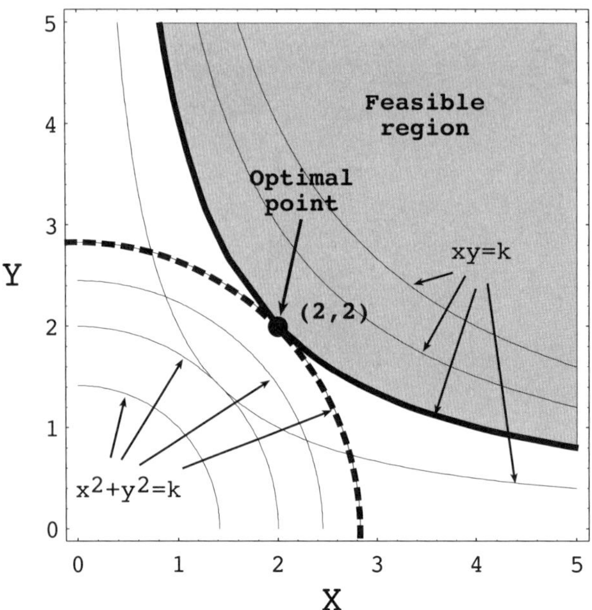

Figure 8.10: Illustration of the minimization problem in Example 8.1.

We have four cases:

Case 1. $\mu_1 = 0; \mu_2 = 0$. In this case, from (8.14) and (8.15) we get $x = y = 0$, which does not satisfy (8.16). Thus, the point $(0, 0)$ is not a KKT point.

Case 2. $\mu_1 \neq 0; \mu_2 = 0$. In this case, (8.14) and (8.15) give $x^2 = y^2$, i.e., $x = \pm y$, which together with (8.18) leads to $y = \pm 2$. Since (8.14) together with (8.17) and (8.20) imply that x and y must be both positive, we conclude that $(2, 2)$ is a KKT point.

Case 3. $\mu_1 = 0; \mu_2 \neq 0$. In this case, (8.15) leads to $y = 0$, and then (8.16) cannot hold.

Case 4. $\mu_1 \neq 0; \mu_2 \neq 0$. In this case, (8.19) leads to $x = 0$, and then (8.16) cannot hold.

The only KKT point $(2, 2)$ is the optimal point, as it is illustrated in Figure 8.10.

■

Example 8.2 (KKTC for the general case). Consider the following optimization problem. Minimize

$$x^2 + y^2$$

subject to

$$\begin{array}{rcl} xy & \geq & 4 \\ x + y & = & 5 \end{array} \tag{8.22}$$

The Lagrangian function is

$$\mathcal{L}(x, y; \mu, \lambda) = x^2 + y^2 + \mu(4 - xy) - \lambda(x + y - 5)$$

and the KKTCs are

$$\frac{\partial \mathcal{L}}{\partial x} = 2x - \mu y - \lambda = 0 \tag{8.23}$$

$$\frac{\partial \mathcal{L}}{\partial y} = 2y - \mu x - \lambda = 0 \tag{8.24}$$

$$x + y = 5 \tag{8.25}$$

$$xy \geq 4 \tag{8.26}$$

$$\mu(4 - xy) = 0 \tag{8.27}$$

$$\mu \geq 0. \tag{8.28}$$

(8.29)

We have two cases:

Case 1. $\mu = 0$. In this case, from (8.23), (8.24) and (8.25) we get $x = y = \frac{5}{2}$, which satisfies (8.26). Thus, the point $(\frac{5}{2}, \frac{5}{2})$ is a KKT point.

Case 2. $\mu \neq 0$. In this case, (8.25) and (8.27) give $x = 4, y = 1$ and $x = 1, y = 4$. However, these two solutions lead, from (8.23) and (8.24) to $\mu < 0$. Thus, we conclude that they are not KKT points.

The only KKT point $(\frac{5}{2}, \frac{5}{2})$ is the optimal point, as is illustrated in Figure 8.11.

■

Example 8.3 (KKTC for the general case). Consider the following optimization problem. Minimize

$$x^2 + y^2$$

subject to

$$\begin{array}{rcl} xy & \geq & 4 \\ x + y & = & 5 \\ (x - 4)^2 + (y - 2)^2 & \leq & 1 \end{array} \tag{8.30}$$

The Lagrangian function is

$$\mathcal{L}(x, y; \mu, \lambda) = x^2 + y^2 + \lambda(x + y - 5) + \mu_1(4 - xy) + \mu_2((x - 4)^2 + (y - 2)^2 - 1)$$

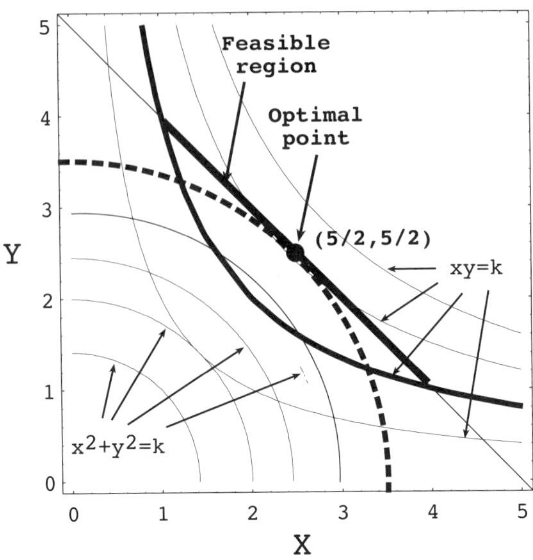

Figure 8.11: Illustration of the minimization problem in Example 8.2.

and the KKTC

$$\begin{aligned}
\frac{\partial \mathcal{L}}{\partial x} &= 2x + \lambda - \mu_1 y + 2\mu_2(x-4) = 0 && (8.31)\\
\frac{\partial \mathcal{L}}{\partial y} &= 2y + \lambda - \mu_1 x + 2\mu_2(y-2) = 0 && (8.32)\\
x + y &= 5 && (8.33)\\
xy &\geq 4 && (8.34)\\
(x-4)^2 + (y-2)^2 &\leq 1 && (8.35)\\
\mu_1(4 - xy) &= 0 && (8.36)\\
\mu_2((x-4)^2 + (y-2)^2 - 1) &= 0 && (8.37)\\
\mu_1 &\geq 0 && (8.38)\\
\mu_2 &\geq 0 && (8.39)
\end{aligned}$$

We have four cases:

Case 1. $\mu_1 = 0; \mu_2 = 0$. In this case, from (8.31)–(8.33) we get $x = y = \frac{5}{2}$, which does not satisfy (8.35). Thus, the point $(\frac{5}{2}, \frac{5}{2})$ is not a KKT point.

Case 2. $\mu_1 \neq 0; \mu_2 = 0$. In this case, (8.33) and (8.36) lead to the candidate points $(4,1)$ and $(1,4)$. However, $(1,4)$ does not satisfies (8.35), and $(4,1)$ with (8.31) and (8.32) leads to $\mu_1 = -2 < 0$, i.e., is not a KKT point.

Case 3. $\mu_1 = 0; \mu_2 \neq 0$. In this case, (8.33) and (8.35) lead to $(3,2)$ and $(4,1)$.

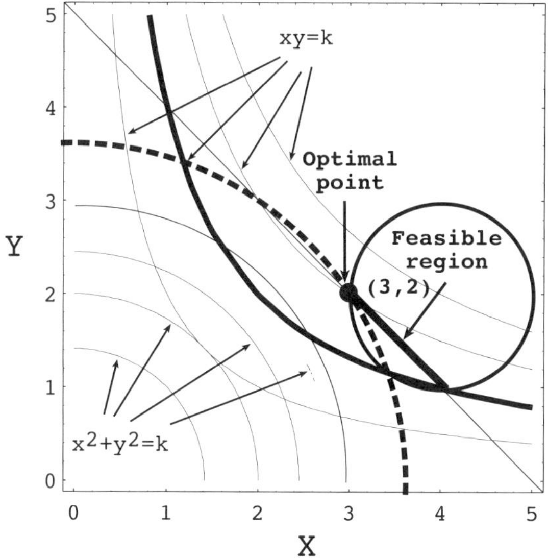

Figure 8.12: Illustration of the minimization problem in Example 8.3.

From (8.31) and (8.32) with $(4,1)$, we get $\mu_1 < 0$; thus it is not a KKT point. Similarly, with $(3,2)$ we get $\mu_1 = 1 > 0$; thus it is a KKT point.

Case 4. $\mu_1 \neq 0; \mu_2 \neq 0$. In this case, (8.33), (8.36), and (8.37) lead to the point $(4,1)$. Then, (8.31) and (8.32) lead to $\mu_1 = \lambda + 8$ and $\mu_2 = -15 - 3\lambda/2$, which cannot satisfy (8.38) and (8.39) simultaneously.

The only KKT point $(3,2)$ is the optimal point, as is illustrated in Figure 8.12.

■

Example 8.4 (KKTCs). In this example we compute the points that satisfy the KKTCs of the following problem. Minimize

$$Z = -x_1 + x_2$$

subject to

$$\begin{aligned} -x_1^2 + x_2 &= 0 \\ x_1^2 + x_2^2 - 4 &\leq 0 \\ -x_1 &\leq 0 \\ -x_2 &\leq 0 \end{aligned}$$

We must introduce one multiplier for each constraint. We denote as μ_1, μ_2, and μ_3 the multipliers for the inequality constraints, and as λ the multiplier for the equality constraint.

The KKTC can be summarized as follows:

1. The stationary condition of the Lagrangian is stated as

$$\begin{pmatrix} -1 \\ 1 \end{pmatrix} + \lambda \begin{pmatrix} -2x_1 \\ 1 \end{pmatrix} + \mu_1 \begin{pmatrix} 2x_1 \\ 2x_2 \end{pmatrix} + \mu_2 \begin{pmatrix} -1 \\ 0 \end{pmatrix} + \mu_3 \begin{pmatrix} 0 \\ -1 \end{pmatrix} = \begin{pmatrix} 0 \\ 0 \end{pmatrix} \tag{8.40}$$

2. The primal feasibility conditions are

$$\begin{aligned} x_1^2 + x_2^2 - 4 &\leq 0 \\ -x_1 &\leq 0 \\ -x_2 &\leq 0 \\ -x_1^2 + x_2 &= 0 \end{aligned} \tag{8.41}$$

3. The slackness conditions are

$$\mu_1(x_1^2 + x_2^2 - 4) = 0 \tag{8.42}$$

$$\mu_2(-x_1) = 0 \tag{8.43}$$

$$\mu_3(-x_2) = 0 \tag{8.44}$$

4. The dual feasibility conditions are

$$\mu_1, \ \mu_2, \ \mu_3 \geq 0 \tag{8.45}$$

To solve this system of equalities and inequalities, we consider the following cases (full enumeration of possibilities):

Case 1. $\mu_2 \neq 0$. If $\mu_2 \neq 0$, then using (8.43), $x_1 = 0$, and (8.40) implies

$$\begin{aligned} -1 - \mu_2 &= 0 \\ 1 + 2x_2\mu_1 - \mu_3 + \lambda &= 0 \end{aligned}$$

Since $\mu_2 = -1$ and the non negativity condition (8.45) does not hold, then any KKT points must satisfy $\mu_2 = 0$.

Case 2. $\mu_3 \neq 0$, and $\mu_2 = 0$. If $\mu_3 \neq 0$ then using (8.44), $x_2 = 0$, and using the relationship $-x_1^2 + x_2 = 0$ of (8.41) we obtain $x_1 = 0$, and using (8.40), we obtain the contradiction

$$-1 = 0$$

This shows that any KKT point must have $\mu_3 = 0$.

Case 3. $\mu_1 \neq 0$, and $\mu_2 = \mu_3 = 0$. If $\mu_1 \neq 0$, then using (8.42), we obtain $x_1^2 + x_2^2 - 4 = 0$, and using the feasibility condition, we form the following system of equations

$$\begin{aligned} x_1^2 + x_2^2 - 4 &= 0 \\ -x_1^2 + x_2 &= 0 \end{aligned} \tag{8.46}$$

The only solution satisfying condition (8.46) is $\bar{x} = (\sqrt{\delta}, \delta)$, where $\delta = \dfrac{(-1+\sqrt{17})}{2}$ and, using (8.40), we get

$$\begin{aligned} -1 + 2\sqrt{\delta}\mu_1 - 2\sqrt{\delta}\lambda &= 0 \\ 1 + 2\delta\mu_1 + \lambda &= 0 \end{aligned}$$

with solution

$$\begin{aligned} \mu_1 &= \frac{2\delta - \sqrt{\delta}}{2\delta(1+2\delta)} > 0 \\ \lambda &= \frac{-1}{2\sqrt{\delta}} + \frac{2\delta - \sqrt{\delta}}{2\delta(1+2\delta)} \end{aligned}$$

which is a KKT point.

Case 4. The last case is $\mu_1 = \mu_2 = \mu_3 = 0$. Using (8.40), we obtain the system of equations

$$\begin{aligned} -1 - 2\lambda x_1 &= 0 \\ 1 + \lambda &= 0 \end{aligned}$$

and we obtain the solution $\lambda = -1$ and $x_1 = 1/2$. Using now (8.41), we get $x_2 = x_1^2 = 1/4$. Since this point is feasible, it is also a KKT point.

■

Next, we shall justify why the KKTC are necessary for most NLPP. To this end, we need the following definition.

Definition 8.6 (Active constraint). *Let $\bar{\mathbf{x}} \in \mathbb{R}^n$, and let $j \in \{1, \ldots, m\}$. The inequality constraint $g_j(\mathbf{x}) \leq 0$ is said to be an active or binding constraint at the point $\bar{\mathbf{x}}$ if $g_j(\bar{\mathbf{x}}) = 0$; it is said to be inactive or nonbinding if $g_j(\bar{\mathbf{x}}) < 0$. We denote the set of indices of active constraints as $I(\bar{\mathbf{x}})$:*

$$I(\bar{\mathbf{x}}) = \{j | g_j(\bar{\mathbf{x}}) = 0\}$$

■

The main point in distinguishing between active and nonactive constraints is that a nonactive constraint at a point allows a whole neighborhood of it to satisfy the same constraint, while at an active constraint this may not, and typically will not, be true.

Lemma 8.1 *KKTCs are necessary for a local optimum of "most" NLPP.*

Proof. We adopt an intuitive approach at this point. We divide our analysis into two steps, examining first the case of equality constraints, and afterward the general case.

1. *Sketch of the proof of KKTC for only equality constraints.* Consider the following NLPP. Minimize

$$Z = f(\mathbf{x}) \tag{8.47}$$

subject to

$$\mathbf{h}(\mathbf{x}) \quad = \quad \mathbf{0} \tag{8.48}$$

where $f : \mathbb{R}^n \to \mathbb{R}$, $\mathbf{h}: \mathbb{R}^n \to \mathbb{R}^\ell$, $\mathbf{h}(\mathbf{x}) = (h_1(\mathbf{x}), \dots, h_\ell(\mathbf{x}))^T$ are continuously differentiable functions over the feasible region $S = \{\mathbf{x} | \mathbf{h}(\mathbf{x}) = \mathbf{0}\}$. Assume that $\bar{\mathbf{x}}$ is a local minimum of the problem. Let

$$\sigma : (-\epsilon, \epsilon) \to \mathbb{R}^n, \quad \epsilon > 0$$

be a curve entirely contained in the feasible region S such that

$$\sigma(0) = \bar{\mathbf{x}}, \quad \sigma'(0) = \mathbf{d}$$

Since σ is contained in S, we must have

$$\mathbf{h}(\sigma(t)) = \mathbf{0}$$

for all t in the interval $(-\epsilon, \epsilon)$. The derivative of this composition at 0 must vanish because $\mathbf{h}(\sigma(t))$ is constant, and by the chain rule

$$\nabla \mathbf{h}(\bar{\mathbf{x}})^T \mathbf{d} = 0$$

On the other hand, the composition $f(\sigma(t))$ must have a local minimum at $t = 0$, and consequently its derivative must vanish as well:

$$\nabla f(\bar{\mathbf{x}})^T \mathbf{d} = 0$$

Since $\mathbf{d} = \sigma'(0)$ runs through the tangent vectors to the feasible set S at $\bar{\mathbf{x}}$, the previous statements can be summarized by saying that

$$\nabla \mathbf{h}(\bar{\mathbf{x}})^T \mathbf{d} = 0$$

implies

$$\nabla f(\bar{\mathbf{x}})^T \mathbf{d} = 0$$

and this can happen only if $\nabla f(\bar{\mathbf{x}})$ belongs to the linear subspace spanned by the vectors $\nabla \mathbf{h}(\bar{\mathbf{x}})$. This dependence gives rise to the multipliers λ_k in (8.3).

2. *Sketch of the proof of KKTC for the general case.* Minimize

$$Z = f(\mathbf{x}) \tag{8.49}$$

subject to

$$\begin{array}{rcl} \mathbf{h}(\mathbf{x}) & = & \mathbf{0} \\ \mathbf{g}(\mathbf{x}) & \leq & \mathbf{0} \end{array} \tag{8.50}$$

where $f : \mathbb{R}^n \to \mathbb{R}$, $\mathbf{h}: \mathbb{R}^n \to \mathbb{R}^\ell$, $\mathbf{g}: \mathbb{R}^n \to \mathbb{R}^m$ with $\mathbf{h}(\mathbf{x}) = (h_1(\mathbf{x}), \ldots, h_\ell(\mathbf{x}))^T$ and $\mathbf{g}(\mathbf{x}) = (g_1(\mathbf{x}), \ldots, g_m(\mathbf{x}))^T$ are continuously differentiable functions over the feasible region $S = \{\mathbf{x}|\mathbf{h}(\mathbf{x}) = \mathbf{0}, \mathbf{g}(\mathbf{x}) \leq \mathbf{0}\}$. Our strategy is to analyze this situation by considering an auxiliary nonlinear programming problem associated with the particular local minimum we have, $\bar{\mathbf{x}}$.

Again let $\bar{\mathbf{x}}$ be a local minimum for our NLPP. Let $I(\bar{\mathbf{x}})$ be its corresponding set of active constraints. Consider the following auxiliary problem. Minimize

$$Z = f(\mathbf{x}) \tag{8.51}$$

subject to

$$\begin{array}{rcl} g_i(\mathbf{x}) & = & 0, \quad \forall i \in I(\bar{\mathbf{x}}) \\ \mathbf{h}(\mathbf{x}) & = & \mathbf{0} \end{array} \tag{8.52}$$

It is easy to argue that $\bar{\mathbf{x}}$ is also a local minimum for this new problem. The validity of this assertion is closely related to the set of active constraints in $\bar{\mathbf{x}}$. Therefore, by our previous step applied to this auxiliary problem, we conclude the existence of multipliers $\bar{\mu}$ and $\bar{\lambda}$ where $\mu_j = 0$ for j not belonging to $I(\bar{\mathbf{x}})$ such that

$$\nabla f(\bar{\mathbf{x}}) + \sum_{k=1}^{\ell} \lambda_k \nabla h_k(\bar{\mathbf{x}}) + \sum_{j=1}^{m} \mu_j \nabla g_j(\bar{\mathbf{x}}) \quad = \quad \mathbf{0} \tag{8.53}$$

The only remaining requirement to be shown is the nonnegativity of the multipliers μ_j for j in $I(\bar{\mathbf{x}})$. To this aim, we fix one such index s in $I(\bar{\mathbf{x}})$ and consider a second auxiliary nonlinear programming problem as follows. Minimize

$$Z = f(\mathbf{x}) \tag{8.54}$$

subject to

$$\begin{array}{rcl} \mathbf{h}(\mathbf{x}) & = & \mathbf{0} \\ g_i(\mathbf{x}) & = & 0, \quad \forall i \in I(\bar{\mathbf{x}}), i \neq s \\ g_s(\mathbf{x}) & < & 0 \end{array} \tag{8.55}$$

Choose a curve

$$\sigma : [0, \epsilon) \to \mathbb{R}^n$$

entirely contained in the feasible region S for this last problem, such that

$$\sigma(0) = \bar{\mathbf{x}}, \quad \sigma'(0) = \mathbf{d}, \quad \nabla g_s(\bar{\mathbf{x}})^T \mathbf{d} < 0$$

Notice that the last requirement is possible because the composition

$$g_s(\sigma(t)), \quad t \in [0, \epsilon)$$

has a maximum at $t = 0$, because $g_s(\sigma(0)) = g_s(\bar{\mathbf{x}}) = 0$ and $g_s(\sigma(h)) \leq 0, \forall h \neq 0$, and its derivative at this point is precisely

$$\nabla g_s(\bar{\mathbf{x}})^T \mathbf{d}$$

On the other hand, the composition $f(\sigma(t))$ has a minimum at $t = 0$ while

$$g_i(\sigma(t)), i \in I(\bar{\mathbf{x}}), i \neq s$$

are constant. Consequently

$$\nabla f(\bar{\mathbf{x}})^T \mathbf{d} \geq 0, \quad \nabla g_i(\bar{\mathbf{x}})^T \mathbf{d} = 0$$

If we perform the inner product of (8.53) by $\mathbf{d}$, many terms vanish and we obtain

$$\nabla f(\bar{\mathbf{x}})^T \mathbf{d} + \mu_s \nabla g_s(\bar{\mathbf{x}})^T \mathbf{d} = 0$$

The first term is nonnegative and the factor for μ_s is negative. If μ_s were also negative, the preceding equality would be impossible. Hence we must have

$$\mu_s \geq 0$$

By repeating this argument for each index $s \in I(\bar{\mathbf{x}})$, we have the conclusion on the sign of the multipliers. ■

There are some technicalities involved in the previous paragraphs, especially those related to the curves σ with specific properties.

Example 8.5 (Global optimum). In this example we compute the global optima (maximum and minimum) of the objective function

$$Z = x_1^3 + x_2^3 + x_3^3$$

over the feasible set determined by the two constraints

$$\begin{array}{rcl} x_1^2 + x_2^2 + x_3^2 - 4 & = & 0 \\ x_1 + x_2 + x_3 & \leq & 1 \end{array}$$

It is elementary to realize that this feasible set is closed and bounded so that any continuous objective function attains its maximum and minimum values on that set. The candidates to such extreme values are the solutions of the KKT conditions. Our first task consists in finding all such solutions.

In this particular case the optimality conditions become

$$\begin{array}{rcl} \begin{pmatrix} 3x_1^2 \\ 3x_2^2 \\ 3x_3^2 \end{pmatrix} + \mu \begin{pmatrix} 1 \\ 1 \\ 1 \end{pmatrix} + \lambda \begin{pmatrix} 2x_1 \\ 2x_2 \\ 2x_3 \end{pmatrix} & = & \begin{pmatrix} 0 \\ 0 \\ 0 \end{pmatrix} \\ x_1^2 + x_2^2 + x_3^2 - 4 & = & 0 \\ \mu(x_1 + x_2 + x_3 - 1) & = & 0 \\ x_1 + x_2 + x_3 & \leq & 1 \end{array}$$

where as usual μ is the multiplier for the inequality constraint and λ the one for the equality constraint. Notice that the admissible candidates for minimum will have $\mu \geq 0$ and those for maximum will correspond to $\mu \leq 0$.

To solve this system of equalities and inequalities, we consider the following cases (full enumeration of possibilities):

Case 1. $\mu = 0$. In this case the system to be solved simplifies to

$$\begin{array}{rcl} 3x_1^2 + \lambda 2x_1 & = & 0 \\ 3x_2^2 + \lambda 2x_2 & = & 0 \\ 3x_3^2 + \lambda 2x_3 & = & 0 \\ x_1^2 + x_2^2 + x_3^2 - 4 & = & 0 \\ x_1 + x_2 + x_3 & \leq & 1 \end{array}$$

Because of the invariance of this equation with respect to permutations of the variables, it suffices to study the following cases:

Case 1a. $x_1 = x_2 = x_3 = 0$. This is impossible because of the constraint

$$x_1^2 + x_2^2 + x_3^2 - 4 = 0$$

Case 1b. $x_1 = x_2 = 0$, $x_3 \neq 0$. In this case, the two constraints we must enforce lead to the only solution $(0, 0, -2)^T$ and the ones obtained by permutations $(0, -2, 0)^T$ and $(-2, 0, 0)^T$.

Case 1c. $x_1 = 0$, $x_2, x_3 \neq 0$. The equations on the multiplier λ imply $x_2 = x_3$ and then the two constraints lead to the only solution $(0, -\sqrt{2}, -\sqrt{2})^T$ and those obtained by permutations $(-\sqrt{2}, 0, -\sqrt{2})^T$, $(-\sqrt{2}, -\sqrt{2}, 0)^T$.

Case 1d. $x_1, x_2, x_3 \neq 0$. The equations on the multiplier λ lead to $x_1 = x_2 = x_3$, and the constraints yield the only candidate:

$$\left(-\tfrac{2}{\sqrt{3}}, -\tfrac{2}{\sqrt{3}}, -\tfrac{2}{\sqrt{3}}\right)^T.$$

In principle, all these points could correspond to the maximum and minimum since the multiplier μ vanishes.

Case 2. $\mu \neq 0$. This time the system has the form

$$\begin{array}{rcl} 3x_1^2 + \lambda 2x_1 & = & 0 \\ 3x_2^2 + \lambda 2x_2 & = & 0 \\ 3x_3^2 + \lambda 2x_3 & = & 0 \\ x_1^2 + x_2^2 + x_3^2 - 4 & = & 0 \\ x_1 + x_2 + x_3 & = & 1 \end{array}$$

Those solutions with $\mu > 0$ will be candidates to the minimum and those with $\mu < 0$ to the maximum. The first three equations express that the vectors $(3x_1^2, 3x_2^2, 3x_3^2)^T$, $(1, 1, 1)^T$, and $(2x_1, 2x_2, 2x_3)^T$ are linearly dependent, so that the determinant of these three vectors ought to vanish. This determinant is a Vandermonde determinant, whose value is, up to the sign, $(x_1 - x_2)(x_1 - x_3)(x_2 - x_3)$. Hence, by eliminating the multipliers in this way, we should solve the equivalent system

$$\begin{aligned} (x_1 - x_2)(x_1 - x_3)(x_2 - x_3) &= 0 \\ x_1^2 + x_2^2 + x_3^2 - 4 &= 0 \\ x_1 + x_2 + x_3 &= 1 \end{aligned}$$

Once again, because of the symmetry of the situation, it is enough to consider the particular case $x_1 = x_2$. Replacing this condition in the other two equations we get

$$\begin{aligned} x_3 + 2x_1 &= 1 \\ 2x_1^2 + (1 - 2x_1)^2 &= 4 \end{aligned}$$

From this, we obtain the two solutions

$$\frac{1}{6}(2 + \sqrt{22},\ \ 2 + \sqrt{22},\ \ 2 - 2\sqrt{22})^T$$

and

$$\frac{1}{6}(2 - \sqrt{22},\ \ 2 - \sqrt{22},\ \ 2 + 2\sqrt{22})^T$$

On the other hand, if we add the first three equations and keep in mind the constraints, we arrive at

$$12 + 3\mu + 2\lambda = 0$$

which provides the relationship between the two multipliers. If we solve for μ here and take this expression into the first equation we get

$$3x_1^2 + \mu - x_1(12 + 3\mu) = 0$$

and thus

$$\mu = \frac{3x_1(4 - x_1)}{1 - 3x_1}$$

Taking the value for x_1 we have previously obtained ($x_1 = \frac{(2-\sqrt{22})}{6}$), it is easy to check that in both cases $\mu < 0$ so that these two points for case II are candidates for the maximum.

Once the maximum and minimum candidates have been identified, it is a matter of computing the value of the objective function on each of them and deciding those extreme values. In this particular example, the minimum is attained at $(-2, 0, 0)^T, (0, -2, 0)^T, (0, 0, -2)^T$ and its value is -8, and the maximum is achieved at $\frac{1}{6}(2 + \sqrt{22},\ \ 2 + \sqrt{22},\ \ 2 - 2\sqrt{22})^T$, and its value is $\frac{68-11\sqrt{22}}{18}$. ∎

8.3 Optimality Conditions: Sufficiency and Convexity

8.3.1 Convexity

Because the differentiability is a local concept (it depends only on the value of the function on a neighborhood of the point), it is used to characterize a local minimum. It is not possible, however, to characterize a global minimum of a function based only on the differentiability of the function f. Thus, we must introduce a new concept, which depends on the behavior of the function over its domain of definition, that guarantees it. To facilitate the treatment we use *convex* functions. Then, we can demonstrate mathematically the coincidence of local and global minimum. Bazaraa et al. [9] have dealt with generalizations of convex functions and the relationships among them.

Definition 8.7 (Convex function). *Let $f : S \to \mathbb{R}$, where S is a nonempty convex set in $\mathbb{R}^n$. The function f is said to be convex on S if for any two points $\mathbf{x}^1$ and $\mathbf{x}^2$ and any scalar λ such that $0 \le \lambda \le 1$, we have*

$$f\left(\lambda \mathbf{x}^1 + (1-\lambda)\mathbf{x}^2\right) \le \lambda f(\mathbf{x}^1) + (1-\lambda) f(\mathbf{x}^2). \tag{8.56}$$

If strict inequality holds in (8.56), f is said to be strictly convex. Similarly, a function f is concave if (8.56) holds with the inequality reversed, that is, if $(-f)$ is convex. ■

Figure 8.13 shows three examples of convex, concave, and neither convex nor concave functions.

Before proceeding to better understand convexity we must convince ourselves that it is indeed a property that guarantees the nonexistence of local minima that are not global.

Theorem 8.2 (Global and local optima). *Consider the convex function $f : S \to \mathbb{R}$, where S is a nonempty convex set in $\mathbb{R}^n$. Every local minimum of f in S is also a global minimum of f in S. Moreover, if f is strictly convex, there is at most one single global minimum.* ■

Proof. Let $\bar{\mathbf{x}}$ be a local minimum of f in S, so that there exists $\epsilon > 0$ such that

$$f(\bar{\mathbf{x}}) \le f(\mathbf{x}), \quad \forall \mathbf{x} \in S, \|\mathbf{x} - \bar{\mathbf{x}}\| < \epsilon$$

For any $\mathbf{y} \in S$, since S is convex we have

$$\mathbf{z} = (1-\lambda)\bar{\mathbf{x}} + \lambda \mathbf{y} \in S, \quad \forall \lambda \in [0,1]$$

In particular, for λ sufficiently small, $\|\mathbf{z} - \mathbf{x}\| < \epsilon$, and hence, by using the convexity of f, we obtain

$$f(\bar{\mathbf{x}}) \le f(\mathbf{z}) \le (1-\lambda) f(\bar{\mathbf{x}}) + \lambda f(\mathbf{y})$$

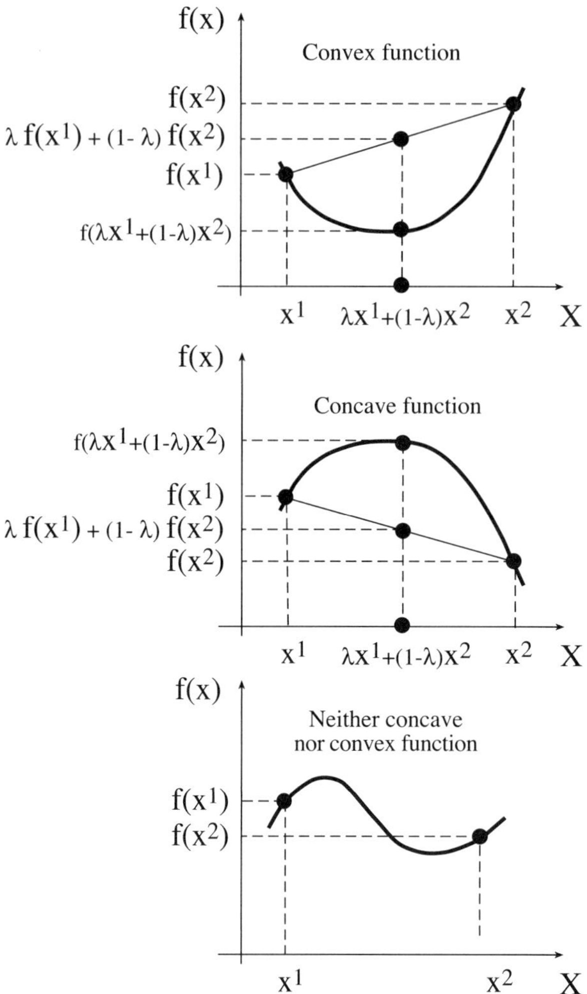

Figure 8.13: Convex, concave, and neither convex nor concave functions in one dimension.

Reorganizing terms, we arrive at

$$0 \leq \lambda\left(f(\mathbf{y}) - f(\bar{\mathbf{x}})\right)$$

and because of the nonnegativity of λ, we conclude that

$$f(\bar{\mathbf{x}}) \leq f(\mathbf{y}).$$

The arbitrariness of $\mathbf{y} \in S$ implies that $\bar{\mathbf{x}}$ is really a global minimum of f in S.

The uniqueness part will be proved by contradiction. Assume that there exists two different optima, $\mathbf{x}^1 \neq \mathbf{x}^2$, such that $f(\mathbf{x}^1) = f(\mathbf{x}^2)$, and let $\lambda \in (0, 1)$.

Since f is strictly convex, we have

$$f(\lambda \mathbf{x}^1 + (1-\lambda)\mathbf{x}^2) < \lambda f(\mathbf{x}^1) + (1-\lambda) f(\mathbf{x}^2) = f(\mathbf{x}^1) = f(\mathbf{x}^2)$$

which contradicts the assumption that $f(\mathbf{x}^1)$ and $f(\mathbf{x}^2)$ are optima. Therefore we actually have uniqueness. ■

Once we know that convexity is a desirable property for the functions involved in a nonlinear programming problem, especially with respect to the issue of global minima, we need a better understanding of this concept, although the reader may already have some experience for functions of one variable.

Let us start by considering the geometric interpretation of convex and concave functions (see Figure 8.13). Let $\mathbf{x}^1$ and $\mathbf{x}^2$ be two distinct points in S, and consider the point $\lambda \mathbf{x}^1 + (1-\lambda)\mathbf{x}^2 \in S$ for $\lambda \in (0,1)$. Note that $\lambda f(\mathbf{x}^1) + (1-\lambda) f(\mathbf{x}^2)$ gives the weighted average of $f(\mathbf{x}^1)$ and $f(\mathbf{x}^2)$, while $f(\lambda \mathbf{x}^1 + (1-\lambda)\mathbf{x}^2)$ gives the value of f at the point $\lambda \mathbf{x}^1 + (1-\lambda)\mathbf{x}^2$. So, for a convex function f, the value of f at the point $\lambda \mathbf{x}^1 + (1-\lambda)\mathbf{x}^2$ is less than or equal to the height of the chord joining the points $[\mathbf{x}^1, f(\mathbf{x}^1)]$ and $[\mathbf{x}^2, f(\mathbf{x}^2)]$. For a concave function, the chord is below the function itself.

When a function of a single variable is differentiable, convexity can be understood in terms of the derivative of the function. It essentially amounts to demanding that its derivative be increasing. When the function is twice differentiable, convex functions can be identified requiring the second derivative to be positive or nonnegative. For functions of several variables, we also have these two criteria that we discuss below.

Theorem 8.3 (Convexity and differentiability). *Let $S \subset \mathbb{R}^n$ be a convex set, and let $f : \mathbb{R}^n \to \mathbb{R}$ be differentiable on S. Then f is a convex function on S if and only if*

$$f(\mathbf{x}^1) - f(\mathbf{x}^2) \geq \nabla f(\mathbf{x}^1)^T(\mathbf{x}^2 - \mathbf{x}^1); \ \ \forall \mathbf{x}^1, \mathbf{x}^2 \in S \tag{8.57}$$

Further, f is said to be a strictly convex function on S if and only if

$$f(\mathbf{x}^1) - f(\mathbf{x}^2) > \nabla f(\mathbf{x}^1)^T(\mathbf{x}^2 - \mathbf{x}^1); \ \ \forall \mathbf{x}^1, \mathbf{x}^2 \in S \ \textit{such that}\ \mathbf{x}^1 \neq \mathbf{x}^2 \tag{8.58}$$

■

To deal with the criterion based on the second derivative, the reader should be reminded about the concepts of twice differentiable functions and positive semidefinite matrices.

Definition 8.8 (Twice differentiable function). *We say that $f : \mathbb{R}^n \to \mathbb{R}$ is twice differentiable at the point $\mathbf{x}$ if there exists a column vector $\nabla f(\mathbf{x})$, and a $n \times n$ matrix $\nabla^2 f(\mathbf{x})$, such that*

$$\lim_{\mathbf{y} \to \mathbf{x}} \frac{f(\mathbf{y}) - f(\mathbf{x}) - \nabla f(\mathbf{x})^T(\mathbf{y} - \mathbf{x}) - \frac{1}{2}(\mathbf{y} - \mathbf{x})^T \nabla^2 f(\mathbf{x})(\mathbf{y} - \mathbf{x})}{\|\mathbf{y} - \mathbf{x}\|^2} = 0.$$

■

The matrix $\nabla^2 f(\mathbf{x})$ is called the *Hessian matrix* of f at the point $\mathbf{x}$. The ijelement of $\nabla^2 f(\mathbf{x})$ is the second-order partial derivative $\partial^2 f(\mathbf{x})/\partial x_i \partial x_j$. Moreover, if these partial derivatives are continuous, then f is called twice continuously differentiable and then, $\nabla^2 f(\mathbf{x})$ is a symmetric matrix.

Definition 8.9 (Positive semidefinite matrix). *A symmetric matrix* $\mathbf{A}$ *is positive semidefinite if*

$$\mathbf{x}^T \mathbf{A} \mathbf{x} \geq 0$$

for any vector $\mathbf{x}$. *Moreover, if the equality*

$$\mathbf{x}^T \mathbf{A} \mathbf{x} = 0$$

occurs only when $\mathbf{x} = \mathbf{0}$, *then* $\mathbf{A}$ *is called positive definite.* ■

Equivalently, the matrix $\mathbf{A}$ is positive semi-definite if all its eigenvalues are non negative; that is, all the values of λ satisfying the equation

$$\det(\mathbf{A} - \lambda \mathbf{I}) = 0$$

are nonnegative. $\mathbf{A}$ is positive definite if all eigenvalues are positive. The eigenvalues can be calculated by any standard package such as Mathematica, Maple, or Matlab. In the case of 2×2 matrices a direct useful criterion is that the 2×2 symmetric matrix $\mathbf{A} = (a_{ij})$ is positive definite if $a_{11} > 0$ and $a_{11}a a_{22} - a_{12}^2 > 0$.

The criterion on convexity based on the Hessian matrix reads as follows.

Theorem 8.4 (Convex functions). *Let* $S \subset \mathbb{R}^n$ *be an open convex set and let* $f : \mathbb{R}^n \to \mathbb{R}$ *be twice differentiable on* S. *Then, the following properties hold:*

1. f *is a convex function on* S *if and only if* $\nabla^2 f(\tilde{\mathbf{x}})$ *is a positive semi-definite matrix for every* $\tilde{\mathbf{x}} \in S$:

$$\mathbf{y}^T \nabla^2 f(\tilde{\mathbf{x}}) \mathbf{y} \geq 0; \;\; \forall \mathbf{y} \in \mathbb{R}^n$$

2. f *is a strictly convex function on* S *if and only if* $\nabla^2 f(\tilde{\mathbf{x}})$ *is a positive definite matrix for every* $\tilde{\mathbf{x}} \in S$:

$$\mathbf{y}^T \nabla^2 f(\tilde{\mathbf{x}}) \mathbf{y} > 0; \;\; \forall \mathbf{y} \in \mathbb{R}^n$$

■

Some relevant convex functions are

1. **Affine functions**: Let $f(\mathbf{x}) = \mathbf{b}^T \mathbf{x} + a$, where $a \in \mathbb{R}$ and $\mathbf{b}$, $\mathbf{x} \in \mathbb{R}^n$, be an affine function. Then, the Hessian matrix is $\nabla^2 f(\tilde{\mathbf{x}}) = 0$ for all $\tilde{\mathbf{x}} \in \mathbb{R}^n$. Since this function satisfies Theorem 8.4, it is a convex function. Note that f is also a concave function because $\nabla^2(-f)(\tilde{\mathbf{x}})$ is also equal to zero.

2. **Quadratic forms**: Let $f(\mathbf{x}) = \frac{1}{2}\mathbf{x}^T\mathbf{C}\mathbf{x} + \mathbf{b}^T\mathbf{x} + a$ be a quadratic form. If the matrix C is positive semidefinite, then f is a convex function, because the Hessian matrix of f is $\nabla^2 f(\mathbf{x}) = \mathbf{C}$ for all $\mathbf{x}$.

When dealing with convex functions, we often combine several of them in some way. It is important to know whether that particular process maintains the convexity property. Some operations that preserve convexity are

1. **Nonnegative linear combinations of convex functions**. Let $f_i(\mathbf{x})$, $i = 1, \ldots, k$ be convex functions, and let λ_i $i = 1, \ldots, k$ be positive scalars. Consider the function $h(\mathbf{x}) = \sum_{i=1}^{k} \lambda_i f_i(\mathbf{x})$. Since the sum of semidefinite matrices is also a semidefinite matrix, h is a convex function.

2. **First composition**. If h is convex and $\mathbf{T}$ is a linear (or more generally affine) transformation, the composition $g = h(\mathbf{T})$ is also convex.

3. **Second composition**. If g is convex, and h is a convex, nondecreasing function of one variable, the composition $f = h(g)$ is also convex.

4. **Supremum**. The supremum of any family of convex functions is also a convex function.

This is straightforward to check.

8.3.2 Sufficiency of the Karush–Kuhn–Tucker Conditions

The KKTCs are not sufficient for a local minimum. For example, the problem of minimizing

$$Z = x^3$$

subject to

$$x \in \mathbb{R}$$

has a KKT point at $x = 0$, but it is not a local minimum. Even if we know that a certain vector corrresponds to a local minimum, there is no way to guarantee that it is a global minimum.

The next lemma gives a necessary and sufficient condition for a global optimum for an important class of NLPP with an implicitly defined constrained set.

Lemma 8.2 (Convex programming problem) *Consider the convex programming problem (CPP). Minimize*

$$Z = f(\mathbf{x})$$

subject to

$$\mathbf{x} \in S$$

where f is convex and differentiable and S is a convex set. Then $\mathbf{x}^* \in S$ is a global optimum if and only if

$$\nabla f(\mathbf{x}^*)^T(\mathbf{x} - \mathbf{x}^*) \geq 0, \ \forall \mathbf{x} \in S$$

Proof. Let $\mathbf{x}$ be a feasible solution of CPP. Then

$$\mathbf{x}^* + \lambda(\mathbf{x} - \mathbf{x}^*) = \lambda\mathbf{x} + (1-\lambda)\mathbf{x}^* \in S, \forall\lambda \in [0,1]$$

because S is a convex set. This implies that $\mathbf{x}^* + \lambda\mathbf{d} \in S, \forall\lambda \in (0,1)$, where $\mathbf{d} = \mathbf{x} - \mathbf{x}^*$ is a feasible direction. Consider the function of one variable $\lambda \in [0,1]$

$$\varphi(\lambda) = f(\mathbf{x}^* + \lambda\mathbf{d}).$$

Since, by assumption, $\mathbf{x}^*$ is a global minimum of f in S, φ has a minimum at $\lambda = 0$. Therefore, by the chain rule, we obtain

$$0 \leq \varphi'(0) = \nabla f(\mathbf{x}^*)^T\mathbf{d}$$

and the implication is proved.

Conversely, using Theorem 8.3, from the convexity and differentiability of f on S we have

$$f(\mathbf{x}) - f(\mathbf{x}^*) \geq \nabla f(\mathbf{x}^*)^T(\mathbf{x} - \mathbf{x}^*) \geq 0 \quad \forall\mathbf{x} \in S.$$

■

The following theorem gives sufficient conditions to ensure that the KKT points are also global minima.

Theorem 8.5 (Sufficiency of the Karush–Kuhn–Tucker Conditions). *Consider the NLPP problem involving minimization of*

$$Z = f(\mathbf{x})$$

subject to

$$\begin{array}{rcl} \mathbf{h}(\mathbf{x}) & = & \mathbf{0} \\ \mathbf{g}(\mathbf{x}) & \leq & \mathbf{0} \end{array}$$

Assume that there exists a triplet $(\bar{\mathbf{x}}, \bar{\boldsymbol{\mu}}, \bar{\boldsymbol{\lambda}})$ satisfying the KKTC. Let $K^+ = \{k|\bar{\lambda}_k > 0\}$ and $K^- = \{k|\bar{\lambda}_k < 0\}$. Suppose that $f(\mathbf{x})$, $g_i(\mathbf{x})$ for all $i \in I(\bar{\mathbf{x}})$ and $h_k(\mathbf{x})$ for all $k \in K^+$ are convex functions on $\mathbb{R}^n$, and $h_k(\mathbf{x})$ for all $k \in K^-$ are concave functions on $\mathbb{R}^n$. Then $\bar{\mathbf{x}}$ is a global solution to the NLPP. ■

Proof. We shall show that $\bar{\mathbf{x}}$ is an optimal solution using Lemma 8.2 . Let $\mathbf{x}$ be any feasible solution to our NLPP. Then, since $g_i(\mathbf{x}) \leq 0$ and $g_i(\bar{\mathbf{x}}) = 0$, for $i \in I(\bar{\mathbf{x}})$, we have $g_i(\mathbf{x}) \leq g_i(\bar{\mathbf{x}})$, and by the convexity of g_i on S it follows that

$$g_i(\bar{\mathbf{x}} + \lambda(\mathbf{x} - \bar{\mathbf{x}})) = g_i(\lambda\mathbf{x} + (1-\lambda)\bar{\mathbf{x}}) \leq \lambda g_i(\mathbf{x}) + (1-\lambda)g_i(\bar{\mathbf{x}}) \leq g_i(\bar{\mathbf{x}})$$

for all $\lambda \in (0,1)$. This implies that g_i does not increase when moving from $\bar{\mathbf{x}}$ along the direction $\mathbf{d} = \mathbf{x} - \bar{\mathbf{x}}$. Thus, we must have

$$\nabla g_i(\bar{\mathbf{x}})^T(\mathbf{x} - \bar{\mathbf{x}}) \leq \mathbf{0} \tag{8.59}$$

Similarly, since h_k is convex for all $k \in K^+$ and h_k is concave for all $k \in K^-$, we have

$$\nabla h_k(\bar{\mathbf{x}})^T(\mathbf{x} - \bar{\mathbf{x}}) \leq \mathbf{0}, \quad \forall k \in K^+ \tag{8.60}$$

$$\nabla h_k(\bar{\mathbf{x}})^T(\mathbf{x} - \bar{\mathbf{x}}) \geq \mathbf{0}, \quad \forall k \in K^- \tag{8.61}$$

Multiplying (8.59), (8.60), and (8.61) by $\bar{\mu}_i \geq 0$, $\bar{\lambda}_k > 0$, and $\bar{\lambda}_k < 0$, respectively, and adding up, we get

$$\left[\sum_{k \in K^+ \cup K^-} \bar{\lambda}_k \nabla h_k(\bar{\mathbf{x}}) + \sum_{i \in I(\bar{\mathbf{x}})} \mu_i \nabla g_i(\bar{\mathbf{x}}) \right]^T (\mathbf{x} - \bar{\mathbf{x}}) \leq 0 \tag{8.62}$$

By assumption, $(\bar{x}, \bar{\mu}, \bar{\lambda})$ is a KKT point. Multiplying condition (8.53) by $\mathbf{x} - \bar{\mathbf{x}}$ and noting that $\bar{\mu}_i = 0$ for all $i \notin I(\bar{\mathbf{x}})$, then (8.62) implies that

$$\nabla f(\bar{\mathbf{x}})^T(\mathbf{x} - \bar{\mathbf{x}}) \geq 0$$

and using Lemma 8.2, the proof is completed. ∎

Example 8.6 (Special case: unconstrained problems). For all unconstrained problems the KKTC are necessary optimality conditions. Let $f : \mathbb{R}^n \to \mathbb{R}$ be convex on $\mathbb{R}^n$ and differentiable at $\mathbf{x}^*$. Then $\mathbf{x}^*$ is a global optimal solution of the problem. We minimize

$$Z = f(\mathbf{x}), \quad \mathbf{x} \in \mathbb{R}^n$$

if and only if $\nabla f(\mathbf{x}^*) = 0$. ∎

The most important class of NLPP for which KKT conditions are always necessary and sufficient for global minima are, by far, the so-called convex programs (CP). A CP consists of finding the minimum of a convex objective function over a set of constraints where inequality constraints are associated with convex functions and equality restrictions are formulated as affine ones. Specifically, we are concerned with the following problem. Minimize

$$Z = f(\mathbf{x})$$

subject to

$$\mathbf{g}(\mathbf{x}) \leq \mathbf{0}$$

where f and $\mathbf{g}$ are convex and continuously differentiable functions. We denote as S the feasible region determined by the constraints.

A slightly more general representation of this problem is obtained taking into account that the function $\mathbf{h}(\mathbf{x}) = \mathbf{A}\mathbf{x} + \mathbf{b}$, where $\mathbf{A}$ is an $m \times n$ matrix, and $\mathbf{b}$ an $n \times 1$ vector, is simultaneously a concave and a convex function; thus $\mathbf{h}$ and $-\mathbf{h}$ are convex functions. The function $\mathbf{h}$ is called an *affine function* (a linear function plus a constant term). Constraints of the form $\mathbf{h}(\mathbf{x}) = \mathbf{0}$ where $\mathbf{h}$ is an affine function are common in convex programs.

The most general form of a convex program can then be written as follows. Minimize

$$Z = f(\mathbf{x})$$

subject to

$$\begin{aligned} \mathbf{h}(\mathbf{x}) &= \mathbf{0} \\ \mathbf{g}(\mathbf{x}) &\leq \mathbf{0} \end{aligned}$$

where the functions f and $\mathbf{g}$ are convex and continuously differentiable, and $\mathbf{h}$ is an affine function. An important particular case is the linear programming problem, thus the discussion of this chapter can be fully applied to linear programming problems.

Example 8.7 (Least-squares fit). In many instances in science and technology, we have a series of experimental data relating arbitrary functions $r(x)$ and $s(y)$ of two different variables x and y, respectively. The relationship between these quantities is unknown, and we have information only about a data pairs set $\{(r(x_i), s(y_i))|i = 1, 2, \ldots, n\}$. Often we would like to find the best curve of a certain class fitting these pairs in the sense that the mean-squared error of the $s(y)$ coordinates is minimized. If we are looking for a linear model to fit the set of given data, we would like to find the slope α and the independent term β so that they minimize the objective function

$$f(\alpha, \beta) = \sum_i (\alpha r(x_i) + \beta - s(y_i))^2$$

Since both parameters α and β cannot be constrained in sign or in any other way, it turns out to be an unconstrained NLPP. Since the objective function f tends to infinity as $(\alpha, \beta) \to \infty$ and, on the other hand, $f \geq 0$, the NLPP admits optimal solutions. Furthermore, f is a convex function of the pair (α, β), even more, strictly convex. Consequently, the optimization problem has a unique optimal solution that can be found by looking at the KKT conditions for the problem. In the case of an unconstrained problem, we only have to impose that the partial derivatives of f with respect to α and β should vanish:

$$\begin{aligned} \frac{\partial f(\alpha, \beta)}{\partial \alpha} &= \sum_i (\alpha r(x_i) + \beta - s(y_i)) r(x_i) = 0 \\ \frac{\partial f(\alpha, \beta)}{\partial \alpha} &= \sum_i (\alpha r(x_i) + \beta - s(y_i)) = 0 \end{aligned}$$

or after rearranging terms, we arrive at the linear system

$$\begin{aligned} \sum_i r(x_i)^2 \alpha + \sum_i r(x_i) \beta &= \sum_i r(x_i) s(y_i) \\ \sum_i r(x_i) \alpha + n\beta &= \sum_i s(y_i) \end{aligned}$$

The solution of this system is easily found as

$$\begin{aligned}\alpha &= \frac{n\sum_i r(x_i)s(y_i) - \sum_i r(x_i)\sum_i s(y_i)}{(n\sum_i r(x_i)^2 - (\sum_i r(x_i))^2)}\\ \beta &= \frac{\sum_i r(x_i)^2\sum_i s(y_i) - \sum_i r(x_i)\sum_i r(x_i)s(y_i)}{(n\sum_i r(x_i)^2 - (\sum_i r(x_i))^2)}\end{aligned}$$

■

Example 8.8 (Cobb–Douglas Problem). The level of production of a good depends on the level of investment, denoted as x_1, and the level of work, denoted as x_2. The Cobb–Douglas's function, which gives the level of production as a function of x_1 and x_2, is:

$$f(x_1, x_2) = x_1^\alpha x_2^\beta$$

where $\alpha > 0$, $\beta > 0$, and $\alpha + \beta \leq 1$. We assume a budget constraint. Let C be the total amount of available money, and let a be the cost per unit of work, this constraint can be written as

$$x_1 + ax_2 = C. \tag{8.63}$$

Since the decision maker wishes to obtain the maximum level of production, we are led to the following optimization problem. Maximize

$$Z = x_1^\alpha x_2^\beta$$

subject to

$$\begin{array}{rrcl} x_1 & +ax_2 & = & C\\ -x_1 & & \leq & 0\\ & -x_2 & \leq & 0\end{array}$$

Using the fact that

$$\max_{\mathbf{x}\in S} f(\mathbf{x}) = -\min_{\mathbf{x}\in S}(-f(\mathbf{x}))$$

the previous problem can be formulated as follows. Minimize

$$Z = -x_1^\alpha x_2^\beta$$

subject to

$$\begin{array}{rrcl} x_1 & +ax_2 & = & C\\ -x_1 & & \leq & 0\\ & -x_2 & \leq & 0\end{array}$$

This is a CPP because the constraint functions are linear and the objective function is convex. We shall show this using Theorem 8.4. The gradient vector is

$$\nabla f(x_1, x_2) = \begin{pmatrix} -\alpha x_1^{\alpha-1}x_2^\beta\\ -\beta x_1^\alpha x_2^{\beta-1}\end{pmatrix}$$

and the Hessian matrix is

$$\nabla^2 f(x_1, x_2) = \begin{pmatrix} -\alpha(\alpha-1)x_1^{\alpha-2}x_2^{\beta} & -\alpha\beta x_1^{\alpha-1}x_2^{\beta-1} \\ -\alpha\beta x_1^{\alpha-1}x_2^{\beta-1} & -\beta(\beta-1)x_1^{\alpha}x_2^{\beta-2} \end{pmatrix}$$

Using the criterion on the determinant according to Definition 8.9, we shall prove that the Hessian matrix is a positive semidefinite matrix. By assumption, the parameter α satisfies $0 < \alpha < 1$ and $-\alpha(\alpha - 1) > 0$; in the feasible region, $x_1 \geq 0$ and $x_2 \geq 0$, thus $x_1^{\alpha-2}x_2^{\beta} \geq 0$. In addition, $\det(\nabla^2 f(x_1, x_2)) = \alpha\beta(1-\alpha-\beta)x_1^{2(\alpha-1)}x_2^{2(\beta-1)}$, and it is nonnegative in the feasible region. Thus, $\nabla^2 f(x_1, x_2)$ is positive definite and Theorem 8.5 justifies that any KKT point is an optimal solution. The stationary condition of the Lagrangian can be stated as

$$\begin{pmatrix} -\alpha x_1^{\alpha-1}x_2^{\beta} \\ -\beta x_1^{\alpha}x_2^{\beta-1} \end{pmatrix} + \mu_1 \begin{pmatrix} 1 \\ 0 \end{pmatrix} + \mu_2 \begin{pmatrix} 0 \\ 1 \end{pmatrix} + \lambda \begin{pmatrix} 1 \\ a \end{pmatrix} = \begin{pmatrix} 0 \\ 0 \end{pmatrix}$$

We assume that $x_1 > 0$ and $x_2 > 0$, and by the complimentary slackness conditions $\mu_1 = \mu_2 = 0$. Then, the system of equations

$$\begin{aligned} -\alpha x_1^{\alpha-1}x_2^{\beta} + \lambda &= 0 \\ -\beta x_1^{\alpha}x_2^{\beta-1} + \lambda a &= 0 \end{aligned}$$

leads to

$$-\alpha x_1^{\alpha-1}x_2^{\beta} = -\frac{\beta}{a}x_1^{\alpha}x_2^{\beta-1}$$

Dividing by $x_1^{\alpha-1}x_2^{\beta-1} > 0$, we get

$$x_2 = \frac{\beta}{a\alpha}x_1$$

and using (8.63), we finally obtain

$$\begin{aligned} x_1 &= \frac{\beta}{\alpha+\beta}C \\ x_2 &= \frac{\alpha}{a(\alpha+\beta)}C \end{aligned}$$

■

8.4 Duality Theory

Duality plays a crucial role in the theory and computational algorithms of nonlinear programming (see, for example, Bazaraa et al. [9] and Luenberger [68]). In this section the results obtained for linear programming problems are generalized to convex programming problems, and in the next chapter these results are used to develop computational methods to approximate optimal solutions for NLPP.

Consider the following general nonlinear *primal problem* (P). Minimize

$$Z_P = f(\mathbf{x}) \tag{8.64}$$

subject to

$$\begin{array}{rcl} \mathbf{h}(\mathbf{x}) & = & \mathbf{0} \\ \mathbf{g}(\mathbf{x}) & \leq & \mathbf{0} \end{array} \tag{8.65}$$

where $f : \mathbb{R}^n \to \mathbb{R}$, $\mathbf{h} : \mathbb{R}^n \to \mathbb{R}^\ell$, $\mathbf{g} : \mathbb{R}^n \to \mathbb{R}^m$. The *dual problem* requires the introduction of the dual function defined as

$$\theta(\boldsymbol{\lambda}, \boldsymbol{\mu}) \quad = \quad \text{Inf}_{\mathbf{x}} \ \{f(\mathbf{x}) + \boldsymbol{\lambda}^T \mathbf{h}(\mathbf{x}) + \boldsymbol{\mu}^T \mathbf{g}(\mathbf{x})\} \tag{8.66}$$

The *dual problem* (D) is then defined as follows. Maximize

$$Z_D = \theta(\boldsymbol{\lambda}, \boldsymbol{\mu})$$

subject to

$$\boldsymbol{\mu} \geq \mathbf{0}$$

Using the Lagrangian

$$\mathcal{L}(\mathbf{x}, \boldsymbol{\lambda}, \boldsymbol{\mu}) = f(\mathbf{x}) + \boldsymbol{\lambda}^T \mathbf{h}(\mathbf{x}) + \boldsymbol{\mu}^T \mathbf{g}(\mathbf{x})$$

we can rewrite D as

$$\max_{\boldsymbol{\lambda}, \boldsymbol{\mu}; \boldsymbol{\mu} \geq 0} \left(\inf_{\mathbf{x}} \ \mathcal{L}(\mathbf{x}, \boldsymbol{\lambda}, \boldsymbol{\mu}) \right) \tag{8.67}$$

We always assume that $f, \mathbf{h}$, and $\mathbf{g}$ are such that the infimum of the Lagrangian function is always attained at some $\mathbf{x}$, so that the "infimum" operation in (8.66) and (8.67) can be replaced by the "minimum" operation. Then, problem (8.67) is referred to as the *max–min dual problem.*

In this situation, if we denote by $\mathbf{x}(\boldsymbol{\lambda}, \boldsymbol{\mu})$ a point where the minimum of the Lagrangian is attained (considered as a function of the multipliers), we can write

$$\theta(\boldsymbol{\lambda}, \boldsymbol{\mu}) = f(\mathbf{x}(\boldsymbol{\lambda}, \boldsymbol{\mu})) + \boldsymbol{\lambda}^T \mathbf{h}(\mathbf{x}(\boldsymbol{\lambda}, \boldsymbol{\mu})) + \boldsymbol{\mu}^T \mathbf{g}(\mathbf{x}(\boldsymbol{\lambda}, \boldsymbol{\mu}))$$

On the other hand, under the assumption of convexity of f and $\mathbf{g}$, affineness of $\mathbf{h}$ and positivity of $\boldsymbol{\mu}$, the Lagrangian is convex in $\mathbf{x}$, and therefore at a point of global minimum its gradient should vanish

$$\nabla \theta(\boldsymbol{\lambda}, \boldsymbol{\mu}) = \nabla f(\mathbf{x}(\boldsymbol{\lambda}, \boldsymbol{\mu})) + \boldsymbol{\lambda}^T \nabla \mathbf{h}(\mathbf{x}(\boldsymbol{\lambda}, \boldsymbol{\mu})) + \boldsymbol{\mu}^T \nabla \mathbf{g}(\mathbf{x}(\boldsymbol{\lambda}, \boldsymbol{\mu})) = \mathbf{0} \tag{8.68}$$

These two last identities can be utilized to derive an expression for the gradient of the dual function as well as its Hessian. Indeed, according to the chain rule

$$\begin{aligned} \nabla_{\boldsymbol{\mu}} \theta(\boldsymbol{\lambda}, \boldsymbol{\mu}) = {} & [\nabla f(\mathbf{x}(\boldsymbol{\lambda}, \boldsymbol{\mu})) + \boldsymbol{\lambda}^T \nabla \mathbf{h}(\mathbf{x}(\boldsymbol{\lambda}, \boldsymbol{\mu})) + \boldsymbol{\mu}^T \nabla \mathbf{g}(\mathbf{x}(\boldsymbol{\lambda}, \boldsymbol{\mu}))]^T \nabla_{\boldsymbol{\mu}} \mathbf{x}(\boldsymbol{\lambda}, \boldsymbol{\mu}) \\ & + \mathbf{g}(\mathbf{x}(\boldsymbol{\lambda}, \boldsymbol{\mu})) \end{aligned}$$

and by (8.68)

$$\nabla_{\boldsymbol{\mu}}\theta(\boldsymbol{\lambda}, \boldsymbol{\mu}) = \mathbf{g}(\mathbf{x}(\boldsymbol{\lambda}, \boldsymbol{\mu})) \tag{8.69}$$

Similarly

$$\nabla_{\boldsymbol{\lambda}}\theta(\boldsymbol{\lambda}, \boldsymbol{\mu}) = \mathbf{h}(\mathbf{x}(\boldsymbol{\lambda}, \boldsymbol{\mu})) \tag{8.70}$$

that is, the gradient of the dual function is the vector of missmatches of the corresponding constraints.

Concerning the Hessian, notice that by differentiating (8.69) and (8.70) with respect to $\boldsymbol{\mu}$ and $\boldsymbol{\lambda}$, respectively, we obtain

$$\nabla^2_{\boldsymbol{\mu}}\theta(\boldsymbol{\lambda}, \boldsymbol{\mu}) = \nabla_{\mathbf{x}}\mathbf{g}(\mathbf{x}(\boldsymbol{\lambda}, \boldsymbol{\mu})) \;\; \nabla_{\boldsymbol{\mu}}\mathbf{x}(\boldsymbol{\lambda}, \boldsymbol{\mu}) \tag{8.71}$$

and

$$\nabla^2_{\boldsymbol{\lambda}}\theta(\boldsymbol{\lambda}, \boldsymbol{\mu}) = \nabla_{\mathbf{x}}\mathbf{h}(\mathbf{x}(\boldsymbol{\lambda}, \boldsymbol{\mu})) \;\; \nabla_{\boldsymbol{\lambda}}\mathbf{x}(\boldsymbol{\lambda}, \boldsymbol{\mu}) \tag{8.72}$$

To obtain an expression for $\nabla_{\boldsymbol{\mu}}\mathbf{x}(\boldsymbol{\lambda}, \boldsymbol{\mu})$ and $\nabla_{\boldsymbol{\lambda}}\mathbf{x}(\boldsymbol{\lambda}, \boldsymbol{\mu})$, we differentiate (8.68) with respect to $\boldsymbol{\mu}$ and $\boldsymbol{\lambda}$, respectively. The result is

$$\nabla^2_{\mathbf{x}}\mathcal{L}(\mathbf{x}(\boldsymbol{\lambda}, \boldsymbol{\mu}), \boldsymbol{\lambda}, \boldsymbol{\mu}) \; \nabla_{\boldsymbol{\mu}}\mathbf{x}(\boldsymbol{\lambda}, \boldsymbol{\mu}) + \nabla_{\mathbf{x}}\mathbf{g}(\mathbf{x}(\boldsymbol{\lambda}, \boldsymbol{\mu})) = \mathbf{0}$$

and a similar equation exchanging $\boldsymbol{\lambda}$ by $\boldsymbol{\mu}$ and $\mathbf{h}$ by $\mathbf{g}$. Taking these formulas into (8.71) and (8.72), we arrive at

$$\nabla^2_{\boldsymbol{\mu}}\theta(\boldsymbol{\lambda}, \boldsymbol{\mu}) = -\nabla_{\mathbf{x}}\mathbf{g}(\mathbf{x}(\boldsymbol{\lambda}, \boldsymbol{\mu}))[\nabla^2_{\mathbf{x}}\mathcal{L}(\mathbf{x}(\boldsymbol{\lambda}, \boldsymbol{\mu}), \boldsymbol{\lambda}, \boldsymbol{\mu})]^{-1} \; \nabla_{\mathbf{x}}\mathbf{g}(\mathbf{x}(\boldsymbol{\lambda}, \boldsymbol{\mu}))$$

and the parallel formula for $\nabla^2_{\boldsymbol{\lambda}}\theta(\boldsymbol{\lambda}, \boldsymbol{\mu})$. Finally, if we are interested in the mixed second partial derivatives

$$\nabla^2_{\boldsymbol{\lambda}\,\boldsymbol{\mu}}\theta(\boldsymbol{\lambda}, \boldsymbol{\mu})$$

the same computations lead to

$$\nabla^2_{\boldsymbol{\lambda}\,\boldsymbol{\mu}}\theta(\boldsymbol{\lambda}, \boldsymbol{\mu}) = -\nabla_{\mathbf{x}}\mathbf{g}(\mathbf{x}(\boldsymbol{\lambda}, \boldsymbol{\mu}))[\nabla^2_{\mathbf{x}}\mathcal{L}(\mathbf{x}(\boldsymbol{\lambda}, \boldsymbol{\mu}), \boldsymbol{\lambda}, \boldsymbol{\mu})]^{-1} \; \nabla_{\mathbf{x}}\mathbf{h}(\mathbf{x}(\boldsymbol{\lambda}, \boldsymbol{\mu}))$$

and similarly

$$\nabla^2_{\boldsymbol{\lambda}\,\boldsymbol{\mu}}\theta(\boldsymbol{\lambda}, \boldsymbol{\mu}) = -\nabla_{\mathbf{x}}\mathbf{h}(\mathbf{x}(\boldsymbol{\lambda}, \boldsymbol{\mu}))[\nabla^2_{\mathbf{x}}\mathcal{L}(\mathbf{x}(\boldsymbol{\lambda}, \boldsymbol{\mu}), \boldsymbol{\lambda}, \boldsymbol{\mu})]^{-1} \; \nabla_{\mathbf{x}}\mathbf{g}(\mathbf{x}(\boldsymbol{\lambda}, \boldsymbol{\mu}))$$

All these formulas are relevant in computational methods (see next chapter).

The following theorem shows that any value of the dual problem is a lower bound of the optimal value of the primal problem. This may be used to terminate computations in an iterative algorithm where such points are available.

Theorem 8.6 (Weak duality). *For any feasible solution* $\mathbf{x}$ *of the primal problem (8.65) and for any feasible solution of the dual problem (8.66), the following holds*

$$f(\mathbf{x}) \geq \theta(\boldsymbol{\lambda}, \boldsymbol{\mu}) \tag{8.73}$$

■

Proof. From the definition of θ, for every feasible $\mathbf{x}$ and for every $\boldsymbol{\mu} \geq 0$ and $\boldsymbol{\lambda} \in \mathbb{R}^{\ell}$, we have

$$\theta(\boldsymbol{\lambda}, \boldsymbol{\mu}) = \text{Inf}_{\mathbf{y}} \left[f(\mathbf{y}) + \boldsymbol{\lambda}^T \mathbf{h}(\mathbf{y}) + \boldsymbol{\mu}^T \mathbf{g}(\mathbf{y}) \right] \leq f(\mathbf{x}) + \boldsymbol{\lambda}\mathbf{h}(\mathbf{x}) + \boldsymbol{\mu}^T \mathbf{g}(\mathbf{x}) \leq f(\mathbf{x})$$

■

If the feasible regions of both primal an dual are nonempty, then (8.73) implies that $\sup \theta$ and $\inf f$, both taken over their respective constraint sets, are finite. Of course, neither the *supremum* nor the *infimum* need to be attained at any feasible point. If they are, then (8.73) yields the result that if both primal an dual are feasible, both have optimal solutions.

A corollary of high practical interest is stated below.

Corollary 8.2 (Primal and dual optimality)
1. The following holds

$$\sup\{\theta(\boldsymbol{\lambda}, \boldsymbol{\mu}) | \boldsymbol{\mu} \geq \mathbf{0}\} \leq \inf\{f(\mathbf{x}) | \mathbf{h}(\mathbf{x}) = \mathbf{0}, \mathbf{g}(\mathbf{x}) \leq \mathbf{0}\} \tag{8.74}$$

2. If $f(\mathbf{x}^*) = \theta(\boldsymbol{\lambda}^*, \boldsymbol{\mu}^*)$ for some feasible solution $\mathbf{x}^*$ of the primal problem (8.65) and for some feasible solution $(\boldsymbol{\lambda}^*, \boldsymbol{\mu}^*)$ of the dual problem (8.67), then $\mathbf{x}^*$ and $(\boldsymbol{\lambda}^*, \boldsymbol{\mu}^*)$ are optimal solutions of the primal and dual problems, respectively.

3. If $\sup\{\theta(\boldsymbol{\lambda}, \boldsymbol{\mu}) : \boldsymbol{\mu} \geq \mathbf{0}\} = +\infty$, then the primal problem has no feasible solution.

4. If $\inf\{f(\mathbf{x}) : \mathbf{h}(\mathbf{x}) = \mathbf{0}, \mathbf{g}(\mathbf{x}) \leq \mathbf{0}\} = -\infty$, then $\theta(\boldsymbol{\lambda}, \boldsymbol{\mu}) = -\infty$ for every $\boldsymbol{\lambda} \in \mathbb{R}^{\ell}$ and $\boldsymbol{\mu} \geq \mathbf{0}$.

If a multiplier vector solves the dual problem and the *duality gap* does not occur, that is, if the relationship (8.74) is not strictly satisfied, then, the solutions of the Lagrangian problem associated with this multiplier vector are optimal solutions of the primal. This result introduces a way to solve the primal problem by means of the dual problem. The main result consists of obtaining conditions that guarantee the nonexistence of the duality gap. This is the situation when we deal with convex programs.

Theorem 8.7 (Duality theorem for convex programs). *Consider a convex programming problem as (8.64)–(8.65). If $\mathbf{x}^*$ solves the primal problem, its associated vector of multipliers $(\boldsymbol{\lambda}^*, \boldsymbol{\mu}^*)$ solves the dual problem, and $\boldsymbol{\mu}^* \mathbf{g}(\mathbf{x}^*) = 0$. Conversely, if $(\boldsymbol{\lambda}^*, \boldsymbol{\mu}^*)$ solves the dual problem, and there exists a solution, $\mathbf{x}^*$, of the Lagrangian problem associated to this vector of multipliers such that $\boldsymbol{\mu}^* \mathbf{g}(\mathbf{x}^*) = 0$, then $\mathbf{x}^*$ is an optimal solution of the primal.* ■

Proof. If $\mathbf{x}^*$ solves the primal problem and $(\boldsymbol{\mu}^*, \boldsymbol{\lambda}^*)$ is its associated vector of multipliers, we know that they should be a solution of the KKT conditions

$$\nabla f(\mathbf{x}^*) + \boldsymbol{\lambda}^* \nabla \mathbf{h}(\mathbf{x}^*) + \boldsymbol{\mu}^* \nabla \mathbf{g}(\mathbf{x}^*) = \mathbf{0}$$

$$\begin{aligned} \boldsymbol{\mu}^* \mathbf{g}(\mathbf{x}^*) &= \mathbf{0} \\ \boldsymbol{\mu}^* &\geq \mathbf{0} \\ \mathbf{g}(\mathbf{x}^*) &\leq \mathbf{0} \\ \mathbf{h}(\mathbf{x}^*) &= \mathbf{0} \end{aligned}$$

Let us look at the minimization problem defining $\theta(\boldsymbol{\lambda}^*, \boldsymbol{\mu}^*)$

$$\theta(\boldsymbol{\lambda}^*, \boldsymbol{\mu}^*) = \text{Inf}_{\mathbf{x}} \ \{f(\mathbf{x}) + \boldsymbol{\lambda}^{*T}\mathbf{h}(\mathbf{x}) + \boldsymbol{\mu}^{*T}\mathbf{g}(\mathbf{x})\}$$

The condition coming from the above KKT system

$$\nabla f(\mathbf{x}^*) + \boldsymbol{\lambda}^* \nabla \mathbf{h}(\mathbf{x}^*) + \boldsymbol{\mu}^* \nabla \mathbf{g}(\mathbf{x}^*) = \mathbf{0}$$

together with the convexity of the objective function (notice $\boldsymbol{\mu}^* \geq \mathbf{0}$)

$$f(\mathbf{x}) + \boldsymbol{\lambda}^{*T}\mathbf{h}(\mathbf{x}) + \boldsymbol{\mu}^{*T}\mathbf{g}(\mathbf{x})$$

imply that $\mathbf{x}^*$ is truly a global minimizer for the problem determining $\theta(\boldsymbol{\lambda}^*, \boldsymbol{\mu}^*)$ and therefore

$$\theta(\boldsymbol{\lambda}^*, \boldsymbol{\mu}^*) = f(\mathbf{x}^*)$$

and by Corollary 8.2 the pair $(\boldsymbol{\lambda}^*, \boldsymbol{\mu}^*)$ is an optimal solution of the dual. Notice that the condition

$$\boldsymbol{\mu}^* \mathbf{g}(\mathbf{x}^*) = \mathbf{0}$$

also holds.

Conversely, assume that the pair $(\boldsymbol{\lambda}^*, \boldsymbol{\mu}^*)$ is optimal for the dual problem, and that $\mathbf{x}^*$ is a corresponding solution for the associated Lagrangian problem such that $(\boldsymbol{\mu}^*)^T\mathbf{g}(\mathbf{x}^*) = 0$. Applying the KKT conditions to the dual, it is not hard to conclude that

$$\nabla_{\boldsymbol{\mu}} \theta(\boldsymbol{\lambda}^*, \boldsymbol{\mu}^*) \leq \mathbf{0}, \quad \nabla_{\boldsymbol{\lambda}} \theta(\boldsymbol{\lambda}^*, \boldsymbol{\mu}^*) = \mathbf{0}$$

But observe that by (8.69) and (8.70), we obtain

$$\mathbf{g}(\mathbf{x}^*) \leq \mathbf{0}; \ \ \mathbf{h}(\mathbf{x}^*) = \mathbf{0}$$

This implies that $\mathbf{x}^*$ is feasible for the primal. Then

$$\theta(\boldsymbol{\lambda}^*, \boldsymbol{\mu}^*) = f(\mathbf{x}^*) + \boldsymbol{\lambda}^{*T}\mathbf{h}(\mathbf{x}^*) + \boldsymbol{\mu}^{*T}\mathbf{g}(\mathbf{x}^*) = f(\mathbf{x}^*)$$

and by Corollary 8.2 we have our conclusion. ■

Example 8.9 (Dual problem). Consider the following (primal) problem (P). Minimize

$$Z_P = x_1^2 + x_2^2$$

subject to

$$\begin{aligned} x_1 + x_2 &\leq 4 \\ -x_1 &\leq 0 \\ -x_2 &\leq 0 \end{aligned}$$

The unique optimal solution to this problem is $(x_1^*, x_2^*) = (0, 0)$ with the optimal fuction value equal to 0. This is a trivial convex programming problem. Now let us formulate the dual problem (D). We need first to find the dual function:

$$\begin{aligned}\theta(\mu_1, \mu_2, \mu_3) &= \min_{\mathbf{x} \in \mathbf{R}^2} \{x_1^2 + x_2^2 + \mu_1(x_1 + x_2 - 4) + \mu_2(-x_1) + \mu_3(-x_2)\} \\ &= \min_{x_1}\{x_1^2 + (\mu_1 - \mu_2)x_1\} + \min_{x_2}\{x_2^2 + (\mu_1 - \mu_3)x_2\} - 4\mu_1\end{aligned}$$

After a few elementary computations we explicitly find the Lagrangian function

$$\theta(\boldsymbol{\mu}) = -\frac{1}{4}[(\mu_1 - \mu_2)^2 + (\mu_1 - \mu_3)^2] - 4\mu_1$$

an the dual problem (D) becomes

$$\sup\{\theta(\boldsymbol{\mu}) | \boldsymbol{\mu} \geq 0\}$$

Notice how the dual function is concave. The dual problem optimum is attained when $\mu_1^* = \mu_2^* = \mu_3^* = 0$ and $Z_D^* = \theta(\boldsymbol{\mu}^*) = 0$. ■

It is relevant at this point to stress that in some instances it may be important to look at duality partially, in the sense that some but not all constraints are dualized. Those restrictions not dualized are incorporated as such in the definition of the dual function. This is, for example, the underlying driving motivation for the Lagrangian decomposition techniques to be discussed at the end of Chapter 9.

8.5 Practical Illustration of Duality and Separability

To illustrate the relationship between primal and dual problems, consider a given commodity that is produced and supplied under the following conditions:

1. Different producers compete in the market
2. Demand in every time period of a study horizon has to be strictly satisfied
3. No storage capacity is available
4. Producers have different production cost functions and production constraints

We will consider two different approaches to this problem (i) the centralized or *primal* approach and (ii) the competitive market or *dual* approach.

8.5.1 Centralized or Primal Approach

A centralized (primal) approach to this problem is as follows. Producers elect a system operator which is in charge of guaranteeing the supply of demand every time period. To this end, the system operator

1. Is entitled to know the actual production cost function of every producer and its production constraints
2. Knows the actual commodity demand in every time period
3. Has the power to dictate the actual production of every producer

Assume that the main elements of this problem are

1. **Data.**

$c_{ti}(p_{ti})$: production cost of producer i in period t to produce p_{ti},

Π_i: producer i operation feasible region

d_t: demand in period t

T: number of time periods

I: number of producers

2. **Variables.**

p_{ti}: amount of commodity produced by producer i during period t

3. **Constraints.**

$$\sum_{i=1}^{I} p_{ti} = d_t, \qquad t = 1, \dots, T \tag{8.75}$$

$$p_{ti} \in \Pi_i, \qquad t = 1, \dots, T, i = 1, \dots, I \tag{8.76}$$

This enforces for every time period the strict supply of demand, and the operational restrictions of every producer.

4. **Function to be optimized.** Since the operator has the objective of minimizing total production cost, he has to solve the following problem. Minimize

$$\sum_{t=1}^{T} \sum_{i=1}^{I} c_{ti}(p_{ti}), \tag{8.77}$$

which is the sum of the production costs of all producers over all time periods.

Thus, the primal problem is as follows. Minimize

$$\sum_{t=1}^{T} \sum_{i=1}^{I} c_{ti}(p_{ti}) \tag{8.78}$$

subject to

$$\sum_{i=1}^{I} p_{ti} = d_t, \qquad t = 1, \ldots, T \tag{8.79}$$

$$p_{ti} \in \Pi_i, \qquad t = 1, \ldots, T, i = 1, \ldots, I \tag{8.80}$$

The solution of this problem provides the optimal level of production of every producer. Then, the system operator communicates its optimal production to every producer, who actually produces this optimal value.

Note that the approach explained is a centralized one. The system operator has knowledge of the production cost functions and the production constraints of every producer, can determine the actual (optimal) production of every producer, and has the power to force every producer to produce its centrally determined optimal value.

Applying Duality

We analyze in this section the structure of the dual problem of the cost minimization problem (primal problem) (8.77)–(8.80).

If partial duality over constraints (8.79) is considered, the Lagrangian function has the form

$$\mathcal{L}(p_{ti}, \lambda_t) = \sum_{t=1}^{T} \sum_{i=1}^{I} c_{ti}(p_{ti}) + \sum_{t=1}^{T} \lambda_t \left(d_t - \sum_{i=1}^{I} p_{ti} . \right) \tag{8.81}$$

Arranging terms, this Lagrangian function can be expressed as

$$\mathcal{L}(p_{ti}, \lambda_t) = \sum_{i=1}^{I} \left[\sum_{t=1}^{T} (c_{ti}(p_{ti}) - \lambda_t \, p_{ti}) \right] + \sum_{t=1}^{T} \lambda_t \, d_t \tag{8.82}$$

The evaluation of the dual function for a given value of the Lagrange multipliers $\lambda_t = \widehat{\lambda}_t$, $t = 1, \ldots, T$, is achieved by solving the following problem. Maximize

$$\mathcal{L}(p_{ti}, \widehat{\lambda}_t) \tag{8.83}$$

subject to

$$p_{ti} \in \Pi_i, \qquad t = 1, \ldots, T, i = 1, \ldots, I \tag{8.84}$$

Using the explicit expression (8.82) for the Lagrangian function, the problem presented above now entails maximization of

$$\sum_{i=1}^{I} \left[\sum_{t=1}^{T} \left(c_{ti}(p_{ti}) - \widehat{\lambda}_t \, p_{ti} \right) \right] + \sum_{t=1}^{T} \widehat{\lambda}_t \, d_t \tag{8.85}$$

subject to

$$p_{ti} \in \Pi_i, \qquad t = 1, \ldots, T, i = 1, \ldots, I \tag{8.86}$$

Taking into account that the expression $\sum_{t=1}^{T} \widehat{\lambda}_t \, d_t$ is constant, the previous problem reduces to maximization of

$$\sum_{i=1}^{I} \left[\sum_{t=1}^{T} \left(c_{ti}(p_{ti}) - \widehat{\lambda}_t \, p_{ti} \right) \right] \tag{8.87}$$

subject to

$$p_{ti} \in \Pi_i, \qquad t = 1, \ldots, T, i = 1, \ldots, I \tag{8.88}$$

This problem has the very remarkable property of being separable; that is, it decomposes by producer. The individual problem of producer j is therefore maximization of

$$\sum_{t=1}^{T} \left(\widehat{\lambda}_t p_{tj} - c_{tj}(p_{tj}) \right) \tag{8.89}$$

subject to

$$p_{tj} \in \Pi_j, \qquad t = 1, \ldots, T \tag{8.90}$$

Therefore, the evaluation of the dual function can be achieved by solving problem (8.89)-(8.90) for every producer, namely, I times.

What is again particularly remarkable is the interpretation of problem (8.89)–(8.90); it represents the maximization of the profit of producer j subject to its production restrictions. Lagrange multipliers λ_t, $t = 1, \ldots, T$ are naturally interpreted as the selling prices of the commodity in periods $t = 1, \ldots, T$.

It can be concluded that to evaluate the dual function, every producer has to solve its own profit maximization problem, and then, the operator adds the maximum profit of every producer and the constant term $\sum_{t=1}^{T} \widehat{\lambda}_t \, d_t$.

Duality Theorem and Dual Problem Solution

The duality theorem states that, under certain convexity assumptions, the solutions of the primal and the dual problems are the same in terms of the objective function value. It also states that the evaluation of the dual function for the optimal values of Lagrangian multipliers λ_t^*, $t = 1, \ldots, T$, results in optimal values for productions p_{ti}^*, $t = 1, \ldots, T$, $i = 1, \ldots, I$. As a consequence, the solution of the primal problem can be obtained solving the dual one.

Duality theory also states that under very general conditions in the structure of the primal problem, the dual function is concave, and therefore, it can be maximized using a simple steepest-ascent procedure.

To use a steepest-ascent procedure, it is necessary to know the gradient of the dual function. Duality theory provides us with a simple expression for the gradient of the dual function. For productions p_{ti}, $t = 1, \ldots, T$, $i = 1, \ldots, I$, the components of the gradient of the dual function are

$$d_t - \sum_{i=1}^{I} p_{ti} \qquad t = 1, \ldots, T \tag{8.91}$$

That is, the component t of the gradient of the dual function is the actual demand imbalance in period t.

A steepest-ascent procedure to solve the dual problem (and therefore the primal one) includes the following steps:

1. The system operator sets up initial values of Lagrange multipliers (selling prices).

2. Using the Lagrange multipliers presented above, every producer maximizes its own profit meeting its own production constraints.

3. The operator calculates the demand imbalance in every time period (component of the gradient of the dual function).

4. If the norm of the gradient of the dual function is not sufficiently small, a steepest-ascent iteration is performed and the procedure continues in 2. Otherwise stop, the dual solution has been attained.

8.5.2 Competitive Market or Dual Approach

The preceding four-step algorithm can be interpreted in the framework of a competitive market. This is done below.

1. The market operator broadcasts to producers the initial ($k = 1$) value of selling price for every period, $\lambda_t^{(k)}$, $t = 1, \ldots, T$.

2. Every producer j solves its profit maximization problem by maximizing

$$\sum_{t=1}^{T} \left(\lambda_t^{(k)} p_{tj} - c_{tj}(p_{tj}) \right) \tag{8.92}$$

subject to

$$p_{tj} \quad \in \quad \Pi_j, \qquad \forall t \tag{8.93}$$

and sends to the market operator its optimal production schedule, $p_{tj}^{(k)}$, $t = 1, \ldots, T$, $j = 1, \ldots, I$.

3. The market operator computes the demand imbalance in every time period

$$d_t - \sum_{i=1}^{I} p_{ti}^{(k)} \qquad t = 1, \ldots, T \tag{8.94}$$

and update prices proportionally to imbalances

$$\lambda_t^{(k+1)} = \lambda_t^{(k)} + K\left(d_t - \sum_{i=1}^{I} p_{ti}^{(k)}\right); \qquad t = 1, \ldots, T \tag{8.95}$$

where K is a proportionality constant.

4. If prices remain constant in two consecutive rounds, *stop*; the optimal solution (market equilibrium) has been attained. Otherwise, go to step 2.

This dual solution algorithm can be interpreted as the functioning of a multiround auction. In fact, economists call this iterative procedure a Walrasian auction.

A Walrasian auction preserves the privacy of the corporate information (cost function and production constraints) of every producer and allows them to maximize their respective own profits.

8.5.3 Conclusion

Some comments are in order. The primal approach is centralized, and the system operator has complete information of the production cost function and operation constraints of every producer. Moreover, it is given the power of deciding how much should produce every producer.

As opposed to the primal approach, the dual approach is decentralized in nature and every producer maintains the privacy of its production cost function and operation constraints. Furthermore, every producer decides how much to produce to maximize its own benefits.

Duality theory guarantees that both approaches yield identical results.

8.6 Constraint Qualifications

We have seen in previous sections the significance and relevance of the KKT conditions for solving NLPP. However, they are not always necessary conditions for optimality. In fact, only if certain regularity is assumed, KKT conditions are necessary. Hypotheses ensuring this regularity are usually referred to as "constraint qualifications." This section explains these issues and specifies the class of NLPP for which KKT conditions hold. We again consider a typical NLPP as follows. Minimize

$$Z = f(\mathbf{x}) \tag{8.96}$$

subject to

$$\begin{array}{rcl} \mathbf{h}(\mathbf{x}) & = & \mathbf{0} \\ \mathbf{g}(\mathbf{x}) & \leq & \mathbf{0} \end{array} \tag{8.97}$$

Since in most of the NLPPs one encounters in practice some constraint qualification conditions (CQDs), that hold, and a full discussion of these would exceed the objective of this text, we simply quote here some of those CQDs

- **Slater's constraint qualification**. The functions $g_i(\mathbf{x})$ for all $i \in I(\bar{\mathbf{x}})$ are convex, and $\nabla h_k(\bar{\mathbf{x}})$ for all $k = 1, \ldots, \ell$ are linearly independent. Furthermore, there exists $\tilde{\mathbf{x}} \in \mathbb{R}^n$ such that $g_i(\tilde{\mathbf{x}}) < 0$ for all $i \in I(\bar{\mathbf{x}})$ and $h_k(\tilde{\mathbf{x}}) = 0$ for all $k = 1, \ldots, \ell$.

- **Linear independence constraint qualification**. The gradients of the binding constraints are linearly independent, namely, $\nabla g_i(\bar{\mathbf{x}})$ for all $i \in I(\bar{\mathbf{x}})$ and $\nabla h_k(\bar{\mathbf{x}})$ for all $k = 1, \ldots, \ell$ are linearly independent.

Example 8.10 (Special cases). In this example we analyze the following cases:

1. **Unconstrained problems**. In this case $S = \mathbb{R}^n$ and the CQC is trivially satisfied. This theorem is formulated as: If $\bar{\mathbf{x}}$ is a local minimum and f is differentiable at $\bar{\mathbf{x}}$ then $\nabla f(\bar{\mathbf{x}}) = 0$.

2. **Problems with only equality constraints: Lagrange multipliers**. In this case the CQC becomes $\nabla h_k(\bar{\mathbf{x}})$ for $k = 1, \ldots, \ell$ which are linearly independent.

3. **Problems with only inequality constraints**. In this case requirements for the constraints $\mathbf{h}(\mathbf{x}) = \mathbf{0}$ do not exist.

■

Exercises

8.1 Solve the following problems by applying KKTC.

(a) Maximize

$$Z = x_1 - \exp(-x_2)$$

subject to

$$\begin{array}{rcl} -\sin x_1 + x_2 & \leq & 0 \\ x_1 & \leq & 3 \end{array}$$

(b) Minimize

$$Z = (x_1 - 4)^2 + (x_2 - 3)^2$$

subject to

$$\begin{array}{rcl} x_1^2 - x_2 & \leq & 0 \\ x_2 & \leq & 4 \end{array}$$

(c) Maximize

$$Z = x_2$$

subject to

$$\begin{aligned} x_1^2 - x_2^2 &\leq 4 \\ -x_1^2 + x_2 &\leq 0 \\ x_1,\ x_2 &\geq 0 \end{aligned}$$

(d) Minimize

$$Z = -x_1^3 + x_2$$

subject to

$$\begin{aligned} \max\{0, x_1\} - x_2 &\leq 0 \\ x_1^2 + 3x_2 &\leq 4 \end{aligned}$$

(e) Maximize

$$Z = x_1$$

subject to

$$\begin{aligned} x_1^4 + x_1^3 - 3x_1^2 - 5x_1 - x_2 &\geq 2 \\ x_1 &\leq 1 \\ x_2 &\geq 0 \end{aligned}$$

(f) Minimize

$$Z = -x_1$$

subject to

$$\begin{aligned} x_1^2 + x_2^2 &\leq 1 \\ (x_1 - 1)^3 - x_2 &\leq 0 \end{aligned}$$

(g) Maximize

$$Z = x_1$$

subject to

$$\begin{aligned} x_1^2 - x_2 &\leq 0 \\ x_1 - x_2 &\leq -1 \end{aligned}$$

(h) Minimize

$$Z = -x_1 + x_2^2$$

subject to

$$\begin{aligned} \exp(-x_1) - x_2 &\leq 0 \\ -\exp(-x_1^2) + x_2 &\leq 0 \end{aligned}$$

(i) Minimize

$$Z = 4x_1^2 + 2x_2^2 + 4x_1x_2 + 3x_1 + \exp(x_1 + 2x_2)$$

subject to

$$\begin{aligned} x_1^2 + x_2^2 &\leq 10 \\ \sqrt{x_2} &\geq \frac{1}{2} \\ x_2 &\geq 0 \\ x_1 + 2x_2 &= 0 \end{aligned}$$

(j) Maximize

$$Z = 3x_1 - x_2 + x_3^2$$

subject to

$$\begin{aligned} x_1 + x_2 + x_3 &\leq 0 \\ -x_1 + 2x_2 + x_3^2 &= 0 \end{aligned}$$

(k) Minimize

$$Z = x_1 + x_2$$

subject to

$$\begin{aligned} x_1 + x_2^2 &= 4 \\ -2x_1 - x_2 &\leq 4 \end{aligned}$$

(l) Minimize

$$Z = (x_1 + 1)^2 + (x_2 - 1)^2$$

subject to

$$\begin{aligned} (x_1 - 1)(4 - x_1^2 - x_2^2) &\geq 0 \\ x_2 - \frac{1}{2} &= 0 \\ 100 - 2x_1^2 - x_2^2 &= 0 \end{aligned}$$

(m) Maximize

$$Z = x_2 - x_1$$

subject to

$$\begin{array}{rcl} x_1^2 + x_2^2 & \leq & 4 \\ -x_1^2 + x_2 & = & 0 \\ x_1,\ x_2 & \geq & 0 \end{array}$$

8.2 Consider the following problem. Minimize

$$Z = x_1$$

subject to

$$x_1^2 + x_2^2 = 1$$

Compute the dual function and show that is a concave function. Draw it. Find the optimal solutions for the dual and primal problems by means of comparing their objective functions.

8.3 Consider the following problem. Minimize

$$Z = e^{-x}$$

subject to

$$x \geq 0$$

Solve the primal problem. Compute explicitly the dual function and solve the dual problem.

8.4 Consider the following problem. Minimize

$$Z = 2x_1^2 + x_2^2 - 2x_1x_2 - 6x_2 - 4x_1$$

subject to

$$\begin{array}{rcl} x_1^2 + x_2^2 & = & 1 \\ -x_1 + 2x_2 & \leq & 0 \\ x_1 + x_2 & \leq & 8 \\ x_1,\ x_2 & \geq & 0 \end{array}$$

(a) Solve the problem by KKTCs

(b) Compute explicitly the dual function.

(c) Does a duality gap exist?

8.5 Find the minimum of the objective function

$$f(x_1, x_2, x_3) = 12(x_1^2 + x_2^2 + x_3^2)$$

subject to

$$x_1 x_2 + x_1 x_3 + x_2 x_3 = 1$$

8.6 Find the minimum length of a ladder that must lean against a wall if a box of dimensions a and b is placed right at the corner of that same wall.

8.7 Consider the function

$$\begin{aligned} f(y_1, y_2, y_3) = \quad & \min \{y_1 x_1 + y_2 x_2 + y_3 x_3 | x_1 + x_2 + x_3 = 1, \\ & x_1 - x_2 - x_3 = 0, x_i \geq 0\} \end{aligned}$$

Solve the mathematical programming problem. Maximize

$$Z = f(y_1, y_2, y_3)$$

subject to

$$\begin{aligned} y_1^2 + y_2^2 + y_3^2 &\leq 1 \\ y_3 - y_2 &\leq -\frac{1}{2} \end{aligned}$$

8.8 Find the maximum of the integral

$$J = \int_{x_1}^{x_2} (e^{-t} - e^{-2t})\, dt$$

with respect to the limits of integration subject to the constraint $x_2 - x_1 = c$, where c is a constant.

8.9 Find the maximum and the minimum of the function

$$f(x_1, x_2) = \int_{x_1}^{x_1 + x_2} \log(1 + t + t^2)\, dt$$

over the region determined by the inequalities

$$\begin{aligned} (x_1 + x_2)^2 &\leq 4 \\ -2 \leq x_2 &\leq 2 \end{aligned}$$

8.10 Find the closest point of the surface $xyz = 1$ to the origin.

8.11 Consider the function

$$f(x_1, x_2, x_3) = x_1^2 + x_2^2 + x_3^2 + x_1 + x_2 + x_3$$

Check that f is convex and find both its maximum and minimum subject to the constraints

$$\begin{aligned} x_1^2 + x_2^2 + x_3^2 &= 4 \\ x_3 &\leq 1 \end{aligned}$$

8.12 Determine the maximum and minimum values of

$$f(x_1, x_2, x_3) = x_1^4 x_2 + x_3^4$$

under the restriction

$$x_1^4 + x_2^4 + x_3^4 \leq 1$$

8.13 Same as in the previous exercise with the function

$$f(x_1, x_2) = x_1^3 - x_1^2 x_2^2 + x_1^2 + x_2^2$$

and constraints

$$\begin{array}{rcl} x_1^2 + x_2^2 & \leq & 4 \\ x_2 & \geq & x_1 - 1 \end{array}$$

8.14 Find the maximum length of a beam that can be pushed through the door of a room assuming that the height of the door is h and is located a distance d from the facing wall. The height of the room can be assumed infinite.

8.15 Consider the following problem. Maximize

$$x_2 - x_1$$

subject to

$$\begin{array}{r} x_1^2 + x_2^2 \leq 4 \\ -x_1^2 + x_2 = 0 \\ x_1, x_2 \geq 0 \end{array}$$

(a) Write the KKT conditions, and solve for the point $\mathbf{x}$ satisfying these conditions.

(b) Discuss if KKT conditions are necessary or sufficient for the optimality of this problem.

8.16 Consider the following problem. Minimize

$$Z = (x_1 - 3)^2 + (x_2 - 5)^2$$

subject to

$$\begin{array}{rrcl} x_1^2 & -x_2 & \leq & 0 \\ x_1 & +2x_2 & \leq & 10 \\ -x_1 & & \leq & 1 \\ x_1, & x_2 & \geq & 0 \end{array}$$

(a) Solve the problem using the KKTCs.

(b) Let $X = \{\mathbf{x} \in \mathbb{R}^2 | x_1 + 2x_2 \leq 10\}$. Formulate the dual problem.

(c) Prove that the dual function is differentiable at $\mathbf{0}$.

(d) Verify that the gradient at $\mathbf{0}$ is an ascent direction.

(e) Apply the gradient method to solve the dual problem, selecting as the initial guess $\mu^0 = \mathbf{0}$.

8.17 Consider the following problem. Minimize

$$f(\mathbf{x})$$

subject to:

$$\mathbf{x} \in X$$
$$\mathbf{A}\mathbf{x} = \mathbf{b}$$

where X is a polytope, and f is a concave function.

(a) Write the dual function associated with the set X.

(b) Prove that there exists an extreme point of X such that minimize the Lagrangian function.

(c) Show that the dual function is concave and piecewise.

8.18 Joan writes books and Sebastian publishes and sells books. They live in an unidimensional world where the number of copies sold of a given book depends only on the selling price of that book. Sebastian has computed the selling price most convenient for him, and Joan did the same for herself. Their respective selling prices differ by the quantity

$$\frac{c}{2(1-\gamma)}$$

where c is the cost of publishing the book and γ the author fee per book sold in per unit of the selling price. Have they both, Joan and Sebastian, rightly computed their respective most convenient selling prices? Why are they different?

Chapter 9

Computational Methods for Nonlinear Programming

In this chapter we deal with computational algorithms for obtaining solutions of the NLPP. In general, these methods generate a sequence of points such that its limit point is an optimal solution to the problem under consideration. To ensure convergence, we must assume that the NLPP is a differentiable convex program. However, in practice, these algorithms are applied even when the requirements for convergence are not satisfied.

The stopping criteria are normally based on the KKTC optimality conditions. When an iterate of the sequence satisfies, within a given tolerance, those conditions, the procedure is stopped and the corresponding iterate is taken as the local minimum.

Constrained optimizations problems are sometimes solved by converting them to unconstrained ones. Thus, many computational methods for the constrained problem have been developed based on existing methods for the unconstrained case, which, formally speaking, is a special case of the constrained optimization problem, where the domain of definition is $S = \mathbb{R}^n$.

In this chapter we deal first with unconstrained methods in Section 9.1, and in Section 9.2 with the constrained case.

To begin unconstrained optimization, we discuss the minimization of a function of one variable. In Section 9.1.2 we study the descent directions methods to minimize a function of several variables. These methods transform the problem of minimizing a function of several variables in a sequence of one-dimensional minimization problems.

In Section 9.2 we discuss several strategies to deal with nonlinear problems with equality and inequality constraints. The first strategy consists of avoiding the constraints (or simplifying them) by solving the dual problem. The second strategy, which leads to the *penalty and barrier methods*, consists of converting the constrained problem into an equivalent sequence of unconstrained problems.

It is worthwhile mentioning that the GAMS package is a useful tool to formulate mathematical programming problems that save us from worrying about all these matters. GAMS uses various optimizers for NLPP that are somehow based on the methods described in this chapter.

9.1 Unconstrained Optimization Algorithms

9.1.1 Line Search Methods

We begin with the following one-dimensional problem. Minimize

$$Z = f(x) \tag{9.1}$$

where $x \in \mathbb{R}$ and $f : \mathbb{R} \to \mathbb{R}$ is a differentiable function for any $x \in \mathbb{R}$.

There is a wide collection of procedures to solve this problem. However, they can be divided into two categories:

1. Those that use information on derivatives. They are based on applying numerical methods to solve the necessary condition (KKT conditions)

$$f'(x^*) = 0 \tag{9.2}$$

where x^* is the solution looked for.

2. Those that use evaluations of the objective function only. They use interpolation formulas to iteratively estimate the minimum of a one-dimensional problem.

Line Search Using Derivatives

Note that if f is also a convex function, then, a point x^* is a global optimum if and only if it satisfies Equation (9.2), and problems (9.1) and (9.2) are equivalent. Since this problem leads us to find a root of equation (9.2), we can use root finding methods, such as the Newton, the quasi-Newton, or the bisection methods. In this section we discuss only the first two methods.

First, we remind the reader how to generate the sequence to solve equation $g(x) = 0$ using these methods:

Newton method. Since the value of the function $g(x)$ at the point $x^{(t+1)}$ can be approximated by

$$g(x^{(t+1)}) = g(x^{(t)}) + g'(x^{(t)})(x^{(t+1)} - x^{(t)})$$

and we want this to be zero, we obtain

$$x^{(t+1)} = x^{(t)} - \frac{g(x^{(t)})}{g'(x^{(t)})}$$

Quasi-Newton or Secant method. Replacing in the above expression $g'(x^{(t)})$ by the approximation

$$\frac{g(x^{(t)}) - g(x^{(t-1)})}{x^{(t)} - x^{(t-1)}}$$

we get

$$x^{(t+1)} = x^{(t)} - \frac{g(x^{(t)})}{g(x^{(t)}) - g(x^{(t-1)})}(x^{(t)} - x^{(t-1)})$$

Applying these results to Equation (9.2) we obtain

1. **Newton's method**

$$x^{(t+1)} = x^{(t)} - \frac{f'(x^{(t)})}{f''(x^{(t)})}$$

which use first and second derivatives of the objective function.

2. **Secant method**

$$x^{(t+1)} = x^{(t)} - \frac{f'(x^{(t)})}{f'(x^{(t)}) - f'(x^{(t-1)})}(x^{(t)} - x^{(t-1)})$$

which requires only first derivatives

Example 9.1 (Line search using derivatives). Consider the following problem. Minimize

$$Z = f(x) = (x-1)^4 \tag{9.3}$$

It is clear that its optimal solution is achieved at $x^* = 1$. However, we illustrate in detail how the Newton method can be applied. The first and second derivatives are

$$\begin{array}{rcl} f'(x) & = & 4(x-1)^3 \\ f''(x) & = & 12(x-1)^2 \end{array}$$

and the Newton recursive sequence becomes

$$\begin{array}{rcl} x^{(t+1)} & = & x^{(t)} - \dfrac{f'(x^{(t)})}{f''(x^{(t)})} = x^{(t)} - \dfrac{4(x^{(t)}-1)^3}{12(x^{(t)}-1)^2} \\ & = & x^{(t)} - \dfrac{1}{3}(x^{(t)}-1) = \dfrac{2}{3}x^{(t)} + \dfrac{1}{3} \end{array} \tag{9.4}$$

Note that if the sequence converges to the point $\bar{x}$, then taking limits on both sides of the relationship (9.4), we obtain $\bar{x} = \frac{2}{3}\bar{x} + \frac{1}{3}$, and this equation has the unique solution $\bar{x} = x^* = 1$.

Table 9.1: Summary of computations for the Newton and secant methods.

	Newton's method		Secant's method	
t	$x^{(t)}$	$f'(x^{(t)})$	$x^{(t)}$	$f'(x^{(t)})$
1	0.000	-4.000	0.000	-4.000
2	0.333	-1.185	0.333	-1.185
3	0.556	-0.351	0.474	-0.583
4	0.704	-0.104	0.610	-0.238
5	0.803	-0.031	0.703	-0.104
6	0.868	-0.009	0.777	-0.045
7	0.912	-0.003	0.831	-0.019
8	0.942	-8.02×10^{-4}	0.873	-0.008
9	0.961	-2.38×10^{-4}	0.904	-0.003
10	0.974	-7.04×10^{-5}	0.927	-0.002

If we apply the quasi-Newton method, we obtain

$$\begin{aligned} x^{(t+1)} &= x^{(t)} - \frac{f'(x^{(t)})}{\dfrac{f'(x^{(t)}) - f'(x^{(t-1)})}{x^{(t)} - x^{(t-1)}}} \\ &= x^{(t)} - \left[\frac{x^{(t)} - x^{(t-1)}}{(x^{(t)} - 1))^3 - (x^{(t-1)} - 1)^3}\right](x^{(t)} - 1)^3 \\ &= x^{(t)} - \frac{x^{(t)} - x^{(t-1)}}{1 - \left(\dfrac{x^{(t-1)} - 1}{x^{(t)} - 1}\right)^3} \end{aligned} \tag{9.5}$$

Note that the Newton method requires only one initial point, but the quasi-Newton method needs two initial points. Table 9.1 shows the first 10 iterations for both methods. We have used as initial point $x_1 = 0$ for the Newton method, and $x_1 = 0$ and $x_2 = \frac{1}{3}$ for the quasi-Newton method.

■

Line Search Method without Derivatives

In this section we discuss the quadratic fit line search method, that uses a parabolic interpolation based on three points as an approximation to the objective function f. After computing the candidate minimum value we use it, together with two previous points, to repeat the process. More formally, suppose that we are trying to minimize a convex function $f(x)$, and assume that we have three points $a < b < c$ such that (see Figure 9.1)

$$f(a) \geq f(b) \leq f(c) \tag{9.6}$$

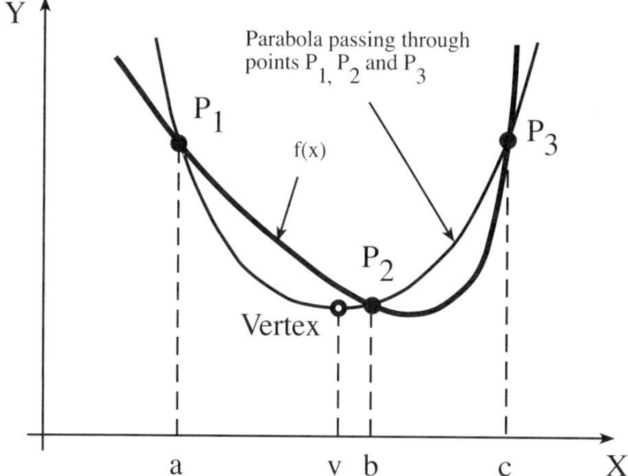

Figure 9.1: Objective function to be minimized and its approximating parabola.

We can assume that at least one of the previous inequalities is strict. Otherwise, because of its convexity, function f is constant and attains its minimum in the interval $[a, c]$. Next, we fit the quadratic curve passing through the points $(a, f(a)), (b, f(b))$ and $(c, f(c))$. The vertex of this parabola is

$$v = b - \frac{1}{2} \frac{(b-a)^2[f(b)-f(c)]^2 - (b-c)^2[f(b)-f(a)]}{(b-a)[f(b)-f(c)] - (b-c)[f(b)-f(a)]} \tag{9.7}$$

If the three points are collinear, the denominator degenerates to zero, and expression (9.7) is not valid. However, this does not occur under at least one of the above mentioned assumptions $f(a) > f(b)$ and $f(b) < f(c)$. Note that the condition (9.6) implies that the vertex of the parabola is a minimum (not a maximum), and that it is located on the interval (a, c).

Now we have a set of four points $\{a, b, c, v\}$ to choose from the new three points to iterate the process. The selection criterion looks for three new points to satisfy condition (9.6). Three cases are to be considered:

Case 1. $v < b$ (see Figure 9.2). There are two possibilities: (a) if $f(b) \geq f(v)$ the minimum of f is in the interval (a, b), thus, the new set of three points is $\{a', b', c'\} = \{a, v, b\}$; (b) on the other hand, if $f(b) < f(v)$, the minimum of f is in the interval (v, c), thus, the new set of three points becomes $\{a', b', c'\} = \{v, b, c\}$.

Case 2. $v > b$. By an argument similar to case 1, if $f(v) \geq f(b)$, we let $\{a', b', c'\} = \{a, b, v\}$, and if $f(v) < f(b)$, we let $\{a', b', c'\} = \{b, v, c\}$.

Case 3. $v = b$. In this case we recover the initial three points, and we are not able to decide which is the new reduced interval. To avoid this problem, we

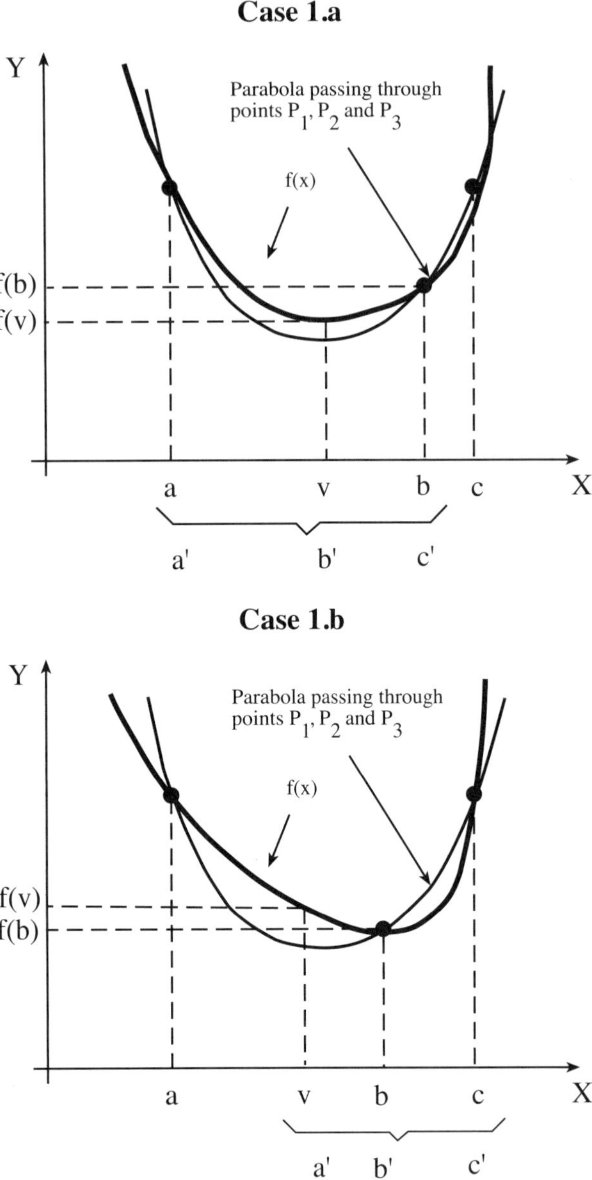

Figure 9.2: Illustration of cases 1a and 1b.

replace b by $(a+b)/2$ and repeat the process again, with the guarantee that the new situation will be in case 1 or case 2.

The quadratic fit technique described above entails global convergence under

Table 9.2: Summary of computations for the quadratic fit line search.

t	a	b	c	v	$f(v)$	$f(b)$
1	0.0000	0.3333	1.500	0.9545	4.268×10^{-6}	0.1975
2	0.3333	0.9545	1.500	1.0720	2.801×10^{-5}	4.268×10^{-6}
3	0.3333	0.9545	1.072	1.0130	3.240×10^{-8}	4.268×10^{-6}
4	0.9545	1.0130	1.072	0.9918	4.509×10^{-9}	3.240×10^{-8}
5	0.9545	0.9918	1.013	1.0020	2.715×10^{-11}	4.509×10^{-9}
6	0.9918	1.0020	1.013	0.9984	6.077×10^{-12}	2.715×10^{-11}
7	0.9918	0.9984	1.002	1.0000	9.785×10^{-15}	6.077×10^{-12}
8	0.9984	1.0000	1.002	0.9997	4.799×10^{-15}	9.785×10^{-15}
9	0.9997	1.0000	1.002	1.0000	3.821×10^{-19}	9.785×10^{-15}
10	0.9997	1.0000	1.000	1.0000	3.193×10^{-19}	3.821×10^{-19}

the hypothesis that f is a convex function. A modification that makes the procedure implementation easier consists of replacing systematically the last generated point by the vertex v. This new procedure has local convergence, which will be dependent on how close the initial points are to the optimal value.

Example 9.2 (Quadratic fit line search numerical example). We have applied the quadratic fit line search to Example 9.1. Table 9.2 shows the different three point sets generated as a function of the values $v, b, f(v)$ and $f(b)$. Note that for this example this method converges more rapidly than the Newton and the secant methods. This situation is due to the fact that x^* is not a simple root of $f'(x)$. In other situations the Newton method is faster.

■

9.1.2 Multidimensional Unconstrained Optimization

In this subsection we consider the problem of minimizing a function of several variables using derivatives. Consider the following problem. Minimize

$$Z = f(\mathbf{x}) \tag{9.8}$$

where $\mathbf{x} \in \mathbb{R}^n$ and $f : \mathbb{R}^n \to \mathbb{R}$ is a differentiable function for any $\mathbf{x} \in \mathbb{R}^n$. This problem can be solved using the *descent method*, described below.

A descent method generates a sequence of points such that the objective function value is sequentially reduced, i.e., a sequence $\{\mathbf{x}^{(t)}\}$ such that $f(\mathbf{x}^{(1)}) > f(\mathbf{x}^{(2)}) > \cdots > f(\mathbf{x}^{(t)}) > \cdots$ is generated until some stopping criterion is satisfied. This process is referred to as the *method of descent direction*. One iteration of the algorithm consists of the following two main steps:

1. **Descent direction generation problem**. Given a point $\mathbf{x}^{(t)}$, the problem consists of obtaining a direction $\mathbf{d}^{(t)}$ such that a slight movement from the point $\mathbf{x}^{(t)}$ in such direction decreases the objective function. Recall

that $\mathbf{d}$ is a descent direction of f at $\mathbf{x}$ if and only if there exist a positive number $\bar{\alpha}$ such that

$$f(\mathbf{x} + \alpha \mathbf{d}) < f(\mathbf{x}) \text{ for all } \alpha \in (0, \bar{\alpha})$$

2. **Line search problem.** Having obtained a descent direction $\mathbf{d}^{(t)}$ of f at the current trial point $\mathbf{x}^{(t)}$, the problem consists of deciding how large is the step length, α_t, such that the new point $\mathbf{x}^{(t)} + \alpha_t \mathbf{d}^{(t)}$ has a value of the objective function smaller than the value at the original point $\mathbf{x}^{(t)}$; thus, choose a $\alpha_t > 0$ such that $f(\mathbf{x}^{(t)} + \alpha_t \mathbf{d}^{(t)}) < f(\mathbf{x}^{(t)})$ to generate the next trial point $\mathbf{x}^{(t+1)} = \mathbf{x}^{(t)} + \alpha_t \mathbf{d}^{(t)}$. In many cases, the step length α_t is computed as the length that minimizes the objective function along the direction $\mathbf{d}^{(t)}$.

The following lemma shows that using the gradient, $\nabla f(\mathbf{x})$, descent directions can be characterized.

Lemma 9.1 *Let $f : \mathbb{R}^n \to \mathbb{R}$ be differentiable at $\mathbf{x} \in \mathbb{R}^n$. Let $\mathbf{d}$ be a vector in $\mathbb{R}^n$. If $\nabla f(\mathbf{x})^T \mathbf{d} < 0$, then $\mathbf{d}$ is a descent direction of f at $\mathbf{x}$.*

Proof. Since f is differentiable at $\mathbf{x}$, it follows that

$$f'(\mathbf{x}; \mathbf{d}) = \lim_{\alpha \to 0^+} \frac{f(\mathbf{x} + \alpha \mathbf{d}) - f(\mathbf{x})}{\alpha} = \nabla f(\mathbf{x})^T \mathbf{d} < 0.$$

Thus, there exist a positive number $\bar{\alpha}$ such that

$$\frac{f(\mathbf{x} + \alpha \mathbf{d}) - f(\mathbf{x})}{\alpha} < 0 \text{ for all } \alpha \in (0, \bar{\alpha})$$

and the lemma holds. ■

Figure 9.3 shows that a descent direction has a positive projection on the minus gradient direction.

Descent directions algorithms have the following structure.

Algorithm 9.1 (Descent direction).

Step 1. (Initial guess). *Choose an initial point $\mathbf{x}^{(1)} \in \mathbb{R}^n$, and let $t = 1$.*

Step 2. (Search direction generation). *Obtain a descent search direction $\mathbf{d}^{(t)}$.*

Step 3. (Optimality check). *If $\mathbf{d}^{(t)} = \mathbf{0}$, then* stop *($\mathbf{x}^{(t)}$ is a KTT point of the problem). Otherwise, continue.*

Step 4. (Line search). *Find a step length, α_t, which solves the onedimensional problem. Minimize*

$$Z_{LS} = f(\mathbf{x}^{(t)} + \alpha \mathbf{d}^{(t)})$$

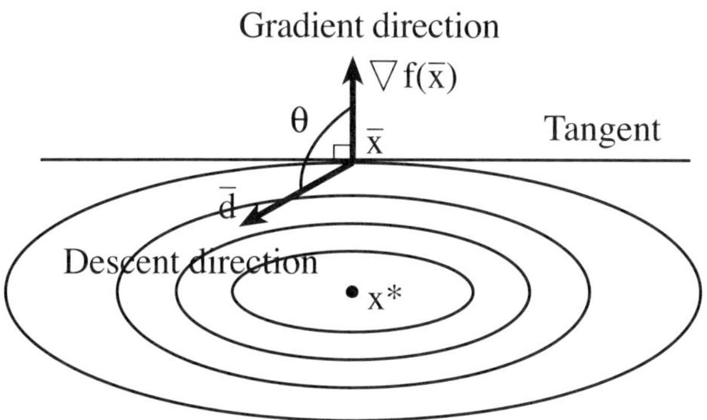

Figure 9.3: Illustration of descent directions in $\mathbb{R}^2$.

subject to

$$\alpha \geq 0$$

Step 5. (Update the trial point). Let $\mathbf{x}^{(t+1)} = \mathbf{x}^{(t)} + \alpha_t \mathbf{d}^{(t)}$.

Step 5. (Stopping criterion). $\|\mathbf{x}^{(t+1)} - \mathbf{x}^{(t)}\| \leq \epsilon$, then *stop*. Otherwise, let $t = t + 1$, and go to step 2.

■

All the methods discussed thus far have searched for a minimum in an n-dimensional space by performing one-dimensional minimizations in a set of directions $\{\mathbf{d}^{(t)}\}$. The Newton and the quasi-Newton line search methods discussed above use the first two derivatives of the univariate function $g(\alpha) = f(\mathbf{x}^{(t)} + \alpha \mathbf{d}^{(t)})$. They can be computed as

$$\begin{aligned} g'(\alpha) &= \nabla f(\mathbf{x}^{(t)} + \alpha \mathbf{d}^{(t)})^T \mathbf{d}^{(t)} \\ g''(\alpha) &= (\mathbf{d}^{(t)})^T \nabla^2 f(\mathbf{x}^{(t)} + \alpha \mathbf{d}^{(t)}) \mathbf{d}^{(t)} \end{aligned} \tag{9.9}$$

Thus, the efficiency of any such procedure depends critically on the efficiency of the method used to solve the line search. In practice, the computation of the step length α_t have to be a tradeoff between the reduction of f and the required computational time. Some practical strategies perform an *inexact line search* to identify α_t, which achieves an adequate reduction in f at a minimum computing cost. The aim of the inexact line searches is to keep the overall convergence to the stationary point, but reducing the amount of evaluations of the objective and the gradient functions. A popular inexact line search is the so called *Armijo's rule*. Let $0 < \varepsilon < 1$ and $\delta > 1$, which respectively manage the acceptable step length from being too large or too small. The first-order Taylor approximation

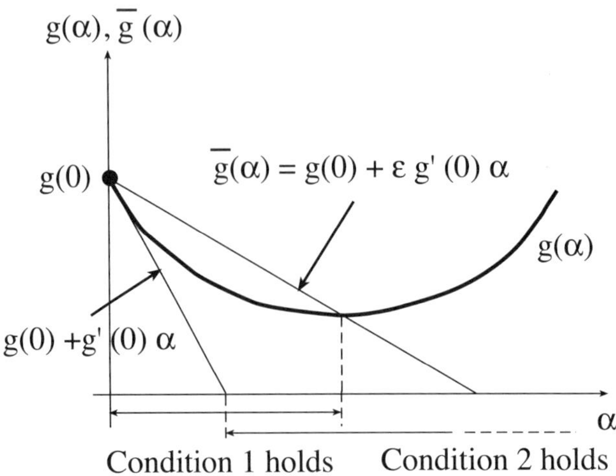

Figure 9.4: Graphical illustration of Armijo's rule.

of g at $\alpha = 0$ is given by $g(0) + \alpha g'(0)$. Now, we perturb the previous straight line, and define

$$\bar{g}(\alpha) = g(0) + \varepsilon g'(0)\alpha = f(\mathbf{x}^{(t)}) + \varepsilon\alpha\nabla f(\mathbf{x}^{(t)})^T\mathbf{d}^{(t)}$$

A step length α_t will be accepted if the following two conditions are satisfied:

$$g(\alpha_t) \le \bar{g}(\alpha_t) \tag{9.10}$$

$$g(\delta\alpha_t) > \bar{g}(\delta\alpha_t) \tag{9.11}$$

Figure 9.4 illustrates Armijo's rule. The condition (9.10) guarantees a sufficient descent in the value of the function g. This condition does not suffice by itself to ensure that the algorithm makes reasonable progress. Figure 9.4 shows that for all sufficiently small values of α, this condition holds. On the other hand, condition (9.11) guarantees a minimum step length.

In its most basic form Armijo's rule proceeds as follows.

Algorithm 9.2 (Armijo's rule).

Step 1(Initial guess). Choose $\hat{\alpha} > 0$, $\varepsilon \in (0, 1)$ and $\delta > 1$. Typical values are $\varepsilon = 0.2$ and $\delta = 2$ or $\delta = 10$. Set $\alpha = \hat{\alpha}$.

Step 2. If $g(\alpha) \le \bar{g}(\alpha)$, then go to step 3. Otherwise go to step 4.

Step 3. If $g(\delta\alpha) > \bar{g}(\delta\alpha)$, then *stop*, and let $\alpha_t = \alpha$. Otherwise, let $\alpha = \delta\alpha$ and go to step 2.

Step 4. If $g(\alpha/\delta) \le \bar{g}(\alpha/\delta)$, then stop. Otherwise, let $\alpha = \alpha/\delta$ and go to step 3. ■

This simple strategy for terminating a line search is appropriate for the Newton method but is not appropriate for the quasi-Newton and the conjugate gradient methods. The interested reader can find a detailed discussion of this problem in Nocedal and Wright [81].

These methods converge to a candidate for local optimum, and the algorithm stops when the iterates are sufficiently close to it. The stopping rule is frequently chosen as $\|\nabla f(\mathbf{x}^{(t)})\| \leq \varepsilon$, that is based on the necessary optimality condition $\nabla f(\mathbf{x}) = \mathbf{0}$ and the continuity of the function $\nabla f(\mathbf{x})$.

To apply a descent direction method, two steps are required. The first step is the selection of the descent direction, and the second one, the determination of the step length along the selected direction. Possible selections of descent directions are:

1. **Steepest descent method**. This method uses $\mathbf{d}^{(t)} = -\nabla f(\mathbf{x}^{(t)})$, as the search direction at $\mathbf{x}^{(t)}$ that, using Lemma 9.1, can be shown to be a descent direction. In fact, if $\nabla f(\mathbf{x}^{(t)}) \neq \mathbf{0}$ the following relation holds

$$\nabla f(\mathbf{x}^{(t)})^T \mathbf{d}^{(t)} = -\nabla f(\mathbf{x}^{(t)})^T \nabla f(\mathbf{x}^{(t)}) = -\left(\|\nabla f(\mathbf{x}^{(t)})\|\right)^2 < 0 \qquad (9.12)$$

Thus, $\mathbf{d}^{(t)}$ is a descent direction of f at $\mathbf{x}^{(t)}$.

2. **Newton Method**. This method chooses

$$\mathbf{d}^{(t)} = -\left[\nabla^2 f(\mathbf{x}^{(t)})\right]^{-1} \nabla f(\mathbf{x}^{(t)})$$

It is worth noting that the search direction $\mathbf{d}^{(t)}$ cannot be calculated if $\nabla^2 f(\mathbf{x}^{(t)})$ is a singular matrix. Even if $\nabla^2 f(\mathbf{x}^{(t)})$ is nonsingular, $\mathbf{d}^{(t)}$ is not necessarily a descent direction when $\nabla^2 f(\mathbf{x}^{(t)})$ is not positive definite. Assuming that $\nabla^2 f(\mathbf{x}^{(t)})$ is a positive definite matrix and $\nabla f(\mathbf{x}^{(t)}) \neq \mathbf{0}$, then its inverse matrix is also positive definite:

$$\nabla f(\mathbf{x}^{(t)})^T \mathbf{d}^{(t)} = \nabla f(\mathbf{x}^{(t)})^T \left(-\left[\nabla^2 f(\mathbf{x}^{(t)})\right]^{-1}\right) \nabla f(\mathbf{x}^{(t)}) < 0 \qquad (9.13)$$

and using Lemma 9.1, $\mathbf{d}^{(t)}$ is a descent direction of f at $\mathbf{x}^{(t)}$. Convex programming is a special relevant case. Using the characterization Theorem 8.4, it can be shown that for convex functions $\nabla^2 f(\mathbf{x}^{(t)})$ is a positive semidefinite matrix, and if it is nonsingular, then it is a positive definite matrix.

3. **Quasi-Newton Methods**. The search direction of Newton's method becomes undefined when the Hessian matrix is singular. Moreover, the computational effort required to obtain the inverse of Hessian matrices may become excessive for problems of even modest size. To overcome these difficulties, $[\nabla^2 f(\mathbf{x}^{(t)})]$ is approximated by a positive definite matrix

$\mathbf{B}^{(t)}$, which is updated successively so that it progressively converges to the exact Hessian matrix. The search direction is computed as

$$\mathbf{d}^{(t)} = -\left[\mathbf{B}^{(t)}\right]^{-1} \nabla f(\mathbf{x}^{(t)})$$

The resultant procedures are called *quasi-Newton methods.* One example of these methods is called *Davidon–Fletcher–Powell's* (DFP) *formula.* With the purpose of describing this method, in the following we denote

$$\mathbf{H}^{(t)} = \left[\mathbf{B}^{(t)}\right]^{-1}$$

The updating formula for this method is

$$\mathbf{H}^{(t+1)} = \mathbf{H}^{(t)} + \frac{\mathbf{p}^{(t)}\left(\mathbf{p}^{(t)}\right)^T}{\left(\mathbf{p}^{(t)}\right)^T\mathbf{q}^{(t)}} - \frac{\mathbf{H}^{(t)}\mathbf{q}^{(t)}\left(\mathbf{q}^{(t)}\right)^T\mathbf{H}^{(t)}}{\left(\mathbf{q}^{(t)}\right)^T\mathbf{H}^{(t)}\mathbf{q}^{(t)}} \tag{9.14}$$

where $\mathbf{p}^{(t)} = \mathbf{x}^{(t+1)} - \mathbf{x}^{(t)}$, $\mathbf{q}^{(t)} = \nabla f(\mathbf{x}^{(t+1)}) - \nabla f(\mathbf{x}^{(t)})$, and $\mathbf{H}^{(1)}$ is the identity matrix $\mathbf{I}_n$. It can be shown that $\mathbf{H}^{(t+1)}$ iteratively approximates the inverse of the Hessian matrix (see Luenberger [68]). This formula was first proposed by Davidon [31] and later improved by Fletcher and Powell [39]. Another example is the *Broyden–Goldfarb–Shanno's* (BFGS) *formula* which updates as follows:

$$\mathbf{H}^{(t+1)} = \left(\mathbf{I}_n - \frac{\mathbf{p}^{(t)}\left(\mathbf{q}^{(t)}\right)^T}{\left(\mathbf{q}^{(t)}\right)^T\mathbf{p}^{(t)}}\right)\mathbf{H}^{(t)}\left(\mathbf{I}_n - \frac{\mathbf{q}^{(t)}\left(\mathbf{p}^{(t)}\right)^T}{\left(\mathbf{q}^{(t)}\right)^T\mathbf{p}^{(t)}}\right) + \frac{\mathbf{p}^{(t)}\left(\mathbf{p}^{(t)}\right)^T}{\left(\mathbf{q}^{(t)}\right)^T\mathbf{p}^{(t)}}. \tag{9.15}$$

4. **A general case**. The previous methods are particular cases of considering a transformation of the minus gradient by means of a positive definite matrix. Let $\mathbf{A}^{(t)}$ be a positive matrix, namely, $\mathbf{x}^T\mathbf{A}^{(t)}\mathbf{x} > 0$ for all $\mathbf{x} \neq \mathbf{0}$. Let $\mathbf{d}^{(t)} = \mathbf{A}^{(t)}[-\nabla f(\mathbf{x}^{(t)})]$, then

$$\nabla f(\mathbf{x}^{(t)})^T\mathbf{d}^{(t)} = \nabla f(\mathbf{x}^{(t)})^T\mathbf{A}^{(t)}[-f(\mathbf{x}^{(t)})] < 0$$

and using Lemma 9.1, $\mathbf{d}^{(t)}$ is a descent direction. An important set of positive definite matrices are the called *projection matrices.* An $n \times n$ matrix is said to be a *projection matrix* if $\mathbf{P}^T = \mathbf{P}$ and $\mathbf{PP} = \mathbf{P}$.

5. **Conjugate gradient method**. This method computes the search direction by

$$\mathbf{d}^{(t)} = -\nabla f(\mathbf{x}^{(t)}) + \beta_{t-1}\mathbf{d}^{(t-1)} \tag{9.16}$$

where $\mathbf{d}^{(1)} = \mathbf{0}$. Using Lemma 9.1 to prove that it is a descent direction, we obtain

$$\nabla f(\mathbf{x}^{(t)})^T\mathbf{d}^{(t)} = -\|\nabla f(\mathbf{x}^{(t)})\|^2 + \beta_{t-1}\nabla f(\mathbf{x}^{(t)})^T\mathbf{d}^{(t-1)}. \tag{9.17}$$

Note that $\mathbf{x}^{(t)} = \mathbf{x}^{(t-1)} + \alpha_t \mathbf{d}^{(t-1)}$ minimizes the function f along the set $x^{(t-1)} + \alpha \mathbf{d}^{(t-1)}$. So the function $g(\alpha) = f(x^{(t-1)} + \alpha \mathbf{d}^{(t-1)})$ achieves a minimum at the value $\alpha = \alpha_{t-1}$ and $g'(\alpha_{t-1}) = 0$. Using (9.9) (with $t-1$ replacing t), we obtain

$$g'(\alpha_{t-1}) = \nabla f(x^{(t-1)} + \alpha_{t-1}\mathbf{d}^{(t-1)})^T \mathbf{d}^{(t-1)} = \nabla f(x^{(t)})^T \mathbf{d}^{(t-1)} = 0$$

As a result, the second term of (9.17) is zero, and $\nabla f(\mathbf{x}^{(t)})^T \mathbf{d}^{(t)} < 0$ if $\nabla f(\mathbf{x}^{(t)}) \neq \mathbf{0}$. There are many variants of the conjugate methods that differ in the choice of the parameter β_{t-1}. The Fletcher–Reeves method uses the formula

$$\beta_{t-1}^{FR} = \begin{cases} 0 & \text{if } t \text{ is a multiple of } n \\ \dfrac{\|\nabla f(\mathbf{x}^{(t)})\|^2}{\|\nabla f(\mathbf{x}^{(t-1)})\|^2} & \text{otherwise} \end{cases}$$

The Polak–Ribière method defines this parameter as follows:

$$\beta_{t-1}^{PR} = \begin{cases} 0 & \text{if } t \text{ is a multiple of } n \\ \dfrac{\nabla f(\mathbf{x}^{(t)})^T \left(\nabla f(\mathbf{x}^{(t)}) - \nabla f(\mathbf{x}^{(t-1)})\right)}{\|\nabla f(\mathbf{x}^{(t-1)})\|^2} & \text{otherwise} \end{cases}$$

To *restart* the iteration at every n steps, and for only one iteration, these methods set $\beta_{t-1} = 0$; that is, they choose a steepest-descent direction. This is so because for quadratic functions only n steps are required to find the optimum. In the nonquadratic case, after n steps, the full procedure is restarted.

These methods have finite convergence for a convex quadratic function after n successive exact line searches. Moreover, both algorithms are identical in this case. Numerical experiments show that the Polak–Ribière method tends to be the most robust and efficient of the two.

Example 9.3 (Numerical example). To illustrate the above methods, consider the following problem. Minimize

$$f(x_1, x_2) = -2x_1x_2 - 2x_2 + x_1^2 + 2x_2^2$$

which optimal solution is $(1,1)^T$.

1. Steepest-descent method. We perform two iterations of the steepest descent method beginning at the point $\mathbf{x}^{(1)} = \mathbf{0}$. At iteration t, this method uses as the search direction the minus gradient function at the point:

$$\mathbf{d}^{(t)} = -\nabla f(\mathbf{x}^{(t)}) = (-2x_2 + 2x_1, -2x_1 - 2 + 4x_2)^T$$

Step 1 (Initial guess). Let $\mathbf{x}^{(1)} = (0,0)^T$. For simplicity, we start from the origin, but in practical cases it can be more convenient to make a wiser selection of the starting point.

Step 2 (Search direction generation). In the first iteration

$$\mathbf{x}^{(1)} = (0,0)^T \text{ and } \mathbf{d}^{(1)} = -\nabla f(0,0) = (0,2)^T$$

Step 3 (Optimality check). As $\mathbf{d}^{(1)} \neq \mathbf{0}$ this is a descent direction.

Step 4 (Line search). To compute the step length, we solve the one-dimensional problem. Minimize

$$Z_{LS} = f(\mathbf{x}^{(1)} + \alpha \mathbf{d}^{(1)})$$

subject to

$$\alpha \geq 0$$

Since

$$\mathbf{x}^{(1)} + \alpha \mathbf{d}^{(1)} = (0,0)^T + \alpha(0,2)^T = (0, 2\alpha)^T$$

the objective function of this problem is

$$f(0, 2\alpha) = -4\alpha + 8\alpha^2$$

As $g(\alpha) = f(\mathbf{x}^{(1)} + \alpha \mathbf{d}^{(1)})$ is a convex function, the sufficient optimality condition is that $g'(\alpha) = 0$. Then, we solve $g'(\alpha) = -4 + 16\alpha = 0$ and obtain $\alpha_1 = \frac{1}{4} > 0$.

Step 5 (Update). Let $\mathbf{x}^{(2)} = \mathbf{x}^{(1)} + \alpha_1 \mathbf{d}^{(1)} = (0,0)^T + \frac{1}{4}(0,2)^T = (0, \frac{1}{2})^T$.

Step 6 (Convergence check). Since the gradient is not small enough, we repeat the iteration with a new point $\mathbf{x}^{(2)} = (0, \frac{1}{2})^T$, and let $t = 1 + 1 = 2$.

Step 2 (Search direction generation). The search direction at $\mathbf{x}^{(2)}$ is

$$\mathbf{d}^{(2)} = -\nabla f(0, \frac{1}{2}) = (1,0)^T$$

Step 3 (Optimality check). As $\mathbf{d}^{(2)} \neq \mathbf{0}$ this is a descent direction.

Step 4 (Line search). Since

$$\mathbf{x}^{(2)} + \alpha \mathbf{d}^{(2)} = (0, \frac{1}{2}) + \alpha(1,0) = (\alpha, \frac{1}{2})^T$$

then

$$f(\mathbf{x}^{(2)} + \alpha \mathbf{d}^{(2)}) = f(\alpha, \frac{1}{2}) = -\alpha + \alpha^2 - \frac{1}{2}.$$

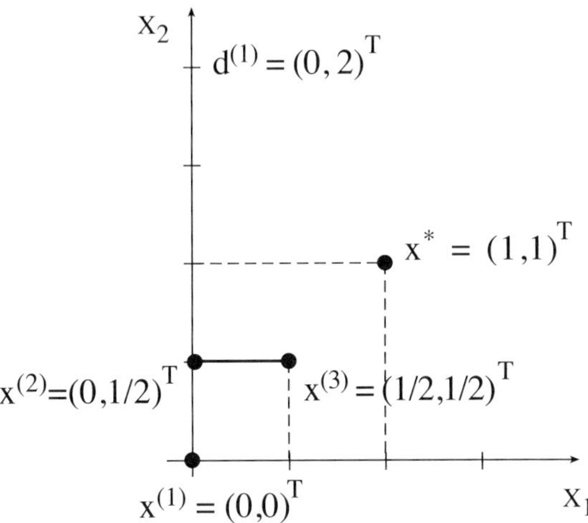

Figure 9.5: Progress of the steepest descent method.

Now, in the line search subproblem it becomes necessary to minimize

$$Z_{LS} = -\alpha + \alpha^2 - \frac{1}{2}$$

subject to

$$\alpha \geq 0$$

As $g(\alpha) = -\alpha + \alpha^2 - \frac{1}{2}$ is a convex function, we solve $g'(\alpha) = -1 + 2\alpha = 0$ and we obtain $\alpha_2 = \frac{1}{2} > 0$.

Step 5 (Update). We let

$$\mathbf{x}^{(3)} = \mathbf{x}^{(2)} + \alpha_2 \mathbf{d}^{(2)} = \left(0, \frac{1}{2}\right)^T + \frac{1}{2}(1, 0)^T = \left(\frac{1}{2}, \frac{1}{2}\right)^T$$

and $t = 2 + 1 = 3$

This procedure continues until convergence is attained. Figure 9.5 shows the evolution of the algorithm until the solution is obtained.

2. Newton's method. This method generates the search direction at the iteration t as

$$\mathbf{d}^{(t)} = -\left(\nabla^2 f(\mathbf{x}^{(t)})\right)^{-1} \nabla f(\mathbf{x}^{(t)})$$

The Hessian matrix is

$$\nabla^2 f(x_1, x_2) = \begin{pmatrix} 2 & -2 \\ -2 & 4 \end{pmatrix}$$

with inverse

$$\left(\nabla^2 f(x_1, x_2)\right)^{-1} = \begin{pmatrix} 1 & \frac{1}{2} \\ \frac{1}{2} & \frac{1}{2} \end{pmatrix}.$$

Step 1 (Initial guess). Let $\mathbf{x}^{(1)} = (0,0)^T$.

Step 2 (Search direction generation). In the first iteration $\mathbf{x}^{(1)} = (0,0)^T$ and the search direction becomes

$$\mathbf{d}^{(1)} = -[\nabla^2 f(\mathbf{x}^{(1)})]^{-1} \nabla f(\mathbf{x}^{(1)}) = \begin{pmatrix} -1 & -\frac{1}{2} \\ -\frac{1}{2} & -\frac{1}{2} \end{pmatrix} \begin{pmatrix} 0 \\ -2 \end{pmatrix} = \begin{pmatrix} 1 \\ 1 \end{pmatrix}$$

Step 3 (Optimality check). As $\mathbf{d}^{(1)} \neq \mathbf{0}$ and f is a strictly convex function this is a descent direction.

Step 4 (Line search). To compute the step length, we solve the one-dimensional problem, Minimize

$$Z_{LS} = f(\mathbf{x}^{(1)} + \alpha \mathbf{d}^{(1)})$$

subject to

$$\alpha \geq 0$$

Since

$$\mathbf{x}^{(1)} + \alpha \mathbf{d}^{(1)} = (0,0)^T + \alpha(1,1)^T = (\alpha, \alpha)^T$$

we get

$$f(\alpha, \alpha) = -2\alpha + \alpha^2.$$

and the optimal value is achieved at $\alpha_1 = 1 > 0$

Step 5 (Update). We let $\mathbf{x}^{(2)} = \mathbf{x}^{(1)} + \alpha_1 \mathbf{d}^{(1)} = (0,0)^T + (1,1)^T = (1,1)^T$.

Step 6 (Check the convergence). Since the condition to be checked is

$$\left| \nabla f(\mathbf{x}^{(2)}) \right| < \varepsilon$$

and we have $\nabla f(\mathbf{x}^{(2)}) = \mathbf{0}$, an optimal solution has been achieved, and the algorithm stops. This is due to the fact that the Newton method computes the descent direction as the solution of a quadratic approximation to the original problem in each iteration. As the original function is indeed a quadratic function the approximation coincides with the original function, and we get the exact solution in a single iteration.

3. Quasi-Newton method. In this example we use the DFP formula to update the approximation to the inverse Hessian matrix.

Step 1 (Initial guess). Let $\mathbf{x}^{(1)} = (0, 0)^T$. The first iteration of the Quasi-Newton method uses as an approximation to the inverse Hessian matrix the identity matrix:

$$\mathbf{H}^{(1)} = \begin{pmatrix} 1 & 0 \\ 0 & 1 \end{pmatrix}$$

and the search direction is $\mathbf{d}^{(1)} = -\mathbf{H}^{(1)}\nabla f(\mathbf{x}^{(1)}) = -\nabla f(\mathbf{x}^{(1)})$. This means that the first iteration coincides with the steepest-descent method. We do not repeat the steepest-descent step here, and assume that we are in iteration $t = 2$ and $\mathbf{x}^2 = (0, \frac{1}{2})^T$.

Step 2 (Search direction generation). To use the updating formula used in the DFP method, we need the following intermediate vector and matrices:

$$\begin{aligned}
\mathbf{x}^{(1)} &= \begin{pmatrix} 0 \\ 0 \end{pmatrix} \\
\mathbf{x}^{(2)} &= \begin{pmatrix} 0 \\ \frac{1}{2} \end{pmatrix} \\
\mathbf{p}^{(1)} &= \mathbf{x}^{(2)} - \mathbf{x}^{(1)} = \begin{pmatrix} 0 \\ \frac{1}{2} \end{pmatrix} \\
\nabla f(\mathbf{x}^{(1)}) &= \begin{pmatrix} 0 \\ -2 \end{pmatrix} \\
\nabla f(\mathbf{x}^{(2)}) &= \begin{pmatrix} -1 \\ 0 \end{pmatrix} \\
\mathbf{q}^{(1)} &= \nabla f(\mathbf{x}^{(2)}) - \nabla f(\mathbf{x}^{(1)}) = \begin{pmatrix} -1 \\ 2 \end{pmatrix} \\
\mathbf{p}^{(1)}\mathbf{p}^{(1)^T} &= \begin{pmatrix} 0 \\ \frac{1}{2} \end{pmatrix} (0, \frac{1}{2}) = \begin{pmatrix} 0 & 0 \\ 0 & \frac{1}{4} \end{pmatrix} \\
\mathbf{p}^{(1)^T}\mathbf{q}^{(1)} &= (0, \frac{1}{2}) \begin{pmatrix} -1 \\ 2 \end{pmatrix} = 1 \\
\mathbf{H}^{(1)}\mathbf{q}^{(1)}\mathbf{q}^{(1)^T}\mathbf{H}^{(1)} &= \begin{pmatrix} 1 & 0 \\ 0 & 1 \end{pmatrix} \begin{pmatrix} -1 \\ 2 \end{pmatrix} (-1, 2) \begin{pmatrix} 1 & 0 \\ 0 & 1 \end{pmatrix} \\
&= \begin{pmatrix} 1 & -2 \\ -2 & 4 \end{pmatrix} \\
\mathbf{q}^{(1)^T}\mathbf{H}^{(1)}\mathbf{q}^{(1)} &= (-1, 2) \begin{pmatrix} 1 & 0 \\ 0 & 1 \end{pmatrix} \begin{pmatrix} -1 \\ 2 \end{pmatrix} = 5
\end{aligned}$$

and obtain the updated approximate inverse Hessian matrix

$$\begin{aligned}\mathbf{H}^{(2)} &= \begin{pmatrix} 1 & 0 \\ 0 & 1 \end{pmatrix} + \frac{1}{1}\begin{pmatrix} 0 & 0 \\ 0 & \frac{1}{4} \end{pmatrix} - \frac{1}{5}\begin{pmatrix} 1 & -2 \\ -2 & 4 \end{pmatrix} \\ &= \begin{pmatrix} \frac{4}{5} & \frac{2}{5} \\ \frac{2}{5} & \frac{9}{20} \end{pmatrix}\end{aligned}$$

and the search direction

$$\mathbf{d}^{(2)} = -\mathbf{H}^{(2)}\nabla f(\mathbf{x}^{(2)}) = \begin{pmatrix} -\frac{4}{5} & -\frac{2}{5} \\ -\frac{2}{5} & -\frac{9}{20} \end{pmatrix}\begin{pmatrix} -1 \\ 0 \end{pmatrix} = \begin{pmatrix} \frac{4}{5} \\ \frac{2}{5} \end{pmatrix}.$$

Step 3 (Optimality check). As $\mathbf{d}^{(2)} \neq \mathbf{0}$ this is a descent direction.

Step 4 (Line search). To compute the step length, we solve the one-dimensional problem by minimizing

$$Z_{LS} = f(\mathbf{x}^{(2)} + \alpha\mathbf{d}^{(2)})$$

subject to

$$\alpha \geq 0$$

Since

$$\mathbf{x}^{(2)} + \alpha\mathbf{d}^{(2)} = \left(0, \frac{1}{2}\right)^T + \alpha\left(\frac{4}{5}, \frac{2}{5}\right)^T = \left(\frac{4\alpha}{5}, \frac{1}{2} + \frac{2\alpha}{5}\right)^T$$

then

$$f\left(\frac{4}{5}\alpha, \frac{1}{2} + \frac{2}{5}\alpha\right) = -\frac{1}{2} - \frac{4}{5}\alpha + \frac{8}{25}\alpha^2.$$

The optimal value of the line search is $\alpha_2 = \frac{5}{4} > 0$.

Step 5 (Update). We let

$$\mathbf{x}^{(3)} = \mathbf{x}^{(2)} + \alpha_2\mathbf{d}^{(2)} = \left(0, \frac{1}{2}\right)^T + \frac{5}{4}\left(\frac{4}{5}, \frac{2}{5}\right)^T = (1, 1)^T.$$

Step 6 (Convergence check). Since $\nabla f(\mathbf{x}^{(3)}) = \mathbf{0}$, we have attained the optimum, and the algorithm stops.

Optimality is obtained after two iterations (the dimension of the Hessian matrix). This is a consequence of its quadratic convex character.

4. Conjugate gradient method. In this example the Fletcher–Reeves and the Polak–Ribière formulas are equivalent because it is a quadratic function.

Step 1 (Initial guess). Let $\mathbf{x}^{(1)} = (0,0)^T$. Again, since the first iteration of the conjugate gradient method coincides with the steepest-descent method, the first search direction is $\mathbf{d}^{(1)} = -\nabla f(\mathbf{x}^{(1)})$, and we assume that the current iteration is $t = 2$ and $\mathbf{x}^2 = (0, \frac{1}{2})^T$.

Step 2 (Search direction generation). We obtain

$$\begin{aligned}
\nabla f(\mathbf{x}^{(1)}) &= \begin{pmatrix} 0 \\ -2 \end{pmatrix}; \ \nabla f(\mathbf{x}^{(2)}) = \begin{pmatrix} -1 \\ 0 \end{pmatrix} \\
\|\nabla f(\mathbf{x}^{(1)})\|^2 &= 0^2 + (-2)^2 = 4 \\
\|\nabla f(\mathbf{x}^{(2)})\|^2 &= 1^2 + 0^2 = 1 \\
\beta_1^{FR} &= \frac{\|\nabla f(\mathbf{x}^{(2)})\|^2}{\|\nabla f(\mathbf{x}^{(1)})\|^2} = \frac{1}{4}
\end{aligned}$$

and compute the search direction as

$$\mathbf{d}^{(2)} = \nabla f(\mathbf{x}^{(2)}) + \beta_1^{FR}\mathbf{d}^{(1)} = \begin{pmatrix} -1 \\ 0 \end{pmatrix} + \frac{1}{4}\begin{pmatrix} 0 \\ 2 \end{pmatrix} = \begin{pmatrix} -1 \\ \frac{1}{2} \end{pmatrix}$$

Step 3 (Optimality check). As $\mathbf{d}^{(2)} \neq \mathbf{0}$ this is a descent direction.

Step 4 (Line search). To compute the step length, we solve the one-dimensional problem. Minimize

$$Z_{LS} = f(\mathbf{x}^{(2)} + \alpha \mathbf{d}^{(2)})$$

subject to

$$\alpha \geq 0$$

Since

$$\mathbf{x}^{(2)} + \alpha \mathbf{d}^{(2)} = (0, \frac{1}{2})^T + \alpha(1, \frac{1}{2})^T = (\alpha, \frac{1}{2} + \frac{\alpha}{2})^T$$

we get

$$f(\alpha, \frac{1}{2} + \frac{\alpha}{2}) = -1/2 - \alpha + \frac{\alpha^2}{2}$$

The optimal value of the line search is $\alpha_2 = 1 > 0$.

Step 5 (Update). We let

$$\mathbf{x}^{(3)} = \mathbf{x}^{(2)} + \alpha_2 \mathbf{d}^{(2)} = (0, \frac{1}{2})^T + 1(1, \frac{1}{2})^T = (1,1)^T$$

Step 6 (Convergence check). Since the iterate $\mathbf{x}^{(3)}$ satisfies $\nabla f(\mathbf{x}^{(3)}) = \mathbf{0}$, we have achieved the optimum, and the algorithm stops.

For convex quadratic functions the optimum is achieved after a number of iterations equal to the dimension of the Hessian matrix. ■

9.2 Constrained Optimization Algorithms

In this section we address the constrained optimization problem. Those problems can be solved using mainly the following methods:

1. *Dual methods* that solve the dual problem instead of the primal one.

2. *Penalty methods* that transform the constrained problem into a new problem where the constraints are incorporated into the objective function by means of an adequate selection of penalty parameters. These algorithms transform the original constrained problem into a sequence of unconstrained optimization problems.

3. *Augmented Lagrangian or multipliers method* that are quadratic penalty methods using the augmented Lagrangian function as the objective function.

4. *Feasible direction methods* that extend to constrained problems the unconstrained optimization basic descent direction algorithm. Now the directions are forced to be feasible.

5. *Sequential quadratic programming methods* that solve a sequence of quadratic programming problems that approximate the original problem.

In this book only the first two methods are discussed. The reader interested in other methods is referred to Bazaraa et al. [9].

9.2.1 Dual Methods

This section deals with methods for solving the NLPP based on the solution of the dual problem. The main motivation is that the dual problem in the majority of the cases requires the maximization of a concave function over a very simple convex set, and thus, it has no local maximum different from the global maximum. Two main questions must be addressed:

1. How to obtain the optimal solution of the primal problem solving the dual one.

2. How to solve the dual problem.

Consider the following problem. Minimize

$$Z_P = f(\mathbf{x})$$

subject to

$$\begin{array}{rcl}\mathbf{h}(\mathbf{x}) & = & \mathbf{0}\\ \mathbf{g}(\mathbf{x}) & \leq & \mathbf{0}.\end{array}$$

where $f : \mathbb{R}^n \rightarrow \mathbb{R}$, $\mathbf{h} : \mathbb{R}^\ell \rightarrow \mathbb{R}$ and $\mathbf{g} : \mathbb{R}^n \rightarrow \mathbb{R}^m$ are continuous. In its dual problem we maximize

$$Z_D = \theta(\boldsymbol{\lambda}, \boldsymbol{\mu})$$

subject to

$$\boldsymbol{\mu} \geq \mathbf{0}$$

where the dual function is defined by

$$\theta(\boldsymbol{\lambda}, \boldsymbol{\mu}) = \text{Infimum}_{\mathbf{x}} \left(f(\mathbf{x}) + \sum_{k=1}^{\ell} \lambda_k h_k(\mathbf{x}) + \sum_{j=1}^{m} \mu_j g_j(\mathbf{x}) \right)$$

with domain of definition

$$D = \{(\boldsymbol{\lambda}, \boldsymbol{\mu}) | \boldsymbol{\mu} \geq \mathbf{0}, \text{ Minimum } \mathcal{L}(\mathbf{x}, \boldsymbol{\lambda}, \boldsymbol{\mu}) \text{ exists}\},$$

that is, D is the set of vectors $(\boldsymbol{\lambda}, \boldsymbol{\mu})$ such that $\boldsymbol{\mu}$ is nonnegative for which $\mathcal{L}(\mathbf{x}, \boldsymbol{\lambda}, \boldsymbol{\mu})$ has a finite infimum. For simplicity, we assume that

$$D = \mathbb{R}^\ell \times (\mathbb{R}^m)^+ = \{(\boldsymbol{\lambda}, \boldsymbol{\mu}) | \boldsymbol{\mu} \geq \mathbf{0}\}$$

To evaluate the dual function for $\boldsymbol{\lambda}^{(t)}$ and $\boldsymbol{\mu}^{(t)}$, it is necessary to solve the problem

$$\min_{\mathbf{x}} \ \mathcal{L}(\mathbf{x}, \boldsymbol{\lambda}^{(t)}, \boldsymbol{\mu}^{(t)})$$

which is called *the Lagrangian problem.*

To discuss the first question, we define the set

$$X(\boldsymbol{\lambda}, \boldsymbol{\mu}) = \{\mathbf{x} | \mathbf{x} \text{ minimizes } \mathcal{L}(\mathbf{x}, \boldsymbol{\lambda}, \boldsymbol{\mu})\}$$

If the duality gap does not exist, i.e., $Z_D^* = Z_P^*$, and we are able to solve the dual problem and obtain as optimal values $(\boldsymbol{\lambda}^*, \boldsymbol{\mu}^*)$, then any feasible point $\mathbf{x}^* \in X(\boldsymbol{\lambda}^*, \boldsymbol{\mu}^*)$ solves the primal problem. Theorem 9.1 justifies that for convex programs the duality gap does not exist, and therefore we can obtain a primal solution based on the dual solutions. It gives a condition based on the differentiability of the dual function to validate the procedure for obtaining the primal solution based on the dual problem.

Theorem 9.1 (Sufficient condition for a primal solution based on a dual solution). *Let* $(\boldsymbol{\lambda}^*, \boldsymbol{\mu}^*)$ *solve the dual problem, and assume that* $\theta(\boldsymbol{\lambda}, \boldsymbol{\mu})$ *is differentiable at* $(\boldsymbol{\lambda}^*, \boldsymbol{\mu}^*)$*. Then any element* $\mathbf{x}^* \in X(\boldsymbol{\lambda}^*, \boldsymbol{\mu}^*)$ *solves the primal problem.* ■

Theorem 9.2 (Differentiability of the dual function). *The dual objective function is differentiable at a point* $(\boldsymbol{\lambda}^*, \boldsymbol{\mu}^*)$ *if and only if each* h_k *and* g_j *are constant over* $X(\boldsymbol{\lambda}^*, \boldsymbol{\mu}^*)$*. In this case the partial derivatives of* θ *are given by*

$$\left.\frac{\partial \theta(\boldsymbol{\lambda}, \boldsymbol{\mu})}{\partial \lambda_k}\right|_{(\boldsymbol{\lambda}^*, \boldsymbol{\mu}^*)} = h_k(\mathbf{x}), \quad \textit{for any } \mathbf{x} \in X(\boldsymbol{\lambda}^*, \boldsymbol{\mu}^*)$$

$$\left.\frac{\partial \theta(\boldsymbol{\lambda}, \boldsymbol{\mu})}{\partial \mu_j}\right|_{(\boldsymbol{\lambda}^*, \boldsymbol{\mu}^*)} = g_j(\mathbf{x}), \quad \textit{for any } \mathbf{x} \in X(\boldsymbol{\lambda}^*, \boldsymbol{\mu}^*).$$

■

A condition guaranteeing that the constraint functions are constant over $X(\boldsymbol{\lambda}^*, \boldsymbol{\mu}^*)$ is that the set $X(\boldsymbol{\lambda}^*, \boldsymbol{\mu}^*)$ contains only a single element; that is, $\mathcal{L}(\mathbf{x}, \boldsymbol{\lambda}^*, \boldsymbol{\mu}^*)$ is miniminized at the unique point $\mathbf{x}(\boldsymbol{\lambda}^*, \boldsymbol{\mu}^*)$. For convex programming problems, this uniqueness can be guaranteed. Since the objective function for these problems is strictly convex, the Lagrangian function $\mathcal{L}(\mathbf{x}, \boldsymbol{\lambda}, \boldsymbol{\mu})$ is strictly convex for all $(\boldsymbol{\lambda}, \boldsymbol{\mu})$. Thus, the Lagrangian problem has only one solution (if it exists) by Theorem 8.1.

According to the previous results, under the assumptions of $X(\boldsymbol{\lambda}, \boldsymbol{\mu})$ being a singleton, the dual problem is unconstrained, except for the conditions $\boldsymbol{\mu} \geq \mathbf{0}$, and $\theta(\boldsymbol{\lambda}, \boldsymbol{\mu})$ differentiable in D. Thus, a steepest-ascent method is appropriate for maximizing the dual function (this addresses the second question). Since the constraints $\boldsymbol{\mu} \geq \mathbf{0}$ imply that the ascent direction

$$\mathbf{d}^{(t)} = (\mathbf{d}_{\boldsymbol{\lambda}}^{(t)}, \mathbf{d}_{\boldsymbol{\mu}}^{(t)}) = (\nabla_{\boldsymbol{\lambda}} \theta(\boldsymbol{\lambda}^{(t)}, \boldsymbol{\mu}^{(t)}), \nabla_{\boldsymbol{\mu}} \theta(\boldsymbol{\lambda}^{(t)}, \boldsymbol{\mu}^{(t)}))$$

could be an infeasible direction, the projection of the gradient onto the nonnegative set $\{(\boldsymbol{\lambda}, \boldsymbol{\mu}) \in \mathbb{R}^{\ell} \times \mathbb{R}^m | \boldsymbol{\mu} \geq \mathbf{0}\}$ can be used as the search direction, obtaining both, an ascent and a feasible direction. This method is called *Rosen's gradient projection method for problems with linear constraints* (see Rosen [96]) particularized for the constraints $\boldsymbol{\mu} \geq \mathbf{0}$.

The steepest-ascent algorithm, modified to handle the constraints $\boldsymbol{\mu} \geq \mathbf{0}$, to maximize $\theta(\boldsymbol{\lambda}, \boldsymbol{\mu})$ is described below.

Algorithm 9.3 (Modified steepest-ascent algorithm).

Step 1 (Initial guess). Choose an initial point $(\boldsymbol{\lambda}^{(1)}, \boldsymbol{\mu}^{(1)}) \in D$, and let t=1.

Step 2 (Lagrangian Problem). Solve the Lagrangian problem with $\boldsymbol{\lambda} = \boldsymbol{\lambda}^{(t)}$ and $\boldsymbol{\mu} = \boldsymbol{\mu}^{(t)}$, obtaining $x(\boldsymbol{\lambda}^{(t)}, \boldsymbol{\mu}^{(t)})$.

Step 3 (Evaluation). Evaluate the dual function

$$\theta(\boldsymbol{\lambda}^{(t)}, \boldsymbol{\mu}^{(t)}) = \mathcal{L}(\mathbf{x}(\boldsymbol{\lambda}^{(t)}, \boldsymbol{\mu}^{(t)}), \boldsymbol{\lambda}^{(t)}, \boldsymbol{\mu}^{(t)})$$

and its gradient (see Chapter 8)

$$\begin{array}{rcl} \nabla_{\boldsymbol{\mu}}\theta(\boldsymbol{\lambda}^{(t)}, \boldsymbol{\mu}^{(t)}) & = & \mathbf{g}(\mathbf{x}(\boldsymbol{\lambda}^{(t)}, \boldsymbol{\mu}^{(t)})) \\ \nabla_{\boldsymbol{\lambda}}\theta(\boldsymbol{\lambda}^{(t)}, \boldsymbol{\mu}^{(t)}) & = & \mathbf{h}(\mathbf{x}(\boldsymbol{\lambda}^{(t)}, \boldsymbol{\mu}^{(t)})) \end{array}$$

Step 4 (Optimality check). If

$$\nabla_{\boldsymbol{\mu}}\theta(\boldsymbol{\lambda}^{(t)}, \boldsymbol{\mu}^{(t)}) = \mathbf{0} \text{ and } \nabla_{\boldsymbol{\lambda}}\theta(\boldsymbol{\lambda}^{(t)}, \boldsymbol{\mu}^{(t)}) = \mathbf{0}$$

then *stop* and exit, $(\boldsymbol{\lambda}^{(t)}, \boldsymbol{\mu}^{(t)})$ solves the dual problem. Otherwise, continue.

Step 5 (Search direction generation). Define a search direction $\mathbf{d}^{(t)} = (\mathbf{d}_{\boldsymbol{\lambda}}^{(t)}, \mathbf{d}_{\boldsymbol{\mu}}^{(t)})$, by

$$\begin{array}{rcl} \mathbf{d}_{\mu_j}^{(t)} & = & \begin{cases} \left.\dfrac{\partial\theta(\boldsymbol{\lambda}, \boldsymbol{\mu})}{\partial\mu_j}\right|_{(\boldsymbol{\lambda}^{(t)}, \boldsymbol{\mu}^{(t)})} & \text{if } \mu_j^{(t)} > 0 \\ \max\left\{0, \left.\frac{\partial\theta(\boldsymbol{\lambda}, \boldsymbol{\mu})}{\partial\mu_j}\right|_{(\boldsymbol{\lambda}^{(t)}, \boldsymbol{\mu}^{(t)})}\right\} & \text{if } \mu_j^{(t)} = 0 \end{cases} \\ \mathbf{d}_{\boldsymbol{\lambda}}^{(t)} & = & \nabla_{\boldsymbol{\lambda}}\theta(\boldsymbol{\lambda}^{(t)}, \boldsymbol{\mu}^{(t)}) \end{array}$$

Step 6 (Line search). Find a step length, α_t, which solves the following one-dimensional problem. Maximize with respect to α

$$\theta\left((\boldsymbol{\lambda}^{(t)}, \boldsymbol{\mu}^{(t)}) + \alpha\mathbf{d}^{(t)}\right)$$

subject to

$$\begin{array}{rcl} (\boldsymbol{\lambda}^{(t)}, \boldsymbol{\mu}^{(t)}) + \alpha\mathbf{d}^{(t)} & \in & D \\ \alpha & \geq & 0 \end{array}$$

Step 7 (Update). Let $(\boldsymbol{\lambda}^{(t+1)}, \boldsymbol{\mu}^{(t+1)}) = (\boldsymbol{\lambda}^{(t)}, \boldsymbol{\mu}^{(t)}) + \alpha_t\mathbf{d}^{(t)}$, and $t = t + 1$.

Step 8 (Convergence check). If the following conditions hold

$$||\boldsymbol{\lambda}^{(t+1)} - \boldsymbol{\lambda}^{(t)}|| \leq \epsilon \text{ and } ||\boldsymbol{\mu}^{(t+1)}) - \boldsymbol{\mu}^{(t)})|| \leq \epsilon$$

then *stop* the process and exit. Otherwise go to step 2.

∎

Note that to evaluate the dual function we must solve the Lagrangian problem, and that the line search is an intensive computational procedure. It is necessary to choose the step size α_t such that an increase in the dual function be produced; thus we select a step size α_t such that

$$\theta\left((\boldsymbol{\lambda}^{(t)}, \boldsymbol{\mu}^{(t)}) + \alpha_t \mathbf{d}^{(t)}\right) > \theta(\boldsymbol{\lambda}^{(t)}, \boldsymbol{\mu}^{(t)}) \tag{9.18}$$

If θ is differentiable, unless $(\boldsymbol{\lambda}^{(t)}, \boldsymbol{\mu}^{(t)})$ maximizes θ, there exists α_t satisfying (9.18). In some cases, the line search becomes simple thanks to the special structure of the dual function (see Example 9.4).

It should be noted that the Rosen gradient projection method converges to a global solution for concave differentiable objective functions with the simple constraints $\boldsymbol{\mu} \geq 0$ (see [33]).

However, it should be noted that in most practical cases the dual function is not differentiable.

Example 9.4 (Dual method). Consider the following problem. Minimize

$$Z = \sum_{i=1}^{m}\sum_{j=1}^{n} f_{ij}(T_{ij}) = \sum_{i=1}^{m}\sum_{j=1}^{n} T_{ij}(\frac{\log T_{ij}}{t_{ij}} - 1) \tag{9.19}$$

subject to

$$\begin{aligned} \sum_{j=1}^{n} T_{ij} &= r_i, \quad i = 1, \ldots, m && (9.20)\\ \sum_{i=1}^{m} T_{ij} &= c_j, \quad j = 1, \ldots, n && (9.21)\\ T_{ij} &\geq 0, \quad i = 1, \ldots, m, \; j = 1, \ldots, n && (9.22) \end{aligned}$$

where t_{ij}, r_i, and c_j are constants, and T_{ij} are the problem variables. Problem (9.19)–(9.22) is the primal problem. Below we formulate the dual problem. We consider that $T = \{\mathbf{T} | T_{ij} \geq 0\}$, and we introduce the multipliers $\lambda_i^r; i = 1, \ldots, m$ for constraints (9.20) and $\lambda_j^c; j = 1, \ldots, n$ for constraints (9.21).

The Lagrangian function becomes

$$\begin{aligned} \mathcal{L}(\mathbf{T}, \boldsymbol{\lambda}^r, \boldsymbol{\lambda}^c) &= \sum_{i=1}^{m}\sum_{j=1}^{n} f_{ij}(T_{ij}) + \sum_{i=1}^{m} \lambda_i^r \left(\sum_{j=1}^{n} T_{ij} - r_i\right)\\ &\quad + \sum_{j=1}^{n} \lambda_j^c \left(\sum_{i=1}^{m} T_{ij} - c_j\right)\\ &= \sum_{i=1}^{m}\sum_{j=1}^{n} \left(f(T_{ij}) + (\lambda_i^r + \lambda_j^c) T_{ij}\right)\\ &\quad - \left(\sum_{i=1}^{m} r_i \lambda_i^r + \sum_{j=1}^{n} c_j \lambda_j^c\right). \end{aligned} \tag{9.23}$$

Note that the second term does not depend on the T_{ij} variables.

To compute the dual function we must solve the following problem. Minimize with respect to $\mathbf{T}$

$$Z = \mathcal{L}(\mathbf{T}, \boldsymbol{\lambda}^r, \boldsymbol{\lambda}^c) \tag{9.24}$$

subject to

$$\mathbf{T} \in T \tag{9.25}$$

Then, the dual function is

$$\theta(\boldsymbol{\lambda}^r, \boldsymbol{\lambda}^c) = \min_{\mathbf{T} \in T} \mathcal{L}(\mathbf{T}, \boldsymbol{\lambda}^r, \boldsymbol{\lambda}^c) \tag{9.26}$$

This problem is separable in the T_{ij} variables, and therefore the solution of (9.24)–(9.25) can be obtained by solving the following $n \times m$ one dimensional problems. Minimize

$$Z_{ij} = f_{ij}(T_{ij}) + \lambda_{ij} T_{ij}$$

subject to

$$T_{ij} \geq 0$$

where $\lambda_{ij} = \lambda_i^r + \lambda_j^c$.

This is a convex programming problem because the objective function is the sum of two convex functions, and the constraints are linear. A sufficient condition for the optimality of a feasible solution T_{ij} is the gradient to be zero. This leads to the equation

$$f'_{ij}(T^*_{ij}) = -\lambda_{ij}$$

whose solution is

$$T^*_{ij} \quad = \quad t_{ij} \exp(-\lambda_{ij}) > 0 \tag{9.27}$$

and then, the optimal objective function value is

$$\begin{aligned} Z^*_{ij} &= f_{ij}(T^*_{ij}) + \lambda_{ij} T^*_{ij} \\ &= T^*_{ij} \left(\log(t_{ij} \exp(-\lambda_{ij})/t_{ij}) - 1\right) + \lambda_{ij} T^*_{ij} \\ &= -t_{ij} \exp(-\lambda_{ij}) \end{aligned} \tag{9.28}$$

Substituting T^*_{ij} in the Lagrangian function, we obtain

$$\begin{aligned} \theta(\boldsymbol{\lambda}^r, \boldsymbol{\lambda}^c) &= \mathcal{L}(T^*_{ij}, \lambda_i^r, \lambda_j^c) = \sum_{i=1}^{m} \sum_{j=1}^{n} Z^*_{ij} - \left(\sum_{i=1}^{m} r_i \lambda_i^r + \sum_{j=1}^{n} c_j \lambda_j^c \right) \\ &= -\sum_{i=1}^{m} \sum_{j=1}^{n} t_{ij} \exp(-\lambda_{ij}) - \sum_{i=1}^{m} r_i \lambda_i^r - \sum_{j=1}^{n} c_j \lambda_j^c \end{aligned}$$

and the dual problem can be stated as follows. Maximize

$$\theta(\boldsymbol{\lambda}^r, \boldsymbol{\lambda}^c)$$

Table 9.3: Data for the matrix balancing problem

i or j	(r_i)	(c_j)
1	12000	6750
2	10500	7300
3	3800	10000
4	7700	9950

	t_{ij}			
	1	2	3	4
1	–	60	275	571
2	50	–	410	443
3	123	61	–	47
4	205	265	75	–

Table 9.4: Dual results for Example 9.4

(t)	$(\boldsymbol{\lambda}^r)^{(t)}$				$\nabla_{\boldsymbol{\lambda}^r}\theta = h_r(\mathbf{T}^{(t)})$			
1	0	0	0	0	-11094.00	-9597.00	-3569.00	-7155.00
2	-1.54	-1.33	-0.50	-0.99	2507.84	1329.54	-2784.66	-3760.61
3	-1.31	-1.21	-0.76	-1.34	-3338.84	-2237.14	-2113.47	158.99
100	-1.68	- 1.52	-1.15	-0.92	0.00	0.00	0.00	0.00
(t)	$(\boldsymbol{\lambda}^c)^{(t)}$				$\nabla_{\boldsymbol{\lambda}^c}\theta = h_c(\mathbf{T}^{(t)})$			
1	0	0	0	0	-6372.00	-6914.00	-9240.00	-8889.00
2	-0.88	-0.96	-1.28	-1.23	-4461.50	-4439.23	960.95	5231.90
3	-1.30	-1.37	-1.19	-0.75	-2282.42	-1891.17	-1181.18	-2175.70
100	-1.78	-1.82	-1.04	-0.64	0.00	0.00	0.00	0.00

Table 9.5: Additional dual results for Example 9.4

(t)	$\|\nabla\theta\|_t$	α_t	$\theta((\boldsymbol{\lambda}^r, \boldsymbol{\lambda}^c)^{(t+1)})$
1	23063.15	1.3e-004	-2585.00
2	9892.49	9.3e-005	48793.09
3	5963.18	7.3e-005	53273.57
100	0.00	4.8e-005	55709.91

In what follows we apply this method to solve problem (9.19)–(9.22) using the data provided in Table 9.3.

A summary of the computations and the solution are shown in Table 9.4, which provides dual variable values, the gradient of the dual function, the result of the line search, including the norm of the dual function gradient, the step length, and the actual value of the dual function. Tables 9.4 and 9.5 provide the dual function, the gradient of dual function and the step length used in each step. Finally, Table 9.6 gives the solution of the problem (values of the primal variables). Note that the gradient of the dual function approaches zero, and the dual function increases and reaches the optimum (the gradient reaches zero). ∎

Table 9.6: Primal solution of Example 9.4

	T_{ij}			
	1	2	3	4
1	–	1997.89	4176.46	5825.64
2	1364.38	–	5293.38	3842.24
3	2322.86	1195.02	–	282.12
4	3062.76	4107.08	530.16	–

9.2.2 Penalty Methods

Penalty methods transform the original constrained problem into a sequence of unconstrained problems through the use of penalty functions. The idea of converting a constrained optimization problem into a sequence of appropriately built unconstrained problems is very appealing, since unconstrained problems can be solved both efficiently and reliably. Since all penalty methods treat equalities in the same way, penalty function methods are classified according to the procedure employed for handling inequality constraints, There are two types of methods:

1. *Exterior* point methods. The sequence of solutions of the unconstrained problems contains only infeasible points.

2. *Interior or barrier methods.* The sequence of solutions of the unconstrained problems contains only feasible points. A particular case of these methods is the LP *interior point method* that is analyzed at the end of this section.

Exterior Penalty Method

In the *exterior penalty method*, the penalty imposed on the objective function at point $\mathbf{x}$ increases as it deviates from the feasible set. Once again, as the penalty parameters are updated, the corresponding sequence of minimum points of the penalty problems converges to an optimal solution. In this case the sequence of points lies outside the feasible region.

We consider the following optimization problem. Minimize

$$Z = f(\mathbf{x})$$

subject to

$$\begin{array}{rcl} \mathbf{h}(\mathbf{x}) & = & \mathbf{0} \\ \mathbf{g}(\mathbf{x}) & \leq & \mathbf{0} \end{array} \tag{9.29}$$

where $f : \mathbb{R}^n \to \mathbb{R}$, $\mathbf{g} : \mathbb{R}^n \to \mathbb{R}^m$ and $\mathbf{h} : \mathbb{R}^\ell \to \mathbb{R}$ are continuous. We denote by S the feasible region.

The *exterior penalty function* is

$$P(\mathbf{x};r) = f(\mathbf{x}) + r\psi(\mathbf{h}(\mathbf{x})\mathbf{g}(\mathbf{x}))$$

where r is the *penalty parameter* and $\psi : \mathbb{R}^{\ell+m} \to \mathbb{R}$ is the *penalty function* (a function of the constraints), which satisfies

$$\begin{aligned} \psi(\mathbf{h}(\mathbf{x}), \mathbf{g}(\mathbf{x})) &= 0; \quad \forall \mathbf{x} \in S \\ \psi(\mathbf{h}(\mathbf{x}), \mathbf{g}(\mathbf{x})) &> 0; \quad \forall \mathbf{x} \notin S \end{aligned}$$

Theorem 9.3 (Convergence of the exterior method). *Let $\{r_t\}$ be a sequence of positive numbers diverging to $+\infty$ and $\psi : \mathbb{R}^{\ell+m} \to \mathbb{R}$ be an exterior continuous penalty function. We define the sequence*

$$\mathbf{x}^{(t)} = \textit{arg minimize} \left\{f(\mathbf{x}) + r_t\psi\left(\mathbf{h}(\mathbf{x}), \mathbf{g}(\mathbf{x})\right)\right\}, \; t = 1, 2, \ldots \tag{9.30}$$

and assume that for each r_t, there exists an optimal solution of this problem.

Every limit point of the sequence $\{\mathbf{x}^{(t)}\}$ generated by this exterior method is a global minimum of the original constrained problem. In particular, if the original problem has a unique optimal solution, then it is the limit point of $\{\mathbf{x}^{(t)}\}$. ■

A proof of this theorem can be found in Luenberger [68].

The most common penalty function is

$$P_{\alpha,\beta}(\mathbf{x};r) = f(\mathbf{x}) + r\left[\sum_{k=1}^{\ell} |h_k(\mathbf{x})|^{\alpha} + \sum_{j=1}^{m} |g_j(\mathbf{x})|_+^{\beta}\right]; \; \alpha, \beta \geq 1$$

where $|x|_+ = \max\{x, 0\}$ is called the *nonnegative operator*. Two particular cases are

1. Absolute-value penalty function

$$P_1(\mathbf{x};r) = f(\mathbf{x}) + r\left[\sum_{k=1}^{\ell} |h_k(\mathbf{x})| + \sum_{j=1}^{m} |g_j(\mathbf{x})|_+\right]$$

2. Quadratic penalty function

$$P_2(\mathbf{x};r) = f(\mathbf{x}) + r\left[\sum_{k=1}^{\ell} |h_k(\mathbf{x})|^2 + \sum_{j=1}^{m} |g_j(\mathbf{x})|_+^2\right]$$

Algorithms based on exterior penalty methods have the following structure.

Algorithm 9.4 (Exterior penalty method).

Step 1 (Initialization). Choose an initial point $\mathbf{x}^{(0)}$ and an initial penalty parameter $r_1 > 0$ and set $t = 1$. Let $\varepsilon > 0$ be a given tolerance, and $\eta > 1$ be a fixed number.

Step 2 (Subproblem). Solve by an appropriate descent method, where $\mathbf{x}^{(t-1)}$ is an initial point, the problem. Minimize

$$f(\mathbf{x}) + r_t \psi(\mathbf{h}(\mathbf{x}), \mathbf{g}(\mathbf{x})) \tag{9.31}$$

whose solution is $\mathbf{x}^{(t)}$.

Step 3 (Stopping criteria). If $\|\mathbf{x}^{(t)} - \mathbf{x}^{(t-1)}\| < \varepsilon$, then *stop* the process and exit. Otherwise, go to step 4.

Step 4 (Increase counter). Set $r_{t+1} = \eta r_t$. Set $t = t + 1$, and go to step 2. ■

Penalty methods have the following properties:

1. $P(\mathbf{x}(r), r)$ and $f(x(r))$, where $x(r)$ minimizes $P(\mathbf{x}, r)$, are nondecreasing functions in r.
2. $\psi\left(h(\mathbf{x}(r))\right)$ is a nonincreasing function in r.
3. The following limits condition holds:

$$\lim_{t \to +\infty} r_t \psi\left(h(\mathbf{x}^{(t)})\right) = 0$$

Those properties are analyzed in detail in Luenberger [68].

Example 9.5 (Exterior penalty method). Consider the following problem. Minimize

$$Z = (x_1 - 6)^2 + (x_2 - \frac{17}{2})^2$$

subject to

$$x_1^2 - x_2 = 0$$

Note that at iteration t, the problem to be solved for obtaining $x^{(t)}$ consists of minimization of

$$Z = (x_1 - 6)^2 + (x_2 - 17/2)^2 + r_t |x_1^2 - x_2|^\beta$$

Table 9.7 summarizes the computations using the penalty function method for the values $\beta = 1$ and $\beta = 2$. The starting point is taken as $\mathbf{x}^{(0)} = (0, 0)$. The initial value of the penalty parameter is taken as $r_1 = 0.001$ and $\eta = 10$. Figure 9.6 illustrates how the algorithm progresses for $\beta = 1$ and $\beta = 2$.

■

Table 9.7: Results for Example 9.5

β	t	r_t	$x_1^{(t)}$	$x_2^{(t)}$	$f(x^{(t)})$	$P(\mathbf{x}^{(t)}, r_t)$	$r_t\psi(h(\mathbf{x}^{(t)}))$
2	1	0.001	5.7227	8.5242	0.07746	0.6643	5.8687×10^{-1}
	2	0.010	4.7164	8.6361	1.6662	3.5180	1.8518×10^{-0}
	3	0.100	3.5184	8.8527	6.2826	7.5263	1.2437×10^{-0}
	4	1.000	3.0744	8.9758	8.7858	9.0122	2.2640×10^{-1}
	5	10.00	3.0078	8.9974	9.2004	9.2251	2.4740×10^{-2}
	6	100.0	3.0008	8.9997	9.2450	9.2475	2.4978×10^{-3}
	7	1,000	3.0001	9.0000	9.2495	9.2498	2.4997×10^{-4}
	8	10,000	3.0000	9.0000	9.2499	9.2500	2.5023×10^{-5}
1	1	0.001	5.9940	8.5005	0.0000	0.02746	2.7428×10^{-2}
	2	0.010	5.9406	8.5050	0.0003	0.27141	2.6786×10^{-1}
	3	0.100	5.4545	8.5500	0.3000	2.4202	2.1202×10^{-0}
	4	1.000	3.0000	9.0000	9.2500	9.2500	1.3051×10^{-5}
	5	10.00	3.0000	9.0000	9.2500	9.2500	6.2022×10^{-9}
	6	100.0	3.0000	9.0000	9.2500	9.2500	2.1316×10^{-12}
	7	1,000	3.0000	9.0000	9.2500	9.2500	0
	8	10,000	3.0000	9.0000	9.2500	9.2500	0
Exact	∞		3	9	9.25	9.25	0

Interior or Barrier Methods

Let us consider the following inequality constrained optimization problem. Minimize

$$Z = f(\mathbf{x})$$

subject to

$$\mathbf{g}(\mathbf{x}) \leq \mathbf{0}$$

where $f : \mathbb{R}^n \to \mathbb{R}, \mathbf{g} : \mathbb{R}^n \to \mathbb{R}^m$ are continuous functions, and $S = \{\mathbf{x}|\mathbf{g}(\mathbf{x}) \leq \mathbf{0}\}$ is the feasible region. We assume that $S^- = \{\mathbf{x}|\mathbf{g}(\mathbf{x}) < \mathbf{0}\}$ is not empty, and the closure of S^- is the set S.

We consider the objective function

$$P(\mathbf{x}; r) = f(\mathbf{x}) + r\phi\left(\mathbf{g}(\mathbf{x})\right)$$

where r is the *penalty parameter* and ϕ is the *interior penalty function*, that is, a function of the constraints. This function forms an infinite barrier along the boundary of the feasible region, that tends to favor the selection of feasible points over infeasible ones when using the unconstrained search method starting at a feasible point.

Interior penalty functions are functions defined on $(\mathbb{R}^m)^- = \{\mathbf{y} \in \mathbb{R}^m|\mathbf{y} < \mathbf{0}\}$, which satisfy the following conditions:

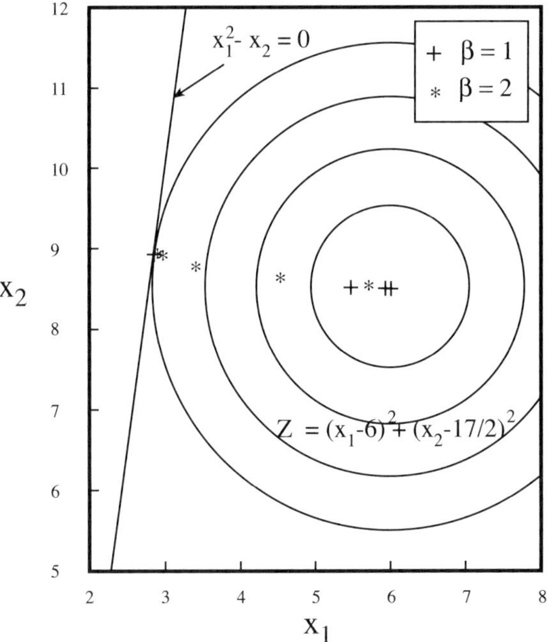

Figure 9.6: Illustration of the progress of the exterior penalty method for $\beta = 1$ and $\beta = 2$.

1. It must be positive in the interior of the feasible region

$$\phi\left(\mathbf{g}(\mathbf{x})\right) > 0; \quad \forall \mathbf{x} \in S^{-} \tag{9.32}$$

2. It must tend to ∞ in the boundary of the feasible region

$$\lim_{\mathbf{x} \to S - S^{-}} \phi(\mathbf{g}(\mathbf{x})) = +\infty \tag{9.33}$$

Actual examples of the penalty term, valid for the feasible region $\mathbf{g}(\mathbf{x}) \le \mathbf{0}$, are

1. The *logarithmic barrier penalty*

$$\phi(\mathbf{g}(\mathbf{x})) = -\sum_{j=1}^{m} \log(-g_j(\mathbf{x}))$$

2. The *inverse barrier penalty*

$$\phi(\mathbf{g}(\mathbf{x})) = \sum_{j=1}^{m} \frac{-1}{g_j(\mathbf{x})}$$

Theorem 9.4 (Convergence of barrier methods). *Let $\{r_t\}$ be a sequence of positive numbers converging to zero, and let $\phi : \mathbb{R}^m \to \mathbb{R}$ be an interior penalty function, we define*

$$\mathbf{x}^{(t)} = \textit{arg minimize}_{\mathbf{x} \in S^-} \left\{ f(\mathbf{x}) + r_t \phi(g(\mathbf{x})) \right\}, \ t = 1, 2, \ldots \tag{9.34}$$

Every limit point of a sequence $\{\mathbf{x}^{(t)}\}$ generated by a barrier method by means of (9.34) is a global minimum of the original constrained problem. In particular, if the original problem has a unique optimal solution, then it is the limit point of $\{\mathbf{x}^{(t)}\}$. ■

A proof of this theorem can be found in Luenberger [68].

A prototype of a barrier algorithm has the following structure.

Algorithm 9.5 (Barrier method).

Step 1 (Initialization). Choose an initial point $\mathbf{x}^{(0)}$ such that $\mathbf{x}^{(0)} \in S^-$, an initial penalty parameter $r_1 > 0$ and set $t = 1$. Let $\varepsilon > 0$ be a specified small number, and $\eta \in (0, 1)$ be a fixed number.

Step 2 (Unconstrained problem solution). Solve, by an appropriate descent method, where $\mathbf{x}^{(t-1)}$ is the initial point, the following problem. Minimize

$$Z = \{ f(\mathbf{x}) + r_t \phi(\mathbf{g}(\mathbf{x})) \}$$

subject to

$$\mathbf{x} \in S^-$$

whose optimal solution is $\mathbf{x}^{(t)}$.

Step 3 (Stopping Criteria). If $\|\mathbf{x}^{(t)} - \mathbf{x}^{(t-1)}\| < \varepsilon$, then stop the process and exit (alternative stopping criteria exist). Otherwise, go to step 4.

Step 4 (Increase counter). Set $r_{t+1} = \eta r_t$. Set $t = t + 1$, and go to step 2. ■

Example 9.6 (Barrier method). Consider the following problem. Minimize

$$Z = -5x_1 + x_1^2 - 8x_2 + 2x_2^2$$

subject to

$$3x_1 + 2x_2 \le 6$$

The logarithmic barrier function is

$$P(\mathbf{x}, r) = -5x_1 + x_1^2 - 8x_2 + 2x_2^2 - r \log(6 - 3x_1 + 2x_2)$$

The stationary points of $P(\mathbf{x}, r)$ as functions of r are

$$\begin{aligned}\frac{\partial P}{\partial x_1} &= 2x_1 - 5 + \frac{3r}{6 - 3x_1 - 2x_2} = 0\\ \frac{\partial P}{\partial x_2} &= 4x_2 - 8 + \frac{2r}{6 - 3x_1 - 2x_2} = 0\end{aligned}$$

which leads to $2x_1 - 6x_2 + 7 = 0$. Therefore

$$2x_1 - 5 - \frac{3r}{6 - 3x_1 - 2(-7/6 - x_1/3)} = 0$$

$$4x_2 - 8 - \frac{2r}{6 - 3(7/2 + 3x_2) - 2x_2} = 0$$

which is equivalent to

$$\begin{aligned}22x_1^2 - 77x_1 - 9r + 55 &= 0\\ 22x_2^2 - 77x_2 - r + 66 &= 0\end{aligned} \tag{9.35}$$

The solution of (9.35) produce the desired stationary values as a function of r:

$$x_1(r) = \frac{77 - \sqrt{99(11 + 8r)}}{44}$$

$$x_2(r) = \frac{77 - \sqrt{11(11 + 8r)}}{44}$$

where the other roots have been rejected because they are infeasible. Figure 9.7 shows the contours of the objective function as well as the feasible region. Note that the optimal solution is reached at the boundary of the feasible region, and that the trajectory toward the optimum belongs to the feasible region.

Note that

$$\begin{aligned}\lim_{r \to 0^+} x_1(r) &= 1\\ \lim_{r \to 0^+} x_2(r) &= \frac{3}{2}\\ \lim_{r \to 0^+} r\phi(\mathbf{x}) &= 0\end{aligned}$$

■

Example 9.7 (Numerical barrier method). Consider the following problem. Minimize

$$Z = 9 - 8x_1 - 6x_2 - 4x_3 + 2x_1^2 + 2x_2^2 + x_3^2 + 2x_1x_2 + 2x_1x_3$$

subject to

$$\begin{aligned}x_1 + x_2 + 2x_3 &\leq 3\\ x_i &\geq 0, \quad i = 1, 2, 3\end{aligned}$$

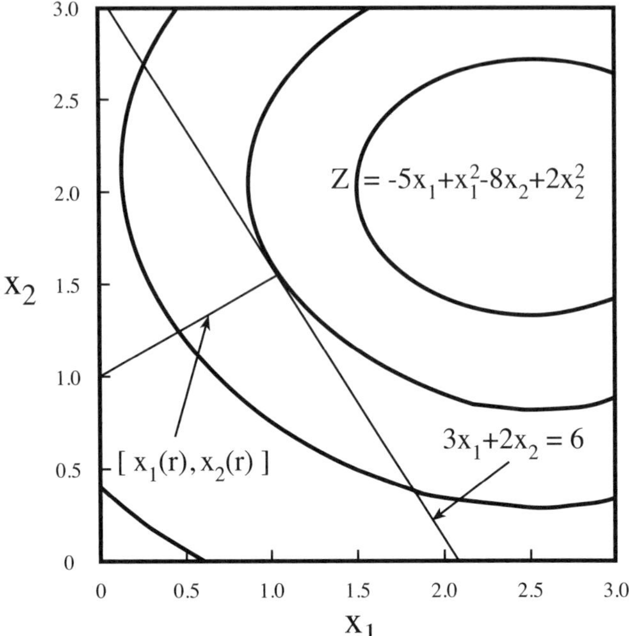

Figure 9.7: Illustration of the log barrier penalty method.

To compute the solution of this problem by means of the interior penalty function method, we have used the MATLAB package. This program uses a quasi-Newton method to solve the unconstrained problems. It uses the BFGS formula for updating the approximation of the Hessian matrix. The line search algorithm is a (cubic) polynomial interpolation method.

The logarithmic barrier function is not defined for negative x. However, it is possible to generate an infeasible point using such a barrier function, for instance, with an overly ambitious line search. Therefore, a special logic must be included to detect, prevent, and/or recover from such a situation. In this example and for inverse penalty and log barrier penalty methods, infeasible points are generated. To avoid this possibility, the following modified penalty term has been used:

$$\hat{\phi}(\mathbf{g}) = \phi(\hat{\mathbf{g}}) \text{ where } \hat{g}_i = \min\{g_i, -\varepsilon\}$$

for some small value of the tolerance ε.

The results of the numerical optimization are summarized in Table 9.8.

■

Table 9.8: Results for Example 9.7

	t	r_t	$x_1^{(t)}$	$x_2^{(t)}$	$x_3^{(t)}$	$f(x^{(t)})$	$P(\mathbf{x}^{(t)}, r_t)$	$r_t\phi(g(\mathbf{x}^{(t)}))$
Inverse	1	1.00	1.123	0.775	0.598	0.703	7.415	6.712
penalty	2	10^{-1}	1.292	0.697	0.331	0.232	1.043	0.810
	3	10^{-2}	1.379	0.737	0.351	0.154	0.259	0.104
	4	10^{-3}	1.379	0.795	0.351	0.125	0.145	0.020
	5	10^{-4}	1.373	0.799	0.374	0.115	0.121	0.005
	6	10^{-5}	1.335	0.776	0.440	0.112	0.114	0.001
	7	10^{-6}	1.336	0.775	0.442	0.111	0.112	0.000
Log	1	1.00	1.164	0.732	0.314	0.354	2.415	2.061
penalty	2	10^{-1}	1.372	0.736	0.333	0.168	0.426	0.258
	3	10^{-2}	1.344	0.769	0.422	0.120	0.160	0.040
	4	10^{-3}	1.334	0.776	0.442	0.112	0.118	0.006
Solution	∞		$\frac{12}{9}$	$\frac{7}{9}$	$\frac{4}{9}$	$\frac{1}{9}$	$\frac{1}{9}$	0

9.2.3 The Interior Point Method

For decades the simplex method dominated the solution algorithms for solving linear programming problems. However, from the theoretical point of view, the computing time required by the simplex algorithm may grow exponentially with the size of the problem defined as the number of variables plus the number of constraints. This exponential growth, however, corresponds to the worst case. In most practical cases the actual growth is slower than exponential. During those decades of simplex predominance, many researchers and practitioners tried to develop algorithms whose computing time requirements grew in a polynomial way with the problem size. All trials failed, either from the theoretical or the practical point of view, until 1984, where a distinguished researcher at AT&T, Karmarkar [57] proposed an algorithm with polynomial computational complexity that proved to be highly competitive with the simplex in terms of solution time when solving large linear programming problems. Pioneering interior point research works, previous to Karmarkar algorithm, include those of Frisch [45], Fiacco and McCormick [38], and Khachiyan [63]. The Karmarkar algorithm originated a lot of research work around his original idea. This idea has evolved and matured and it has been improved in many respects. One of the most fruitful derivations is the primal–dual logarithmic barrier algorithm introduced by Meggido [72] and explained in this chapter following Torres and Quintana [90] and Medina et al. [71]. Further and detailed information on primal-dual interior-point algorithms can be found in Wright [105].

Algorithm Foundation

This section provides the elements on which the primal–dual logarithmic barrier interior-point algorithm is based.

Consider a primal LP problem expressed in standard form:
Minimize

$$Z = \mathbf{c}^T \mathbf{x} \tag{9.36}$$

subject to

$$\begin{array}{rcl} \mathbf{A}\mathbf{x} & = & \mathbf{b} \\ \mathbf{x} & \geq & \mathbf{0} \end{array} \tag{9.37}$$

where $\mathbf{x}, \mathbf{c} \in \mathbb{R}^n$, $\mathbf{b} \in \mathbb{R}^m$ and $\mathbf{A} \in \mathbb{R}^m \times \mathbb{R}^n$.

The dual of the above problem, as previously stated, is as follows. Maximize

$$Z = \mathbf{b}^T \mathbf{y} \tag{9.38}$$

subject to

$$\mathbf{A}^T \mathbf{y} \leq \mathbf{c} \tag{9.39}$$

and converting inequalities to equalities using dual slack variables, the preceding dual problem becomes maximization of

$$Z = \mathbf{b}^T \mathbf{y} \tag{9.40}$$

subject to

$$\begin{array}{rcl} \mathbf{A}^T \mathbf{y} + \mathbf{z} & = & \mathbf{c} \\ \mathbf{z} & \geq & \mathbf{0} \end{array} \tag{9.41}$$

Dual variables are therefore $\mathbf{y} \in \mathbb{R}^m$ and $\mathbf{z} \in \mathbb{R}^n$.

To eliminate the nonnegativity constraints, logarithmic barrier functions are used; thus we deal with the following LPP. Maximize

$$Z = \mathbf{b}^T \mathbf{y} + \mu \sum_{j=1}^{n} \ln z_j \tag{9.42}$$

subject to

$$\mathbf{A}^T \mathbf{y} + \mathbf{z} = \mathbf{c} \tag{9.43}$$

It should be noted that z_j will never be zero and therefore the logarithmic barrier term is well posed.

The interior point method addresses the solution of the preceding problem for different values of the barrier parameter μ. This parameter is defined so that $\mu_0 > \mu_1 > \mu_2 > \ldots > \mu_\infty = 0$. The theorem below shows why parameter μ is chosen in that manner.

Theorem 9.5 (Convergence). *The sequence of parameters $\{\mu_t\}_{t=0}^{\infty}$ generates a sequence of problems of the type (9.42)–(9.43). The sequence of solutions of these problems, $\{x_t\}_{t=0}^{\infty}$, converges to the solution of the problem (9.40)–(9.41).* ■

The proof of this theorem can be found, for instance, in Wright [105].

The Lagrangian function of problem (9.42)-(9.43) has the form

$$\mathcal{L}(\mathbf{x}, \mathbf{y}, \mathbf{z}, \mu) = \mathbf{b}^T\mathbf{y} + \mu \sum_{j=1}^{m} \ln z_j - \mathbf{x}^T(\mathbf{A}^T\mathbf{y} + \mathbf{z} - \mathbf{c}) \tag{9.44}$$

Note that the Lagrangian multipliers, $\mathbf{x}$, are the variables of the original (primal) problem.

Using the Lagrangian above, the first order optimality conditions of the problem (9.42)–(9.43) are

$$\left\{ \begin{array}{l} \nabla_{\mathbf{x}}\mathcal{L}(\mathbf{x}, \mathbf{y}, \mathbf{z}, \mu) = \mathbf{A}^T\mathbf{y} + \mathbf{z} - \mathbf{c} = \mathbf{0} \\ \nabla_{\mathbf{y}}\mathcal{L}(\mathbf{x}, \mathbf{y}, \mathbf{z}, \mu) = \mathbf{A}\mathbf{x} - \mathbf{b} = \mathbf{0} \\ \nabla_{\mathbf{z}}\mathcal{L}(\mathbf{x}, \mathbf{y}, \mathbf{z}, \mu) = \mathbf{X}\mathbf{Z}\mathbf{e} - \mu\mathbf{e} = \mathbf{0} \end{array} \right. \tag{9.45}$$

where

$$\begin{array}{rcl} \mathbf{X} & \equiv & \mathrm{diag}(x_1, x_2, \ldots, x_n) \\ \mathbf{Z} & \equiv & \mathrm{diag}(z_1, z_2, \ldots, z_m) \\ \mathbf{e} & = & (1, 1, \ldots, 1)^T \end{array} \tag{9.46}$$

and the dimension of $\mathbf{e}$ is $n \times 1$.

The reader is encouraged to explore the compact notation given above, which uses diagonal matrices. It is illustrated in the Example 9.8.

Example 9.8 (Illustrative example). Consider the following primal problem. Minimize

$$Z = \begin{pmatrix} -3 & -5 & 0 & 0 & 0 \end{pmatrix} \begin{pmatrix} x_1 \\ x_2 \\ x_3 \\ x_4 \\ x_5 \end{pmatrix} \tag{9.47}$$

subject to

$$\begin{pmatrix} 1 & 0 & 1 & 0 & 0 \\ 0 & 1 & 0 & 1 & 0 \\ 3 & 2 & 0 & 0 & 1 \end{pmatrix} \begin{pmatrix} x_1 \\ x_2 \\ x_3 \\ x_4 \\ x_5 \end{pmatrix} = \begin{pmatrix} 4 \\ 6 \\ 18 \end{pmatrix} \tag{9.48}$$

$$x_1, x_2, x_3, x_4, x_5 \geq 0 \tag{9.49}$$

The feasible region of this problem, expressed in canonical form, is provided in Figure 9.8.

In its dual problem we maximize

$$Z = \begin{pmatrix} 4 & 6 & 18 \end{pmatrix} \begin{pmatrix} y_1 \\ y_2 \\ y_3 \end{pmatrix} \tag{9.50}$$

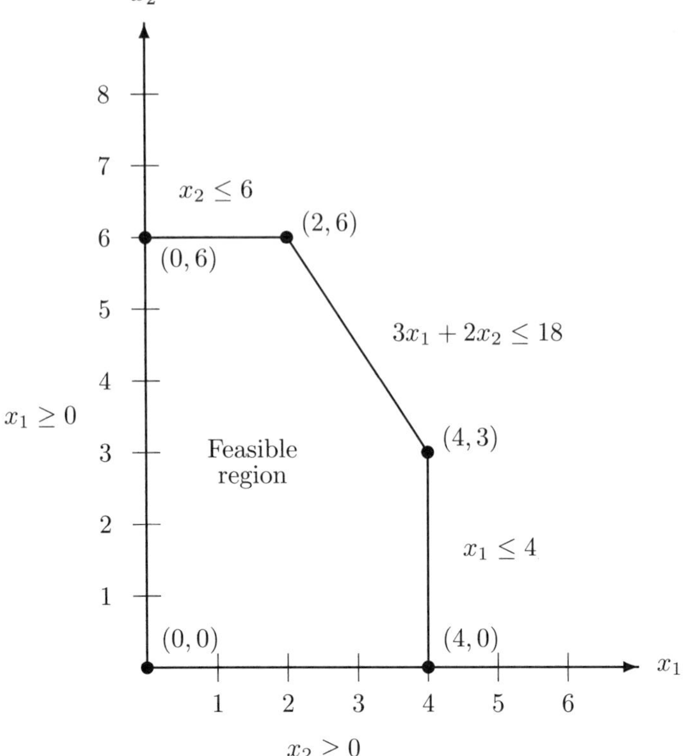

Figure 9.8: Example in canonical form.

subject to

$$\begin{pmatrix} 1 & 0 & 3 \\ 0 & 1 & 2 \\ 1 & 0 & 0 \\ 0 & 1 & 0 \\ 0 & 0 & 1 \end{pmatrix} \begin{pmatrix} y_1 \\ y_2 \\ y_3 \end{pmatrix} + \begin{pmatrix} z_1 \\ z_2 \\ z_3 \\ z_4 \\ z_5 \end{pmatrix} = \begin{pmatrix} -3 \\ 5 \\ 0 \\ 0 \\ 0 \end{pmatrix} \tag{9.51}$$

$$z_1, z_2, z_3, z_4, z_5 \geq 0 \tag{9.52}$$

The Lagrangian function is

$$\mathcal{L}(\mathbf{x}, \mathbf{y}, \mathbf{z}, \mu) = \begin{pmatrix} 4 & 6 & 18 \end{pmatrix} \begin{pmatrix} y_1 \\ y_2 \\ y_3 \end{pmatrix} + \mu \sum_{j=1}^{5} \ln z_j$$

$$
-\begin{pmatrix} x_1 \\ x_2 \\ x_3 \\ x_4 \\ x_5 \end{pmatrix}^T \left\{ \begin{pmatrix} 1 & 0 & 3 \\ 0 & 1 & 2 \\ 1 & 0 & 0 \\ 0 & 1 & 0 \\ 0 & 0 & 1 \end{pmatrix} \begin{pmatrix} y_1 \\ y_2 \\ y_3 \end{pmatrix} + \begin{pmatrix} z_1 \\ z_2 \\ z_3 \\ z_4 \\ z_5 \end{pmatrix} - \begin{pmatrix} -3 \\ 5 \\ 0 \\ 0 \\ 0 \end{pmatrix} \right\} \tag{9.53}
$$

The optimality conditions [Equations (9.45)] are

$$
\begin{pmatrix} 1 & 0 & 3 \\ 0 & 1 & 2 \\ 1 & 0 & 0 \\ 0 & 1 & 0 \\ 0 & 0 & 1 \end{pmatrix} \begin{pmatrix} y_1 \\ y_2 \\ y_3 \end{pmatrix} + \begin{pmatrix} z_1 \\ z_2 \\ z_3 \\ z_4 \\ z_5 \end{pmatrix} - \begin{pmatrix} -3 \\ 5 \\ 0 \\ 0 \\ 0 \end{pmatrix} = \begin{pmatrix} 0 \\ 0 \\ 0 \\ 0 \\ 0 \end{pmatrix} \tag{9.54}
$$

$$
\begin{pmatrix} 1 & 0 & 1 & 0 & 0 \\ 0 & 1 & 0 & 1 & 0 \\ 3 & 2 & 0 & 0 & 1 \end{pmatrix} \begin{pmatrix} x_1 \\ x_2 \\ x_3 \\ x_4 \\ x_5 \end{pmatrix} - \begin{pmatrix} 4 \\ 6 \\ 18 \end{pmatrix} = \begin{pmatrix} 0 \\ 0 \\ 0 \end{pmatrix} \tag{9.55}
$$

$$
\begin{pmatrix} x_1 & & & & \\ & x_2 & & & \\ & & x_3 & & \\ & & & x_4 & \\ & & & & x_5 \end{pmatrix} \begin{pmatrix} z_1 & & & & \\ & z_2 & & & \\ & & z_3 & & \\ & & & z_4 & \\ & & & & z_5 \end{pmatrix} \begin{pmatrix} 1 \\ 1 \\ 1 \\ 1 \\ 1 \end{pmatrix} - \begin{pmatrix} \mu \\ \mu \\ \mu \\ \mu \\ \mu \end{pmatrix} = \begin{pmatrix} 0 \\ 0 \\ 0 \\ 0 \\ 0 \end{pmatrix} \tag{9.56}
$$

■

To solve the system of equations (9.45) by the Newton method, $\mathbf{x}$, $\mathbf{y}$, and $\mathbf{z}$ are substituted respectively by $\mathbf{x} + \Delta\mathbf{x}$, $\mathbf{y} + \Delta\mathbf{y}$, and $\mathbf{z} + \Delta\mathbf{z}$, which neglecting second order terms leads to the following incremental system:

$$
\begin{array}{rcl} \mathbf{A}\Delta\mathbf{x} & = & \mathbf{0} \\ \mathbf{A}^T\Delta\mathbf{y} + \Delta\mathbf{z} & = & \mathbf{0} \\ \mathbf{Z}\Delta\mathbf{x} + \mathbf{X}\Delta\mathbf{z} & = & \mu\mathbf{e} - \mathbf{XZe} \end{array} \tag{9.57}
$$

The primal–dual search directions are obtained solving (9.57) for $\Delta\mathbf{y}$, $\Delta\mathbf{z}$, and $\Delta\mathbf{x}$:

$$
\begin{array}{rcl} \Delta\mathbf{y} & = & -(\mathbf{AXZ}^{-1}\mathbf{A}^T)^{-1}\mathbf{AZ}^{-1}\mathbf{v}(\mu) \\ \Delta\mathbf{z} & = & -\mathbf{A}^T\Delta\mathbf{y} \\ \Delta\mathbf{x} & = & \mathbf{Z}^{-1}\mathbf{v}(\mu) - \mathbf{XZ}^{-1}\Delta\mathbf{z} \end{array} \tag{9.58}
$$

where $\mathbf{v}(\mu) = \mu\mathbf{e} - \mathbf{XZe}$.

A Newton iteration, including step lengths, is performed to obtain a new point

$$
\begin{array}{rcl} \mathbf{x}^{t+1} & = & \mathbf{x}^t + \alpha_p\Delta\mathbf{x} \\ \mathbf{y}^{t+1} & = & \mathbf{y}^t + \alpha_d\Delta\mathbf{y} \\ \mathbf{z}^{t+1} & = & \mathbf{z}^t + \alpha_d\Delta\mathbf{z} \end{array} \tag{9.59}
$$

where $0 \leq \alpha_p \leq 1$, $0 \leq \alpha_d \leq 1$ are respectively the primal and dual step lengths. The primal step length is applied to primal variables $\mathbf{x}$, whereas the dual step length is applied to the dual variables $\mathbf{y}$ and $\mathbf{z}$.

The selection of α_p and α_d is done so that $\mathbf{x}$ and $\mathbf{z}$ remain strictly positive. It should be noted that $\mathbf{y}$ is not required to be positive. To ensure positiveness, the parameter σ $(0 < \sigma < 1)$ is used. Therefore

$$\begin{aligned} \alpha_x &= \min\left\{\frac{-x_i}{\Delta x_i} \text{ so that } \Delta x_i < -\delta\right\} \\ \alpha_p &= \min\{1, \sigma\alpha_x\} \end{aligned} \tag{9.60}$$

$$\begin{aligned} \alpha_z &= \min\left\{\frac{-z_j}{\Delta z_j} \text{ so that } \Delta z_j < -\delta\right\} \\ \alpha_d &= \min\{1, \sigma\alpha_z\} \end{aligned} \tag{9.61}$$

where δ is a tolerance (e.g., $\delta = 0.0001$) and σ is typically set to 0.99995.

It should be noted that these step length limitations become apparent for large negative values of variable increments Δx_i and Δz_j.

Duality Gap

For feasible $\mathbf{x}$ and $\mathbf{y}$, the duality gap is given by

$$\mathbf{c}^T\mathbf{x} - \mathbf{b}^T\mathbf{y}. \tag{9.62}$$

This duality gap is actually used as a measure of the proximity to optimality.

Initial Infeasibility

If the available initial point is not feasible, applying the Newton method to system (9.45) (first- order optimality conditions) yields the incremental system below

$$\begin{aligned} \mathbf{Z}\Delta\mathbf{x} + \mathbf{X}\Delta\mathbf{z} &= \mu\mathbf{e} - \mathbf{XZe} \\ \mathbf{A}\Delta\mathbf{x} &= \mathbf{b} - \mathbf{Ax} \\ \mathbf{A}^T\Delta\mathbf{y} + \Delta\mathbf{z} &= \mathbf{c} - \mathbf{A}^T\mathbf{y} - \mathbf{z} \end{aligned} \tag{9.63}$$

whose solution is

$$\begin{aligned} \Delta\mathbf{y} &= -(\mathbf{AXZ}^{-1}\mathbf{A}^T)^{-1}(\mathbf{AZ}^{-1}\mathbf{v}(\mu) - \mathbf{AXZ}^{-1}r_d - r_p) \\ \Delta\mathbf{z} &= -\mathbf{A}^T\Delta\mathbf{y} + r_d \\ \Delta\mathbf{x} &= \mathbf{Z}^{-1}\mathbf{v}(\mu) - \mathbf{XZ}^{-1}\Delta\mathbf{z} \end{aligned} \tag{9.64}$$

where

$$\begin{aligned} r_p &= \mathbf{b} - \mathbf{Ax} \\ r_d &= \mathbf{c} - \mathbf{A}^T\mathbf{y} - \mathbf{z} \end{aligned} \tag{9.65}$$

are respectively the primal and dual residuals.

If the initial point is not feasible, feasibility and optimality are achieved simultaneously as the interior-point algorithm progresses; in other words, the residuals and the duality gap approach zero simultaneously. If, on the other hand, the initial point is feasible, system (9.63) reduces to system (9.45).

Barrier Parameter Updating

A possible selection of μ is given below. From the optimality conditions (9.45), the following equality is obtained:

$$\mu = \frac{\mathbf{z}^T\mathbf{x}}{n} \tag{9.66}$$

To achieve a "central path," to avoid a premature approach to the feasibility boundary, the parameter ρ is introduced. Then

$$\mu = \rho\frac{\mathbf{z}^T\mathbf{x}}{n} \tag{9.67}$$

Values for the parameter ρ are tuned up experimentally by trial and error. A reasonable choice is [100]

$$\rho = \begin{cases} 0.1 & \text{if} \quad \mathbf{c}^T\mathbf{x} > \mathbf{b}^T\mathbf{y} \\ 0.2 & \text{if} \quad \mathbf{c}^T\mathbf{x} < \mathbf{b}^T\mathbf{y} \end{cases} \tag{9.68}$$

If $\mathbf{c}^T\mathbf{x} = \mathbf{b}^T\mathbf{y}$, the optimal solution has been found.

Stopping Criterion

The algorithm stops when the duality gap is small enough, that is, when

$$\frac{|\mathbf{c}^T\mathbf{x}^t| - |\mathbf{b}^T\mathbf{y}^t|}{\max\{1, |\mathbf{c}^T\mathbf{x}^t|\}} < \varepsilon \tag{9.69}$$

where ε is a per unit tolerance.

The numerator of the fraction above represents the actual duality gap, whereas the denominator is the maximum between 1 and the primal objective function value. This denominator form prevents a division by 0.

Bounds

Consider that the original primal problem includes upper and lower bounds on the $\mathbf{x}$ variables; Minimize

$$Z = \mathbf{c}^T\mathbf{x} \tag{9.70}$$

subject to

$$\begin{aligned} A\mathbf{x} &= \mathbf{b} \\ \mathbf{l} \le \mathbf{x} &\le \mathbf{u} \end{aligned} \tag{9.71}$$

where some of the components of $\mathbf{u}$ can be infinite and some of the components of $\mathbf{l}$ can be zero.

Using surplus and slack variables $\mathbf{s}$ and $\mathbf{v}$, the problem above can be written as follows. Minimize

$$Z = \mathbf{c}^T\mathbf{x} \tag{9.72}$$

subject to

$$\begin{array}{rcl} \mathbf{Ax} & = & \mathbf{b} \\ \mathbf{x}+\mathbf{s} & = & \mathbf{u} \\ \mathbf{x}-\mathbf{v} & = & \mathbf{l} \\ \mathbf{s} & \geq & \mathbf{0} \\ \mathbf{v} & \geq & \mathbf{0} \end{array} \tag{9.73}$$

and using logarithmic barrier parameters, the problem above now involves minimization of

$$Z = \mathbf{c}^T\mathbf{x} - \mu\left(\sum_{i=1}^{n}(\ln s_i + \ln v_i)\right) \tag{9.74}$$

subject to

$$\begin{array}{rcl} \mathbf{Ax} & = & \mathbf{b} \\ \mathbf{x}+\mathbf{s} & = & \mathbf{u} \\ \mathbf{x}-\mathbf{v} & = & \mathbf{l} \end{array} \tag{9.75}$$

and its Lagrangian has the form

$$\begin{array}{rcl} \mathcal{L}(\mathbf{x},\mathbf{y},\mathbf{z},\mu) & = & \mathbf{c}^T\mathbf{x} - \mu\left(\sum_{i=1}^{n}(\ln s_i + \ln v_i)\right) \\ & & -\mathbf{y}^T(A\mathbf{x}-\mathbf{b}) + \mathbf{w}^T(\mathbf{x}+\mathbf{s}-\mathbf{u}) - \mathbf{z}^T(\mathbf{x}-\mathbf{v}-\mathbf{l}) \end{array} \tag{9.76}$$

where the dual variables associated with constraints $\mathbf{Ax} = \mathbf{b}$ are $\mathbf{y}$, the dual variables associated with constraints $\mathbf{x}+\mathbf{s} = \mathbf{u}$ are $\mathbf{w}$, and the dual variables associated with constraints $\mathbf{x}-\mathbf{v} = \mathbf{l}$ are $\mathbf{z}$.

The optimality conditions are

$$\left\{\begin{array}{rcl} \nabla_{\mathbf{x}}\mathcal{L}(\mathbf{x},\mathbf{y},\mathbf{z},\mu) & = & \mathbf{A}^T\mathbf{y}+\mathbf{z}-\mathbf{w}-\mathbf{c}=\mathbf{0} \\ \nabla_{\mathbf{y}}\mathcal{L}(\mathbf{x},\mathbf{y},\mathbf{z},\mu) & = & \mathbf{Ax}-\mathbf{b}=\mathbf{0} \\ \nabla_{\mathbf{z}}\mathcal{L}(\mathbf{x},\mathbf{y},\mathbf{z},\mu) & = & \mathbf{x}-\mathbf{v}-\mathbf{l}=\mathbf{0} \\ \nabla_{\mathbf{s}}\mathcal{L}(\mathbf{x},\mathbf{y},\mathbf{z},\mu) & = & \mathbf{SWe}-\mu\mathbf{e}=\mathbf{0} \\ \nabla_{\mathbf{v}}\mathcal{L}(\mathbf{x},\mathbf{y},\mathbf{z},\mu) & = & \mathbf{VZe}-\mu\mathbf{e}=\mathbf{0} \\ \nabla_{\mathbf{w}}\mathcal{L}(\mathbf{x},\mathbf{y},\mathbf{z},\mu) & = & \mathbf{x}+\mathbf{s}-\mathbf{u}=\mathbf{0} \end{array}\right. \tag{9.77}$$

For the usual case of $\mathbf{l} = \mathbf{0}$, which implies $\mathbf{x} = \mathbf{v}$, these conditions become

$$\begin{array}{rcl} \mathbf{Ax} & = & \mathbf{b} \\ \mathbf{x}+\mathbf{s} & = & \mathbf{u} \\ \mathbf{A}^T\mathbf{y}+\mathbf{z}-\mathbf{w} & = & \mathbf{c} \\ \mathbf{XZe} & = & \mu\mathbf{e} \\ \mathbf{SWe} & = & \mu\mathbf{e} \end{array} \tag{9.78}$$

The search directions resulting from applying the Newton method to the system above are

$$\begin{array}{rcl} \Delta\mathbf{y} & = & (\mathbf{A}\Theta\mathbf{A})^{-1}\left[(\mathbf{b}-\mathbf{Ax}) + \mathbf{A}\Theta((\mathbf{c}-\mathbf{A}^T\mathbf{y}-\mathbf{z}+\mathbf{w})+\boldsymbol{\rho}(\mu))\right] \\ \Delta\mathbf{x} & = & \Theta\left[\mathbf{A}^T\Delta\mathbf{y}-\boldsymbol{\rho}(\mu)-(\mathbf{c}-\mathbf{A}^T\mathbf{y}-\mathbf{z}+\mathbf{w})\right] \\ \Delta\mathbf{z} & = & \mu X^{-1}\mathbf{e}-\mathbf{Ze}-\mathbf{X}^{-1}\mathbf{Z}\Delta\mathbf{x} \\ \Delta\mathbf{w} & = & \mu\mathbf{S}^{-1}\mathbf{e}-\mathbf{We}+\mathbf{S}^{-1}\mathbf{W}\Delta\mathbf{x} \\ \Delta\mathbf{s} & = & -\Delta\mathbf{x} \end{array} \tag{9.79}$$

where

$$\begin{aligned}\Theta &= (\mathbf{X}^{-1}\mathbf{Z} + \mathbf{S}^{-1}\mathbf{W})^{-1} \\ \boldsymbol{\rho}(\mu) &= \mu(\mathbf{S}^{-1} - \mathbf{X}^{-1})\mathbf{e} - (\mathbf{W} - \mathbf{Z})\mathbf{e}\end{aligned} \tag{9.80}$$

It should be noted that, although equations become slightly more complicated including bounds than without including them, the basic computation work remains the same.

Initial Point Selection

If no feasible initial point is available, the heuristic below, due to Vanderbei [100], usually provide a good starting point.

First, vector $\hat{\mathbf{x}}$ is computed as

$$\hat{x}_j = \frac{1}{\|A_j\|_2 + 1} \tag{9.81}$$

where A_i is the column j of the constraint matrix $\mathbf{A}$, and $\|\cdot\|_2$ is the quadratic norm.

Then, the scalar β is computed as

$$\beta = \frac{\|\mathbf{b}\|_2 + 1}{\|\mathbf{A}\hat{\mathbf{x}}\|_2 + 1} \tag{9.82}$$

The initial vector $\mathbf{x}^{(0)}$ is finally computed as

$$\mathbf{x}^{(0)} = 10\,\beta\,\hat{\mathbf{x}}. \tag{9.83}$$

In any component of vector $\mathbf{x}$ is larger than half of its upper bound, it is actually set to half of that upper bound. The dual variable vector $\mathbf{y}$ is set to zero. The dual variable vector $\mathbf{z}$ is initialized as a function of vector $\mathbf{x}$ as follows.

If x_j has a finite upper bound, then

$$z_j^0 = \begin{cases} 1 & \text{if } c_j < 0 \\ c_j + 1 & \text{otherwise} \end{cases} \tag{9.84}$$

If, on the other hand, x_j upper bound is infinite, then

$$z_j^0 = \begin{cases} 1 & \text{if } c_j < 1 \\ c_j & \text{otherwise} \end{cases} \tag{9.85}$$

Algorithm 9.6 (The primal–dual logarithmic barrier algorithm).

Input. A linear programming problem.

Output. Its solution or a message indicating that it is unbounded or has no solution.

Step 0. Set $t = 0$ and select an initial point $(\mathbf{x}^t, \mathbf{y}^t, \mu_t)$, using Equation (9.83). This initial point may be (primal and/or dual) feasible or not. Calculate the dual slack vector $\mathbf{z}^t = \mathbf{c} - \mathbf{A}^T\mathbf{y}^t$.

Step 1. Calculate the diagonal matrices $\mathbf{X}^t = \text{diag}(x_1^t, x_2^t, \ldots, x_n^t)$ and $\mathbf{Z}^t = \text{diag}(z_1^t, z_2^t, \ldots, z_m^t)$.

Step 2. Calculate the vector $\mathbf{v}(\mu_t) = \mu_t\mathbf{e} - \mathbf{X}^t\mathbf{Z}^t\mathbf{e}$, and the residual vectors $r_p^t = \mathbf{A}\mathbf{x}^t - \mathbf{b}$ and $r_d^t = \mathbf{c} - \mathbf{A}^T\mathbf{y}^t - \mathbf{z}^t$. If the current point is primal and dual feasible, the residuals are zero.

Step 3. Solve the Newton system for the search directions $\Delta\mathbf{y}^t$, $\Delta\mathbf{z}^t$, and $\Delta\mathbf{x}^t$, using Equations (9.64).

Step 4. Compute the step lengths α_p^t and α_d^t, using Equations (9.60) and (9.61).

Step 5. Update primal and dual variables, using equations (9.59).

Step 6. If the duality gap is small enough [Equation (9.69)], *stop*, an ε optimum has been found. Otherwise, set $t \leftarrow t + 1$ and continue.

Step 7. Update barrier parameter μ_t [Equations (9.67) and (9.68)] and go to step 1.

■

Exercises

9.1 Find the minimum of $f(x) = \exp(x - 1) - x$ by each of the following procedures:

(a) Secant's method

(b) Newton's method

(c) Quadratic fit line search

(d) Armijo's rule for inexact line search

9.2 Repeat the previous exercise for the functions

$$f(x) = x^4 - 6x^2$$

and

$$f(x) = x \ln x$$

9.3 Losses in a superconducting power generator depend on the temperature and the pressure of the cooling fluid as

$$f(x_1, x_2) = 100(x_2 - x_1^2)^2 + (1 - x_1)^2$$

where $f(x_1, x_2)$ are the losses, x_1 the pressure, and x_2 the temperature. To determine the working condition with minimum losses, use

(a) The steepest-descent method

(b) The Newton method

(c) A quasi-Newton method

(d) A conjugate gradient method

9.4 Consider the following problem. Minimize

$$Z = (x_1 - 2)^2 + 4(x_2 - 6)^2$$

subject to

$$\begin{aligned} 12x_1 + 6(x_2 + 1)^2 &\leq 4 \\ x_1, x_2 &\geq 0. \end{aligned}$$

(a) Derive the optimal solution using the KKT conditions.

(b) Use a conjugate gradient method to solve the unconstrained problem, starting at the origin. Perform at least three iterations.

9.5 Select one of the possible methods to solve the following problem. Minimize

$$Z = x_1^2 + 2x_2^2 + x_1 x_2 - 4x_1 - 6x_2$$

subject to:

$$\begin{aligned} x_1 + 2x_2 &\leq 4 \\ x_1, x_2 &\geq 0 \end{aligned}$$

and justify the selection.

9.6 Solve the following problem. Minimize

$$Z = (x - 2)^2$$

subject to

$$x \leq 1$$

by each of the following procedures:

(a) Logarithmic barrier penalty method

(b) Inverse barrier penalty method

9.7 Solve the following problem. Minimize

$$Z = x^2$$

subject to

$$x = 1$$

by each of the following exterior penalty methods:

(a) Absolute-value penalty function

(b) Quadratic penalty methods

9.8 Consider the following problem. Minimize

$$Z = -2x_1 + 8x_2 + x_1x_2 + x_1^2 + x_1^2 + 6x_2^2$$

subject to:

$$\begin{aligned} x_1 + 2x_2 &\leq 4 \\ 2x_1 + x_2 &\leq 5 \\ x_1, x_2 &\geq 0 \end{aligned}$$

(a) Find a local minimum of the unrestricted problem. Discuss the optimality conditions, and justify whether the found point is an optimal solution of the unrestricted problem.

(b) Use a method of descent directions to solve the unrestricted problem. Start the method at the origin $(0, 0)$. Justify the choice of the method and the obtained results.

(c) Solve the problem (b) using one of the algorithms in this chapter. Perform at least two iterations.

(d) Derive the dual problem and solve it using the steepest-ascent method.

9.9 Consider the folowing problem. Minimize

$$Z = x_1^2 + x_2^2$$

subject to

$$(x_1 - 2)^2 + (x_2 - 2)^2 \leq 1$$

Solve it by each of the following exterior penalty methods:

(a) Absolute-value penalty function

(b) Quadratic penalty methods

9.10 Consider the following problem. Minimize

$$-x_1 - x_2$$

subject to

$$\begin{aligned} \exp(x_1) + \exp(x_2) &\leq 20 \\ x_1, x_2 &\geq 0 \end{aligned}$$

(a) Derive an optimal solution using the KKTs conditions.

(b) Solve this problem using the dual method.

9.11 Consider the following strictly convex program *(Coordinate descent method and dualization)*. Minimize

$$Z = \sum_{i=1}^{m} f_i(x_i)$$

subject to

$$\begin{aligned} \sum_{i=1}^{m} g_{ij}(x_i) &\leq 0, \ j = 1, \ldots, r \\ x_i &\in X_i, \ i = 1, \ldots, m \end{aligned}$$

where $f_i : \mathbb{R}^{n_i} \to \mathbb{R}$ are strictly convex functions, x_i are the components of x, g_{ij} are given convex functions on $\mathbb{R}^{n_i}$, and X_i are given convex and compact subsets of $\mathbb{R}^{n_i}$.

(a) Compute the dual problem.

(b) *(Coordinate descent algorithm)*. The coordinate descent method is used for minimizing differentiable functions without the use of derivatives. Here the cost is minimized along one coordinate direction at each iteration. The order in which coordinates are chosen may vary in the course of the algorithm. In the case where this order is cyclical, given $\mathbf{x}^{(t)}$, the ith coordinate of $\mathbf{x}^{(t+1)}$ is determined by

$$x_i^{(t+1)} = \arg\min_{\zeta \in \mathbb{R}} \ f(\mathbf{x}_1^{(t)}, \ldots, \mathbf{x}_{i-1}^{(t)}, \zeta, \mathbf{x}_{i+1}^{(t)}, \ldots, \mathbf{x}_n^{(t)})$$

Take into account the special structure of the dual problem to apply the coordinate descent method by means of a parallel computation.

(c) For continuously differentiable cost functions, it can be shown to generate sequences whose limit points are stationary; that is, if $\{\mathbf{x}^{(t)}\}_{t \in T} \to \bar{\mathbf{x}}$, then $\nabla f(\bar{\mathbf{x}}) = 0$. Show that the coordinate descent method converges to the optimal solution of the dual problem.

9.12 Solve the LP problem. Minimize

$$Z = x_1 - x_2$$

subject to:

$$\begin{aligned} x_1 &\leq 1 \\ x_2 &\leq 1 \\ x_1, x_2 &\geq 0 \end{aligned}$$

using the interior point method.

Part III

Software

Chapter 10

The GAMS Package

10.1 Introduction

GAMS [53] (general algebraic modeling system) is a computer environment to define, analyze and solve optimization problems. It is a terse and powerful language. Some effort is required to comprehend its features, but once its structures are understood, a highly versatile and powerful tool becomes available to solve mathematical programming problems. The reader should notice that GAMS cannot solve more than what the mathematical programming state of the art allows.

Packages similar to GAMS are AMPL [41, 54] and AIMMS [13, 55]. Most of the features of any of these tools are shared by the others and, in general, there is no particular reason to choose one versus the others. GAMS is used in this book because authors happen to be familiar with it.

The most relevant features of GAMS are

1. Its ability to solve small problems (dozens of variables and constraints) and large problems (thousands of variables and constraints) writing basically the same amount of code. A compact and efficient index feature allows one to write blocks of similar constraints by writing just "one constraint."

2. The separation of the problem modeling process from the solution technique used. A GAMS user takes care of consistently formulate a problem and, once expressed in GAMS notation, one of the many available GAMS solvers generates the solution for it. As a result, the GAMS user focuses on problem modeling, without being disturbed by solution algorithm technicalities. This makes it possible a rich, evolving, and changing modeling process that is neither messy nor painful.

3. GAMS almost mimics the mathematical description of a mathematical programming problem. As a result, the GAMS code is almost self-explanatory for readers with some optimization background.

4. GAMS also provides mechanisms to solve collections of structured optimization problems, such as those arising from decomposition techniques.

The reader should be aware not to violate GAMS "grammatical" rules. A single violation can produce many errors.

Some references to the GAMS language are the GAMS manual [53], the second chapter of that manual, which is self-contained, and reference [24], which provides an engineering view of GAMS.

In what follows, an example is analyzed and the basic features of the GAMS language are reviewed.

10.2 Illustrative Example

A transportation problem has been selected to highlight some of the GAMS features. It is formulated below.

Consider a set of markets and production plants. Given (1) the demand of every market, (2) the maximum production capacity of every plant, and (3) the transportation cost from every plant to every market, the transportation problem consists of determining the amount of commodity to be shipped from every plant to every market so that the transportation cost is minimum.

The mathematical programming formulation of the preceding problem is as follows. Minimize

$$\sum_i \sum_j c_{ij} x_{ij}$$

subject to

$$\begin{array}{rcl} \sum_j x_{ij} & \leq & a_i, \ \forall i \\ \sum_i x_{ij} & \geq & b_j, \ \forall j \\ x_{ij} & \geq & 0, \ \forall i, j \end{array}$$

where the indices are

i: for production plants

j: for markets

The data are

a_i: the maximum production capacity of plant i in tons

b_j: the demand of market j in tons

c_{ij}: the transportation cost from plant i to market j in dollars per ton and kilometer

The decision variables are

x_{ij}: the amount of commodity to ship from plant i to market j in tons

Table 10.1: Distances in kilometers from production plants to markets

	Markets		
Plants	m_1	m_2	m_3
p1	2.0	1.6	1.8
p2	2.5	1.2	1.4

Two production plants and three markets are considered. Plant maximum capacities are respectively 300 and 500 tons. Market demands are respectively 100, 200, and 300 tons. Distances in kilometers from production plants to markets are given in Table 10.1.

The transportation cost in dollars per ton and kilometer is 0.09.

In GAMS, the above problem is expressed, by default, in a file with extension '.gms', with the content given below (the reader should not be too concerned, at this time, in understanding all commands, because they will be explained later in the chapter, but just to get a global view of a GAMS input file):

```
$Title  The Transportation Problem
* Simple transportation example

  Sets
       i    production plants     / p1, p2 /
       j    markets               / m1*m3 /;

  Table d(i,j)  distance in kilometers
                m1          m2           m3
      p1        2.0         1.6          1.8
      p2        2.5         1.2          1.4;

  Scalar f  freight (dollars per ton and km)  /0.09/;

  Parameters
       a(i)  capacity of plant i in tons
         /    p1     300
              p2     500  /
       b(j)  demand at market j in tons
         /    m1     100
              m2     200
              m3     300  /
       c(i,j)  transportation cost in dollars per ton;

  c(i,j) = f * d(i,j);

  Variables
       x(i,j)  shipment quantities in tons
       z       total transportation costs in dollars;
```

```
Positive Variable x;

Equations
     cost        objective function
     supply(i)   meet supply limit at plant i
     demand(j)   satisfy demand at market j;

cost ..        z  =e=  sum((i,j), c(i,j)*x(i,j));
supply(i) ..   sum(j, x(i,j))  =l=  a(i);
demand(j) ..   sum(i, x(i,j))  =g=  b(j);

Model transport /all/;

Solve transport using lp minimizing z;

Display x.l;
```

The reader may observe, in this problem, the close parallelism between the mathematical programming formulation and the GAMS code (specifically the equation definitions).

The first line is used to define the title of the problem, and the second line is a comment. In GAMS, any line starting with the "*" symbol (asterisk) is interpreted as a comment.

The command `Sets` is used to define two things: the indices names used below, and the sets of ordered values that those indices can take. In our example, we use the index `i` to refer to production plants `p1` and `p2`. Similarly, index `j` refers to markets `m1`, through (indicated by the asterisk) `m3`.

The command `Table` is used to define matrices and assign them the corresponding values (in the example, the matrix given in Table 10.1 is represented using this command). Note that those values are given in tabular form.

The `Scalar` command can be used to define scalars and assign them values, as the scalar `f` in the example.

The command `Parameters` is used to define vectors and assign them particular values. In our example we use two vectors: `a(i)` and `b(j)`, referring to the plant capacities, and the market demands, respectively. Their corresponding elements are listed using the command `parameter`. To define the transportation cost `c(i,j)` as a data matrix, the command `parameter` is also used. The elements of this matrix are not enumerated, they are computed in an assignment sentence in terms of previously defined data. It is possible to give numerical values to those vectors by listing them or by an assignment sentence.

Optimization variables are defined by the command `Variables`. They are the unknowns of our problem, which values must be determined by the solver. Since the variable `x(i,j)` is positive, we should inform GAMS on that.

The command `Equations` is used to declare the objective function and the constraints. First, we name them and, if needed, indicate (using sets) the number of constraints following the same pattern. Then, using the ".." symbol, the constraints are defined as they are formulated mathematically. Sums, products, different functions, etc. are some of the allowed expressions when defining constraints

Finally, we should name the model, which in our example is `transport`, and tell GAMS to solve the problem using one of the available solvers (`lp` in our example, because it is a linear programming problem).

Optionally, we can indicate GAMS what variables or information to be displayed in the output file. Here, `Display x.l` indicates that the optimal values of `x(i,j)` should be displayed.

Once the input file is ready, the GAMS solver can be run, and the solution, or a list of errors obtained.

Since our GAMS code contains no errors, GAMS gives the solution in the output file. Parts of the GAMS solution file (".lst") associated with the problem above is given below.

```
---- EQU SUPPLY       meet supply limit at plant i
       LOWER     LEVEL      UPPER    MARGINAL
p1      -INF    100.000    300.000       .
p2      -INF    500.000    500.000    -0.036

---- EQU DEMAND       satisfy demand at market j
       LOWER     LEVEL      UPPER    MARGINAL
m1    100.000   100.000     +INF       0.180
m2    200.000   200.000     +INF       0.144
m3    300.000   300.000     +INF       0.162
```

The minimum (`LOWER`), optimal (`LEVEL`), maximum (`UPPER`), and marginal (`MARGINAL`) values for every `SUPPLY` and `DEMAND` constraint are provided. Marginal values contain information about how much a change in the RHS of a constraint affect the objective function value. In the output file, dots indicate zeroes.

The lower and upper bounds on constraints depend on their type. If the constraints specified in the GAMS input file are "greater than or equal to" constraints (as in `SUPPLY` equations), the lower bounds of these constraints are $-\infty$ (`-INF` in the GAMS output file), and their upper bounds are their corresponding constant right-hand-sides. If the equations specified are "less than or equal to" equations (as in `DEMAND` equations), the lower bounds of these equations are their corresponding constant right-hand-sides and their upper bounds are $+\infty$ (`+INF` in the GAMS output file). When dealing with equality constraints, their lower and upper bounds are both equal to their corresponding constant RHSs.

The "level" of an equation is the result of evaluating its variable left-hand side after solving the problem. The "marginal" value of an equation is also determined by the solver and shown in the GAMS output file.

The GAMS output variable information is provided below.

```
---- VAR X            shipment quantities in tons
```

Table 10.2: Some commands of the GAMS package

Command	Purpose
Set(s)	Name the indices and their possible values
Scalar(s)	Name the scalars and assign them values
Parameter(s)	Name the vectors and assign them values
Table(s)	Name arrays and assign them values
Variable(s)	Declare variables, assign them a type (optional), and give them upper and lower bounds
Equation(s)	Name the function to be optimized and the constraints
Model	Name models and the list of associated constraints
Solve	Tell GAMS the solver to be used and solve the model
Display	Tell GAMS the elements to be listed in the output report

```
            LOWER      LEVEL      UPPER     MARGINAL
p1.m1         .       100.000      +INF        .
p1.m2         .          .         +INF       EPS
p1.m3         .          .         +INF        .
p2.m1         .          .         +INF      0.081
p2.m2         .       200.000      +INF        .
p2.m3         .       300.000      +INF        .

                        LOWER      LEVEL      UPPER     MARGINAL
---- VAR Z               -INF      77.400      +INF
  Z          total transportation costs in dollars
```

Lower, level (optimal), upper, and marginal values are provided for every variable. If the lower (upper) bound is not stated in the GAMS input file, it is considered to be $-\infty$ $(+\infty)$. `EPS` is used to indicate a very small (nonzero) number. The level (optimal) and marginal values of a variable are determined after solving the problem.

10.3 Language Features

The basic GAMS commands are described in the next sections. Extensive information can be found in the GAMS manual [53].

To use GAMS we need to prepare an input file in which a complete definition of the problem to be solved is given.

A GAMS data file is structured in parts, and each part deals with a different component. In Table 10.2 we show the different components (associated with different commands) of a GAMS file.

Before using GAMS, it is important to know some general rules used by GAMS. The most important are:

1. For GAMS, upper and lowercase letters are identical. Thus, the user is free to choose the typing style.

2. Every command must end with a semicolon. If the user forgets the semicolon GAMS runs into problems.

3. Commands can be used in any order with the only constraint that any element must be defined before it is used.

4. GAMS, as any programming language, has some reserved words that must be used with care. When GAMS reads a line of text looks for these words and acts consequently.

5. Some commands are identified by their first letters, no matter whether they end with or without "s". This option has been selected by GAMS to facilitate the understanding of the code, as normal English commands. For example, we can use `Set` or `Sets` because GAMS makes no distinction.

6. Commands can be written in any style. For example, multiple commands per line and multiple lines per command are allowed. The number of blanks or spaces is indifferent to the GAMS compiler (only one is considered).

7. One single command is valid to declare or define one or several elements of the same type. There is no need to repeat the command name.

8. Documentation can and should be incorporated into the GAMS code. Any line starting with an asterisk, in its first column, is ignored by the GAMS compiler and is considered as a comment line. Other comments can be inserted within some declarative commands such as: `set`, `scalar`, `parameter`, `table`, `equation`, and `model`. The inserted comments (optionally in between inverted commas) make the declared element more easily understandable.

9. Most commands are used for declaration to name the elements and optionally to give them particular values. Declare an element (scalar, parameter, matrix, etc.) means convert it in a valid element to be used by GAMS. Elements are declared by giving them a name with any of the declaration commands (sets, scalar, parameter, table, variables, equations, and model).

10. The valid names for GAMS elements must start with a letter and can be followed by up to nine more letters or digits.

10.3.1 Sets

The reserved word `Set` or `Sets` is used in GAMS to declare indices, and to specify the set of their possible values (this explains the name of this command). An index should be declared and then its possible values listed, as in

```
Sets
    i   production plants    / p1, p2 /
    j   markets              / m1*m3 /;
```

The two indices defined here are `i` and `j`. Strings after the set name (`production plants` and `markets`) are comments, ignored by the GAMS compiler. The set assignment is specified in between two "/" symbols. The symbol "*" helps to define a set in a compact way, for instance, `/m1*m3/` is equivalent to `/m1, m2, m3/`. However, GAMS treats set values as character strings. The command concludes with a semicolon symbol.

The above sets are known as *static* or *constant* sets; their elements never change throughout the program execution. In GAMS it is also possible to define dynamic sets that contain some of the elements of a given static set. The elements in dynamic sets may change at some point throughout the program execution. A brief explanation of dynamic sets is given in Section 10.3.12. Additionally, multi-dimensional sets can be defined to establish a mapping between elements that belong to different sets. For instance, to model a network of nodes in GAMS, a two-dimensional set is defined including the valid links between two different nodes as it follows:

```
Sets N        set of nodes       /N1*N4/
     MAP(N,N) mapping set of N   /N1.N2,N1.N3,N2.N4,N3.N4/
```

This code defines a network containing 4 nodes and 4 links. To check for a link between any two given nodes (n and n'), the following GAMS code can be used (see Table 10.8 to know more details about the `if` statement):

```
Alias(N,NP);
If(NP$MAP(N,NP), ...; );
```

In the preceding example, the command `alias` is necessary to refer to different nodes that belong to the same set. With this command the sets `N` and `NP` are identical (note that the set is not duplicated). In general, aliases are used to deal with sums, products, an so on, when two or more indices belonging to the same set are referenced and need to vary independently.

In GAMS there are six admissible operators for dealing with ordered sets: `card`, `ord`, `+`, `-`, `++` and `--`. Ordered sets are one-dimensional static sets comprised of an ordered sequence of elements. There are different ways to obtain an ordered set:

- If its elements do not belong to any previously defined set
- If its elements belong to other sets but the relative position between the shared elements is kept the same as in previously defined ordered sets

An example of ordered and unordered sets is given below:

```
Sets A1 /Tuesday, Wednesday, Friday, Thursday/
     A2 /Sunday, Thursday, Saturday, Friday/
     A3 /Wednesday, Friday, Saturday, Monday/;
```

Table 10.3: Examples of using some operators for the given ordered set A1

Operation	Result
`card(A1)`	4
`ord('Wednesday')`	2
`'Wednesday'+1`	Friday
`'Thursday'++1`	Tuesday
`'Tuesday'--1`	Thursday

In this example, the set A1 is ordered because its elements do not belong to any previously defined set. The set A2 is not ordered because the relative position between Thursday and Friday is different when it is compared with the previously defined ordered set A1. The last defined set is ordered because the relative position of its elements is the same as in the ordered set A1, regardless of whether the relative position between Friday and Saturday is different from the unordered set A2.

The possible operators to be used on the ordered set A1 are illustrated in Table 10.3. The operator card returns the number of elements belonging to a set (this operator can also be used when dealing with dynamic or any kind of unordered set). The operator ord returns the position of an element in a set. Given an element of a set, it is possible to refer to its next or previous element in this ordered set by using the operators +1 or -1. With these two operators, the set is not considered to be a circular list of elements, that is, there is no element that follows its last element or any previous for its first element. The operators ++ and -- allows us to consider a set as a circular list of ordered elements; therefore it always exists the next and the previous of a given element in the set.

10.3.2 Scalars

In GAMS scalars are declared, and optionally assigned a value, using the reserved word Scalar or Scalars. After the name, comments can be included, as it is stated in next sentence.

```
Scalar f  freight (dollars per ton and km)  /0.09/;
```

The scalar f value is declared and the value 0.09 is assigned to it within two "/" symbols. The semicolon ends the command.

10.3.3 Parameters and Tables

Parameters and tables are GAMS structures used to define data arrays having two or more dimensions. Parameters and tables are equivalent, except for the case of defining one-dimensional arrays. In this case, only the parameter structure is allowed. In this section, we start explaining the use of the parameter command, and next, we deal with tables. Finally, both commands are compared.

The reserved word `Parameter` or `Parameters` is used to declare vectors of data. Optionally, this statement allows to assign values to each element of the vector and to include self-explanatory comments as in

```
Parameters
     a(i)  capacity of plant i in tons
       /    p1    300
            p2    500  /;
```

In this example the parameter `a(i)` is defined in terms of set `i`. For every set element (`p1,p2`), a parameter value is provided (`300,500`). The parameter assignment takes place in between two "/" symbols. A semicolon concludes the command. The following GAMS rules must be taken into consideration when assigning values to vectors:

1. The optional list of possible indices and their corresponding parameter values must be enclosed in slashes "`/.../`" and separated by commas or carriage returns (different lines).

2. The index value pairs can be given in any order.

3. The default value for the parameters is zero. Thus, only values different from zero must be given.

4. GAMS automatically checks that the given index values are valid set values for the corresponding index. Otherwise, GAMS lists the corresponding errors.

To define a data matrix (or vector) as a function of previously defined data, the parameter command can be used. For instance

```
Parameter c(i,j)  transportation cost in dollars per ton;
          c(i,j) = f * d(i,j);
```

It should be noted that GAMS allows a very compact data entry. `c(i,j)` is a two-dimensional matrix defined as a function of `f`, a previously defined scalar, and `d(i,j)`, which is also an already defined two-dimensional matrix.

The preceding data matrix definition is a compact version of the following code:

```
Parameter c(i,j)  transportation cost in dollars per ton
              /  p1.m1  0.180
                 p2.m1  0.225
                 p1.m2  0.144
                 p2.m2  0.108
                 p1.m3  0.162
                 p2.m3  0.126  /;
```

Matrices of data can be defined in GAMS using either tables or two-dimensional parameters (depending on more than one set). The reserved word `Table` or `Tables` identifies the table command. Tables are defined using two (or more) indices. Comments can be included after the table name, as in

```
Table d(i,j)  distance in km
              m1          m2          m3
     p1       2.0         1.6         1.8
     p2       2.5         1.2         1.4;
```

In this example the matrix `d(i,j)` is defined using indices `i` and `j`. For every crossed couple of set elements (`p1.m1, p1.m2, p1.m3, p2.m1, p2.m2, p2.m3`), a table value is specified (`2.0, 1.6, 1.8, 2.5, 1.2, 1.4`).

Finally, to show the equivalence between table and parameter commands the matrix `c(i,j)` is defined using the following table command:

```
Table c(i,j)  transportation cost in dollars per ton
          m1     m2     m3
     p1   0.180  0.144  0.162
     p2   0.225  0.108  0.126;
```

Notice that this is not the most compact notation for representing the `c(i,j)` array.

The structure of tables with more than two sets is explained through an example. Suppose that we have a table containing the number of people sharing different features (religion, race, study level). A code used to obtain this kind of table could be

```
Sets re types of religions /re1*re3/
     sl types of study levels /sl1*sl2/
     ra types of races /ra1*ra4/;

Table c(re,sl,ra)  number of people with the same features
                    ra1         ra2       ra3      ra4
    re1*re2.sl1      2           4        10        2
    re3.sl1          6           2         5        7
    re1*re3.sl2      3           9         7        8 ;

Display c;
```

This code allows us to define a three-dimensional array in a compact way (the asterisk facility is explained in Section 10.3.9). The display command allows us to show the table above with an extended format.

```
----      11 PARAMETER C   number of people with the same features
                 ra1          ra2          ra3           ra4
re1.sl1        2.000        4.000       10.000         2.000
re1.sl2        3.000        9.000        7.000         8.000
re2.sl1        2.000        4.000       10.000         2.000
re2.sl2        3.000        9.000        7.000         8.000
re3.sl1        6.000        2.000        5.000         7.000
re3.sl2        3.000        9.000        7.000         8.000
```

10.3.4 Mathematical Expression Rules in Assignments

To assign values using mathematical expressions, the following rules must be observed:

1. The assignment of values to vectors or matrices can be done for all its elements by writing just one sentence and avoiding the typical structures using loops. The following code substitutes the old values of the matrix c_{ij} by 3.

   ```
   c(i,j)=3;
   ```

2. A specific assignment can be done by specifying the indices values in quotes, as in

   ```
   c('p1','m1')=0.180;
   ```

3. Scalars, parameters, and tables can be assigned values more than once. The new values override the old ones.

4. Mathematical expressions can incorporate standard mathematical functions and operations. Table 10.4 lists some of the available mathematical functions. For instance, to model a random data vector $x_i, \forall i = 1 \cdots 10$ obtained from the uniform distribution $U(0, 0.5)$, the following code is used:

   ```
   Set I /I1*I10/; Parameter x(I);
   x(I)=uniform(0,1)*0.5;
   ```

10.3.5 Variables

Variables are declared in GAMS in a simple way:

```
Variables
     x(i,j)  shipment quantities in tons
     z       total transportation costs in dollars;
```

The reserved word `Variable` or `Variables` is used to declare the optimization variables. When declaring a variable, the dimensions of the referenced variable should be specified. A variable (`z` in the transportation example) for representing the objective function value must always be declared in any GAMS program. After the variable name, comments can be included. The semicolon ends the command.

Characteristic properties of the variables can be declared, too:

```
Positive Variable x;
```

Table 10.4: Some functions of the GAMS package

Function name	Description
abs(x)	Absolute value of x
arctan(x)	Inverse of the tangent function (in radians)
ceil(x)	Smallest integer greater than or equal to x
cos(x)	Cosine function (x in radians)
errorf(x)	Cumulative distribution function of the normal $N(0,1)$ distribution at x
exp(x)	Exponential function
floor(x)	Largest integer less than or equal to x
log(x)	Natural logarithm of x
log10(x)	Logarithm in base 10 of x
mapval(x)	Mapping function
max(x_1,x_2,...)	Largest value in the list
min(x_1,x_2,...)	Smallest value in the list
mod(x,y)	Remainder obtained after dividing x by y
normal(x,y)	Random number from a normal random variable with mean x and standard deviation y
power(x,y)	Power function x^y (where y must be an integer)
$x**y$	Power function x^y (where x must be positive)
round(x)	Round x to the nearest integer
round(x,y)	Rounds x to y decimal places
sign(x)	Sign of x, 1 if positive, -1 if negative, and 0 if null.
sin(x)	Sine function (in radians)
sqr(x)	Square of x
sqrt(x)	Square root of x
trunc(x)	Is equal to sign(x) * floor(abs(x))
uniform(x,y)	Random number from a uniform distribution $U(x,y)$

This GAMS command establishes a lower bound for variable `x(i,j)`, namely, $x_{ij} \geq 0$. In this case, it is not necessary to indicate the dimension of the variable; this is due to the fact that the dimension has already been specified in a previous sentence.

It is also possible to declare `Free` (no bounds by default), `Binary`, `Integer`, or `Negative` variables. The variable representing the objective function value should always be declared as `Free`.

The valid types of variables and the associated possible and default ranges are given in Table 10.5.

It is also possible to set lower and upper bounds for variables using GAMS suffixes. For instance, the following mathematical expression

$$2.0 \leq r \leq 5.0$$

Table 10.5: Types of valid variables in GAMS, associated ranges and default ranges

Variable type	Range	Default range
Binary	$\{0,1\}$	$\{0,1\}$
Free (default)	$(-\infty,\infty)$	$(-\infty,\infty)$
Integer	$\{0,1,\ldots,n\}$	$\{0,1,\ldots,100\}$
Negative	$(-\infty,0)$	$(-\infty,0)$
Positive	$(0,\infty)$	$(0,\infty)$

is written in GAMS as

```
Positive variable r;
r.lo = 2.0;  r.up = 5.0;
```

where the suffix `lo` indicates lower bound on the variable and the suffix `up` denotes its upper bound. Both suffixes can be used to modify the default range of any variable. For instance, the following code changes the default upper limit of an integer variable `i` from 100 to 1000:

```
Integer variable i; i.up=1000;
```

When the upper and lower bounds on a variable are the same, it is used the suffix `fx`. For instance, the expression

$$y_i = 3.0, \quad \forall i$$

is specified in GAMS as

```
y.fx(i) = 3.0;
```

It should be noted that it is only necessary one GAMS sentence to assign the same value to all the components of a vector (the same for matrices), regardless the number of components they have.

Sometimes, when dealing with nonlinear programming problems it is necessary to state initial values for some variables. For instance, the following matrix assignment (notice that the next sentence does not fix the elements of matrix z_{ij} to value 3.0, it just states 3.0 as initial value)

$$s_{ij} \leftarrow 3.0, \quad \forall i, \forall j$$

can be done by means of the GAMS code as

```
s.l(i,j) = 3.0;
```

where the suffix `l` directs the solver to use the specified value as the starting point.

10.3.6 Equations

The term *Equations* is used in GAMS in a wide sense. In addition to equations it also allows to define any kind of constraints.

The reserved word `Equation` or `Equations` is used to declare the constraints of a given problem (see lines 2–4 in the example below). Once the constraints are declared they are defined using the symbol " ..". This symbol couples the constraint names with their definitions (see lines 5–7). Standard mathematical notation can be used to model many different constraints. A semicolon should be included after the definition of every equation.

```
Equations
     cost         objective function
     supply(i)    meet supply limit at plant i
     demand(j)    satisfy demand at market j;
cost ..         z  =e=  sum((i,j), c(i,j)*x(i,j));
supply(i) ..    sum(j, x(i,j))  =l=  a(i);
demand(j) ..    sum(i, x(i,j))  =g=  b(j);
```

It should be noted that the summations are expressed in GAMS as `sum(i, x(i,j))`, which indicates $\sum_i x_{ij}$. Similarly to summation, product operation is defined in GAMS as `prod(i, x(i,j))` representing $\Pi_i x_{ij}$.

Equations symbols are as follows:

- `=e=` indicates "is equal to",
- `=l=` indicates "is less than or equal to"
- `=g=` indicates "is greater than or equal to".

When an "equation" is defined as a function of a set [as `supply(i)` or `demand(j)`], this "equation" is equivalent to as many equations as the number of elements in the corresponding set.

Therefore, the following definition of the constraint `supply(i)`

```
supply(i) ..    sum(j, x(i,j))  =l=  a(i);
```

is equivalent to the following two equations:

```
supply1 ..   sum(j,x('p1',j))  =l=  a('p1');
supply2 ..   sum(j,x('p2',j))  =l=  a('p2');
```

The reader should be aware of the power of this compact notation.

It should also be noted that the objective function is also declared using the command `Equation`.

10.3.7 Model

The command `Model` is used to define the equations to be included in the model to be solved. The `Model` command below declares that the problem considered includes all previously defined equations.

```
Model transport /all/;
```

This sentence can also be written as

```
Model transport /cost,supply,demand/;
```

Alternatively, it is possible to build a model without including all the defined equations. For instance, the following model `transport1` does not include the `demand` constraints:

```
Model transport1 "Demand constraint not included" /cost,supply/;
```

Note that some comments have been inserted after the model name in between inverted quotes.

10.3.8 Solve

The `Solve` command orders GAMS to solve the formulated problem. The `Solve` command below directs GAMS to solve the `transport` problem using a linear programming solver (`lp`) and minimizing the variable `z`.

```
Solve transport using lp minimizing z;
```

The reserved word `lp` is used for linear programming. This and other options are indicated in Table 10.6.

Table 10.6: List of available solvers in GAMS

Solver name	Purpose
`lp`	Linear programming
`nlp`	Nonlinear programming
`dnlp`	Nonlinear programming with discontinuous derivatives
`mip`	Mixed-integer programming
`rmip`	Relaxed mixed-integer programming
`minlp`	Mixed-integer nonlinear programming
`rminlp`	Relaxed mixed-integer nonlinear programming
`mcp`	Mixed complementary problems
`mpec`	Mathematical problems with equilibrium constraints
`cns`	Constrained nonlinear systems

In GAMS, some suffixes associated with the problem modeling provide useful information once the problem is solved. Remarkable suffixes among many others are: `modelstat` to check the model status, `solvestat` to check the solver status and `resusd` to check the time (expressed in CPU seconds) spent by the solver. An example including the suffix `resusd` is given in Section 10.3.10. The next example shows a possible use of the suffix `modelstat`.

```
If (transport.modelstat ne 1,
     Solve transport1 using lp minimizing z;
);
```

Table 10.7: List of possible values for modelstat and solvestat suffixes

Code	modelstat	solvestat
1	Optimal (lp, rmip)	Normal completion
2	Locally optimal (nlp)	Iteration interrupt
3	Unbounded (lp, rmip, nlp)	Resource interrupt
4	Infeasible (lp, rmip)	Terminated by solver
5	Locally infeasible (nlp)	Evaluation error limit
6	Intermediate infeasible (nlp)	Error
7	Intermediate non-optimal (nlp)	-
8	Integer solution (mip)	-
9	Intermediate non-integer (mip)	-
10	Integer infeasible (mip)	-
11	-	-
12	Error unknown	-
13	Error no solution	-

This code checks whether the solution of the model transport is optimal or not using the command if (explained later in this chapter) and the suffix modelstat. If the solution is not optimal, an alternative model transport1 is solved. The suffix solvestat is handled similarly.

The code values of the model status suffix are dependent on the type of models to be solved. The reader can check in Table 10.7 the meaning of the possible code values associated with the suffixes modelstat and solvestat. In parentheses, it is specified which type of mathematical programming problem each code is related to.

The codes explained subsequently indicate that the solution attained is optimal for the input model. If the problem is formulated as lp or rmip, the code obtained after running the GAMS program should be 1, but if the model is nlp or mip, the codes should be 2 or 8, respectively.

For unbounded lp, rmip, or nlp models, the code of the model status suffix is 3. When the user obtains the unbounded code, it is advisable to look for the string UNBND in the GAMS output file and run again the program after modifying the limits of the problematical variables.

Infeasible models are identified with the following codes of the model status suffix: 4 for lp and rmip models, 5 for nlp models, and 10 for mip models. In any of these three cases, the user should look for the string INFES in the output file and, according to his/her knowledge of the problem, modify the input data and/or the structure of the model.

All of the codes explained above match with the code value 1 of the solver status suffix. This value means that the solver has not found difficulties in solving the problem. Occasionally, difficulties arise because the solver exceeds its maximum number of iterations; that is, the solver status code value is 2, but these difficulties are easily overcome if the user increases this limit (1000 iterations by default) modifying the option `iterlim` before the `solve` statement appears in the GAMS program. The sentence to modify the default value of this option is as follows:

```
  Option iterlim=1e8;
```

The lack of machine resources is identified with the code value 3 of the solver status suffix. This value can be changed to 1 if the user modifies in the program the option `reslim` (1000 seconds by default) before the statement `solve`.

```
  Option reslim=1e10;
```

When solving difficult `mip` models it is common to reach either the iteration or the resource limit. A possible way to avoid this difficulty is to decrease the accuracy of the attained integer solution. Any of the two following options can be stated for this purpose (the default values are showed):

```
  Option optcr=0.1;
** OR **
  Option optca=0;
```

The previous statements determine the stopping criterion when solving `mip` models. The first option `optcr` is a relative criterion expressed in percentage, while the second is an absolute criterion. Both options are used to compare any of the intermediate integer solutions with the value of the best possible solution (determined by the solver). The default value of the option `optcr` means that the difference between any final solution and the best one should be less than 0.1%. The default value of the option `optca` means that the final solution is optimal.

Other possible reason why the code value of the solver status suffix differs from 1 is the inability of the solver to continue on solving the problem (code value is equal to 4). In this case, the user is recommended to read the explanatory message in the GAMS output file.

The code values 2, 3, and 4 of the solver status lead to intermediate code values of the model status: 6, 7, and 9. Code values 6 and 7 are both related to `nlp` models but they differ in the "quality" of the obtained solution. Code value 6 means that the found solution is not feasible, while code value 7 means that it would be a feasible solution if the solver could finish the current execution. Similarly, the code value 9 refers to an infeasible solution of a `mip` problem when the solver was not able to complete the current execution.

Additionally, a code value 5 of solver status is typically found when solving difficult `nlp` problems that need too many evaluations of the nonlinear terms. Some help is provided in the output file to deal with this specific case. Finally, code values from 6 upward indicate that there are major difficulties in solving any kind of model. It could be due to a preprocessor error, a setup failure, a solver failure, an internal solver error, a postprocessor error, a system failure or, simply, the error is unknown.

10.3.9 Asterisk Facility

In this section, the different uses of the asterisk symbol are summarized. It can be used

- To denote comments. The GAMS input file should contain an asterisk in the first column of each line of comments
- In set definitions to list the elements of the set in a compact way, as was explained in Section 10.3.1,
- To mark errors in the output file, four asterisks are shown at the beginning of each error line
- To mark, in the output file, that nonlinear constraints are infeasible at the starting point. In this case, three asterisks are shown in the equation listing at the end of the nonlinear constraint,
- As the product "*" and the exponential "**" operator,
- To represent the so-called wild set used in the definition of some GAMS structures such as sets, parameters, tables, variables, or equations. This is meaningful when there is no need to declare a set that groups miscellaneous information (labels). For a better understanding, see the following example:

```
Set A this a set of articles /a1,a2/;

Table g(A,*)  features of the A articles
              height    width     weight
*             (cm)      (cm)       (kg)
       a1     1.0        2.7        3.5
       a2     2.4        1.8        4.4;
```

This code could replace the next one:

```
Sets A set of articles /a1,a2/
     B set of features /height,width,weight/;

Table g(A,B)  features of the A articles
              height    width     weight
*             (cm)      (cm)       (kg)
       a1     1.0        2.7        3.5
       a2     2.4        1.8        4.4;
```

10.3.10 Display

In the GAMS standard output file (a file with extension ".lst") additional information is obtained through the command `display`, for example

```
Display x.l;
```

For the transport problem, this command produces the following output

```
----     44 VARIABLE  X.L          shipment quantities in tons

           m1          m2          m3

p1      100.000
p2                  200.000     300.000
```

This output information gives the optimal values for all the components of matrix x_{ij}. Notice that the suffix .l is needed only to obtain the value of a variable after the model is solved, it is not necessary when displaying any kind of data structure. When using the display command, blanks are equivalent to default values, which in this case are zeros.

Other useful information can be obtained using the display command. The next sentence shows how to obtain the computing time spent in solving the transportation example.

```
Display  transport.resusd;
```

It should be noticed that the above sentence is only valid after the corresponding statement solve.

A less restricted format for output data is briefly explained in Section 10.3.14.

10.3.11 Conditional Statements

The "$" symbol can be used to generate convenient subsets of original ordered sets (the ordered sets were explained in Section 10.3.1). The GAMS sentence

```
demand(j)$(ord(j) gt 1) ..   sum(i, x(i,j))  =g=  b(j);
```

is equivalent to

```
demand2..   sum(i, x(i,'m2'))  =g=  b('m2');
demand3..   sum(i, x(i,'m3'))  =g=  b('m3');
```

Note that $(ord(j) gt 1) indicates that only the set elements with ordinal number greater than 1 should be included. The use of the operator ord as well as the operator card when dealing with sets was explained in Section 10.3.1.

Similarly, conditions can be stated at both sides of a constraint. The following code illustrates this case:

```
supply(i) ..   sum(j$(ord(j) lt card(j)), x(i,j))  =l=
               a(i)$(ord(i) eq 1);
```

the above code is equivalent to

```
supply1..  x('p1','m1') + x('p1','m2') =l= b('p1');
supply2..  x('p2','m1') + x('p2','m2') =l= 0 ;
```

Table 10.8: Examples showing the equivalence between the "$" operator and the "if-then-else" statement

$ expression	if-then-else expression

`c('p2','m1')$(z.l ge 70)=0.01;`	`If( z.l ge 70,`
	`c('p2','m1')=0.01;`
	` );`

`c('p2','m1')=0.01$(z.l ge 70);`	`If( z.l ge 70,`
	`c('p2','m1')=0.01;`
	`else`
	` c('p2','m1')=0.0;`
	` );`

Conditions are also possible in other GAMS structures such as data assignments or put statements (see Section 10.3.14). In those structures, the "$" operator can be replaced by an "if-then-else" statement. The examples in Table 10.8 are conditional assignments that use both structures. Although the '$' notation is widely used, the reader should be aware of the importance of its location in any GAMS structure.

Some examples of the use of the conditional `if` statement can be seen in Sections 12.1, 12.2.1, and 12.7.

Typical operators in other programming languages are also valid in GAMS conditional expressions are

- not, and, or, xor as logical operators;
- $<$ (lt), $<=$ (le), $=$ (eq),$<>$ (ne), $>=$ (ge), $>$ (gt) as relational operators

10.3.12 Dynamic Sets

One powerful feature of GAMS is the possibility of defining dynamic sets. Using them, it is possible to modify the elements of a previously defined set througout the GAMS program execution. Dynamic sets are always defined as a subset of a previously defined static set. The value of those variables, parameters, tables, and equations, which depend on a dynamic set can be modified each time the dynamic set is updated. Table 10.9 shows examples of the equivalence between mathematical and GAMS expressions using dynamic sets.

In the first row of Table 10.9, the static set `k` is defined, and then the dynamic subset `s(k)` is defined as a function of `k`. To be able to use a dynamic set it is necessary to define its initial state. A possible initial state, the empty set, is defined in the third row. The fourth row shows how to assign the last element of set `k` to set `s(k)`, this assignment is done using two operators `ord(k)`' and `card(k)`. The '`card`' operator returns the number of elements in any (static or dynamic) set, and the `ord` operator returns the position of an element in a

Table 10.9: Examples of the dynamic sets usage

Mathematical expression	GAMS expression
$k = \{a, b, c\}$ $s \subseteq k$	`Set k /a,b,c/` `s(k);`
$s = \{a, b, c\}$	`s(k)=yes;`
$s = \emptyset$	`s(k)=no;`
$s = \{c\}$	`s(k)=no;` `s(k)$(ord(k) eq card(k))=yes;`
$s = \{b, c\}$	`s(k)=no;` `s(k)$(ord(k) gt 1)=yes;`

Table 10.10: Examples of dynamic set operations

Mathematical expression	GAMS expression

$A = \{a1, a2, a3, a4, a5\}$
$b \subseteq A,\ b = \{a1, a2, a5\}$
$c \subseteq A,\ c = \{a2, a3, a5\}$
	`set A static set /a1*a5/;`
	`set B(A) subset /a1,a2,a5/;`
	`set C(A) subset /a2,a3,a5/;`

$b \cup c = \{a1, a2, a3, a5\}$	`set UN(A) dynamic subset;` `UN(A)=B(A)+C(A);`
$b \cap c = \{a2, a5\}$	`set IN(A) dynamic subset;` `IN(A)=B(A)*C(A);`
$\bar{b} = \{a3, a4\}$	`set COMP(A) dynamic subset;` `COMP(A)=not B(A);`
$b - c = \{a1\}$	`set DIFF(A) dynamic subset;` `DIFF(A)=B(A)-C(A);`

static set (the `ord` operator is only valid for static sets). The last row indicates how to assign two different elements to the `s(k)` set.

In GAMS, set operations should be implemented using dynamic sets. The examples in Table 10.10 show typical set operations using dynamic sets.

Some examples of dynamic sets can be seen in Section 12.2.1.

10.3.13 Iterative Structures

Loops allow the user to solve a collection of related mathematical programming problems. Loops are specially important when solving a mathematical programming problem using decomposition techniques. These techniques usually solve iteratively a set of similar models. Next code, based on the Example 4.1, illustrates how to solve a problem iteratively by means of the loop sentence combined with dynamic sets:

```
Sets K /1*3/
     DYN(K);

Parameter m(K) the slope of C2; m('1')=1;
Positive variables x1,x2; Free variable z; x1.up=3;

Equations
 Cost
 C1
 C2(K)
 C4
 C6;

Cost.. z=e=3*x1+x2; C1.. -x1+x2=l=2; C2(DYN).. x1+m(DYN)*x2=l=6;
           C4.. 2*x1-x2=l=4; C6.. -x1-x2=l=-1;

Model exchap4 /all/;

loop(K, DYN(K)=yes; Solve exchap4 using lp maximizing z; m(K+1)=m(K)+1;
);
```

This loop directs GAMS to solve three related mathematical programming problems and allows a parametric analysis on the binding constraint C2 (a new constraint is added at every iteration changing the slopes of this straight line). The following set of equations does not change along the iterations:

```
C1..      - x1 + x2 =l=  2 ;
C4..      2*x1 - x2 =l=  4 ;
C6..      - x1 - x2 =l= -1 ;
```

of the original problem) and the solver considers equation C2 as

```
C2(1)..    x1 + x2 =l=  6 ;
```

In the second iteration (K = 2), the value of the objective function is 10 and equation C2 is modeled as

```
C2(1)..    x1 + x2 =l=  6 ;
C2(2)..    x1 + 2*x2 =l=  6 ;
```

In the third iteration (K = 3), the value of the objective function is 8.9 and equation C2 is considered as

```
C2(1)..    x1 + x2 =l=  6 ;
C2(2)..    x1 + 2*x2 =l=  6 ;
C2(3)..    x1 + 3*x2 =l=  6 ;
```

It should be noted that these C2 equations cannot be directly used as GAMS code. If equations are not defined generically using a set, they should have different names.

Other iterative structures typical in the majority of programming languages, such as for and while, are also valid in GAMS. Both structures are useful when

Table 10.11: Structure of the statements `for` and `while`

`scalar itecount iteration counter;`
`for(itecount= 1 to 100,`
`...`
`...`
`);`
	`scalar itecount; itecount=1;`
	`while(itecount lt 100,`
	`...`
	`itecount=itecount+1;`
	`);`

the controlling index of the loop is not a set but a scalar. Table 10.11 shows these structures. The statement `while` is typically used when the criterion to stop the iterations (stopping criterion) depends on a conditional statement. On the other hand, the statement `for` is typically used when the exact number of iterations is known in advance and does not depend on any condition.

The following code is an example of how to use the statement `for` in the transportation problem:

```
...
scalar itecount "iteration counter";

for(itecount= 1 to 3,
 Solve transport using lp minimizing z ;
 b(j)=b(j)+ord(j);
);
```

The following code shows the same example as above but using the statement `while`:

```
...
scalar itecount; itecount=1;

while(itecount le 3,
 Solve transport using lp minimizing z ;
 b(j)=b(j)+ord(j); itecount=itecount+1;
);
```

Both codes are equivalent and direct GAMS to solve three related mathematical programming problems. In the first iteration (`itecount = 1`) the GAMS code considers the following sets of `demand` equations

```
demand(m1)..  x(p1,m1) + x(p2,m1) =G= 100 ;
demand(m2)..  x(p1,m2) + x(p2,m2) =G= 200 ;
demand(m3)..  x(p1,m3) + x(p2,m3) =G= 300 ;
```

For the second iteration (`itecount = 2`) the RHS values of the above equations change from 100, 200 and 300 to 101, 202, and 303, respectively. And finally, in the third iteration (`itecount = 3`) the corresponding RHS values are 102, 204, and 306.

10.3.14 Writing Output Files

To let GAMS users write the output information in nonstandard GAMS files facilities the `put` is provided. This statement is useful to write related information into different output files. Although this output format is less restrictive than the `display` statement, its use is more difficult. To generate the same output (see Section 10.3.10) as the command `display` generates in the transportation example, and write it in a nonstandard output file called `transport.out`, the following GAMS code can be used:

```
file out /transport.out/;
put out;
put "----     46 VARIABLE  X.L  shipment quantities in tons"//;
put "              ";
loop(j,put j.tl:12);
put //;
loop(i,
     put i.tl:2;
     loop(j,
          put$(x.l(i,j) ne 0)  x.l(i,j):12:3;
          put$(x.l(i,j) eq 0) "            ";
     );
     put /;
);
```

In the first line of this code, the output file is specified using the `file` command. The name `out` is arbitrarily selected by the user. It refers to the GAMS internal file associated with the user actual output file (in this case it is `transport.out`). Each user output file should have different internal file names.

While the command `display` handles the variable indices automatically, the `put` command requires the use of loops.

In the above example, the put command is followed by

- A GAMS internal file name (as it appears in the second line). This directs GAMS to print in the associated output file what the following put sentences indicate. To write in a different output file, a new internal file name should be defined.

- Text in quotation marks. The user writes the text just as it will be printed in the output file. This is shown in lines 3, 4, and 11 of the previous code.

- Slash symbol "/". This is used to produce a new line in the output file, as is shown in lines 3, 6, and 13.

- Set, parameter and/or variable names with their respective suffixes. Suffix `tl` is used to display the label of the set (see line 5); suffix `l` is used to display the level of the variables (see line 11).

- Conditional statement. If data writing depends on a specific condition, this can be implemented in GAMS using the statement "`$`" (see lines 10 and 11) or the statement `if-then-else`.

10.3.15 Output File: Nonlinear Equation Listing

The GAMS standard output file provides the user with helpful information to check possible errors in their implemented problem. This file is self-explanatory except for non-linear equation listings. An easy example should help the reader understand it.

The next code models the problem of minimizing $e^{-x_1}x_2^3$ taking as starting points $x_1^0 = 1$ and $x_2^0 = 0.5$:

```
POSITIVE VARIABLE X1,X2;
FREE VARIABLE z;
 X1.L=1; X2.L=0.5;

EQUATION
 COST    This is the objective function;

COST..   exp(-X1)*power(X2,3)=e=z;

MODEL nonlinear /all/;
SOLVE nonlinear USING nlp MINIMIZING z;
```

The related part of equation listing is

```
COST..- (0.046)*X1+(0.2759)*X2-Z =E= 0; (LHS = 0.046, INFES = 0.046
***)
```

In this expression, the coefficients enclosed in parentheses represent the first derivative of the COST constraint with respect to each variable evaluated at their current level values (starting point); that is, $-e^{-x_1^0} \cdot (x_2^0)^3 = -0.046$ and $e^{-x_1^0} \cdot 3 \cdot (x_2^0)^2 = 0.2759$. After the semicolon symbol, two equality expressions appear. The first expression is the evaluation of the left-hand side (LHS) of the COST function at the starting point: $e^{-x_1^0} \cdot (x_2^0)^3 = 0.046$. The second expression indicates that this constraint is infeasible at the starting point, but it does not mean that the problem is infeasible at every point.

Chapter 11

Some Examples Using GAMS

11.1 Introduction

In this chapter we explain how some of the examples given in previous chapters can be solved using GAMS. To avoid repetition, we give a detailed explanation of the GAMS commands in the first example, and in the remaining examples we emphasize only the new interesting features.

For the sake of a better understanding, we write GAMS reserved words, data and equation names in capital letters, and the rest of code, in lowercase letters.

11.2 Linear Programming Examples

This section deals with the linear programming problems analyzed in previous chapters.

11.2.1 The Transportation Problem

The transportation problem was analyzed in detail in Section 1.2, and Example 1.1. A particular example can be stated as follows. Minimize

$$Z = \sum_{i=1}^{m} \sum_{j=1}^{n} c_{ij} x_{ij} \tag{11.1}$$

subject to

$$\begin{array}{rcll} \sum\limits_{j=1}^{n} x_{ij} & = & u_i; & \forall i = 1, \ldots, m \\ \sum\limits_{i=1}^{m} x_{ij} & = & v_j; & \forall j = 1, \ldots, n \\ x_{ij} & \geq & 0; & \forall i = 1 \ldots m;\ \forall j = 1, \ldots, n \end{array} \tag{11.2}$$

where $m = n = 3$ and

$$\mathbf{C} = \begin{pmatrix} 1 & 2 & 3 \\ 2 & 1 & 2 \\ 3 & 2 & 1 \end{pmatrix}, \quad \mathbf{u} = \begin{pmatrix} 2 \\ 3 \\ 4 \end{pmatrix}, \text{ and } \mathbf{v} = \begin{pmatrix} 5 \\ 2 \\ 2 \end{pmatrix}$$

From a GAMS-user point of view, the following considerations should be taken into account:

- This problem presents three kinds of equations: (1) a single equation that represents the objective function, (2) a group of three equations associated with the three origins, and (3) a group of three equations associated with the three destinations. Therefore, two different sets for origins and destinations should be defined to represent, in a compact way, the last two groups of constraints. Notice the use of the symbol '*' to list the members of a set.

- There are three different data arrays: c_{ij}, u_i, v_j. Data u_i and v_j should be declared using the parameter command, because both of them are one-dimensional arrays. The c_{ij} array can be declared using either the parameter or the table command. In this case, the table command has been used.

- The optimization variable `z` is declared to represent the value of the objective function to be minimized.

- The array x_{ij} is the array of the optimization variables. There are as many variables as the number of origins times the number of destinations.

- After declaring a variable, it is convenient to define its type. The x_{ij} variables are positive in sign, and `z` is not restricted in sign (free by default).

- Once the necessary sets, data, and optimization variables have been defined, the constraints are declared and defined. The objective function is declared as a single equation (it does not depend on any set). In addition, two groups of constraints are declared using two different names. The first group, `SHIP(I)`, represents as many constraints as the number of origins, and therefore the corresponding set is indicated. The second group, `RECEIVE(J)`, represents as many constraints as the number of destinations, and therefore the corresponding set is indicated.

- Once the constraints have been declared, they should be defined. Since the GAMS notation almost mimics mathematical notation, this is an easy task. Notice the use of the `SUM(J, x(I,J))`, which is the GAMS operator to represent $\sum_{j=1}^{n} x_{ij}$.

- After the definition of equations, we name the model using the term `transport`. This model includes the `COST`, `SHIP(I)`, and `RECEIVE(J)` constraints. In this statement, equation indices are not allowed. Notice that the names enclosed within slashes can be replaced by the reserved word `all`.

- Once the model has been defined, it must be solved. This is a linear programming problem, therefore the reserved word `lp` is used. We also indicate that the objective function should be minimized.

An input GAMS file to solve this problem is

```
$title THE TRANSPORTATION PROBLEM

**  First, indices are declared and defined.
**  Index I is used to refer to the three origins.
**  Index J is used to refer to the three destinations.
**  Note how the symbol '*' is used to list sets members in a compact way.

SETS
 I index of shipping origins       /I1*I3/
 J index of shipping destinations  /J1*J3/;

**  Vectors of data (U(I) and V(J)) are defined as parameters
**  Data are assigned to vector elements

PARAMETERS
 U(I)  the amount of good to be shipped from origin I
       /I1 2
        I2 3
        I3 4/
 V(J)  the amount of good to be received in destination J
       /J1 5
        J2 2
        J3 2/;

**  The C(I,J) data matrix is defined as a table
**  Data are assigned to matrix elements

TABLE C(I,J) cost of sending a unit from origin I to destination J
      J1    J2   J3
I1     1     2    3
I2     2     1    2
I3     3     2    1;

**  The optimization variables are declared.
**  First, the objective function variable (z) is declared.
**  Next, the remaining variables with controlling indices are declared.

VARIABLES
 z       objective function variable
 x(I,J) the amount of product to be shipped from origin I to destination J;

**  The types of variables are given in the following sentence.
**  In the transportation problem, all the variables are positive except
**  the objective function variable.

POSITIVE VARIABLE x(I,J);

**  The objective function equation is declared.
**  The remaining six equations are declared, in a compact way,
**  as two index-dependent equations.
```

```
EQUATIONS
 COST        objective function equation
 SHIP(I)     shipping equation
 RECEIVE(J) receiving equation;

**  The following sentences formulate the above declared equations.
**  All the constraints are equality constraints (=E=).
**  The first one defines the objective function equation as a summation.
**  The second equation group represents three
**  different single equations depending on index I.
**  The left-hand-side is a sum in J of the unknowns x(I,J), and
**  the right-hand-side is a previously defined vector of data.
**  Similarly to SHIP constraints, the RECEIVE constraints are defined.

COST ..          z   =E=  SUM((I,J), C(I,J)*x(I,J)) ;
SHIP(I) ..       SUM(J, x(I,J))  =E=  U(I) ;
RECEIVE(J) ..    SUM(I, x(I,J))  =E=  V(J) ;

**  The next sentence names the model and list its constraints.

MODEL transport /COST,SHIP,RECEIVE/;

**  The next sentence directs GAMS to solve the transportation model using
**  a linear programming solver lp to minimize the objective function.

SOLVE transport USING lp MINIMIZING z;
```

Part of the output file is

```
---- VAR X            the amount of product to be shipped from origin I to
                      destination J
          LOWER     LEVEL     UPPER    MARGINAL
I1.J1       .       2.000     +INF        .
I1.J2       .         .       +INF      2.000
I1.J3       .         .       +INF      4.000
I2.J1       .       1.000     +INF        .
I2.J2       .       2.000     +INF        .
I2.J3       .         .       +INF      2.000
I3.J1       .       2.000     +INF        .
I3.J2       .         .       +INF       EPS
I3.J3       .       2.000     +INF        .

                        LOWER     LEVEL     UPPER    MARGINAL
---- VAR Z               -INF     14.000     +INF        .
  Z           objective function variable
```

In this output file, the optimal values of the variables are listed in the column identified by LEVEL. Therefore, the solution of the transportation problem is

$$Z = 14, \qquad \mathbf{X} = \begin{pmatrix} 2 & 0 & 0 \\ 1 & 2 & 0 \\ 2 & 0 & 2 \end{pmatrix}$$

11.2.2 Production Scheduling Problem 1

The production scheduling problem was analyzed in detail in Section 1.3, and Example 1.2. A particular case can be stated as follows. Maximize

$$Z = \sum_{t=1}^{n}(a_t y_t - b_t x_t - c_t s_t) \tag{11.3}$$

subject to

$$\begin{array}{rcl} s_{t-1} + x_t - s_t & = & y_t, \ \forall t = 1, \ldots, n \\ s_t, x_t, y_t & \geq & 0 \end{array} \tag{11.4}$$

where $a_t = b_t = c_t = 1, \forall t = 1 \ldots n;\ \ n = 4;\ \ s_0 = 2$ and $\mathbf{y} = (2, 3, 6, 1)^T$.

Some significant considerations of the GAMS code for this example are

- The use of the `fx` suffix to fix an element of a variable array to a single value.
- In the objective function equation, the $ command is used to sum specific (not all) elements of an array . This feature is also used to define a group of equations in terms of a subset.

The input GAMS file for this problem is

```
$title PRODUCTION_SCHEDULING PROBLEM

**  The index T is declared. It has 5 elements which are defined
**  using the symbol '*'.

SET
 T   The month index /0*4/;

**  Once the set T has been defined, the data vector Y(T) is
**  declared and all its elements are assigned except
**  the first one (element '0'), which, by default, is set to 0.

PARAMETER

 Y(T) demand in month T
   /1 2
    2 3
    3 6
    4 1/
 A(T)
 B(T)
 C(T);

** A unit value is assigned to all elements of variables A, B and C
** (using index T).

A(T)=1; B(T)=1; C(T)=1;
```

```
**  The optimization variables are declared.
**  First, the objective function variable (z) is declared.
**  This variable does not depend on any
**  index because it is a single-valued variable.
**  The remaining two variables depend on the index T, so, they
**  are declared as functions of it.

VARIABLES

 z    objective function variable
 x(T) number of units produced in month T
 s(T) number of units in storage in month T;

**  With the exception of the objective function variable,
**  which, by default, is not restricted in sign, the other
**  two variables in this problem are defined as positive.

POSITIVE VARIABLES x(T),s(T);

**  The initial value for variable s(T) is set in the next sentence.
**  In month '0', the variable s(T) takes on value 2.

s.fx('0')=2;

**  The objective function, and five constraints, INOUT equations, are
declared.

EQUATIONS
 COST     objective function
 INOUT(T) input and output balance;

**  Once the equations have been declared, they need to be defined.
**  The first equation is the objective function.
**  This equation has a conditional ($) sum.
**  Only the last four members of the ordered index T are
**  considered in the sum when using '$(ord(T) gt 1)'.
**  Next, we define the four INOUT equations.
**  These equations were declared as functions of index T.
**  Since the index T-1 appears in the term s(T-1) in the right hand side
**  of equation, it is necessary to begin with the second member of the set T.
**  To state that equations INOUT do not model the 'month 0',
**  the '$(ord(T) gt 1)' conditional statement is used after its name.

COST..z=E= SUM(T$(ord(T) gt 1), A(T)*Y(T)-B(T)*x(T)-C(T)*s(T));
INOUT(T)$(ord(T) gt 1)..s(T)=E= s(T-1)+x(T)-Y(T);

**  The next sentence names the model and states that all the previously
**  declared and defined constraints must be included in the optimization
**  problem.

MODEL scheduling /ALL/;

**  The next sentence directs GAMS to solve the scheduling model using linear
**  programming and maximizing the objective function variable z.

SOLVE scheduling USING lp MAXIMIZING z;
```

Part of the GAMS output file is as follows:

```
                          LOWER     LEVEL     UPPER    MARGINAL
---- VAR Z                -INF      2.000     +INF

  Z           objective function variable

---- VAR X             number of units produced in month T
     LOWER     LEVEL     UPPER    MARGINAL
1      .         .       +INF        .
2      .       3.000     +INF        .
3      .       6.000     +INF        .
4      .       1.000     +INF        .

---- VAR S             number of units in storage in month T
     LOWER     LEVEL     UPPER    MARGINAL
0     2.000    2.000     2.000     1.000
1      .         .       +INF     -1.000
2      .         .       +INF     -1.000
3      .         .       +INF     -1.000
4      .         .       +INF     -2.000
```

Therefore, the solution of the production scheduling problem is

$$Z = 2, \qquad \mathbf{x} = (0,3,6,1)^T, \qquad \mathbf{s} = (2,0,0,0,0)^T$$

11.2.3 The Diet Problem

The diet problem was analyzed in Section 1.4, and Example 1.3. A particular case can be stated as follows. Minimize

$$Z = \sum_{j=1}^{n} c_j x_j \tag{11.5}$$

subject to

$$\begin{array}{rcll} \sum\limits_{j=1}^{n} a_{ij}x_j & \geq & b_i; & i = 1,\ldots,m \\ x_j & \geq & 0; & j = 1,\ldots,n. \end{array}$$

where $m = 4$, $n = 5$ and

$$\mathbf{A} = \begin{pmatrix} 78.6 & 70.1 & 80.1 & 67.2 & 77.0 \\ 6.50 & 9.40 & 8.80 & 13.7 & 30.4 \\ 0.02 & 0.09 & 0.03 & 0.14 & 0.41 \\ 0.27 & 0.34 & 0.30 & 1.29 & 0.86 \end{pmatrix}, \quad \mathbf{b} = \begin{pmatrix} 74.2 \\ 14.7 \\ 0.14 \\ 0.55 \end{pmatrix}, \quad \mathbf{c} = \begin{pmatrix} 1 \\ 0.5 \\ 2 \\ 1.2 \\ 3 \end{pmatrix}$$

This example does not use any additional GAMS feature. For a careful understanding, the reader should examine the comments included in the following code:

```
$title DIET PROBLEM

**  The sets of indices are defined:
**  Index I is used to refer to the four nutrients.
**  Index J is used to refer to the five foods.

SET
 I    set of nutrients   /DN,DP,Ca,Ph/
 J    set of foods       /Corn,Oats,Milo,Bran,Linseed/;

**  Vectors of data are defined as parameters.
**  Data are assigned to vector elements.

PARAMETERS
 B(I) the minimum required amount of nutrient I
     /DN 74.2
      DP 14.7
      Ca 0.14
      Ph 0.55/
 C(J)  cost of one unit of food J
      /Corn    1
       Oats    0.5
       Milo    2
       Bran    1.2
       Linseed 3/;

**  The matrix of data is defined as a table.
**  Data are assigned to matrix elements.

TABLE A(I,J) the amount of nutrient I in one unit of food J
     Corn  Oats  Milo  Bran  Linseed
DN   78.6  70.1  80.1  67.2   77.0
DP    6.5   9.4   8.8  13.7   30.4
Ca   0.02  0.09  0.03  0.14   0.41
Ph   0.27  0.34  0.30  1.29   0.86;

**  The next step declares the optimization variables.
**  First, the objective function variable is declared.
**  Then, the remaining variables with their associated indices are declared.

VARIABLES
  z    objective function variable
  x(J) the amount of food J to be purchased ;

**  The types of variables are established in the next sentence.
**  In the diet problem, all the variables except
**  the objective-function variable are positive.

POSITIVE VARIABLE x(J);

**  The problem constraints are declared.
**  First, the objective function constraint (COST).
**  Next, the four remaining constraints (NUTFOOD) are
**  declared as functions of index I.

EQUATIONS
 COST      objective function
```

```
 NUTFOOD(I) nutrients and food relation;

**  The objective function is an equality constraint (=E=).
**  The NUTFOOD constraints are inequality constraints (=G=).

COST ..         z  =E=  SUM(J, C(J)*x(J));
NUTFOOD(I) ..   SUM(J, A(I,J)*x(J))  =G=  B(I);

**  The following sentences define the diet model with all the
**  declared constraints and direct GAMS to solve the problem.

MODEL diet /ALL/;
SOLVE diet USING lp MINIMIZING z;
```

Part of the GAMS output file, which provides the optimal solution for the diet problem, is

```
                           LOWER      LEVEL      UPPER     MARGINAL
---- VAR Z                 -INF       0.793      +INF         .
  Z           objective function variable

---- VAR X              the amount of food J to be purchased
              LOWER      LEVEL      UPPER    MARGINAL
Corn            .          .        +INF      0.634
Oats            .        1.530      +INF        .
Milo            .          .        +INF      1.543
Bran            .        0.023      +INF        .
Linseed         .          .        +INF      1.525
```

The optimal solution for the diet problem is

$$Z = 0.793, \qquad \mathbf{x} = (0, 1.53, 0, 0.023, 0)^T \tag{11.6}$$

11.2.4 The Network Flow Problem

The network flow problem, discussed in Section 1.5, and in Example 1.4, can be stated as follows. Minimize

$$Z = \sum_{ij} c_{ij} x_{ij}$$

subject to

$$\begin{array}{rcll} \sum_j (x_{ij} - x_{ji}) & = & f_i; & i = 1, \dots, n \\ -f_{ij} \leq & x_{ij} & \leq f_{ij}; & \forall i < j \end{array}$$

where $n = 4$, we assume that $f_{ij} = 4, \forall i, j$, and $(f_1, f_2, f_3, f_4) = (7, -4, -1, -2)$ and $c_{ij} = 1; \forall i, j$.

In this example two new GAMS features have been used:

- The definition of a subset that allows definition of the different connections between nodes
- The use of subsets to limit the equation index scope in such a way that only the connected nodes are affected

A GAMS input file for this problem is

```
$title NETWORK FLOW PROBLEM

**  First, sets are declared:
**  Set I has four elements.
**  Subset CONEX is defined as a subset of set I.
**  The subset CONEX establishes the valid connections between nodes I.

SET
 I          set of nodes in the network  /I1*I4/
 CONEX(I,I) set of node connections  /I1.I2,I1.I3,I1.I4,I2.I4,I3.I4/;

**  The set of nodes I must be duplicated to refer to
**  its different elements in the same constraint.

 ALIAS(I,J)

**  Vectors of data are defined as parameters.
**  Data are assigned to vector elements.
**  FMAX(I,J) is a data matrix, declared as a parameter, where all
**  its elements have the same value (4). This is a compact way
**  to declare a matrix (instead of using a TABLE).

PARAMETERS
 F(I)   the input or output flow at node I
      /I1   7
       I2  -4
       I3  -1
       I4  -2/
 FMAX(I,J) maximum flow capacity of conduction going from I to J;

FMAX(I,J)=4;

**  Optimization variables are declared.

VARIABLES
  z      objective function variable
  x(I,J) the flow going from node I to node J;

**  Limits are stated on variables using previously defined data.

x.lo(I,J)=-FMAX(I,J);
x.up(I,J)=FMAX(I,J);

** Constraints are declared.

EQUATIONS
 COST        objective function
 BALANCE(I)  conservation of flow conditions;

**  The objective function is a function of flows between connected nodes.
**  This requirement is stated using the $CONEX(I,J) condition.
**  The four BALANCE equations only consider the flow between
**  connected nodes and again the $CONEX(I,J) condition is necessary.

COST ..          z   =E=  SUM(CONEX(I,J),x(I,J)) ;
BALANCE(I) ..   SUM(J$CONEX(I,J),x(I,J))-SUM(J$CONEX(J,I),x(J,I))  =E=  F(I) ;
```

```
**  The next two sentences define the netflow model, considering all the above
**  constraints, and direct GAMS to solve the problem using the lp solver.

MODEL netflow /ALL/;
SOLVE netflow USING lp MINIMIZING z;
```

The GAMS solution of this example is

```
                        LOWER     LEVEL     UPPER    MARGINAL
---- VAR Z              -INF      5.000     +INF        .
  Z           objective function variable

---- VAR X            the flow going from node I to node J
         LOWER     LEVEL     UPPER    MARGINAL
I1.I2    -4.000      .       4.000       .
I1.I3    -4.000    3.000     4.000       .
I1.I4    -4.000    4.000     4.000    -1.000
I2.I4    -4.000   -4.000     4.000      EPS
I3.I4    -4.000    2.000     4.000       .
```

So, the solution of the problem is

$$Z = 5, \;\; x_{12} = 0, \;\; x_{13} = 3, \;\; x_{14} = 4, \;\; x_{24} = -4, \;\; x_{34} = 2$$

The Water Supply Network Problem

This problem was analyzed in Section 1.5, and it can be stated as follows. Minimize

$$Z = \sum_{ij} |x_{ij}|$$

subject to

$$\begin{array}{rcll} \sum_j (x_{ij} - x_{ji}) & = & f_i; & i = 1, \ldots, n \\ -f_{ij} \leq x_{ij} & \leq & f_{ij}; & \forall i < j \end{array}$$

where we assume that $f_{ij} = 8, \forall i, j$, and $(f_1, f_2, f_3, f_4) = (20, -3, -10, -7)$.

Remarkable considerations for this example are

- To avoid the use of the absolute value function (leading to a nonlinear problem), we take the difference of two positives variables to represent a variable not restricted in sign. Therefore, this linear version is stated as follows. Minimize

  $$Z = \sum_{ij} (x_{ij}^+ + x_{ij}^-)$$

  subject to

  $$\begin{array}{rcll} \sum_j \left[(x_{ij}^+ - x_{ij}^-) - (x_{ji}^+ - x_{ji}^-) \right] & = & f_i; & i = 1, \ldots, n \\ -f_{ij} \leq x_{ij}^+ - x_{ij}^- & \leq & f_{ij}; & \forall i < j \end{array}$$

  where we assume that $x_{ij}^+ > 0, \forall i, j$, and $x_{ij}^- > 0, \forall i, j$

- Through conditional statements on subsets, the equations UPLIMIT and LOLIMIT are restricted to be satisfied only by the connected nodes.

```
$title WATER SUPPLY NETWORK (linear)

**  Sets are declared in first place:
**  Set I has four elements.
**  CONEX is defined as a subset of set I.
**  The subset CONEX establishes the valid connections between nodes of I.

SET
 I          set of nodes in the network  /I1*I4/
 CONEX(I,I) set of connections between nodes /I1.I2,I1.I3,I1.I4,I2.I4,I3.I4/;

**  The set of nodes I should be duplicated to refer to its
**  different elements within the same constraint.

 ALIAS(I,J)

**  Vectors of data are defined as parameters.
**  Data are assigned to each vector element.
**  FMAX(I,J) is a data matrix declared as a parameter.
**  All its elements have the same value (8).
**  This is a compact way to declare a matrix (instead of using a TABLE).

PARAMETERS
 X(I,J) total flow going from node I to node J
 F(I)   the input or output flow at node I
      /I1  20
       I2  -3
       I3  -10
       I4  -7/
 FMAX(I,J) maximum flow capacity of conduction going from I to J;

FMAX(I,J)=8;

**  Optimization variables are declared.

VARIABLES
  z       objective function variable
  xp(I,J) positive part of x
  xn(I,J) negative part of x;

**  The flow from I to J is a positive variable.

POSITIVE VARIABLES xp(I,J),xn(I,J);

** Constraints are declared.

EQUATIONS
 COST          objective function
 BALANCE(I)    conservation of flow conditions
 UPLIMIT(I,J)  upper limit of flow going from node I to J
 LOLIMIT(I,J)  lower limit of flow going from node I to J;

**  The objective function is the sum of the absolute values of the flows
**  between all connected nodes. This requirement is stated using the
**  $CONEX(I,J) condition.
```

```
**  The four BALANCE equations only consider the flow between
**  two connected nodes and again the $CONEX(I,J) condition is necessary.
**  The flow going from node I to node J is limited by the maximum flow
**  capacity of conduction using the UPLIMIT(I,J) and LOLIMIT(I,J) constraints.


COST ..          z   =E=  SUM((I,J)$CONEX(I,J),xp(I,J)-xn(I,J)) ;
BALANCE(I) ..    SUM(J$CONEX(I,J),xp(I,J))-SUM(J$CONEX(J,I),xp(J,I))-
                 SUM(J$CONEX(I,J),xn(I,J))+SUM(J$CONEX(J,I),xn(J,I))  =E=
F(I) ;
UPLIMIT(I,J)$(CONEX(I,J)) .. xp(I,J) - xn(I,J) =l=  FMAX(I,J);
LOLIMIT(I,J)$(CONEX(I,J)) .. xp(I,J) - xn(I,J) =g= -FMAX(I,J);

**  The next two sentences define the model with all the above constraints,
**  and direct GAMS to solve the problem using a linear programming solver.

MODEL wsn /ALL/;
SOLVE wsn USING lp MINIMIZING z;

**  The next sentence assigns to x(I,J) the difference between
**  its positive part and its negative part. This is used to
**  model in a linear way, the absolute value function.

x(I,J)=xp.L(I,J)-xn.L(I,J);

**  To print in the listing file (*.lst) the value of
**  x(I,J) the next sentence is used.

 DISPLAY x;
```

Part of the GAMS output file is

```
                          LOWER      LEVEL      UPPER     MARGINAL
---- VAR Z                 -INF      23.000      +INF
  Z           objective function variable

---- VAR XP           positive part of x
         LOWER     LEVEL     UPPER    MARGINAL
I1.I2      .        4.000     +INF       .
I1.I3      .        8.000     +INF       .
I1.I4      .        8.000     +INF       .
I2.I4      .        1.000     +INF       .
I3.I4      .          .       +INF      2.000

---- VAR XN           negative part of x
         LOWER     LEVEL     UPPER    MARGINAL
I1.I2      .          .       +INF      2.000
I1.I3      .          .       +INF      2.000
I1.I4      .          .       +INF      2.000
I2.I4      .          .       +INF      2.000
I3.I4      .        2.000     +INF       .

----     84 PARAMETER X             total flow going from node I to node J
            I2          I3          I4
I1       4.000       8.000       8.000
I2                               1.000
I3                               2.000
```

The solution of the problem is:

$$Z = 23, \quad x_{12} = 4, \quad x_{13} = 8, \quad x_{14} = 8, \quad x_{24} = 1, \quad x_{34} = 2$$

11.2.5 The Portfolio Problem

This problem was analyzed in Section 1.6, and can be stated as follows. Maximize:

$$Z = \sum_j d_j(b_j + x_j) \tag{11.7}$$

subject to

$$\begin{array}{rcl} b_i + x_i & \geq & 0 \\ r(\sum_j v_j(b_j + x_j)) & \leq & v_i(b_i + x_i) \\ \sum_j v_j x_j & = & 0 \\ \sum_j w_j(b_j + x_j) & \geq & (1+s)\sum_j v_j b_j \end{array} \tag{11.8}$$

Consider the particular case in which we have shares of three stocks, 75 of A_1, 100 of A_2, and 35 of A_3, with values \$20, \$20, and \$100, respectively. In addition, we have the information: A_1 will pay no dividends with a new value of \$18, A_2 will pay \$3 per share and the new value will be \$23, and A_3 will pay \$5 per share with a new value of \$102. If we take the proportion r to be 0.25 and s, 0.03, all the preceding constraints become

$$\begin{array}{rcl} x_A & \geq & -75 \\ x_B & \geq & -100 \\ x_C & \geq & -35 \\ 0.25\,[20(75 + x_A) + 20(100 + x_B) + 100(35 + x_C)] & \leq & 20(75 + x_A) \\ 0.25\,[20(75 + x_A) + 20(100 + x_B) + 100(35 + x_C)] & \leq & 20(100 + x_B) \\ 0.25\,[20(75 + x_A) + 20(100 + x_B) + 100(35 + x_C)] & \leq & 100(35 + x_C) \\ 20x_A + 20x_B + 100x_C & = & 0 \\ 18(75 + x_A) + 23(100 + x_B) + 102(35 + x_C) & \geq & 1.03(20(175) + 3500) \end{array} \tag{11.9}$$

The GAMS code to solve this example is

```
$title THE PORTFOLIO PROBLEM

**  Set I is declared with three elements.

SET
 I          set of stocks  /A1,A2,A3/;

**  The set of stocks I should be duplicated in
**  order to refer to its different elements in
**  the same constraint.

ALIAS(I,J);

SCALARS  r  percentage        /0.25/
         s  percentage        /0.03/;
```

```
TABLE  data(I,*)
         B     V    D    W
*             ($)  ($)  ($)
 A1     75    20    0   18
 A2    100    20    3   23
 A3     35   100    5  102;

VARIABLES
  z    objective function variable
  x(I) number of shares of stock I;
POSITIVE VARIABLE x(I);

x.lo(I)=-data(I,'B');

EQUATIONS
 COST         objective function
 NOCHANGE     no change in the current value
 INFLATION    future value must be 3\% greater than the current value
 BALANCE(I)   to avoid excessive reliance on a single stock;

COST ..          z   =E=  SUM(I,data(I,'D')*(x(I)+data(I,'B'))) ;
NOCHANGE ..      SUM(I,data(I,'V')*x(I)) =E= 0;
INFLATION ..     SUM(I,data(I,'W')*(x(I)+data(I,'B')))=G=
                 (1+s)*SUM(I,data(I,'V')*data(I,'B'));
BALANCE(I)..     r*SUM(J,data(J,'V')*(x(J)+data(J,'B')))=L=
                 data(I,'V')*(x(I)+data(I,'B'));

MODEL portfolio /ALL/;
SOLVE portfolio USING lp MAXIMIZING z;
```

Part of the GAMS output file is

```
                         LOWER      LEVEL       UPPER     MARGINAL
---- VAR Z               -INF      612.500      +INF         .
  Z           objective function variable

---- VAR X             number of shares of stock I
       LOWER     LEVEL       UPPER    MARGINAL
A1    -75.000    12.500      +INF         .
A2   -100.000    75.000      +INF         .
A3    -35.000   -17.500      +INF         .
```

The solution for this example is

$$Z = 612.5 \text{ attained at the point } x_A = 12.5, \quad x_B = 75.0, \quad x_C = -17.5$$

11.2.6 The Scaffolding System

This example was dealt with in Section 1.7. A particular example is maximization of

$$Z = \sum_i x_i$$

subject to

$$
\begin{aligned}
T_E + T_F &= x_2 \\
T_C + T_D &= T_F \\
T_A + T_B &= x_1 + T_C + T_D \\
10T_F &= 5x_2 \\
8T_D &= 6T_F \\
10T_B &= 5x_1 + 2T_C + 10T_D
\end{aligned}
$$

The reader must realize that

- The GAMS formulation is valid for any kind of ropes and beams topology.
- The use of subsets is crucial to obtain a compact notation for equations.
- The sign of the resulting loads denotes their directions.

A GAMS input file to solve a linear version of this problem is

```
$title SCAFFOLDING PROBLEM (LINEAR)

**  The indexes are defined in first place:
**  Index B is used to refer to the three beams.
**  Index R is used to refer to the six ropes.
**  Index L is used to refer to the two loads.

SET
 B    set of beams    /B1*B3/
 R    set of ropes    /RA,RB,RC,RD,RE,RF/
 L    set of loads    /L1,L2/

UPP(B,R)
      / B1.(RA,RB)
        B2.(RC,RD)
        B3.(RE,RF)
      /
DOWN(B,R)
      / B1.(RC,RD)
        B2.(RF)
      /
LOAD(B,L)
      / B1.L1
        B3.L2
      /;

**  Vectors of data are defined as parameters.
**  Data are assigned to vector elements.

PARAMETER LMAX(R) maximum load for ropes
  / (RA,RB) 300
    (RC,RD) 200
    (RE,RF) 100
  /;
PARAMETER DL(L) coordinates of load L
  / L1  7
    L2  5
  /;
```

```
PARAMETER DR(R) coordinates of rope R
  / RA  2
    RB 12
    RC  4
    RD 12
    RE  0
    RF 10 /;

VARIABLES
  z    objective function variable
  x(L) the applied load
  t(R) tension on rope R ;

t.up(R) = LMAX(R);

** Problem constraints are declared.

EQUATIONS
 COST       objective function
 FORCES(B)  force equilibrium equation
 MOMENT(B)  moment equilibrium equation;

COST ..          z  =E=  SUM(L, x(L)) ;

FORCES(B)..      SUM(R$UPP(B,R),t(R))=E= SUM(L$LOAD(B,L),x(L))+
                 SUM(R$DOWN(B,R),t(R));

MOMENT(B)..      SUM(R$UPP(B,R),DR(R)*t(R))=E=SUM(L$LOAD(B,L),
                 DL(L)*x(L))+SUM(R$DOWN(B,R),DR(R)*t(R));

MODEL scaffold /COST,FORCES,MOMENT/;
SOLVE scaffold USING lp MAXIMIZING z;
```

Part of the GAMS output file is

```
                         LOWER     LEVEL     UPPER    MARGINAL
---- VAR Z                -INF    640.000    +INF        .
  Z           objective function variable

---- VAR X            the amount of food J to be purchased
      LOWER     LEVEL      UPPER    MARGINAL
L1     -INF    440.000     +INF        .
L2     -INF    200.000     +INF        .

---- VAR T            tension on rope R
      LOWER     LEVEL      UPPER    MARGINAL
RA     -INF    240.000    300.000      .
RB     -INF    300.000    300.000     2.000
RC     -INF     25.000    200.000      .
RD     -INF     75.000    200.000      .
RE     -INF    100.000    100.000     0.400
RF     -INF    100.000    100.000      .
```

Thus, the solution is

$$Z = 640; \quad x_1 = 440, \quad x_2 = 200$$

$$T_A = 240, \quad T_B = 300, \quad T_C = 25, \quad T_D = 75, \quad T_E = 100, \quad T_F = 100$$

11.2.7 Electric Power Economic Dispatch

This problem has been discussed in Section 1.8, and Example 1.7. One example can be stated as follows. Minimize

$$Z = \sum_{i=1}^{n} C_i \, p_i \tag{11.10}$$

subject to

$$\begin{array}{rcll} \delta_k & = & 0 & \\ \sum_{j\in\Omega_i} B_{ij}(\delta_i - \delta_j) + p_i & = & D_i; & i = 1, 2, \dots, n \\ -\overline{P}_{ij} \le B_{ij}(\delta_i - \delta_j) & \le & \overline{P}_{ij}; & \forall j \in \Omega_i, \; i = 1, 2, \dots, n \\ \underline{P}_i \le p_i & \le & \overline{P}_i; & i = 1, 2, \dots, n \end{array} \tag{11.11}$$

where $n = 3$, $k = 3$

$$\underline{\mathbf{p}} = \begin{pmatrix} 0.15 \\ 0.10 \end{pmatrix}, \; \overline{\mathbf{p}} = \begin{pmatrix} 0.6 \\ 0.4 \end{pmatrix}, \; \mathbf{c} = \begin{pmatrix} 6 \\ 7 \end{pmatrix}, \; \mathbf{B} = \begin{pmatrix} 0.0 & 2.5 & 3.5 \\ 2.5 & 0.0 & 3.0 \\ 3.5 & 3.0 & 0.0 \end{pmatrix}, \; \mathbf{D} = \begin{pmatrix} 0.0 \\ 0.0 \\ 0.85 \end{pmatrix}$$

$$\mathbf{P} = \begin{pmatrix} 0.0 & 0.3 & 0.5 \\ 0.3 & 0.0 & 0.4 \\ 0.5 & 0.4 & 0.0 \end{pmatrix}, \; \Omega_1 = \{2,3\}, \;\; \Omega_2 = \{1,3\}, \;\; \Omega_3 = \{1,2\}$$

and the optimization variables are p_1, p_2, δ_1, and δ_2.

Some remarkable considerations to write the GAMS code for this example are:

- The definition of a subset in terms of two different sets. This allows the mapping of elements belonging to different sets (generators and buses).
- The symbol "*" is used in some tables to represent, by columns, the heterogeneous data on generators and lines.
- A column of a table, which has been previously defined in a compact way, can be referred to, in the definition of an equation, by the name given to that column.
- The use of the conditional command on sets that limits the scope of the sum operation.

A GAMS input file for this example is given below.

```
$title THE ECONOMIC DISPATCH PROBLEM

**  The sets G and N are defined.
**  Next, the set MAP is defined as a subset of sets G and N.
**  The subset MAP establishes the valid pairs of G and N elements.
```

```
SETS
  G   index of generators  /G1*G2/
  N   index of buses    /N1*N3/
  MAP(G,N) associates generators with buses /G1.N1,G2.N2/;

**  The data is provided below. The first table contains the columns of data
**  for every generator. The names of these columns have not been previously
**  declared as elements of a set, so they are referred to using the symbol '*'.
**  The same occurs in the second table.

TABLE GDATA(G,*)  generator input data
          PMIN      PMAX      COST
*         (kW)      (kW)     (E/kWh)
  G1      0.15      0.6        6
  G2      0.10      0.4        7;

TABLE LDATA(N,N,*) line input data
           SUS    LIMIT
*          (S)    (kW)
  N1.N2    2.5     0.3
  N1.N3    3.5     0.5
  N2.N3    3.0     0.4;

**  To state that only bus 3 has an electric load, the
**  parameter below is declared and defined.

PARAMETER
  LOAD(N)  load at bus N
           / N3  0.85 /

**  The optimization variables are declared.

VARIABLES
  z              objective function variable
  p(G)           output power for generator G
  d(N)           angle at bus N;

**  Bounds on variables, using previously defined data, are given below.

  p.lo(G)=GDATA(G,'PMIN');
  p.up(G)=GDATA(G,'PMAX');

**  Bus 3 is considered as the reference bus, therefore
**  its corresponding angle is fixed to zero.

  d.fx('N3')=0;

**  Different constraints are declared.

EQUATIONS
  COST           objective function
  MAXPOW(N,N)    maximum line power limit
  MINPOW(N,N)    minimum line power limit
  LOADBAL(N)     load balance equation;

**  The set of buses N should be 'duplicated' to refer
**  to different elements in the same constraint.
```

```
ALIAS(N,NP);

**  The definition of equations begins below.
**  The objective function equation is defined as a summation.
**  The 'cost' index value is used to refer to the column 'cost' of the
**  GDATA table.
**  Next, the inequality constraints MAXPOW and MINPOW are defined.
**  Every equation relates two different buses, therefore they depend on
**  two sets,
**  through N and NP that represent the same set with different names.
**  The last equation group LOADBAL is defined per bus.
**  To ensure that a generator G is connected to bus N the
**  sum is conditioned by $MAP(G,N).

COST..          z =e= SUM(G,GDATA(G,'COST')*p(G));
MAXPOW(N,NP).. LDATA(N,NP,'SUS')*(d(N)-d(NP))=l= LDATA(N,NP,'LIMIT');
MINPOW(N,NP).. LDATA(N,NP,'SUS')*(d(N)-d(NP))=g=-LDATA(N,NP,'LIMIT');
LOADBAL(N)..    SUM(G$MAP(G,N),p(G))+SUM(NP,LDATA(N,NP,'SUS')*(d(N)-d(NP))+
                LDATA(NP,N,'SUS')*(d(N)-d(NP)))=e=LOAD(N);

**  The next sentences define the economic dispatch model
**  with all the declared constraints and direct GAMS to solve the problem.

MODEL ed /COST,MAXPOW,MINPOW,LOADBAL/;
SOLVE ed USING lp MINIMIZING z;
```

The solution of this problem (part of the GAMS output file) is

```
---- VAR Z          objective function variable
      LOWER      LEVEL      UPPER     MARGINAL
      -INF       5.385      +INF        .

---- VAR P          output power for generator G
      LOWER      LEVEL      UPPER     MARGINAL
G1    0.150      0.565      0.600        .
G2    0.100      0.285      0.400        .

---- VAR D          angle at bus N
      LOWER      LEVEL      UPPER     MARGINAL
N1    -INF      -0.143      +INF        .
N2    -INF      -0.117      +INF        .
N3      .          .          .        EPS
```

The optimal solution for the economic dispatch problem is

$$Z = 5.385, \qquad \mathbf{p} = (0.565, 0.285)^T, \qquad \boldsymbol{\delta} = (-0.143, -0.117, 0)^T$$

11.3 Mixed-Integer LPP Examples

This section deals with mixed integer linear programming problems.

11.3.1 The 0–1 Knapsack Example

The knapsack problem was analyzed in detail in Section 2.2. The formulation for this problem is to maximize

$$Z = \sum_{j=1}^{n} c_j x_j$$

subject to

$$\begin{aligned} \sum_{j=1}^{n} a_j x_j &\leq b, \\ x_j &\in \{0,1\} \quad \forall j = 1 \cdots n \end{aligned}$$

where we assume that $a_j = c_j$, $b = 700$, and

$$\mathbf{a} = (100, 155, 50, 112, 70, 80, 60, 118, 110, 55)^T$$

The reader should note that a solution tolerance is specified through OPTION OPTCR=1e-10.

A GAMS input file to solve this problem is

```
$title 0-1 KNAPSACK PROBLEM.

** This knapsack problem is used to determine the maximum loading of a freighter.
** The main characteristic is that A(j)=C(j) for this problem.
** This fact can be taken into account to shorten the GAMS file,
** but the formulation is general for any knapsack problem.
** The next sentence change the termination criterion
** based on the relative gap of the current solution.
** Note that integer programs work with real programs
** and we need to indicate the program what error is
** admitted to consider a real number as an integer number
** If we use a standard value then GAMS stops before
** an optimal solution is achieved.

OPTION OPTCR=1e-10;

SET
 J    set of containers   /c1*c10/;

** Vectors of data A and C are defined as parameters.

PARAMETERS
 C(J) benefit of container J
     /c1     100
      c2     155
      c3      50
      c4     112
      c5      70
      c6      80
      c7      60
      c8     118
      c9     110
      c10     55/
```

```
 A(J) weight of container J;
 A(J) = C(J);

SCALAR  B    maximum capacity of the freighter /700/;

**  Optimization variables are declared.

VARIABLES
 z      objective function variable
 x(J)   binary choice;

** x(J) has the value 1 if the container J is shipped, and 0 otherwise.

BINARY VARIABLE x;

** Constraints are declared.

EQUATIONS
 COST objective function
 CAPA is the loading of the freighter;

** The objective function is the sum of the weights
** of all the containers that are shipped.
** The CAPA equations consider that the loading cannot exceed
** the maximum capacity of the freighter.

COST ..  z=E= SUM(J,C(J)*x(J));
CAPA ..       SUM(J,A(J)* x(J)) =L= B;

** The next two sentences define the model,
** considering all the above constraints, and direct GAMS to solve
** the problem using an integer linear programming solver.

MODEL knapsack /ALL/;
SOLVE knapsack USING mip MAXIMIZING z;
```

A part of the GAMS output file is

```
                          LOWER     LEVEL     UPPER    MARGINAL
---- VAR Z                 -INF    700.000     +INF       EPS
  Z            objective function variable

---- VAR X            binary choice
       LOWER      LEVEL      UPPER     MARGINAL
c1       .        1.000      1.000       EPS
c2       .          .        1.000       EPS
c3       .        1.000      1.000       EPS
c4       .        1.000      1.000       EPS
c5       .        1.000      1.000       EPS
c6       .        1.000      1.000       EPS
c7       .        1.000      1.000       EPS
c8       .        1.000      1.000       EPS
c9       .        1.000      1.000       EPS
c10      .          .        1.000       EPS
```

The solution indicates that the loading consists of the containers

$$c_1, c_3, c_4, c_5, c_6, c_7, c_8, c_9.$$

The optimal value is $Z = 700$ tons, and this means that the ship is full.

11.3.2 Identifying Relevant Symptoms

This example was dealt with in Section 2.3. The optimization problem is formulated as follows. Minimize

$$Z = \sum_{j=1}^{m} x_j$$

subject to

$$\sum_{j=1}^{m} x_j d(c_{ij} - c_{kj}) > a; \quad k \in \{1, 2, \ldots, n\}, \ i \neq k \tag{11.12}$$

where

$$d(c_{ij} - c_{kj}) = \begin{cases} 1 & if \quad c_{ij} \neq c_{kj} \\ 0 & if \quad c_{ij} = c_{kj} \end{cases} \tag{11.13}$$

An input GAMS file to solve this problem is

```
$Title Symptoms

SETS   D set of diseases /D1*D5/
       S set of symptoms /S1*S8/;
ALIAS(D,DP);

SCALAR A discrepancy level /1/;
TABLE  C(D,S)
       S1   S2   S3   S4   S5   S6   S7   S8
 D1     2    3    1    1    1    2    1    2
 D2     1    1    1    1    3    1    2    1
 D3     3    4    2    3    2    2    3    2
 D4     2    2    2    2    2    1    2    3
 D5     1    1    1    2    1    1    1    2;
PARAMETER
DD(D,DP,S) discrepancy measure;
loop((D,DP,S),
    if((C(D,S) ne C(DP,S)),
       DD(D,DP,S)=1;
    else
       DD(D,DP,S)=0;
    );
);
BINARY VARIABLE x(S);
FREE VARIABLE z;
EQUATIONS
 SELECT    number of selected symptoms
 Suff(D,DP) symptoms are sufficient;

SELECT.. z =e= SUM(S,x(S));
Suff(D,DP)$(ord(D) ne ord(DP))..
             SUM(S,x(S)*DD(D,DP,S)) =g= A;

MODEL symptoms /ALL/;

SOLVE symptoms USING mip MINIMIZING z;
```

The resulting solution is that symptoms 1 and 4 are sufficient for identifying all diseases.

11.3.3 The Academy Problem

This example was dealt with in Section 2.3. The optimization problem is formulated as follows. Maximize and minimize

$$Z_j = \sum_{i=1}^{I}\sum_{s=1}^{S} x_{ijs}, \quad \forall j \in \{1,2,\ldots,J\}$$

subject to

$$\begin{array}{rcll}
\sum_{s=1}^{S} x_{ijs} & \leq & 1, & \forall i \in \{1,2,\ldots,I\},\ j \in \{1,2,\ldots,J\} \\
\sum_{j=1}^{J} x_{ijs} & \leq & 1, & \forall i \in \{1,2,\ldots,I\},\ s \in \{1,2,\ldots,S\} \\
\sum_{i=1}^{I}\sum_{s=1}^{S} p_s x_{ijs} & = & c_j, & \forall j \in \{1,2,\ldots,J\} \\
x_{ijs} & \in & \{0,1\}, & \forall i \in \{1,2,\ldots,I\},\ j \in \{1,2,\ldots,J\},\ s \in \{1,2,\ldots,S\}
\end{array}$$

Two special features are used to solve this example via GAMS: loops and dynamic sets. For every candidate we need to solve two problems that differs only in the optimization direction. We define a loop controlled by the candidate index. Inside this loop, the dynamic set is updated with a specific candidate. The objective function equation is conditioned by the dynamic set which states a different candidate in every loop iteration. Further remarks are included in the GAMS file.

An input GAMS file to solve this problem is

```
$title ACADEMY

SETS
 I number of actual members /1*20/
 J number of candidates /1*8/
 DIN(J)
 S the number of different scores that can be assigned /1*4/;

ALIAS(J,J1);

PARAMETER
P(S) the $s$-th score
 /1 10
  2 8
  3 3
  4 1/;

TABLE N(I,J) score assigned to candidate J by actual member I
```

```
          1         2         3         4         5         6         7         8
**************************************************************************************
1         3                   10                  8         1
2         1                   10                  8         3
3                   1                   3         10                  8
4                   3         10                  8         1
5         3                   8                   10                  1
6         1                   10                  8                   3
7         10                  8                   3         1
8         3                   10        1         8
9         8                   3                   10        1
10                  3         10                  1                   8
11        8                   1                   10        3
12                                                10
13                            10                  8
14        10                            1         3                             8
15        3                   10                  8                   1
16        10                  1                   8                   3
17        1         3         10        8
18        1         3         8                   10
19        1                   10                  3         8
20        8         1         10                  3 ;

SCALARS zmin,zmax;

PARAMETER C(J) total score obtained by candidate J;
 C(J)=sum(I,N(I,J));

VARIABLES
 z function to be optimized ;

BINARY VARIABLE
 x(I,J,S) if member I assigns score P(S) to candidate J takes on value 1.
          Otherwise 0.;

EQUATIONS
 OBJ function to be optimized
 L1(I,J) Each member can assign at the most one score to each candidate
 L2(I,S) Each member I can assign score S to at most one candidate
 TOTALSCORE(J) totalscore given;

** The objective function equation depends on the candidate.
** For a specific candidate, the objective function to be
** optimized (minimized and maximized) is the number of members
** that assigns a determined score to this candidate.
** The dynamic set DIN allows to solve
** the same problem for different candidates.
** This set is updated at every iteration
** in the loop of candidates.
** The constraints are identical regardless
** of the candidate.

OBJ(J)$DIN(J)..z=e=sum(I,sum(S,x(I,J,S)));
L1(I,J)..sum(S,x(I,J,S))=l=1;
L2(I,S)..sum(J,x(I,J,S))=l=1;
TOTALSCORE(J)..sum(I,sum(S,P(S)*x(I,J,S)))=e=C(J);
```

```
**  The model includes all the constraints.
MODEL Academy /ALL/;

** The output file is open.
file aux /academy.out/;
put aux;

** First of all, the empty set is assigned to the dynamic set.
DIN(J)=NO;

loop(J1,

** Only the candidate specified by the loop controlling index is considered.
      DIN(J1)=YES;

** The problem is minimized for candidate J1.
      Solve Academy using mip Minimizing z;

** The solution is stored in scalar zmin.
      zmin=z.l;

** The problem is maximized for candidate J1.
      Solve Academy using mip Maximizing z;

** The solution is stored in scalar zmax.
      zmax=z.l;

** Both solutions are written in 'academy.out' file.
      put "J=",J1.tl:3," zmin= ",zmin:3:0," zmax= ",zmax:3:0/;

** Once the problem is optimized in both direction, the empty
** set is assigned to the dynamic set
      DIN(J1)=NO;
);
```

and the content of the output file is

```
J=1    zmin=  8 zmax= 20
J=2    zmin=  3 zmax= 14
J=3    zmin= 15 zmax= 20
J=4    zmin=  2 zmax= 13
J=5    zmin= 15 zmax= 20
J=6    zmin=  2 zmax= 18
J=7    zmin=  3 zmax= 20
J=8    zmin=  1 zmax=  8
```

This solution suggests the following comments:

1. Table $N(I, J)$ is not required, since the scores assigned to the candidates are unknown. However, in this example it is given and used to get the $C(J)$ values.

2. The ranges (upper bound–lower bound) of the candidates with the largest score is smaller than the ranges of the candidates with smaller scores.

3. Even though candidate 6 has a total score (18) larger than that of candidate 2, who has a total score of 14, the lower bound (2) for the first is smaller than the lower bound (3) for the second.

4. Candidates 3 and 5 are not guaranteed to be selected, because the upper bounds (20, 18, and 20) of candidates 1, 6, and 7, respectively, are larger than the lower bound (15) for candidates 3 and 5.

11.3.4 The School Timetable Problem

This problem is discussed in Section 2.5. It can be stated as follows. Minimize

$$\sum_{s\in\Omega}\sum_{c=1}^{n_c}\sum_{h=1}^{n_h}(c+h)\ v(s,c,h)$$

subject to

$$\begin{array}{rcl}
\sum_{s\in\Omega_i}\sum_{c=1}^{n_c}\sum_{h=1}^{n_h} v(s,c,h) & = & n_i,\ \forall i \\
\sum_{s\in\Omega_i}\sum_{c=1}^{n_c} v(s,c,h) & \le & 1,\ \forall h,\ \forall i \\
\sum_{c=1}^{n_c}\sum_{h=1}^{n_h} v(s,c,h) & = & 1,\ \forall s \\
\sum_{s\in\Omega} v(s,c,h) & \le & 1,\ \forall c,\ \forall h \\
\sum_{s\in\Delta_b}\sum_{c=1}^{n_c} v(s,c,h) & \le & 1,\ \forall h,\ \forall b
\end{array}$$

A GAMS input file for this problem is

```
$title CLASS TIMETABLE

SETS
 C   classrooms      /c1*c3/
 H   hours           /h1*h5/
 S   subjects        /s1*s8/
 I   instructors     /i1,i2/
 B   blocks          /b1,b2/
 SI(S,I) maps subjects and instructors /(s1,s2,s8).i1,(s3*s7).i2/
 SB(S,B) maps subjects and blocks      /(s1*s4).b1,(s5*s8).b2/;

VARIABLE z;
BINARY VARIABLE v(S,C,H);

EQUATIONS
 cost         compact the timetable
 const1(I)    every instructor teaches all his (her) subjects
 const2(H,I)  every instructor teaches at most 1 subject every hour
 const3(S)    every subject is taught once
 const4(C,H)  in every classroom-hour pair at most 1 subject is taught
 const5(H,B)  at every hour at most 1 subject of any academic block is taught;

 cost..         SUM((S,C,H),(ord(C)+ord(H))*v(S,C,H)) =e= z ;
 const1(I)..    SUM((S,C,H)$SI(S,I),v(S,C,H)) =e= SUM(S$SI(S,I),1);
```

```
 const2(H,I)..  SUM((S,C)$SI(S,I),v(S,C,H)) =l= 1;
 const3(S)..    SUM((C,H),v(S,C,H)) =e= 1;
 const4(C,H)..  SUM(S,v(S,C,H)) =l= 1;
 const5(H,B)..  SUM((S,C)$SB(S,B),v(S,C,H)) =l= 1;

model timetable /all/;
solve timetable using mip minimizing z;

DISPLAY v.L;
```

Part of the GAMS output file is

```
                          LOWER     LEVEL      UPPER    MARGINAL
---- VAR Z                 -INF     32.000      +INF        .

----       34 VARIABLE  V.L
                H1           H2           H3           H4           H5
S1.C2                     1.000
S2.C2         1.000
S3.C1                                  1.000
S4.C1                                               1.000
S5.C1                                                            1.000
S6.C1                     1.000
S7.C1         1.000
S8.C2                                  1.000
```

The solution in tabular form can be found in Section 2.5.

11.3.5 Models of Discrete Location

This example was considered in Section 2.6. The formulation for this problem is as follows. Maximize

$$Z = \sum_{i \in I} \sum_{j \in J} c_{ij} x_{ij} - \sum_{j \in J} f_j y_j$$

subject to

$$\begin{array}{rcll} \sum\limits_{j \in J} x_{ij} & = & b_i, & \forall i \in I \\ \sum\limits_{i \in I} x_{ij} & \leq & u_j y_j, & \forall j \in J \\ y_j & \in & \{0, 1\}, & \forall j \in J \\ x_{ij} & \geq & 0, & \forall i \in I, \forall j \in J \end{array}$$

where we assume that $u_j = 6 \ \forall j$, $f_j = 10 \ \forall j$ and

$$\mathbf{b} = (1.5, 2.0, 3.0, 4.0, 2.5, 1.0, 2.0)$$

A GAMS input file for this example is

```
$Title MODEL OF DISCRETE LOCATION

** The next sentence change the termination criterion
** based on the relative gap of the current solution.
```

```
** If we use a standard value then GAMS stops before
** an optimal solution is achieved.

OPTION OPTCR=1e-10;

** The indexes are defined in first place.
** Index I is used to refer to the seven cities.
** Index J is used to refer the six available locations
** for the industrial plants.

SET
 I index of cities    /C1*C7/
 J index of locations /L1*L6/;

PARAMETERS

 B(I) demand of a certain good in the city I
  /C1   1.5
   C2   2.0
   C3   3.0
   C4   4.0
   C5   2.5
   C6   1.0
   C7   2.0/

 F(J) amortization cost of an industrial plant at location J
 U(J) maximum production capacity of an industrial plant placed at location J;

 F(J) = 10;
 U(J) = 6;

** The data about the benefit is shown below. The table contains
** the benefit of making a good at the plant located at J, and
** selling it in city I.

TABLE  C(J,I) benefits according to different locations
         C1      C2      C3      C4      C5      C6      C7
 L1      4.0     4.5     2.5     0.5     1.0     0.5    -3.5
 L2      4.0     4.5     2.5     4.2     3.5     1.5    -0.5
 L3      3.5     5.0     4.0     3.5     4.5     1.5     0.0
 L4      1.3     3.0     5.0     3.3     5.5     1.8     1.3
 L5      0.5     1.0     1.5     5.0     4.0     5.5     3.0
 L6     -1.0     0.0     1.5     3.3     4.0     4.5     2.0 ;

** The optimization variables are declared.

VARIABLES
 z      objective function variable
 x(I,J) the amount of good that is made at J and selling at I
 y(J)   location variable. It is equal to 1 if the plant is open at J,
        and 0 otherwise.

** This is a mixed integer programming problem. The variables y(J)
** are binary variables and x(I,J) are positive variables.

POSITIVE VARIABLE x;
BINARY VARIABLE y;
```

```
** Different constraints are declared.

EQUATIONS
 COST    objective function
 SD(I)   satisfying demand of city I
 CAPA(J) capacity of production of the plant I;

** The equation definition is stated below.
** The objective function is the sum of the benefit minus the investment costs.
** The constraint SD(I) is the satisfaction of the demand of the city I.
** The constraint CAPA(J) imposes that the production at the plant J cannot
** exceed its capacity.

COST   .. z=e= SUM((I,J),C(J,I)*x(I,J))-SUM(J,F(J)*y(J));
SD(I)  .. SUM(J,x(I,J)) =e= B(I);
CAPA(J).. SUM(I,x(I,J)) =l= U(J)*y(J);

** The next two sentences define the model, considering all the above
** constraints, and direct GAMS to solve the problem using an integer
** linear programming solver.

MODEL loc /all/;
SOLVE loc USING mip MAXIMIZING z;
DISPLAY x.l;
```

The solution of this problem (part of the GAMS output file) is given below:

```
                          LOWER      LEVEL      UPPER     MARGINAL
---- VAR Z                 -INF      44.450      +INF        EPS
  Z            objective function variable

---- VAR Y            Location variables
       LOWER     LEVEL      UPPER     MARGINAL
l1       .         .        1.000    -10.000
l2       .       1.000      1.000    -10.000
l3       .         .        1.000     -7.000
l4       .       1.000      1.000    -10.000
l5       .       1.000      1.000     -5.200
l6       .         .        1.000    -10.000

----       56 VARIABLE  X.L
              l2          l4          l5
c1          1.500
c2          2.000
c3                      3.000
c4          1.000                   3.000
c5                      2.500
c6                                  1.000
c7                                  2.000
```

The solution of this problem consists of installing three industrial plants in locations L_2, L_4, and L_5, and the production distribution by cities is the following:

Locations	Cities						
	C_1	C_2	C_3	C_4	C_5	C_6	C_7
L_2	1.5	2.0	–	1.0	–	–	–
L_4	–	–	3.0	–	2.5	–	–
L_5	–	–	–	3.0	–	1.0	2.0

11.3.6 Unit Commitment of Thermal Power Units

The unit commitment problem was described in Section 2.7, and Example 2.6. The problem formulation is as follows. Minimize

$$Z = \sum_{k=1}^{K} \sum_{j=1}^{J} [A_j\, v_{jk} + B_j\, p_{jk} + C_j\, y_{jk} + D_j\, z_{jk}] \tag{11.14}$$

subject to

$$\begin{array}{rcll} \underline{P}_j v_{jk} \le p_{jk} & \le & \overline{P}_j v_{jk}; & \forall j, \forall k \\ p_{jk+1} - p_{jk} & \le & S_j; & \forall j, k = 0, \cdots, K-1 \\ p_{jk} - p_{jk+1} & \le & T_j; & \forall j, k = 0, \cdots, K-1 \\ y_{jk} - z_{jk} & = & v_{jk} - v_{jk-1}; & \forall j, k = 1, \cdots, K \\ \sum_{j=1}^{J} p_{jk} & = & D(k); & \forall k \\ \sum_{j=1}^{J} \overline{P}_j\, v_{jk} & \ge & D(k) + R(k); & \forall k \end{array} \tag{11.15}$$

where $K = 4$, $J = 3$, and

$$\underline{\mathbf{P}} = \begin{pmatrix} 50 \\ 80 \\ 40 \end{pmatrix}, \quad \overline{\mathbf{P}} = \begin{pmatrix} 350 \\ 200 \\ 140 \end{pmatrix}, \quad \mathbf{T} = \begin{pmatrix} 300 \\ 150 \\ 100 \end{pmatrix}, \quad \mathbf{S} = \begin{pmatrix} 200 \\ 100 \\ 100 \end{pmatrix}$$

$$\mathbf{A} = \begin{pmatrix} 5 \\ 7 \\ 6 \end{pmatrix}, \quad \mathbf{B} = \begin{pmatrix} 20 \\ 18 \\ 5 \end{pmatrix}, \quad \mathbf{C} = \begin{pmatrix} 0.5 \\ 0.3 \\ 1.0 \end{pmatrix}, \quad \mathbf{E} = \begin{pmatrix} 0.100 \\ 0.125 \\ 0.150 \end{pmatrix}$$

$$\mathbf{D} = \begin{pmatrix} 150 \\ 500 \\ 400 \end{pmatrix}, \quad \mathbf{R} = \begin{pmatrix} 15 \\ 50 \\ 40 \end{pmatrix}$$

One consideration in writing the GAMS code for this example is

- The use of conditional statements to limit the scope of equations. Here, the condition `$(ord(K) GT 1)` forces to satisfy the corresponding equation for every period except for the initial one.

An input GAMS file to solve this problem is given below

```
$title THE UNIT COMMITMENT PROBLEM

** Sets are declared first.
```

```
SETS
  K index of periods of time  /1*4/
  J index of generators /1*3/

**  The data is given below. The first table contains different columns of data
**  for every generator. The column names have not been previously declared
**  as elements of a set, so they should be referenced using the symbol '*'.
**  The same applies to the second table.

TABLE GDATA(J,*) generator input data
        PMIN   PMAX    T      S       A      B      C      E
*       (kW)   (kW)   (kW/h) (kW/h)  (E)    (E)    (E)   (E/kWh)
  1      50    350     300    200     5      20     0.5   0.100
  2      80    200     150    100     7      18     0.3   0.125
  3      40    140     100    100     6       5     1.0   0.150;

TABLE PDATA(K,*) data per period
       D     R
*     (kW) (kW)
  2    150   15
  3    500   50
  4    400   40;

**  The optimization variables are declared.

VARIABLES
  z      objective function variable
  p(J,K) output power of generator j at period k
  v(J,K) is equal to 1 if generator j is committed in period k
  y(J,K) is equal to 1 if generator j is started-up at
         the beginning of period k
  s(J,K) is equal to 1 if generator j is shut-down in period k;

**  The power is a positive variable.

POSITIVE VARIABLES p(J,K);

**  Status decisions are modeled using binary variables.

BINARY VARIABLES v(J,K),y(J,K),s(J,K);

**  Initial values are stated for some variables.

  v.fx(J,'1')=0;
  p.fx(J,'1')=0;

** Constraints are declared.

EQUATIONS
  COST            objective function
  PMAXLIM(J,K)    maximum output power equation
  PMINLIM(J,K)    minimum output power equation
  LOAD(K)         load balance equation
  RESERVE(K)      spinning reserve equation
  LOGIC(J,K)      start-up shut-down and running logic
  RAMPUP(J,K)     maximum up ramp rate limit
  RAMPDOWN(J,K) maximum down ramp rate limit;
```

```
**  The objective function is an equality equation.
**  The remaining constraints are defined for all periods K,
**  except for the initial one.
**  To model this exception the $(ord(K) GT 1) condition is included.

COST.. z =e= SUM((K,J), GDATA(J,'A')*v(J,K)+GDATA(J,'B')*y(J,K)+
                        GDATA(J,'C')*s(J,K)+GDATA(J,'D')*p(J,K));
PMAXLIM(J,K)$(ord(K) GT 1).. p(J,K)=l=GDATA(J,'PMAX')*v(J,K);
PMINLIM(J,K)$(ord(K) GT 1).. p(J,K)=g=GDATA(J,'PMIN')*v(J,K);
LOAD(K)$(ord(K) GT 1)..      SUM(J,p(J,K))=e=PDATA(K,'D');
RESERVE(K)$(ord(K) GT 1)..   SUM(J,GDATA(J,'PMAX')*v(J,K))=g=PDATA(K,'D')
                             +PDATA(K,'R');
LOGIC(J,K)$(ord(K) GT 1)..   y(J,K)-s(J,K)=e=v(J,K)-v(J,K-1);
RAMPUP(J,K)$(ord(K) GT 1).. p(J,K)-p(J,K-1)=l=GDATA(J,'S');
RAMPDOWN(J,K)$(ord(K) GT 1)..p(J,K-1)-p(J,K)=l=GDATA(J,'T');

**  The next sentences define the unit commitment model
**  with all the declared constraints, and direct
**  GAMS to solve the resulting problem using a mixed-integer solver.

MODEL uc /ALL/;
SOLVE uc USING mip MINIMIZING z;
```

Part of the GAMS output file, which provides the optimal solution, is given below.

```
    LOWER     LEVEL     UPPER    MARGINAL
---- VAR Z
    -INF    191.000     +INF        .
    Z          objective function variable

---- VAR P           output power of generator j at period k
       LOWER     LEVEL     UPPER    MARGINAL
1.0      .         .         .        EPS
1.1      .      150.000    +INF        .
1.2      .      350.000    +INF        .
1.3      .      320.000    +INF        .
2.0      .         .         .        EPS
2.1      .         .       +INF       EPS
2.2      .      100.000    +INF        .
2.3      .       80.000    +INF        .
3.0      .         .         .        EPS
3.1      .         .       +INF      0.050
3.2      .       50.000    +INF        .
3.3      .         .       +INF      0.050

---- VAR V  is equal to 1 if generator j is committed in period k
       LOWER     LEVEL     UPPER    MARGINAL
1.0      .         .         .        EPS
1.1      .        1.000     1.000     5.000
1.2      .        1.000     1.000   -12.500
1.3      .        1.000     1.000     5.000
2.0      .         .         .        EPS
2.1      .         .        1.000     7.000
2.2      .        1.000     1.000     7.000
2.3      .        1.000     1.000     9.000
```

```
3.0       .          .          .          EPS
3.1       .          .        1.000      6.000
3.2       .        1.000      1.000      6.000
3.3       .          .        1.000      6.000

---- VAR Y is equal to 1 if generator j is started-up at the
                      beginning of period k
        LOWER      LEVEL      UPPER     MARGINAL
1.1       .        1.000      1.000     20.000
1.2       .          .        1.000     20.000
1.3       .          .        1.000     20.000
2.1       .          .        1.000     18.000
2.2       .        1.000      1.000     18.000
2.3       .          .        1.000     18.000
3.1       .          .        1.000      5.000
3.2       .        1.000      1.000      5.000
3.3       .          .        1.000      5.000

---- VAR S is equal to 1 if generator j is shut-down in period k
        LOWER      LEVEL      UPPER     MARGINAL
1.1       .          .         1.000      0.500
1.2       .          .         1.000      0.500
1.3       .          .         1.000      0.500
2.1       .          .         1.000      0.300
2.2       .          .         1.000      0.300
2.3       .          .         1.000      0.300
3.1       .          .         1.000      1.000
3.2       .          .         1.000      1.000
3.3       .        1.000       1.000      1.000
```

Thus the solution is

$$Z = 191, \quad \mathbf{p} = \begin{pmatrix} 150 & 350 & 320 \\ 0 & 100 & 80 \\ 0 & 50 & 0 \end{pmatrix}, \quad \mathbf{v} = \begin{pmatrix} 1 & 1 & 1 \\ 0 & 1 & 1 \\ 0 & 1 & 0 \end{pmatrix}$$

$$\mathbf{y} = \begin{pmatrix} 1 & 0 & 0 \\ 0 & 1 & 0 \\ 0 & 1 & 0 \end{pmatrix}, \quad \mathbf{s} = \begin{pmatrix} 0 & 0 & 0 \\ 0 & 0 & 0 \\ 0 & 0 & 1 \end{pmatrix}$$

11.4 Nonlinear Programming Examples

In this section we solve using GAMS the nonlinear problems that were stated in Chapter 3.

11.4.1 The Postal Package Example

The postal package problem was analyzed in detail in Section 3.2.1. The formulation for this problem is as follows. Maximize

$$Z = xyz$$

subject to

$$z + 2x + 2y \leq 108$$

$$x, y, z \geq 0$$

A GAMS input file to solve this problem is

```
$title Postal package

POSITIVE VARIABLES x,y,z;
FREE VARIABLE obj;

EQUATIONS
 VOL objective function
 DIM constraint on the dimension and weight;

VOL.. obj =e= x*y*z;
DIM.. z+2*x+2*y =l= 108;

MODEL package /ALL/;

x.l=1; y.l=1; z.l=1;

SOLVE package USING nlp MAXIMIZING obj;
```

The solution of this problem is

$$Z = 11664; \quad x = 18; \quad y = 18; \quad z = 36$$

11.4.2 The Tent Example

The tent problem was analyzed in detail in Section 3.2.2. The formulation for this problem is as follows. Minimize

$$Z = 4(2ab + a\sqrt{h^2 + a^2})$$

subject to

$$\begin{array}{rcl} V & = & 4a^2(b + h/3) \\ H & = & b + h \\ a, b, c & \geq & 0 \end{array}$$

A GAMS input file to solve this problem is

```
$title Tent

SCALAR TV  total tent volume /50/
       TH  total height      /3/;

POSITIVE VARIABLES a,b,h;
FREE VARIABLE z;

EQUATIONS
 SURFACE objective function
 VOLUME  constrains the volume of the tent
 HEIGHT  constrains the height of the tent;
```

```
SURFACE.. z =e= 4*(2*a*b+a*SQRT(SQR(h)+SQR(a)));
VOLUME..  TV =e= 4*SQR(a)*(b+(h/3));
HEIGHT..  TH =e= b+h;

MODEL tent /ALL/;

a.l=100; b.l=100; h.l=100;
SOLVE tent USING nlp MINIMIZING z;
```

The solution of this problem is

$$Z = 58.937; \quad a = 2.633; \quad b = 1.205; \quad h = 1.795$$

11.4.3 The Lightbulb Example

The bulb problem was analyzed in detail in Section 3.2.3. The formulation for this problem is as follows. Minimize

$$Z = k\frac{h^{1/2}}{(h^2 + r^2)^{3/4}}$$

subject to

$$h \geq 0$$

A GAMS input file to solve this problem is

```
$title Bulb

SCALAR k proportionality constant /3/
       r radius  /30/;

POSITIVE VARIABLE h;
FREE VARIABLE z;

EQUATION
 INTENSITY objective function;

INTENSITY.. z =e= k*(SQRT(h)/((SQR(h)+SQR(r))**0.75));

MODEL bulb /all/;

h.l=10;
SOLVE bulb USING nlp MAXIMIZING z;
```

The solution of this problem is

$$Z = 0.062; \quad h = 21.183$$

11.4.4 The Surface Example

The surface problem was analyzed in detail in Section 3.2.4. The formulation for this problem is as follows. Minimize

$$Z = \sqrt{x^2 + y^2 + z^2}$$

subject to

$$\begin{aligned} xyz &= 1 \\ x, y, z &\geq 0 \end{aligned}$$

A GAMS input file to solve this problem is

```
$title Surface

POSITIVE VARIABLES x,y,z;
FREE VARIABLE obj;

EQUATIONS
 DIST objective function
 ON   constraint on the points;

DIST.. obj =e= SQRT(SQR(x)+SQR(y)+SQR(z));
ON..   1 =e= x*y*x;

MODEL surface /all/;

x.l=100; y.l=100; z.l=100;
SOLVE surface USING nlp MINIMIZING obj;
```

The solution of this problem is

$$Z = 1.732; \quad x = 1; \quad y = 1; \quad z = 1$$

11.4.5 The Moving Sand Example

The moving sand problem was analyzed in detail in Section 3.2.5. The formulation for this problem is as follows. Minimize

$$Z = k(3(2xy) + 2(2xz) + 2yz) + 2\frac{50}{xyz}$$

subject to

$$x, y, z \geq 0$$

A GAMS input file to solve this problem is

```
$title sand

SCALAR k proportionality constant /1.5/
       cu cubic units of sand     /50/;

POSITIVE VARIABLES x,y,z;
FREE VARIABLE obj;

EQUATIONS
 COST objective function;

COST.. obj =e= k*(3*(2*x*y)+2*(2*x*z)+2*y*z)+2*(cu/(x*y*z));

MODEL sand /COST/;

x.l=0.5; y.l=0.5; z.l=0.5;
SOLVE sand USING nlp MINIMIZING obj;
```

The solution of this problem is

$$Z = 66.140; \quad x = 0.857; \quad y = 1.715; \quad z = 2.572$$

11.4.6 The Cantilever Beam Example

The cantilever beam problem was analyzed in detail in Section 3.3.1, page 51. The formulation for this problem is as follows. Minimize

$$Z = \gamma L x y$$

subject to

$$\begin{array}{lcl} \dfrac{4FL^3}{Exy^3} & \leq & S \\ x & \geq & 0.5 \\ x, y & \geq & 0 \end{array}$$

A GAMS input file to solve this problem is

```
$title Cantilever beam

SCALARS L length /1/
        E Young modulus /1e6/
        F load at the tip /100/
        S maximum allowable deflection /1 /
        gamma unit weight /100/;


POSITIVE VARIABLES x,y;
FREE VARIABLE obj;

EQUATIONS
 WEIGHT objective function
 SMT constraint given by the strength of materials theory;

WEIGHT.. obj =e= gamma*L*x*y;
SMT..    (4*F*POWER(L,3))/(E*x*POWER(y,3))=l= S;
MODEL beam /ALL/;

x.lo=0.001; y.lo=0.001;
x.l=1; y.l=1;
SOLVE beam USING nlp MINIMIZING obj;
```

The solution of this problem is

$$Z = 4.642; \quad x = 0.5; \quad y = 0.093$$

11.4.7 The Two-Bar Truss Example

The two-bar truss problem was analyzed in detail in Section 3.3.2. The formulation for this problem is as follows. Minimize

$$Z = 2\gamma\sqrt{x^2 + h^2}z$$

subject to

$$
\begin{array}{rcl}
D(x,z) & = & \dfrac{F}{Eh^2 2\sqrt{2}} \dfrac{(h^2+x^2)^{3/2}(h^4+x^4)^{1/2}}{x^2 z} \leq D_0 \\
S^1(x,z) & = & \dfrac{F}{2\sqrt{2}h} \dfrac{(x+h)\sqrt{x^2+h^2}}{xz} \leq S_0 \\
S^2(x,z) & = & \dfrac{F}{2\sqrt{2}h} \dfrac{(h-x)\sqrt{x^2+h^2}}{xz} \leq S_0 \\
x, z & \geq & 0
\end{array}
$$

A GAMS input file to solve this problem is

```
$title Two-bar truss

SCALARS Gamma  unit weight of the bar material /1e2/
        E      Young modulus /1e6/
        F      load at the the fixed joint /15e3/
        S0     maximum admissible stress /6e4/
        D0     maximum admissible displacement of joint 3 /1e-1 /
        h      truss height / 1/;

PARAMETER K constant;
 K= F/(2*SQRT(2)*h);

POSITIVE VARIABLES x,z;
FREE VARIABLE obj;

EQUATIONS
 W      objective function
 D      displacement of the joint 3
 S1     stress at joint 1
 S2     stress at joint 2;


W.. obj =e= 2*Gamma*SQRT(SQR(x)+SQR(h))*z;
D..  K*( (SQR(h)+SQR(x))**(3/2)) * SQRT(h**4+x**4) /(E*h*SQR(x)*z) =l= D0;
S1.. K*((x+h)*SQRT(SQR(x)+SQR(h))) / (x*z) =l= S0;
S2.. K*((h-x)*SQRT(SQR(x)+SQR(h))) / (x*z)=l= S0;

MODEL truss /ALL/;

x.lo=0.05; z.lo=0.001;
x.l=100; z.l=100;
SOLVE truss USING nlp MINIMIZING obj;
```

The solution of this problem is

$$Z = 148.667; \quad x = 0.472; \quad z = 0.329$$

11.4.8 The Column Example

The column problem was analyzed in detail in Section 3.3.3. The formulation for this problem is as follows. Minimize

$$Z = DHxy - \left(\frac{Exy^3}{4H^3 \left(M + \frac{33}{140} DHxy \right)} \right)^{1/2}$$

subject to

$$\begin{array}{rcl} \frac{Mg}{xy} & \leq & S \\ \frac{Mg}{xy} & \leq & \frac{\pi^2 Ey^2}{48H^2} \\ x, y & \geq & 0 \end{array}$$

A GAMS input file to solve this problem is

```
$title Column

SCALARS M  mass supported by the column /100/
        H  height of the column /10/
        D  unit density of the material /100/
        E  Young's modulus of the material  /1e6/
        S  maximum admissible stress /6e4/
        G  acceleration of gravity /9.8/
        Pi /3.141592/;

POSITIVE VARIABLES x,y;
FREE VARIABLE z;

EQUATIONS
 W    objective function
 R1   compression stress constraint
 R2   buckling stress constraint;

W..  z =e= D*H*x*y-SQRT((E*x*(y**3))/(4*(H**3)*(M+(33/140)*D*H*x*y)));
R1.. (M*G)/(x*y) =l= S;
R2.. (M*G)/(x*y) =l= (SQR(Pi)*E*SQR(y))/(48*SQR(H));

MODEL column /ALL/;

x.lo=0.001; y.lo=0.001;
x.l=10; y.lo=10;

SOLVE column USING nlp MINIMIZING z;
```

The solution of this problem is

$$Z = 19989.8; \quad x = 2; \quad y = 10$$

11.4.9 The Scaffolding Example

The ropes–beams example was dealt with in Section 1.7, and consists of maximizing

$$\sum_i x_i$$

subject to

$$\begin{array}{rcl}
\sum\limits_{s\in\Psi_b} t_s & = & \sum\limits_{i\in\Omega_b} x_i + \sum\limits_{x\in\Theta_b} t_s, b \in B \\
\sum\limits_{s\in\Psi_b} dr_s t_s & = & \sum\limits_{i\in\Omega_b} xl_i x_i + \sum\limits_{x\in\Theta_b} dr_s t_s, b \in B \\
0 \le t_s & \le & T_s, s \in S \\
0 \le xl_i & \le & l_b, i \in \Omega_b \\
0 & \le & x_i
\end{array}$$

It should be noted that this formulation is similar to its linear version. Once the corresponding linear code is understood, programming the nonlinear one should be an effortless task.

An input GAMS file to solve the preceding problem is

```
$title SCAFFOLDING (NON-LINEAR)

**  The indexes are defined in first place.
**  Index B is used to refer to the three beams.
**  Index R is used to refer to the six ropes.
**  Index L is used to refer to the two loads.

SET
 B    set of beams    /B1*B3/
 R    set of ropes    /RA,RB,RC,RD,RE,RF/
 L    set of loads    /L1,L2/

UPP(B,R)
      /B1.(RA,RB)
       B2.(RC,RD)
       B3.(RE,RF)/
DOWN(B,R)
      /B1.(RC,RD)
       B2.(RF)/

 LOAD(B,L)
      /B1.L1
       B3.L2/;
**  Vectors of data are defined as parameters.
**  Data are assigned to each vector element.

PARAMETER LMAX(R) maximum load for ropes
    / (RA,RB) 300
      (RC,RD) 200
      (RE,RF) 100/;
```

```
PARAMETER dr(R) coordinates of rope R
    / RA  2
      RB 12
      RC  4
      RD 12
      RE  0
      RF 10/;


**  The next step is to declare the optimization variables.
**  Firstly the objective function variable is declared.
**  Then the remaining variables with controlling indexes
**  are declared.

VARIABLES
  z    objective function variable
  x(L) the amount of food J to be purchased
  T(R) tension on rope R
  d(L) distance from the left end-points of beams;

**  The bounds on variables are established in the next sentences.

POSITIVE VARIABLE d;

T.UP(R) = LMAX(R);


EQUATIONS
 COST       objective function
 FORCES(B)  force equilibrium equation
 MOMENT(B)  moment equilibrium equation;

COST ..          z  =E=  SUM(L, x(L)) ;

FORCES(B)..      SUM(R$UPP(B,R),T(R))=E= SUM(L$LOAD(B,L),x(L))+
                 SUM(R$DOWN(B,R),T(R));

MOMENT(B)..      SUM(R$UPP(B,R),dr(R)*T(R))=E=SUM(L$LOAD(B,L),
                 d(L)*x(L))+SUM(R$DOWN(B,R),dr(R)*T(R));

MODEL ropebeam /COST,FORCES,MOMENT/;
SOLVE ropebeam USING nlp MAXIMIZING z;
```

Part of the GAMS output file is

```
                            LOWER     LEVEL     UPPER    MARGINAL
---- VAR Z                   -INF    700.000     +INF        .
  Z            objective function variable

---- VAR X            the amount of food J to be purchased
      LOWER     LEVEL     UPPER    MARGINAL
L1     -INF    500.000     +INF        .
L2     -INF    200.000     +INF        .

---- VAR T            tension on rope R
      LOWER     LEVEL     UPPER    MARGINAL
RA     -INF    300.000   300.000     1.000
```

```
RB      -INF    300.000   300.000      1.000
RC      -INF     25.000   200.000        .
RD      -INF     75.000   200.000        .
RE      -INF    100.000   100.000      1.000
RF      -INF    100.000   100.000       EPS

---- VAR D            distance from the left end-points of beams
        LOWER      LEVEL     UPPER    MARGINAL
L1        .        6.400     +INF        .
L2        .        5.000     +INF        .
```

and the solution is

$$Z = 700 \text{ attained at the point } x_1 = 500, \;\; x_2 = 200, \;\; d_1 = 6.4, \;\; d_2 = 5.,$$

and the corresponding rope tensions are

$$T_A = 300, \;\; T_B = 300, \;\; T_C = 25, \;\; T_D = 75, \;\; T_E = 100, \;\; T_F = 100$$

11.4.10 Power Circuit State Estimation

We now consider the power circuit state estimation problem, which was stated in Section 3.4, and Example 3.1, as follows. Minimize

$$Z = \sum_{i \in \Omega} \frac{1}{\sigma_i^v}(v_i - \hat{v}_i)^2 + \sum_{k \in \Omega, l \in \Omega_k} \frac{1}{\sigma_{kl}^p}(p_{kl}(\cdot) - \hat{p}_{kl})^2 + \sum_{k \in \Omega, l \in \Omega_k} \frac{1}{\sigma_{kl}^q}(q_{kl}(\cdot) - \hat{q}_{kl})^2$$

subject to no constraint, where $\delta_k = 0$, $k = 2$, $z_{12} = 0.15$, $\theta_{12} = 90°$, and

$$\mathbf{V} = \begin{pmatrix} 1.07 \\ 1.01 \end{pmatrix}, \quad \mathbf{P} = \begin{pmatrix} 0.83 \\ 0.81 \end{pmatrix}, \quad \mathbf{Q} = \begin{pmatrix} 0.73 \\ 0.58 \end{pmatrix}$$

Some remarkable features used in the following GAMS code for this example are:

- The definition of a symmetric matrix by specifying only the values on and above its diagonal. The values below the diagonal are obtained using the conditional `$(ORD(N) GT ORD(NP))` on the corresponding assignment sentence.

- The use of conditional statements to limit the scope of equations. Here, the conditions `$(LINE(N,NP,'Z') NE 0)` are stated in terms of data values instead of using subsets as in other examples. By checking the value of the line impedance, we know whether there is a line connecting two buses.

An input GAMS file to solve this problem is

```
$title THE POWER CIRCUIT ESTIMATION PROBLEM

**  The allowable buses are defined in the next sentence.
```

```
SETS
   N       index of buses       /N1*N2/

** The data is provided below. The first table contains several columns of
   data
** for every line. The names of these columns have not been previously
   declared
** as elements of a set, so they should be referenced using the symbol '*'.
** The same applies to the second table.

TABLE LINE(N,N,*) line input data
              Z       PHI
*           (Ohm)   (degrees)
   N1.N2     0.15      90;

TABLE BUS(N,*) bus measures
           V       P       Q
*         (V)     (W)    (VAr)
   N1    1.07     0.83    0.73
   N2    1.01     0.81    0.58;

** Constant PI is used to convert degrees to radians.

SCALAR
   PI  /3.1416/;

** Optimization variables are declared.

VARIABLES
   z    objective function variable
   V(N) voltage magnitude at bus N
   d(N) voltage angle at bus N;

** Bus 2 is considered as the reference bus, therefore
** its corresponding angle is fixed to zero.

   D.FX('N2')=0;

** The set of buses N should be duplicated to refer to
** different elements in the same constraint.

   ALIAS(N,NP);

** The next sentence converts degrees to radians.

   LINE(N,NP,'PHI')=LINE(N,NP,'PHI')*PI/180;

** Data matrices Z and PHI are defined as symmetric using the
** $(ORD(N) GT ORD(NP)) condition on the sets N and NP.

   LINE(N,NP,'Z')$(ORD(N) GT ORD(NP))=LINE(NP,N,'Z');
   LINE(N,NP,'PHI')$(ORD(N) GT ORD(NP))=LINE(NP,N,'PHI');

** The objective-function constraint is declared in the following lines.

EQUATION
   ERROR    objective function;
```

```
**  The objective function is defined using nonlinear functions
**  SQR, COS and SIN. The $(LINE(N,NP,'Z') NE 0) condition is used
** to check if a line exists between buses N and NP.

ERROR..  z =e= SUM(N, SQR(V(N)-BUS(N,'V')) +SUM(NP$(LINE(N,NP,'Z') NE 0),
               SQR(((1/LINE(N,NP,'Z'))*
              (SQR(V(N))*COS(LINE(N,NP,'PHI'))-V(N)*V(NP)*COS(d(N)-
               d(NP)+LINE(N,NP,'PHI'))))-BUS(N,'P'))+
               SQR(((1/LINE(N,NP,'Z'))*
              (SQR(V(N))*SIN(LINE(N,NP,'PHI'))-V(N)*V(NP)*SIN(d(N)-
               d(NP)+LINE(N,NP,'PHI'))))-BUS(N,'Q'))));

**  The next sentences define the estimation model, and direct GAMS
**  to solve the problem using a nonlinear solver.

MODEL estimation /ERROR/;
SOLVE estimation USING nlp MINIMIZING z;
```

Part of the GAMS output file, that provides the optimal solution is

```
                          LOWER     LEVEL     UPPER    MARGINAL
---- VAR Z                -INF      2.203     +INF        .
  Z            objective function variable

---- VAR V             voltage magnitude at bus N
      LOWER     LEVEL     UPPER    MARGINAL
N1     -INF     1.045     +INF       EPS
N2     -INF     1.033     +INF       EPS

---- VAR D             voltage angle at bus N
      LOWER     LEVEL     UPPER    MARGINAL
N1     -INF     0.002     +INF       EPS
N2       .        .         .        EPS
```

and the solution is

$$Z = 2.203$$

attained at

$$v_1 = 1.045, \;\; v_2 = 1.033, \;\; \delta_1 = 0.002$$

11.4.11 Optimal Power Flow

The optimal power flow problem was stated in Section 3.4.2, page 58, and Example 3.2, as follows. Minimize

$$Z = \sum_{i=1}^{n} C_i p_{Gi}$$

subject to

$$
\begin{array}{rcll}
p_{Gi} - P_{Di} & = & v_i \sum_{k=1}^{n} Y_{ik} v_k \cos(\delta_i - \delta_k - \Theta_{ik}), & \forall i = 1, 2, \ldots, n, \\
q_{Gi} - Q_{Di} & = & v_i \sum_{k=1}^{n} Y_{ik} v_k \sin(\delta_i - \delta_k - \Theta_{ik}), & \forall i = 1, 2, \ldots, n, \\
\underline{V}_i \le v_i & \le & \overline{V}_i, & \forall i = 1, 2, \ldots, n, \\
\underline{P}_{Gi} & \le & p_{Gi} \le \overline{P}_{Gi}, & \forall i = 1, 2, \ldots, n, \\
\underline{Q}_{Gi} & \le & q_{Gi} \le \overline{Q}_{Gi}, & \forall i = 1, 2, \ldots, n, \\
-\pi \le \delta_i & \le & \pi, & \forall i = 1, 2, \ldots, n, \\
\delta_k & = & 0. &
\end{array}
$$

where $n = 3$, and

$$
\underline{P} = \begin{pmatrix} 0 \\ 0 \end{pmatrix}, \ \overline{P} = \begin{pmatrix} 3 \\ 3 \end{pmatrix}, \ \underline{Q} = \begin{pmatrix} -1 \\ -1 \end{pmatrix}, \ \overline{Q} = \begin{pmatrix} 2 \\ 2 \end{pmatrix}, \ \mathbf{C} = \begin{pmatrix} 6 \\ 7 \end{pmatrix}
$$

$$
\underline{V} = \begin{pmatrix} 0.95 \\ 0.95 \\ 0.95 \end{pmatrix}, \ \overline{V} = \begin{pmatrix} 1.13 \\ 1.10 \\ 1.10 \end{pmatrix}, \ \mathbf{Y} = \begin{pmatrix} 22.97 & 12.13 & 10.85 \\ 12.13 & 21.93 & 9.81 \\ 10.85 & 9.81 & 20.65 \end{pmatrix}
$$

$$
\mathbf{P} = \begin{pmatrix} 0 \\ 0 \\ 4.5 \end{pmatrix}, \ \mathbf{Q} = \begin{pmatrix} 0 \\ 0 \\ 1.5 \end{pmatrix}, \ \mathbf{\Theta} = \begin{pmatrix} -1.338 & 1.816 & 1.789 \\ 1.816 & -1.347 & 1.768 \\ 1.789 & 1.768 & -1.362 \end{pmatrix}
$$

The remarks about the GAMS code given in the previous example are also valid for this one.

An input GAMS code to solve the optimal power flow is

```
$title THE OPTIMAL POWER FLOW PROBLEM

**  Sets G and N are defined.
**  The set MAP is defined as a subset of the product of sets G and N.
**  The subset MAP establishes the valid pairs of the elements of G and N.

SETS
    G       index of generators  /G1*G2/
    N       index of buses       /N1*N3/
    MAP(G,N) associates generators with buses /G1.N1,G2.N2/;

**  The data are provided below. The first table contains different columns
**  of data for every generator. The names of these columns have not been
**  previously declared as elements of a set, so they should be referenced
**  using the symbol '*'.
**  The same applies to the second table.

TABLE GDATA(G,*)  generator input data
           PMIN     PMAX     QMIN     QMAX     COST
*          (W)      (W)      (VAr)    (VAr)   (E/Wh)
    G1     0.0      3.0      -1.0     2.0       6
    G2     0.0      3.0      -1.0     2.0       7;
```

```
TABLE LINE(N,N,*) line input data
              Y        PHI
*           (Ohm)     (rad)
    N1.N1   22.97    -1.338
    N2.N2   21.93    -1.347
    N3.N3   20.65    -1.362
    N1.N2   12.13     1.816
    N1.N3   10.85     1.789
    N2.N3    9.81     1.768;

TABLE BUS(N,*)
          VMIN    VMAX   PL    QL
*         (V)     (V)    (W)  (VAr)
    N1    0.95    1.13
    N2    0.95    1.10
    N3    0.95    1.10   4.5   1.5;

**  Constant PI is used to state limits on bus angles.

SCALAR
    PI  /3.1416/;

**  Optimization variables are declared.

VARIABLES
    z    objective function variable
    p(G) active power output for generator G
    q(G) reactive power output for generator G
    v(N) voltage magnitude at bus N
    d(N) voltage angle at bus N;

**  Limits are stated on variables using previously defined data.

    p.lo(G)=GDATA(G,'PMIN');
    p.up(G)=GDATA(G,'PMAX');
    q.lo(G)=GDATA(G,'QMIN');
    q.up(G)=GDATA(G,'QMAX');
    v.lo(N)=BUS(N,'VMIN');
    v.up(N)=BUS(N,'VMAX');
    d.lo(N)=-PI;
    d.up(N)=PI;

**  Bus 3 is considered as the reference bus, therefore
**  its corresponding angle is fixed to zero.

    d.fx('N3')=0;

**  The set of buses N must be duplicated to refer to different elements
**  in the same constraint.

    ALIAS(N,NP);

**  Data matrices Y and PHI are defined as symmetric using
**  the $(ORD(N) GT ORD(NP)) condition on the sets N and NP.

    LINE(N,NP,'Y')$(ORD(N) GT ORD(NP))=LINE(NP,N,'Y');
    LINE(N,NP,'PHI')$(ORD(N) GT ORD(NP))=LINE(NP,N,'PHI');
```

```
** Constraints are declared.

EQUATIONS
    COST        objective function
    PBAL(N)     active power balance constraints
    QBAL(N)     reactive power balance constraints;

**  The objective function is an equality constraint.
**  The remaining constraints are defined using different
**  nonlinear functions. To check if generator
**  G is connected to bus N the sums are conditioned by $MAP(G,N).

    COST..         z =e= SUM(G,GDATA(G,'COST')*p(G));
    PBAL(N)..      SUM(G$MAP(G,N),p(G))-BUS(N,'PL')=e=v(N)*
                   SUM(NP,LINE(N,NP,'Y')*v(NP)*COS(d(N)-d(NP)
                   -LINE(N,NP,'PHI')));
    QBAL(N)..      SUM(G$MAP(G,N),q(G))-BUS(N,'QL')=e= v(N)*
                   SUM(NP,LINE(N,NP,'Y')*v(NP)*SIN(d(N)-d(NP)-LINE(N,NP,'PHI')));

**  The sentences below define the optimal power flow problem with all its
**  constraints, and direct GAMS to solve the problem using the nlp solver.

MODEL opf /COST,PBAL,QBAL/;
SOLVE opf USING nlp MINIMIZING z;
```

The part of the GAMS output file, which provides the optimal solution is

```
                           LOWER      LEVEL     UPPER     MARGINAL
---- VAR Z                  -INF     30.312     +INF         .
  Z          objective function variable

---- VAR P   active power output for generator G
       LOWER      LEVEL      UPPER    MARGINAL
G1       .        3.000      3.000     -0.927
G2       .        1.759      3.000        .

---- VAR Q   reactive power output for generator G
       LOWER      LEVEL      UPPER    MARGINAL
G1    -1.000      1.860      2.000        .
G2    -1.000      0.746      2.000        .

---- VAR V   voltage magnitude at bus N
       LOWER      LEVEL      UPPER    MARGINAL
N1     0.950      1.130      1.130     -0.533
N2     0.950      1.100      1.100     -3.303
N3     0.950      0.979      1.100        .

---- VAR D   voltage angle at bus N
       LOWER      LEVEL      UPPER    MARGINAL
N1    -3.142      0.190      3.142        .
N2    -3.142      0.174      3.142        .
N3       .          .          .         EPS
```

and the solution is

$$\begin{array}{rcl} Z &=& 30.312 \\ p_{G1} &=& 3.000 \\ p_{G2} &=& 1.759 \\ q_{G1} &=& 1.860 \\ q_{G2} &=& 0.746 \\ v_1 &=& 1.130 \\ v_2 &=& 1.100 \\ v_3 &=& 0.979 \\ \delta_1 &=& 0.190 \\ \delta_2 &=& 0.174 \end{array}$$

11.4.12 The Water Supply Network Problem

In the linear version of this example, the cost function was written as

$$Z = \sum_{ij} c_{ij}(x_{ij}^{+} + x_{ij}^{-})$$

which corresponds to

$$Z = \sum_{ij} c_{ij}|x_{ij}|$$

This new function, which now is nonlinear, is used in this example. Because of the lack of derivatives of this function, a special solver must be used. The solver, which deals with nonlinear functions with discontinuous derivatives, is referred to in the code with the reserved word `dnlp`.

An input GAMS file, that can be used to solve the same example with the new nonlinear objective function is

```
$title WATER SUPPLY NETWORK (nonlinear)

**  The sets are declared first. Set I has four elements.
**  Subset CONEX is defined as a subset of set I.
**  The subset CONEX establishes the valid connections between nodes I.

SET
 I          set of nodes in the network  /I1*I4/
 CONEX(I,I) set of node connections   /I1.I2,I1.I3,I1.I4,I2.I4,I3.I4/;

**  The set of nodes I should be duplicated to refer to its
**  different elements in the same constraint.

 ALIAS(I,J)

**  Vectors of data are defined as parameters.
**  Data are assigned to each vector element.
**  FMAX(I,J) is a data matrix declared as a parameter where
**  all its elements have the same value (8).
**  This is a compact way to declare a matrix (instead of using a TABLE).

PARAMETERS
```

```
F(I)   the input or output flow at node I
     /I1   20
      I2  -3
      I3  -10
      I4  -7/

FMAX(I,J) maximum flow capacity of conduction going from I to J;
FMAX(I,J)=8;

**  Optimization variables are declared.

VARIABLES
  z       objective function variable
  x(I,J) the flow going from node I to node J

**  The flow from I to J is a positive magnitude.

POSITIVE VARIABLE x(I,J);

**  Limits are stated on variables using previously defined data.

x.lo(I,J)=-FMAX(I,J);
x.up(I,J)=FMAX(I,J);

** Constraints are declared.

EQUATIONS
 COST         objective function
 BALANCE(I)  conservation of flow conditions;

**  The objective function is the sum of the absolute values of flows between
**  connected nodes. This requirement is stated by the $CONEX(I,J) condition.
**  The four BALANCE equations only consider the flow between
**  two connected nodes and again the $CONEX(I,J) condition is necessary.

COST ..          z   =E=  SUM((I,J)$CONEX(I,J),ABS(x(I,J))) ;
BALANCE(I) ..   SUM(J$CONEX(I,J),x(I,J))-SUM(J$CONEX(J,I),x(J,I))=E=F(I) ;

**  The next two sentences define the model, with all the above constraints,
**  and direct GAMS to solve the problem using a nonlinear programming
**  (with discontinuous derivatives, dnlp) solver.

MODEL wsn_dnlp /ALL/;
SOLVE wsn_dnlp USING dnlp MINIMIZING z;
```

A part of the GAMS output file is

```
                          LOWER      LEVEL      UPPER    MARGINAL
---- VAR Z                 -INF      23.000      +INF        .
  Z           objective function variable

---- VAR X            the flow going from node I to node J
          LOWER      LEVEL      UPPER     MARGINAL
I1.I2    -8.000      4.000      8.000       .
I1.I3    -8.000      8.000      8.000     -2.000
I1.I4    -8.000      8.000      8.000     -1.000
I2.I4    -8.000      1.000      8.000       .
I3.I4    -8.000     -2.000      8.000       .
```

and the solution is

$$Z = 23.000, \quad \mathbf{X} = \begin{pmatrix} 0 & 4 & 8 & 8 \\ -4 & 0 & 0 & 1 \\ -8 & 0 & 0 & -2 \\ -8 & -1 & 2 & 0 \end{pmatrix}.$$

11.4.13 The Matrix Balancing Problem

This example was studied in Section 3.5. It consists of minimizing the following weighted least squares function:

$$\sum_{i=1}^{m}\sum_{j=1}^{n} \omega_{ij}(T_{ij} - t_{ij})^2$$

subject to

$$\begin{aligned} \sum_{j=1}^{n} t_{ij} &= r_i \quad i = 1, \ldots, m \\ \sum_{i=1}^{m} t_{ij} &= c_j \quad j = 1, \ldots, n \\ T_{ij} &\geq 0 \quad i = 1, \ldots, m, \; j = 1, \ldots, n \end{aligned}$$

where $m = n = 4$, and we assume that $\mathbf{r} = (12,000, 10,500, 3800, 7700)^T$, $\mathbf{c} = (6750, 7300, 10,000, 9950)^T$, $\omega_{ij} = 1/T_{ij}$ and

$$\mathbf{T} = \begin{pmatrix} - & 60 & 275 & 571 \\ 50 & - & 410 & 443 \\ 123 & 61 & - & 47 \\ 205 & 265 & 45 & - \end{pmatrix}$$

This example does not use any unusual GAMS feature. For a better understanding, the reader should examine the comments included in the following code:

```
$Title THE MATRIX BALANCING PROBLEM.

** The sets of indices are defined.
** Index I is used to refer to the rows of the updated matrix.
** Index J is used to refer to the columns of the updated matrix.

SET
 I index associated with the rows of the matrix          /R1*R4/
 J index associated with the columns of the matrix       /C1*C4/;

** The vectors of data are defined as parameters.
** Data are assigned to vector elements.

PARAMETERS
 R(I) number of estimated trips from the origin I
  /R1 12000
   R2 10500
```

```
   R3  3800
   R4  7700/
 C(J) number of estimated trips to destination J
  /C1  6750
   C2  7300
   C3 10000
   C4  9950/ ;

** The updated trip matrix is defined as a table.

TABLE OT(I,J) observed (updated) trip matrix
          C1      C2       C3       C4
 R1               60       275      571
 R2        50              410      443
 R3       123     61                 47
 R4       205    265        45        ;

** The next sentences declare the optimization variables.
** First, the objective function variable is declared.
** Then the remaining variables with their associated indices are declared.

VARIABLES
 z       error of good fit
 t(I,J) number of predicted trips from I to J;

** The types of variables are established in the next sentence.
** The variables in the origin-destination trip matrix,
** which define the amount of trips, are positive.

POSITIVE VARIABLES t;

** The problem constraints are declared.
** First, the objective function constraint (COST).
** This function is a weighted least square function
** and gives the distance between observed and estimated trip matrices.
** The remaining constraints mean that the estimated matrix must
** have a certain sum of columns and rows.

EQUATIONS
 COST    objective function
 SO(I)   satisfying origins
 SD(J)   satisfying destinations;

COST..  z=E= SUM((I,J)$(ORD(I) NE ORD(J)),POWER(t(I,J)-OT(I,J),2)/OT(I,J));
SO(I).. SUM(J$(ORD(I) NE ORD(J)),t(I,J)) =e= R(I);
SD(J).. SUM(I$(ORD(I) NE ORD(J)),t(I,J)) =e= C(J);

** The following sentences define the matrix balancing problem
** with all the declared constraints and direct GAMS to solve the problem.

MODEL mbp /ALL/;
SOLVE mbp USING nlp MINIMIZING z;

** The following sentence gives the solution in a table format.

DISPLAY  t.l;
```

The part of the GAMS output file that provides the solution for this matrix balancing model is

```
                         LOWER      LEVEL      UPPER    MARGINAL
---- VAR Z              -INF  441754.6224    +INF         .
  Z           error of goodfit

----      73 VARIABLE  T.L            number of predicted trips from I to J

            C1          C2          C3          C4
R1                  1670.878    4286.405    6042.717
R2     1226.653                 5522.850    3750.497
R3     2386.544     1256.670                 156.786
R4     3136.803     4372.452     190.745
```

The solution of this problem can be found in Section 3.5.

11.4.14 The Traffic Assignment Problem

The traffic problem was analyzed in detail in Section 3.6. The formulation for this problem is as follows. Minimize

$$Z = \sum_{a \in \mathcal{A}} c^0_{a_i} f_{a_i} + \frac{b_{a_i}}{n_{a_i} + 1} \left(\frac{f_{a_i}}{k_{a_i}} \right)^{n_{a_i}+1}$$

subject to

$$\begin{array}{rcl} h_1 + h_2 + h_3 & = & 4000 \\ h_4 + h_5 & = & 2500 \\ h_1 & = & f_1 \\ h_2 + h_4 & = & f_2 \\ h_3 + h_5 & = & f_3 \\ h_2 + h_3 & = & f_4 \\ h_1, \ldots, h_5 & \geq & 0 \end{array}$$

A GAMS input file for solving this problem is

```
$title The Traffic Assignment Problem 4 links x 3 node x 2 commodities

SETS

r routes         /r1*r5/
a links          /a1*a4/
w demands        /w1*w2/;

PARAMETERS

D(w) Demand
/w1 4000
 w2 2500/
K(a) Capacity on link
 /a1 500
  a2 400
  a3 400
  a4 500/
```

```
b(a) Congestion Parameter
 /a1   1.
  a2   1.
  a3   1.
  a4   1./
n(a) Congestion Parameter
 /a1   4.
  a2   4.
  a3   4.
  a4   4./
CO(a) Cost for empty link
  / a1   5
    a2   7
    a3  10
    a4   2/ ;
TABLE DELTAR_W(r,w) Incidence matrix route-demand
         w1      w2
r1       1       0
r2       1       0
r3       1       0
r4       0       1
r5       0       1
TABLE  DELTAR_L(r,a) Incidence matrix route-link
         a1      a2      a3      a4
r1       1       0       0       0
r2       0       1       0       1
r3       0       0       1       1
r4       0       1       0       0
r5       0       0       1       0;

VARIABLES
H(r) Flow on route r
F(a) Flow on link a
Z    Total time on transport network
POSITIVE VARIABLE H, F;

EQUATIONS
COST    objective function
SD(w)   satisfying demand
CF(a)   conservation flow;
SD(w).. SUM(r,H(r)*DELTAR_W(r,w))=e= D(w);
CF(a).. SUM(r,H(r)*DELTAR_L(r,a))=e= F(a);
COST..Z=e=SUM(a,CO(a)*F(a)+(b(a)/(n(a)+1.))*(F(a)/K(a))**(n(a)+1.) );
MODEL TAP /all/;
SOLVE TAP using nlp minimizing Z;
DISPLAY H.L;
DISPLAY F.L;
```

The solution of this problem was given at the end of Example 3.4.

Exercises

11.1 A transport study is being undertaken incorporating three cities A, B, and C. The travel costs between these cities are given in Table 11.1. Note that intraurban movements are excluded from this study.

Table 11.1: Transportation costs (in U.S. dollars) between three cities

	Destination		
Origin	A	B	C
A	–	1.80	2.60
B	1.80	–	1.52
C	2.70	1.52	–

Roadside interviews have been undertaken at several cities and the number of drivers interviewed are shown in Table 11.1 together with their respective origins an destinations. Blank entries indicate no observations.

Table 11.2: Results of the transportation interview

	Destination		
Origin	A	B	C
A	–	6	2
B	8	–	8
C	6	6	–

Assume now a transportation model of functional form

$$T_{ij} = a_i b_j \exp(-\beta c_{ij})$$

where a_i is a constant that depends on the origin i, b_j depends on the destination j, β is a parameter, and c_{ij} is the travel cost between zones. Formulate a least-squares estimation problem to calibrate the parameters of the model. Obtain the solution using the GAMS package.

11.2 Walter builds two types of transformers and has available 6 tons of ferromagnetic material and 28 hour of working time. Transformer 1 requires 2 tons of ferromagnetic material and 7 hours of work, and transformer 2 requires 1 unit of ferromagnetic material and 8 hours of work. Selling prices of transformers 1 and 2 are respectively 120 and 80 thousand Euros (European dollars). How many transformers of each type should Walter manufacture to maximize his benefits? Solve the problem graphically, analytically, and using GAMS.

11.3 Find the optimal solution of the problem. Maximize

$$Z = \sum_{j=1}^{100} \frac{2^j}{j^2} x_j$$

subject to

$$0 \leq \sum_{j=i}^{100} (-1)^{j+i} x_j \leq 1 \text{ for } i = 1, \ldots, 100$$

using GAMS.

11.4 Using the GAMS package, solve the following problem. Minimize

$$Z = x$$

subject to

$$x \ln x \geq 1$$

by each of the following procedures:

(a) Logarithmic barrier penalty method

(b) Inverse barrier penalty method

The aim of this exercise consists of forcing the reader to use loops with the GAMS package, and to illustrate the computational methods above. It is not the aim of this exercise to use GAMS to code the previous methods for solving optimization models, because it is not the natural way to deal with them.

11.5 The bilevel programming model is formulated as follows. Minimize

$$Z_U = F(\mathbf{x}, \hat{\mathbf{y}})$$

subject to

$$\mathbf{x} \in X$$

where $\hat{\mathbf{y}}$ is optimal for minimization of

$$Z_L = f(\mathbf{y}, \mathbf{x})$$

subject to

$$\mathbf{y} \in Y(\mathbf{x})$$

where $F : \mathbb{R}^n \times \mathbb{R}^m \to \mathbb{R}$ and $f : \mathbb{R}^m \times \mathbb{R}^n \to \mathbb{R}$ are given functions, X is a given subset, and $Y(\mathbf{x})$ is a set parameterized by the variable $\mathbf{x}$. The objective $\min_{\mathbf{x} \in X} F(\mathbf{x}, \mathbf{y})$ is referred to as the *upper-level problem*, and the *lower-level problem* is defined as $\min_{\mathbf{y} \in Y(\mathbf{x})} f(\mathbf{y}, \mathbf{x})$ for a fixed $\mathbf{x}$. The following heuristic algorithm four-step [(a)–(d)] has been used to approximate the bilevel mathematical programming for large-scale problems.

(a) Let $\mathbf{x}^{(1)}$ be a feasible point of X. Let T a positive integer. Let $t = 1$ and go to step 2.

(b) Solve the following lower-level subproblem. Minimize

$$Z_L = f(\mathbf{y}, \mathbf{x}^{(\mathbf{t})})$$

subject to

$$\mathbf{y} \in Y(\mathbf{x}^{(t)})$$

and denote any solution point by $\mathbf{y}^{(t)}$.

(c) Solve the following upper-level subproblem. Minimize

$$Z_U = F(\mathbf{x}, \mathbf{y}^{(t)})$$

subject to

$$\mathbf{x} \in X$$

and denote any solution point by $\mathbf{x}^{(t+1)}$.

(d) If $t = T$, then *stop*. Otherwise $t = t + 1$ and go to step 2.

Using the GAMS package and the preceding algorithm to approximate the bilevel problem, minimize

$$Z_U = (x - 8)^2(\hat{y} - 3)^2$$

subject to

$$\begin{array}{rcl} x & \geq & 0 \\ x & \leq & 16 \end{array}$$

where $\hat{y}$ is optimal for minimization of

$$Z_L = y$$

subject to

$$\begin{array}{rrcr} & y & \geq & 0 \\ x & +2y & \geq & 10 \\ x & -2y & \leq & 6 \\ 2x & -y & \leq & 21 \\ x & +2y & \leq & 38 \\ -x & +2y & \leq & 18 \end{array}$$

11.6 Modify the "economic dispatch" example in such a way that the production cost function of every generator is a piecewise linear convex function.

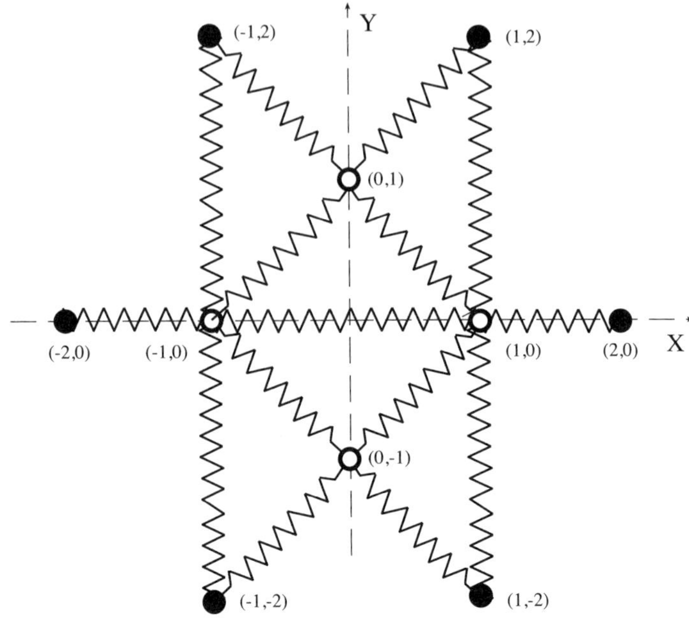

Figure 11.1: System of springs.

11.7 Modify the "unit commitment" example in such a way that the production cost function of every generator is a piecewise linear convex function.

11.8 Modify the "optimal power flow" example considering that all voltage magnitude values are equal to 1. Compare the results obtained using the above simplified model with those obtained using the original formulation.

11.9 In the system of springs of the Figure 11.1, and keeping in mind the explanation given in Exercise 3.8 in Chapter 3, determine the equilibrium state using GAMS.

Part IV

Applications

Chapter 12

Applications

In this chapter we provide some applications to illustrate the power of mathematical programming combined with a tool, such as GAMS, that allows one to solve, in an efficient and easy way, the stated problems. We start in Section 12.1, giving an application to neural and functional networks. In Section 12.2 we present an automatic mesh generation method that can be useful in finite element methods and in all methods that require describing surfaces by means of triangular elements. In Section 12.3 we give some applications to probability. In Section 12.4 we show how to solve some regression models, that are difficult to solve using standard regression techniques. In Section 12.5 we present some applications to continuous optimization problems, as the braquistochrone, the hanging cable, the problems of optimal control of hitting a target and of an harmonic oscillator, and the transportation and the unit commitment problems. In Section 12.6 we provide applications to transportation systems. Finally, in Section 12.7 an application to power systems is provided.

12.1 Applications to Artificial Intelligence

In this section we show how mathematical programming can be applied to deal with neural and functional network problems. This application is based on Castillo et al. [20].

Neural networks consist of one or several layers of neurons connected by links. Each neuron computes a scalar output from a linear combination of inputs, coming from the previous layer, using a given scalar function.

Consider the one-layer neural network in Figure 12.1, where the neural function f is assumed to be invertible.

The typical problem of neural networks consists of learning the thresholds and weights given a set of data

$$D \equiv \{(x_{1s}, x_{2s}, \ldots, x_{Is}, y_s) | s = 1, 2, \ldots, S\}$$

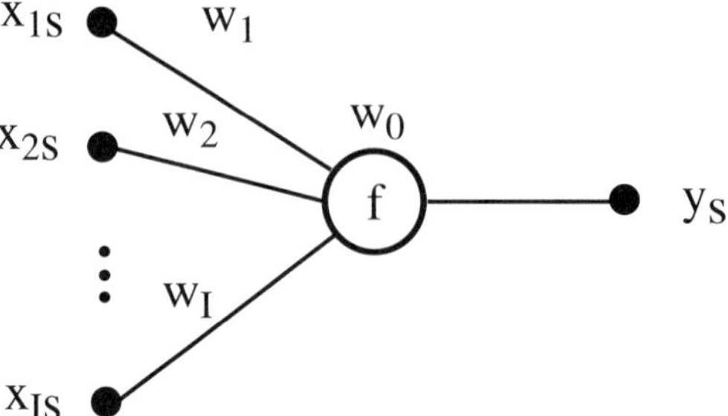

Figure 12.1: A one-layer neural network with its neural function.

which contains the set of inputs $(x_{1s}, x_{2s}, \ldots, x_{Is})$ and the corresponding outputs y_s for a collection of S measurements.

The set of equations giving the outputs as a function of the inputs is

$$y_s = f(w_0 + \sum_{i=1}^{I} w_i x_{is}) + \delta_s; \quad s = 1, 2, \ldots, S \tag{12.1}$$

where w_0 and $w_i; i = 1, 2, \ldots, I$ are the neuron threshold values and weights, respectively, and δ_s measures the error associated with the output y_s.

Three possibilities for learning the thresholds and weights are given below:

Option 1. Minimize

$$Q_1 = \sum_{s=1}^{S} \delta_s^2 = \sum_{s=1}^{S} \left(y_s - f(w_0 + \sum_{i=1}^{I} w_i x_{is}) \right)^2 \tag{12.2}$$

which is the sum of squared errors measured in the output scale (the units of the y_s). Unfortunately, this leads to a nonlinear problem, due to the presence of the usually nonlinear neural function f. Thus, Q_1 is not guaranteed to have a global optima. In fact, it usually has many local optima.

Option 2. Minimize the sum of squared errors:

$$Q_2 = \sum_{s=1}^{S} \epsilon_s^2 \quad = \quad \sum_{s=1}^{S} \left(w_0 + \sum_{i=1}^{I} w_i x_{is} - f^{-1}(y_s) \right)^2, \tag{12.3}$$

This option measures the errors in terms of the input scale (units of the x_{is}). The advantage is that we have a standard least-squares minimization, that can be written as a system of linear equations (linear problem).

Option 3. Alternatively, we can minimize the maximum absolute error:

$$\min_{w} \left\{ \max_{s} \left| w_0 + \sum_{i=1}^{I} w_i x_{is} - f^{-1}(y_s) \right| \right\} \tag{12.4}$$

which can be stated as the following linear programming problem. Minimize ϵ subject to

$$\left.\begin{array}{rcl} w_0 + \sum\limits_{i=1}^{I} w_i x_{is} - \epsilon & \leq & f^{-1}(y_s) \\ -w_0 - \sum\limits_{i=1}^{I} w_i x_{is} - \epsilon & \leq & -f^{-1}(y_s) \end{array}\right\}; \quad s = 1, 2, \ldots, S \tag{12.5}$$

which global optimum can be easily obtained using linear programming techniques.

12.1.1 Learning the Neural Functions

An important improvement is obtained if we learn the f^{-1} function instead of assuming that it is known. When the neural functions are learned instead of being given, we say that we are "in front" of a functional network. More precisely, f^{-1} can be assumed to be a convex combination of a set $\{\phi_1(x), \phi_2(x), , \ldots, \phi_R(x)\}$ of invertible basic functions:

$$f^{-1}(x) = \sum_{r=1}^{R} \alpha_r \phi_r(x) \tag{12.6}$$

where $\{\alpha_r;\ r = 1, 2, \ldots, R\}$ is the corresponding set of coefficients, which has to be chosen for the resulting function f^{-1} to be invertible. Without loss of generality, it can be assumed to be increasing.

Thus, we can consider two more options:

Option 4. Minimize, with respect to $w_i; i = 0, 1, \ldots, I$ and $\alpha_r; r = 1, 2, \ldots, R$, the function

$$Q_4 = \sum_{s=1}^{S} \epsilon_s^2 = \sum_{s=1}^{S} \left(w_0 + \sum_{i=1}^{I} w_i x_{is} - \sum_{r=1}^{R} \alpha_r \phi_r(y_s) \right)^2 \tag{12.7}$$

subject to

$$\sum_{r=1}^{R} \alpha_r \phi_r(y_{s_1}) \leq \sum_{r=1}^{R} \alpha_r \phi_r(y_{s_2}); \quad \forall y_{s_1} < y_{s_2} \tag{12.8}$$

which forces the candidate function f^{-1} to increase in the corresponding interval, and

$$\sum_{r=1}^{R} \alpha_r \phi_r(y_0) = 1 \tag{12.9}$$

which avoids the zero function as the optimum.

Option 5. Alternatively, we can minimize the maximum absolute error:

$$Q_5 = \min_{w,\alpha} \left\{ \max_{s} \left| w_0 + \sum_{i=1}^{I} w_i x_{is} - \sum_{r=1}^{R} \alpha_r \phi_r(y_s) \right| \right\} \tag{12.10}$$

which can be stated as the following linear programming problem. Minimize ϵ subject to

$$
\begin{array}{rcl}
w_0 + \sum\limits_{i=1}^{I} w_i x_{is} - \sum\limits_{r=1}^{R} \alpha_r \phi_r(y_s) - \epsilon & \leq & 0 \\
-w_0 - \sum\limits_{i=1}^{I} w_i x_{is} + \sum\limits_{r=1}^{R} \alpha_r \phi_r(y_s) - \epsilon & \leq & 0 \\
\sum\limits_{r=1}^{R} \alpha_r \phi_r(y_{s_1}) & \leq & \sum\limits_{r=1}^{R} \alpha_r \phi_r(y_{s_2}); \;\; \forall y_{s_1} < y_{s_2} \\
\sum\limits_{r=1}^{R} \alpha_r \phi_r(y_0) & = & 1
\end{array}
\tag{12.11}
$$

which has a global optimum easily obtainable.

To use the network, that is, to calculate the outputs for given inputs, it is necessary to know $f(x)$. However, this method only gives the function f^{-1}. One possible and efficient way to obtain $f(x)$ when $f^{-1}()$ is given is the bisection method.

The four main elements of the neural and functional networks problem are

1. **Data.**

   S: number of data vectors

   I: dimension of data input vectors

   x_{is}: vectors of input data $(x_{1s}, x_{2s}, \dots, x_{Is}); s = 1, 2, \dots, S$

   y_s: output data $y_s; s = 1, 2, \dots, S$

2. **Variables.**

   w_0: neuron threshold value

   w_i: neuron weight for the component i of the input vector

   ϵ: maximum error (it is a nonnegative variable that has sense only for option 3)

3. **Constraints.** For options 1 and 2 there are no constraints. For Option 3 the constraints are

$$
\left.
\begin{array}{rcl}
w_0 + \sum\limits_{i=1}^{I} w_i x_{is} - \epsilon & \leq & f^{-1}(y_s) \\
-w_0 - \sum\limits_{i=1}^{I} w_i x_{is} - \epsilon & \leq & -f^{-1}(y_s)
\end{array}
\right\}; \quad s = 1, \dots, S
\tag{12.12}
$$

   In the case of option 4 we have

$$
\sum_{r=1}^{R} \alpha_r \phi_r(y_{s_1}) \leq \sum_{r=1}^{R} \alpha_r \phi_r(y_{s_2}); \;\; \forall y_{s_1} < y_{s_2}
$$

$$\sum_{r=1}^{R} \alpha_r \phi_r(y_0) = 1$$

and for option 5, to the last two constraints we must add

$$\begin{aligned} w_0 + \sum_{i=1}^{I} w_i x_{is} - \sum_{r=1}^{R} \alpha_r \phi_r(y_s) - \epsilon &\leq 0 \\ -w_0 - \sum_{i=1}^{I} w_i x_{is} + \sum_{r=1}^{R} \alpha_r \phi_r(y_s) - \epsilon &\leq 0 \end{aligned}$$

4. **Function to be optimized.** We have the following cases:

   Option 1. Minimize

$$Q_1 = \sum_{s=1}^{S} \delta_s^2 = \sum_{s=1}^{S} \left(y_s - f(w_0 + \sum_{i=1}^{I} w_i x_{is}) \right)^2 \tag{12.13}$$

   Option 2. Minimize

$$Q_2 = \sum_{s=1}^{S} \epsilon_s^2 = \sum_{s=1}^{S} \left(w_0 + \sum_{i=1}^{I} w_i x_{is} - f^{-1}(y_s) \right)^2 \tag{12.14}$$

   Option 3 and 5. Minimize ϵ.

   Option 2. Minimize

$$Q_4 = \sum_{s=1}^{S} \epsilon_s^2 = \sum_{s=1}^{S} \left(w_0 + \sum_{i=1}^{I} w_i x_{is} - \sum_{r=1}^{R} \alpha_r \phi_r(y_s) \right)^2$$

Once these four elements have been identified, we are ready to solve the problem.

Example 12.1 (Neural network). Consider de data in the first three columns of Table 12.1, which gives the outputs y_s obtained for the corresponding inputs $\{x_{1s}, x_{2s}\}$ for $s = 1, \ldots, 30$. The data have been simulated from the model

$$y_s = (0.3 + 0.3x_{1s} + 0.7x_{2s})^2 + \epsilon_s$$

where $x_{is} \sim U(0, 0.5)$ and $\epsilon_s \sim U(-0.005, 0.005)$.

With the aim of learning the relation $y_s = q(x_1, x_2)$, from these data, we decide to use a one-layer neural network with two inputs $\{x_1, x_2\}$ and one output y, and use the three options with known neuron function, and the two options where the neuron function is learned, described above. For the sake of illustration, a neural function $f(x) = \arctan(x)$, and the following basic functions have been used:

$$\{\phi_1(x), \phi_2(x), \phi_3(x)\} = \{\sqrt{x}, x, x^2\}$$

The following input GAMS code was used to solve this problem. Note that five different models with five different objective functions have been used. Note also the implementation of the bisection method.

```
$title Neural networks (neural2)

SETS
S number of data vectors/1*30/
I dimension of data input vectors/1*2/
J number of different learning functions/1*5/
R number of basic functions/1*3/;

ALIAS(S,S1)

PARAMETERS
X(I,S) input data vectors
Y(S) output data
powers(R) exponents of x in the basic functions
/1 0.5
2  1
3  2/;

X(I,S)=uniform(0,1)*0.5;
Y(S)=sqr(0.3+0.3*X('1',S)+0.7*X('2',S))+(uniform(0,1)-0.5)*0.01;

PARAMETERS
aux auxiliary parameter
x1 auxiliary parameter
x2 auxiliary parameter
x3 auxiliary parameter
f1 auxiliary parameter
f2 auxiliary parameter
f3 auxiliary parameter
ymin minimum value of Y(S)
ymax maximum value of Y(S)
acterror maximum prediction error
maxerror maximum allowed error for the bisection method;

maxerror=0.00001;

VARIABLES
z1 function1 to be optimized
z2 function2 to be optimized
z3 function3 to be optimized
z4 function21 to be optimized
z5 function21 to be optimized
W0 threshold value
W(I) weight associated with component i of input
alfa(R) coefficients of the neural function;

POSITIVE VARIABLES
epsilon error;

EQUATIONS
Q1 definition of z1
Q2 definition of z2
Q3 definition of z3
Q4 definition of z4
Q5 definition of z5
const1(S) upper error bound
const2(S) lower error bound
```

```
const3(S) upper error bound
const4(S) lower error bound
normalized normalized values
increasing(S,S1) the f function must be increasing;

Q1..z1=e=sum(S,sqr(Y(S)-arctan(W0+sum(I,W(I)*X(I,S)))));
Q2..z2=e=sum(S,sqr(W0+sum(I,W(I)*X(I,S))-sin(Y(S))/cos(Y(S))));
Q3..z3=e=epsilon;
Q4..z4=e=sum(S,sqr(W0+sum(I,W(I)*X(I,S))
          -sum(R,alfa(R)*Y(S)**(powers(R)))));
Q5..z5=e=epsilon;
const1(S)..W0+sum(I,W(I)*X(I,S))-epsilon=l=sin(Y(S))/cos(Y(S));
const2(S)..-W0-sum(I,W(I)*X(I,S))-epsilon=l=-sin(Y(S))/cos(Y(S));
const3(S)..W0+sum(I,W(I)*X(I,S))
     -sum(R,alfa(R)*Y(S)**(powers(R)))-epsilon=l=0.0;
const4(S)..-W0-sum(I,W(I)*X(I,S))
     +sum(R,alfa(R)*Y(S)**(powers(R)))-epsilon=l=0.0;
normalized..sum(R,alfa(R)*Y('1')**(powers(R)))=e=1;
increasing(S,S1)$(Y(S)<Y(S1))..sum(R,alfa(R)*Y(S)**(powers(R)))
     =l=sum(R,alfa(R)*Y(S1)**(powers(R)));;

MODEL reg1/Q1/;
MODEL reg2/Q2/;
MODEL reg3/Q3,const1,const2/;
MODEL reg4/Q4,normalized,increasing/;
MODEL reg5/Q5,const3,const4,normalized,increasing/;

file out/neural3.out/;
put out;

put "Data"/;
ymin=0.0;
ymax=0.0;
loop(S,
  loop(I,put X(I,S):12:3," & ";);
  put Y(S):12:3,"\\"/;
  if(ymin<Y(S),ymin=Y(S));
  if(ymax>Y(S),ymax=Y(S));
);
loop(J,
  if(ord(J) eq 1, SOLVE reg1 USING nlp MINIMIZING z1;
        put "z1=",z1.l:12:9/;);
  if(ord(J) eq 2, SOLVE reg2 USING nlp MINIMIZING z2;
        put "z2=",z2.l:12:9/;);
  if(ord(J) eq 3, SOLVE reg3 USING  lp MINIMIZING z3;
        put "z3=",z3.l:12:9/;);
  if(ord(J) eq 4, SOLVE reg4 USING nlp MINIMIZING z4;
        put "z4=",z4.l:12:9/;);
  if(ord(J) eq 5, SOLVE reg5 USING  lp MINIMIZING z5;
        put "z5=",z5.l:12:9/;);
  put "Weights:"/;
  put W0.l:12:3;
  loop(I,put " & ",W.l(I):12:3;);
  put " "/;
  if(((ord(J) eq 4) or (ord(J) eq 5)),
    put "Alfas:"/;
    loop(R,put " & ",alfa.l(R):12:3;);
```

```
    put " "/;
  );
  acterror=-10000;
  put "Data and fitted values:"//;


  loop(S,
    aux=W0.l+sum(I,W.l(I)*X(I,S));
    if(ord(J) le 3,
      f1=arctan(aux);
    else
      x1=ymin-(ymax-ymin)*0.1;
      x2=ymax+(ymax-ymin)*0.1;
      f1=sum(R,alfa.l(R)*abs(x1)**(powers(R)))-aux;
      f2=sum(R,alfa.l(R)*abs(x2)**(powers(R)))-aux;
      if(f1*f2>0.0,
        put "ERROR IN BISECTION ","  S=",S.tl:3," aux=",
                aux:12:3," f1=",f1:12:3," f2=",f2:12:3/;
      else
        while(abs(x1-x2)>maxerror,
          x3=(x1+x2)*0.5;
          f3=sum(R,alfa.l(R)*x3**(powers(R)))-aux;
          if(f3*f1>0.0,
            x1=x3;f1=f3;
          else
            x2=x3;f2=f3;
          );
        );
      );
      f1=((x1+x2)*0.5);
    );
    loop(I,put X(I,S):12:3," & ";);
    aux=(f1-Y(S));
    if(abs(aux)>acterror,acterror=abs(aux));
    put Y(S):12:3," & ",aux:12:3"\\"/;
  );
  put "Maximum error=",acterror:12:9/;
);
```

The corresponding errors are shown in Table 12.1. Finally, Table 12.2 shows the threshold values, the weights, and the α coefficients (only for options 4 and 5), together with the optimal values of the corresponding objective functions.

A simple comparison shows that the errors for the objective functions, Q_4 and Q_5, where the neural functions have been learned, lead to smaller errors, as expected.

■

12.2 Applications to CAD

In this section we discuss an application to automatic mesh generation.

Automatic mesh generation is an area that has attracted the interest of many researchers. The most common mesh generation procedures include triangle, in

Table 12.1: Input and output data for a one-layer neural network

Input		Output	ϵ_s Errors				
x_{1s}	x_{2s}	y_s	Q_1	Q_2	Q_3	Q_4	Q_5
0.086	0.055	0.129	-0.023	-0.024	-0.020	0.009	0.004
0.422	0.251	0.360	0.005	0.007	0.012	0.007	0.008
0.275	0.080	0.193	-0.002	-0.003	0.002	-0.004	-0.004
0.151	0.436	0.426	-0.015	-0.014	-0.006	-0.002	-0.008
0.146	0.133	0.187	0.000	-0.001	0.005	-0.001	-0.003
0.112	0.143	0.183	0.000	-0.001	0.005	0.001	-0.001
0.175	0.297	0.315	0.006	0.006	0.013	-0.002	-0.002
0.428	0.361	0.466	-0.024	-0.022	-0.016	-0.004	-0.008
0.034	0.314	0.280	0.008	0.008	0.015	-0.003	-0.005
0.250	0.232	0.287	0.010	0.010	0.016	-0.001	0.000
0.499	0.207	0.351	0.007	0.008	0.013	0.006	0.009
0.289	0.059	0.181	-0.002	-0.002	0.002	0.000	-0.001
0.496	0.157	0.308	0.013	0.014	0.018	0.005	0.009
0.381	0.023	0.190	-0.007	-0.008	-0.004	-0.007	-0.006
0.065	0.169	0.191	-0.002	-0.003	0.003	-0.003	-0.006
0.320	0.091	0.214	0.000	0.000	0.004	-0.007	-0.006
0.080	0.323	0.300	0.008	0.009	0.016	0.000	-0.002
0.125	0.280	0.281	0.011	0.011	0.018	0.000	0.000
0.334	0.385	0.451	-0.021	-0.019	-0.013	-0.004	-0.008
0.218	0.149	0.216	0.008	0.008	0.013	-0.001	-0.001
0.180	0.331	0.340	0.006	0.007	0.014	0.004	0.002
0.176	0.378	0.376	0.002	0.003	0.010	0.007	0.004
0.066	0.314	0.289	0.009	0.009	0.017	-0.001	-0.003
0.075	0.142	0.178	-0.007	-0.009	-0.003	-0.003	-0.006
0.295	0.043	0.172	-0.003	-0.004	0.000	0.002	0.001
0.415	0.051	0.209	0.007	0.007	0.010	0.000	0.001
0.115	0.321	0.311	0.008	0.008	0.015	0.000	-0.001
0.333	0.273	0.347	0.006	0.007	0.012	0.004	0.005
0.388	0.016	0.181	-0.002	-0.002	0.001	0.000	0.000
0.152	0.396	0.393	-0.009	-0.008	-0.001	-0.003	-0.007
Maximum absolute error			0.024	0.024	0.019	0.009	0.009

two dimensions, and tetrahedral and hexahedral, in three dimensions, generation methods. Most generation problems involve building partitioning elements on arbitrary three-dimensional surfaces. The main surface mesh generation techniques belong to three main groups:

1. **Parametric space based techniques.** Since surfaces in parametric

Table 12.2: Values of the objective functions, threshold values, weights, and coefficients associated with the five options

Objective function	Threshold	Weights		Alphas		
Q	w_0	w_1	w_2	α_1	α_2	α_3
$Q_1 = 0.00292$	0.032	0.353	0.803	–	–	–
$Q_2 = 0.00375$	0.029	0.358	0.812	–	–	–
$Q_3 = 0.01978$	0.034	0.354	0.823	–	–	–
$Q_4 = 0.00086$	0.940	0.369	0.847	4.199	−4.238	2.384
$Q_5 = 0.00988$	0.924	0.415	0.912	4.200	−4.269	2.596

form have an $u-v$ representation, one can use standard techniques to mesh the $u-v$ parametric region, and then move to the associated $x-y-z$ space (see, for example, Cuilliere [29], and Farouki [36]). The main shortcoming of a direct application of this technique is that the resulting elements can be poorly shaped. Thus, some modifications are normally required.

2. **Direct techniques.** These methods proceed by advancing through the surface itself, by generating elements based on previously generated elements, and using surface normals and tangents to compute the direction of the advancing front.

3. **Optimization-based methods.** These methods are based on optimizing a quality function subject to some constraints. Nodes are moved in a direction such that an improvement of the objective function is obtained (Canann et al. [19], Freitag [42] and Li [67]).

In addition to the mesh generation problem, since most mesh generation techniques output meshes that are not optimal, some postprocessing to improve the quality of the elements is normally required. The two main categories of mesh improvement are

1. **Smoothing.** This consists of adjusting node locations without altering connectivities.

2. **Polishing.** This consists of changing the topology or the elements connectivities. Polishing methods use some quality criteria to operate, and can be combined with smoothing techniques (see Freitag [44]).

3. **Refinement.** This consists of reducing the element size. The reduction in size may be required to reproduce better the local reality, or to improve the quality of the elements. In fact, refinement can be seen as a mesh generation. Thus, mesh generation techniques are applicable here.

Figure 12.2: Topology of the selected mesh.

12.2.1 Automatic Mesh Generation

In this section we describe how to generate or improve a mesh using an optimization based method. This application is based on Castillo and Mínguez [23].

The method starts from a given projected mesh topology (the projection of the surface mesh on the X–Y plane), as that shown in Figure 12.1. In this projection, the corner nodes are assumed to be fixed, the boundary nodes can move in the boundary, and the interior nodes are free to move in any direction. The surface triangularization is selected for minimizing the standard deviation or variance of the triangle areas. Thus, we look for one mesh with the maximum similarity in terms of the triangle areas. In other words, we minimize

$$Q = \sum_{i=1}^{m} \frac{[a(i) - \bar{a}]^2}{m}$$

subject to the boundary constraints, where $a(i)$ is the area of the triangle i, $\bar{a}$ is the corresponding mean, and m is the number of triangles.

In summary, we proceed as follows:

Step 1. The topology of the mesh is defined by a projected mesh. A compact definition of one possible and simple mesh is given in Table 12.3, where (x_{min}, y_{min}) and (x_{max}, y_{max}) are the left-top and right-bottom corner coordinates, and n is the number of nodes per column ($n = 12$ for the mesh in Figure 12.2). The resulting mesh has n^2 nodes and $2(n-1)^2$ elements.

Table 12.3: Parametric definition of a rectangular net

	Coordinates of node k
x_k	$x_{min} + floor((k-1)/n))(x_{max} - x_{min})/(n-1)$
y_k	$y_{min} + mod(k-1,n)(y_{max} - y_{min})/(n-1)$
	Nodes of element i for i even
First node	(floor((i-1)/(2(n-1))))n+floor(mod((i-1),(2(n-1)))/2+1)
Second node	(floor((i-1)/(2(n-1))))n+floor(mod((i-1),(2(n-1)))/2+1)+1
Third node	(floor((i-1)/(2(n-1))))n+floor(mod((i-1),(2(n-1)))/2+1)+n+1
	Nodes of element i for i odd
First node	(floor((i-1)/(2(n-1))))n+floor(mod((i-1),(2(n-1)))/2+1)
Second node	(floor(i-1)/(2(n-1))))n+floor(mod((i-1),(2(n-1)))/2+1)+n
Third node	(floor((i-1)/(2(n-1))))n+floor(mod((i-1),(2(n-1)))/2+1)+n+1
	Boundary elements
Lower side	$n(j-1)+1;\ j = 1, 2, \ldots, n$
Upper side	$nj;\ j = 1, 2, \ldots, n$
Right-hand side	$n^2 - n + j;\ j = 1, 2, \ldots, n$
Left-hand side	$j;\ j = 1, 2, \ldots, n$

The following three main components of the mesh are defined:

(a) Node coordinates

(b) Nodes belonging to each element

(c) Boundary nodes

Step 2. The surface to be triangulated is defined by its corresponding equations.

Step 3. The degrees of freedom of all the nodes are defined. Figure 12.3 shows the degrees of freedom of the different sets of nodes. Some nodes corresponding to the corner nodes are fixed, having no degrees of freedom. Some nodes, the boundary no corner nodes, have 1 degree of freedom, which is shown by an arrow indicating the free directions of their possible displacements. Finally, the rest, the interior nodes, have 2 degrees of freedom, marked by two arrows showing their possible displacements.

Step 4. The objective function Q is minimized subject to the boundary constraints, and the final projected and surface meshes are obtained (see Figure 12.4).

To avoid overlapping of triangles, we can force the total area of the projected mesh to remain constant and equal to the initial one. This simple constraint avoids stability problems caused by the overlapping of triangles.

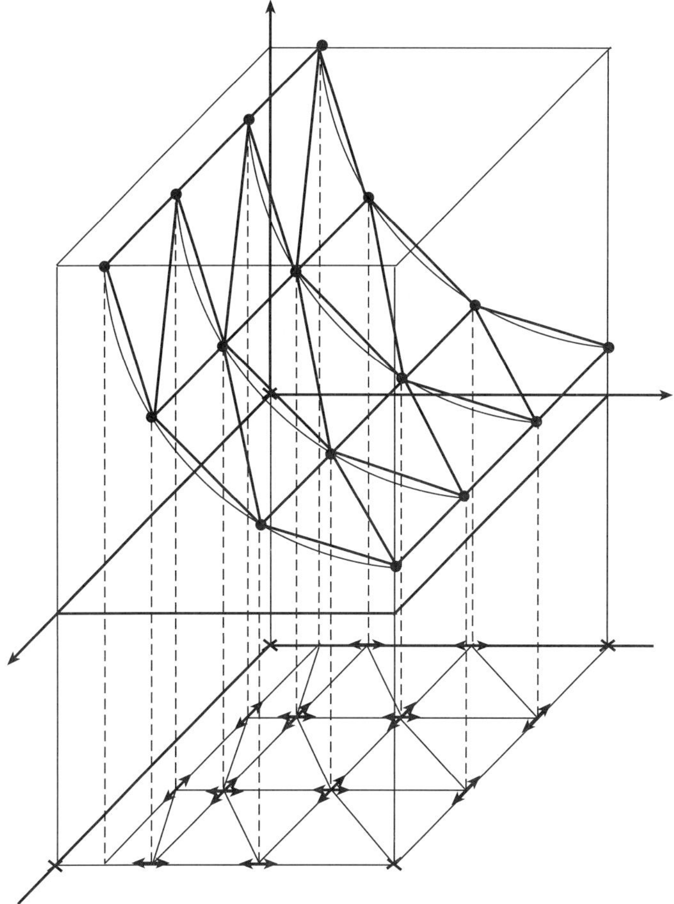

Figure 12.3: Illustration of how the net is generated, showing the degrees of freedom of the nodes.

Example 12.2 (Surface triangulation). Consider the surface with equation

$$z = \frac{xy(x^2 - y^2)}{x^2 + y^2}$$

defined on the square

$$\{(x, y)| - 2 \leq x \leq 2; \ -2 \leq y \leq 2\}$$

Applying the proposed method with a rectangular net of size $n = 12$, and

$$(x_{min}, y_{min}) = (-2, -2); \ \ (x_{max}, y_{max}) = (2, 2)$$

the triangulated surface in Figure 12.4 is obtained. Note that the triangles in the vertical projection have been adapted to get 3D triangles with similar area. ■

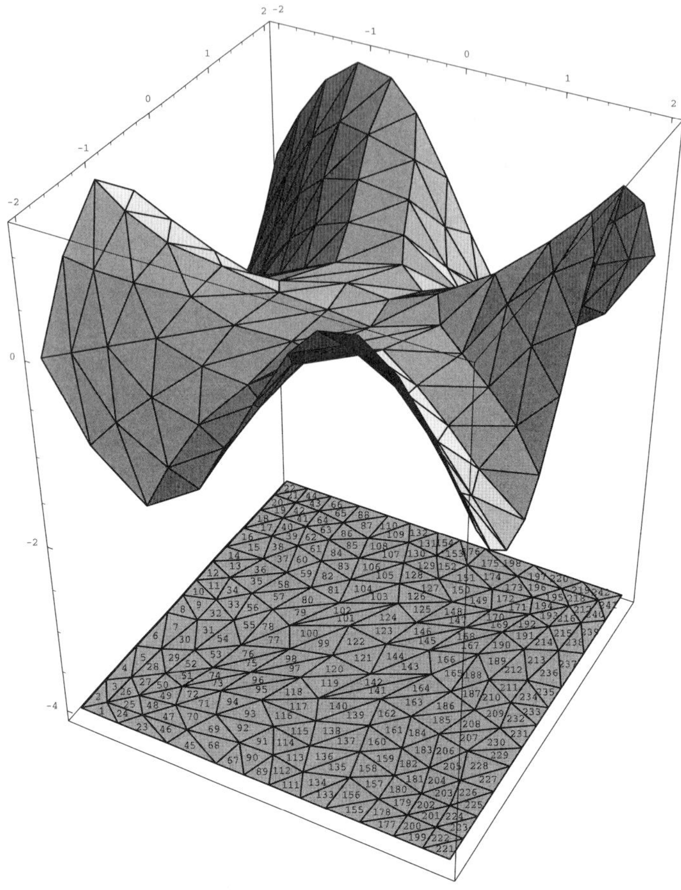

Figure 12.4: Example of a triangulated surface showing the basic mesh and the mesh leading to a minimum variance of triangle areas.

```
$title Mesh

SETS
I number of elements/1*242/
J number of nodes per element/1*3/
K number of nodes/1*144/
N number of nodes per row/1*12/
D space dimension/1*2/
XFIXED(K) nodes with fixed X
YFIXED(K) nodes with fixed Y;

ALIAS(K,K1,K2);
```

```
PARAMETER
t number of squares per dimension
m number of elements per column
xmin minimum value of coordinate x
xmax maximum value of ccordinate x
ymin minimum value of coordinate y
ymax maximum value of ccordinate y
COORD(K,D) coordinates of the initial mesh nodes;

t=card(N);
xmin=-2;
xmax=2;
ymin=-2;
ymax=2;

m=2*t-2;

XFIXED(K)=no;
loop(N,
XFIXED(K)$(ord(K) eq ord(N))=yes;
XFIXED(K)$(ord(K) eq ord(N)+t*t-t)=yes;
YFIXED(K)$(ord(K) eq t*ord(N))=yes;
YFIXED(K)$(ord(K) eq (ord(N)-1)*t+1)=yes;
);

COORD(K,'1')=xmin+floor((ord(K)-1)/t)*(xmax-xmin)/(t-1);
COORD(K,'2')=ymin+mod(ord(K)-1,t)*(ymax-ymin)/(t-1);

PARAMETER

EL(I,J);

loop(I,
if(mod(ord(I),2) eq 0,
EL(I,'1')=(floor((ord(I)-1)/m))*t+floor(mod((ord(I)-1),m)/2+1);
EL(I,'2')=(floor((ord(I)-1)/m))*t+floor(mod((ord(I)-1),m)/2+1)+1;
EL(I,'3')=(floor((ord(I)-1)/m))*t+floor(mod((ord(I)-1),m)/2+1)+t+1;
else
 EL(I,'1')=(floor((ord(I)-1)/m))*t+floor(mod((ord(I)-1),m)/2+1);
EL(I,'2')=(floor((ord(I)-1)/m))*t+floor(mod((ord(I)-1),m)/2+1)+t;
EL(I,'3')=(floor((ord(I)-1)/m))*t+floor(mod((ord(I)-1),m)/2+1)+t+1;
);
);

VARIABLES
z1 function to be optimized
X(K) coordinate of node K
Y(K) coordinate of node K
Z(K) coordinate of node K
area(I) area of element I
mean mean area of elements;

X.l(K)=COORD(K,'1');
Y.l(K)=COORD(K,'2');

X.fx(XFIXED)=COORD(XFIXED,'1');
Y.fx(YFIXED)=COORD(YFIXED,'2');
```

```
EQUATIONS
Q objective function to be optimized
Carea(I)
Cf(K) points are in surface
Cmean;

Q..z1=e=sum(I,sqr(area(I)-mean))/card(I);

Carea(I)..area(I)=e=sum((K,K1,K2)$((EL(I,'1')
   =ord(K)) and (EL(I,'2')=ord(K1)) and (EL(I,'3')=ord(K2))),
sqr((Y(K1)-Y(K))*(Z(K2)-Z(K))-(Y(K2)-Y(K))*(Z(K1)-Z(K)))+
sqr((Z(K1)-Z(K))*(X(K2)-X(K))-(Z(K2)-Z(K))*(X(K1)-X(K)))+
sqr((X(K1)-X(K))*(Y(K2)-Y(K))-(X(K2)-X(K))*(Y(K1)-Y(K))));
Cf(K)..Z(K)=e=(X(K)*Y(K)*(sqr(X(K))-sqr(Y(K))))/(sqr(X(K))+sqr(Y(K)));
Cmean..mean=e=sum(I,area(I))/card(I);

MODEL mallas/ALL/;

SOLVE mallas USING nlp MINIMIZING z1;

file out/mallas3.out/;
put out;


if((card(I) ne (2*(t-1)*(t-1))),
put "Error, card(I) must be equal to ",(2*(t-1)*(t-1)):8:0/;
else
if((card(K) ne t*t),
put "Error, card(K) must be equal to ",(t*t):8:0/;
else
put "z1=",z1.l:12:8," mean=",mean.l:12:8, " cpu=",mallas.resusd:9:2,
     " modelstat=",mallas.modelstat:3:0," solvestat=",mallas.solvestat:3:0/;

loop(K,
put "X(",K.tl:3,")=",COORD(K,'1'):12:8," Y(",K.tl:3,")=",COORD(K,'2'):12:8/;
);

loop(I,
put "I=",i.tl:3," Nodes=",EL(I,'1'):5:0," ",EL(I,'2'):5:0," ",EL(I,'3'):5:0/;
);

loop(K,
put "K=",K.tl:3," X=",X.l(K):12:6," Y=",Y.l(K):12:6," Z=",Z.l(K):12:6/;
);

loop(I,
put "I=",I.tl:3," area=",area.l(I):12:6/;
);

put "COORDINATES"/;
put "{";

loop(K,
if(ord(K)<card(K),
put "{",X.l(K):12:8,",",Y.l(K):12:8,",",Z.l(K):12:8,"},"/;
else
put "{",X.l(K):12:8,",",Y.l(K):12:8,",",Z.l(K):12:8,"}}"/;
```

```
);
);

put "ELEMENTS"/;
put "{";

loop(I,
if(ord(I)<card(I),
put "{",EL(I,'1'):5:0,",",EL(I,'2'):5:0,",",EL(I,'3'):5:0,"},"/;
else
put "{",EL(I,'1'):5:0,",",EL(I,'2'):5:0,",",EL(I,'3'):5:0,"}}"/;
););););
```

12.3 Applications to Probability

In this section we present some applications relevant to probability. We deal with the problem of compatibility of conditional distributions, which arises in many fields of applications, as artificial intelligence and expert systems, for example. The assessment of probabilities is crucial in probability based models arising in these areas. This assessment can be done by considering the joint distributions of the variables involved or, more easily, by considering several sets of conditional distributions and/or marginal ones, because they are of lower dimensions. However, conditionals probabilities are closely interrelated and, consequently, they cannot be freely assessed. Even for a qualified expert, the chances of violating the probability axioms during the assessment process are close to one. So, the problem of compatibility must be dealt with, and the help of a computer is unavoidable in practice. In this section, we show how this problem can be solved using linear programming techniques (see Arnold et al. [1, 2, 3, 4, 5] for a more complete treatment of this problem).

12.3.1 Compatibility of Conditional Probability Matrices

Consider a bidimensional discrete random variable (X, Y), with domain $N = \{(x_i, y_j)|i = 1, 2, \dots, I; j = 1, 2, \dots, J\}$, where X and Y take possible values x_1, x_2, ... x_I and $y_1, y_2, \dots, y_J$, respectively.

Let $\mathbf{P}$ be its joint probability matrix with $p_{ij} = \text{Prob}[X = x_i, Y = y_j]$. The column vectors

$$\begin{array}{rcl} p_{i.} & = & \sum_{j}^{J} p_{ij} \\ p_{.j} & = & \sum_{i}^{I} p_{ij} \end{array}$$

where the dot is used to refer to sum in the corresponding index, give the probabilities of X and Y, respectively, after ignoring the other variable, and are called marginal probabilities of X and Y, respectively.

The conditional probability matrices $\mathbf{A}$ and $\mathbf{B}$ are defined as

$$a_{ij} = \text{Prob}[X = x_i | Y = y_j] = \frac{p_{ij}}{p_{.j}} \tag{12.15}$$

and

$$b_{ij} = \text{Prob}[Y = y_j | X = x_i] = \frac{p_{ij}}{p_{i.}} \tag{12.16}$$

respectively, and give the probabilities of one variable when the value of the other variable has been fixed.

Note that $\mathbf{A}$ and $\mathbf{B}$ exists only if

$$\text{Prob}[Y = y_j] = \sum_{i}^{I} p_{ij} = p_{.j} \neq 0$$

and

$$\text{Prob}[X = x_i] = \sum_{j}^{J} p_{ij} = p_{i.} \neq 0$$

respectively.

From (12.15) and (12.16) it follows that the columns of $\mathbf{A}$ and the rows of $\mathbf{B}$ sum to 1. Thus, valid conditional probability matrices $\mathbf{A}$ and $\mathbf{B}$ must have nonnegative elements such that $\mathbf{A}$ has columns and $\mathbf{B}$ has rows that sum to 1.

If the joint probabilities p_{ij} are given, the conditional probability matrices $\mathbf{A}$ and $\mathbf{B}$ can be obtained using (12.15) and (12.16), respectively. However, obtaining the matrix $\mathbf{P}$ from $\mathbf{A}$ and $\mathbf{B}$ is not a trivial task. In fact, $\mathbf{A}$ and $\mathbf{B}$ can be chosen in such form that there is no $\mathbf{P}$ satisfying (12.15) and (12.16). In this case we say that $\mathbf{A}$ and $\mathbf{B}$ are incompatible. If, on the other hand, this matrix $\mathbf{P}$ exists, we say that $\mathbf{A}$ and $\mathbf{B}$ are compatible.

Consequently, $\mathbf{A}$ and $\mathbf{B}$ are compatible if and only if a joint distribution $\mathbf{P}$ with $\mathbf{A}$ and $\mathbf{B}$ as its associated conditionals exists, that is, such that

$$p_{ij} = a_{ij} p_{.j}; \ \forall i, j \tag{12.17}$$

$$p_{ij} = b_{ij} p_{i.}; \ \forall i, j \tag{12.18}$$

$$p_{ij} \geq 0, \quad \forall i, j, \tag{12.19}$$

$$\sum_{i=1}^{I} \sum_{j=1}^{J} p_{ij} = 1 \tag{12.20}$$

Balow we give three possible methods to determine the compatibility of $\mathbf{A}$ and $\mathbf{B}$:

Method 1 (based on P). Seek a joint probability matrix $\mathbf{P}$ satisfying

$$\begin{array}{rcl} p_{ij} - a_{ij} \sum\limits_{i=1}^{I} p_{ij} & = & 0, \ \ \forall i, j \\ p_{ij} - b_{ij} \sum\limits_{j=1}^{J} p_{ij} & = & 0, \ \ \forall i, j \\ \sum\limits_{i=1}^{I} \sum\limits_{j=1}^{J} p_{ij} & = & 1 \\ p_{ij} & \geq & 0, \ \ \forall i, j \end{array} \tag{12.21}$$

In this method $\mathbf{P}$ is directly sought, and it involves $2|N|+1$ equations in $|N|$ unknowns, where $|N|$ is the cardinality of the domain set N of (X,Y).

Method 2 (based on the two marginals). Seek two probability vectors $\boldsymbol{\tau}$ and $\boldsymbol{\eta}$ satisfying

$$\begin{array}{rcl} \eta_j a_{ij} - \tau_i b_{ij} & = & 0, \ \forall i,j \\ \sum\limits_{i=1}^{I} \tau_i & = & 1 \\ \sum\limits_{j=1}^{J} \eta_j & = & 1 \\ \tau_i \geq 0, \forall i, \ \ \eta_j & \geq & 0, \ \ \forall j \end{array} \tag{12.22}$$

In this method we seek the two marginal probability vectors, $\boldsymbol{\tau}$ and $\boldsymbol{\eta}$, which, combined with $\mathbf{A}$ and $\mathbf{B}$, give $\mathbf{P}$ $(p_{ij} = \eta_j a_{ij} = \tau_i b_{ij})$. This involves $|N|+2$ equations in $I+J$ unknowns.

Method 3 (based on one marginal). Seek one probability vector $\boldsymbol{\tau}$ satisfying

$$\begin{array}{rcl} a_{ij} \sum\limits_{k=1}^{I} \tau_k b_{kj} - \tau_i b_{ij} & = & 0, \ \forall i,j \\ \sum\limits_{i=1}^{I} \tau_i & = & 1 \\ \tau_i & \geq & 0, \ \forall i \end{array} \tag{12.23}$$

In this method, we seek the X-marginal $\boldsymbol{\tau}$ which, combined with $\mathbf{B}$, gives $\mathbf{P}$ $(p_{ij} = \tau_i b_{ij})$. It involves $|N|+1$ equations in I unknowns.

All three methods involve linear equations to be solved subject to non-negativity constraints. However, since system (12.23) is the one that involves a smaller number of variables, it is probably the one we will use to try to solve our problem.

Example 12.3 (Compatibility). Consider the matrices $\mathbf{A}$ and $\mathbf{B}$:

$$\mathbf{A} = \begin{pmatrix} 0.0667 & 0.1905 & 0.3750 & 0.1176 & 0.0769 & 0.1111 \\ 0.1333 & 0.1905 & 0.0000 & 0.1765 & 0.3846 & 0.3333 \\ 0.3333 & 0.1905 & 0.1250 & 0.1765 & 0.0769 & 0.1111 \\ 0.2000 & 0.0952 & 0.2500 & 0.1176 & 0.0769 & 0.1111 \\ 0.2000 & 0.1905 & 0.1250 & 0.1765 & 0.3077 & 0.2222 \\ 0.0667 & 0.1429 & 0.1250 & 0.2353 & 0.0769 & 0.1111 \end{pmatrix} \tag{12.24}$$

$$\mathbf{B} = \begin{pmatrix} 0.0833 & 0.3333 & 0.2500 & 0.1667 & 0.0833 & 0.0833 \\ 0.1176 & 0.2353 & 0.0000 & 0.1765 & 0.2941 & 0.1765 \\ 0.3333 & 0.2667 & 0.0667 & 0.2000 & 0.0667 & 0.0667 \\ 0.2727 & 0.1818 & 0.1818 & 0.1818 & 0.0909 & 0.0909 \\ 0.1765 & 0.2353 & 0.0588 & 0.1765 & 0.2353 & 0.1176 \\ 0.0909 & 0.2727 & 0.0909 & 0.3636 & 0.0909 & 0.0909 \end{pmatrix} \tag{12.25}$$

The following GAMS code generates a random joint probability matrix and the corresponding conditionals. Next, it obtains the joint probability based on A and B using methods 1–3.

```
$title Compatibility Method I

SETS
I number of rows/1*6/
J number of columns/1*6/;

ALIAS(I,I1);
ALIAS(J,J1);

PARAMETER
PP(I,J) Joint probability matrix used to generate conditionals
A(I,J) conditional probability of X given Y
B(I,J) conditional probability of Y given X;

PP(I,J)=round(uniform(0,5));
PP(I,J)=PP(I,J)/sum((I1,J1),PP(I1,J1));
A(I,J)=PP(I,J)/sum(I1,PP(I1,J));
B(I,J)=PP(I,J)/sum(J1,PP(I,J1));

VARIABLE
z value of the objective function;

POSITIVE VARIABLES
P(I,J) Joint probability looked for
TAU(I) Marginal probability of X
ETA(J) Marginal probability of Y;

EQUATIONS
ZDEF1 Function to be optimized in model 1
ZDEF2 Function to be optimized in model 2
ZDEF3 Function to be optimized in model 3
CONST11(I,J)
CONST12(I,J)
CONST2(I,J)
CONST3(I,J)
NORM1 P add up to one
NORM2 TAU add up to one
NORM3 ETA add up to one;

ZDEF1..z=e=P('1','1');
ZDEF2..z=e=TAU('1');
CONST11(I,J)..P(I,J)-A(I,J)*sum(I1,P(I1,J))=e=0;
CONST12(I,J)..P(I,J)-B(I,J)*sum(J1,P(I,J1))=e=0;
CONST2(I,J)..ETA(J)*A(I,J)-TAU(I)*B(I,J)=e=0;
CONST3(I,J)..A(I,J)*sum(I1,TAU(I1)*B(I1,J))-TAU(I)*B(I,J)=e=0;
NORM1..sum((I,J),P(I,J))=e=1;
NORM2..sum(I,TAU(I))=e=1;
NORM3..sum(J,ETA(J))=e=1;

MODEL model1/ZDEF1,CONST11,CONST12,NORM1/;
MODEL model2/ZDEF2,CONST2,NORM2,NORM3/;
MODEL model3/ZDEF2,CONST3,NORM2/;

file out/comp1.out/;
put out;

put "Initial Matrix PP:"/;
```

```
loop(I,loop(J,put PP(I,J):7:4;);put ""/;);
put "Matrix A:"/;
loop(I,loop(J,put A(I,J):7:4;);put ""/;);
put "Matrix B:"/;
loop(I,loop(J,put B(I,J):7:4;);put ""/;);

SOLVE model1 USING lp MINIMIZING z;
put "Final Matrix P Obtained by Model 1:"/;
loop(I,loop(J,put P.l(I,J):7:4;);put ""/;);
put "z=",z.l," modelstat=",model1.modelstat," solvestat=",model1.solvestat/;
SOLVE model2 USING lp MINIMIZING z;
put "Final Matrix P Obtained by Model 2:"/;
loop(I,loop(J,put (TAU.l(I)*B(I,J)):7:4;);put ""/;);
put "z=",z.l," modelstat=",model2.modelstat," solvestat=",model2.solvestat/;
SOLVE model3 USING lp MINIMIZING z;
put "Final Matrix P Obtained by Model 3:"/;
loop(I,loop(J,put (TAU.l(I)*B(I,J)):7:4;);put ""/;);
put "z=",z.l," modelstat=",model3.modelstat," solvestat=",model3.solvestat/;
```

The solution of this problem is the joint probability matrix

$$\mathbf{P} = \begin{pmatrix} 0.0120 & 0.0482 & 0.0361 & 0.0241 & 0.0120 & 0.0120 \\ 0.0241 & 0.0482 & 0.0000 & 0.0361 & 0.0602 & 0.0361 \\ 0.0602 & 0.0482 & 0.0120 & 0.0361 & 0.0120 & 0.0120 \\ 0.0361 & 0.0241 & 0.0241 & 0.0241 & 0.0120 & 0.0120 \\ 0.0361 & 0.0482 & 0.0120 & 0.0361 & 0.0482 & 0.0241 \\ 0.0120 & 0.0361 & 0.0120 & 0.0482 & 0.0120 & 0.0120 \end{pmatrix} \quad (12.26)$$

Since the corresponding programming problem has a solution, the conditional probability matrices **A** and **B** are compatible. ∎

Example 12.4 (Incompatibility). Consider the matrices **A** and **B**:

$$\mathbf{A} = \begin{pmatrix} \frac{1}{2} & \frac{1}{2} & 0 \\ 0 & \frac{1}{2} & \frac{1}{2} \\ \frac{1}{2} & 0 & \frac{1}{2} \end{pmatrix}, \quad (12.27)$$

$$\mathbf{B} = \begin{pmatrix} \frac{1}{3} & \frac{2}{3} & 0 \\ 0 & \frac{1}{3} & \frac{2}{3} \\ \frac{1}{3} & 0 & \frac{2}{3} \end{pmatrix}. \quad (12.28)$$

Since the corresponding programming problem has no solution, the conditional probability matrices **A** and **B** are incompatible.

The reader is encouraged to modify the previous GAMS program to check that this problem is infeasible.

∎

12.3.2 ϵ Compatibility

In Section 12.3.1 we described three linear equation formulations for the search for a matrix **P** compatible with given conditional matrices **A** and **B**. If, instead of exact compatibility, we are willing to accept approximate compatibility, we can replace Methods 1–3 by the following revised versions:

Revised method 1. Seek a probability matrix $\mathbf{P}$ to minimize ϵ subject to

$$\begin{array}{rcl} |p_{ij} - a_{ij} \sum_{i=1}^{I} p_{ij}| & \leq & \epsilon\gamma_{ij}, \ \forall i,j \\ |p_{ij} - b_{ij} \sum_{j=1}^{J} p_{ij}| & \leq & \epsilon\gamma_{ij}, \ \forall i,j \\ \sum_{i=1}^{I} \sum_{j=1}^{J} p_{ij} & = & 1 \\ p_{ij} & \geq & 0, \ \forall i,j \end{array} \tag{12.29}$$

Revised method 2. Seek two probability vectors, $\boldsymbol{\tau}$ and $\boldsymbol{\eta}$, to minimize ϵ subject to

$$\begin{array}{rcl} |\eta_j a_{ij} - \tau_i b_{ij}| & \leq & \epsilon\gamma_{ij}, \ \forall i,j \\ \sum_{i=1}^{I} \tau_i & = & 1 \\ \sum_{j=1}^{J} \eta_j & = & 1 \\ \tau_i \geq 0, \eta_j & \geq & 0, \forall i,j \end{array} \tag{12.30}$$

Revised method 3. Seek one probability vector $\boldsymbol{\tau}$ to minimize ϵ subject to

$$\begin{array}{rcl} |a_{ij} \sum_{i=1}^{I} \tau_i b_{ij} - \tau_i b_{ij}| & \leq & \epsilon\gamma_{ij}, \ \forall i,j \\ \sum_{i=1}^{I} \tau_i & = & 1 \\ \tau_i & \geq & 0, \ \ \forall i \end{array} \tag{12.31}$$

where γ_{ij} are numbers that measure the relative importance of the errors we are willing to accept for the corresponding values p_{ij} of the joint probabilities, and ϵ measures the global degree of incompatibility. Note that if $\epsilon = 0$, we have compatibility; otherwise we do not.

Our aim is to obtain the matrices $\mathbf{P}$, $\boldsymbol{\tau}$ and $\boldsymbol{\eta}$ that minimize ϵ, namely, the degree of incompatibility.

The associated problem is a LPP because each of the constraints involving absolute values in (12.29)–(12.31) can be replaced by two linear inequality constraints (see Chapter 13). So, all constraints can be written as linear constraints.

Example 12.5 (A compatible example). Consider the following matrices **A** and **B**:

$$A = \begin{pmatrix} 0.1429 & 0.3333 & 0.2727 & 0.1667 \\ 0.1429 & 0.0833 & 0.1818 & 0.3333 \\ 0.0000 & 0.2500 & 0.4545 & 0.2500 \\ 0.7143 & 0.3333 & 0.0909 & 0.2500 \end{pmatrix} \tag{12.32}$$

$$B = \begin{pmatrix} 0.1000 & 0.4000 & 0.3000 & 0.2000 \\ 0.1250 & 0.1250 & 0.2500 & 0.5000 \\ 0.0000 & 0.2727 & 0.4545 & 0.2727 \\ 0.3846 & 0.3077 & 0.0769 & 0.2308 \end{pmatrix} \quad (12.33)$$

The GAMS code to build de joint and conditional probability matrices, and solve this problem by the revised methods 1–3 is:

```
$title Compatibility Methods 1, 2 and 3

SETS
I number of rows/1*4/
J number of columns/1*4/;

ALIAS(I,I1);
ALIAS(J,J1);

PARAMETER
PP(I,J) Joint probability matrix used to generate conditionals
A(I,J) conditional probability of X given Y
B(I,J) conditional probability of Y given X;

PP(I,J)=round(uniform(0,5));
PP(I,J)=PP(I,J)/sum((I1,J1),PP(I1,J1));
A(I,J)=PP(I,J)/sum(I1,PP(I1,J));
B(I,J)=PP(I,J)/sum(J1,PP(I,J1));

* The line below modifies the B matrix to get incompatibility

VARIABLE
z value of the objective function
epsilon Maximum error in joint probability;

POSITIVE VARIABLES
P(I,J) Joint probability looked for
TAU(I) Marginal probability of X
ETA(J) Marginal probability of Y;

EQUATIONS
ZDEF1 Function to be optimized in model 1
ZDEF2 Function to be optimized in model 2
ZDEF3 Function to be optimized in model 3
CONST111(I,J)
CONST112(I,J)
CONST121(I,J)
CONST122(I,J)
CONST21(I,J)
CONST22(I,J)
CONST31(I,J)
CONST32(I,J)
NORM1 P add up to one
NORM2 TAU add up to one
NORM3 ETA add up to one;

ZDEF1..z=e=epsilon;
CONST111(I,J)..P(I,J)-A(I,J)*sum(I1,P(I1,J))=l=epsilon;
CONST112(I,J)..-epsilon=l=P(I,J)-A(I,J)*sum(I1,P(I1,J));
```

```
CONST121(I,J)..P(I,J)-B(I,J)*sum(J1,P(I,J1))=l=epsilon;
CONST122(I,J)..-epsilon=l=P(I,J)-B(I,J)*sum(J1,P(I,J1));
CONST21(I,J)..ETA(J)*A(I,J)-TAU(I)*B(I,J)=l=epsilon;
CONST22(I,J)..-epsilon=l=ETA(J)*A(I,J)-TAU(I)*B(I,J);
CONST31(I,J)..A(I,J)*sum(I1,TAU(I1)*B(I1,J))-TAU(I)*B(I,J)=l=epsilon;
CONST32(I,J)..-epsilon=l=A(I,J)*sum(I1,TAU(I1)*B(I1,J))-TAU(I)*B(I,J);
NORM1..sum((I,J),P(I,J))=e=1;
NORM2..sum(I,TAU(I))=e=1;
NORM3..sum(J,ETA(J))=e=1;

MODEL model1/ZDEF1,CONST111,CONST112,CONST121,CONST122,NORM1/;
MODEL model2/ZDEF1,CONST21,CONST22,NORM2,NORM3/;
MODEL model3/ZDEF1,CONST31,CONST32,NORM2/;

file out/comp1.out/;
put out;

put "Initial Matrix PP:"/;
loop(I,loop(J,put PP(I,J):7:4," & ";);put ""/;);
put "Matrix A:"/;
loop(I,loop(J,put A(I,J):7:4," & ";);put ""/;);
put "Matrix B:"/;
loop(I,loop(J,put B(I,J):7:4," & ";);put ""/;);

SOLVE model1 USING lp MINIMIZING z;
put "Final Matrix P Obtained by Model 1:"/;
loop(I,loop(J,put P.l(I,J):7:4," & ";);put ""/;);
put "z=",z.l:12:8," modelstat=",model1.modelstat," solvestat=",model1.solvestat/;
SOLVE model2 USING lp MINIMIZING z;
put "Final Matrix P Obtained by Model 2:"/;
loop(I,loop(J,put (TAU.l(I)*B(I,J)):7:4," & ";);put ""/;);
put "z=",z.l:12:8," modelstat=",model2.modelstat," solvestat=",model2.solvestat/;
SOLVE model3 USING lp MINIMIZING z;
put "Final Matrix P Obtained by Model 3:"/;
loop(I,loop(J,put (TAU.l(I)*B(I,J)):7:4," & ";);put ""/;);
put "z=",z.l:12:8," modelstat=",model3.modelstat," solvestat=",model3.solvestat/;
```

The solution of this problem, which coincides with those for the three methods, is $\epsilon = 0$ and the resulting joint probability matrix

$$\mathbf{P} = \begin{pmatrix} 0.0238 & 0.0952 & 0.0714 & 0.0476 \\ 0.0238 & 0.0238 & 0.0476 & 0.0952 \\ 0.0000 & 0.0714 & 0.1190 & 0.0714 \\ 0.1190 & 0.0952 & 0.0238 & 0.0714 \end{pmatrix} \tag{12.34}$$

■

Example 12.6 (An incompatible example). If we modify the first row of the B matrix in Example 12.5 [in (12.33)], by removing the asterisk in the comment line, we get the output file:

```
Initial Matrix PP:
 0.0238 &  0.0952 &  0.0714 &  0.0476 &
 0.0238 &  0.0238 &  0.0476 &  0.0952 &
```

```
 0.0000 &  0.0714 &  0.1190 &  0.0714 &
 0.1190 &  0.0952 &  0.0238 &  0.0714 &
Matrix A:
 0.1429 &  0.3333 &  0.2727 &  0.1667 &
 0.1429 &  0.0833 &  0.1818 &  0.3333 &
 0.0000 &  0.2500 &  0.4545 &  0.2500 &
 0.7143 &  0.3333 &  0.0909 &  0.2500 &
Matrix B:
 0.2500 &  0.2500 &  0.2500 &  0.2500 &
 0.1250 &  0.1250 &  0.2500 &  0.5000 &
 0.0000 &  0.2727 &  0.4545 &  0.2727 &
 0.3846 &  0.3077 &  0.0769 &  0.2308 &
Final Matrix P Obtained by Model 1:
 0.0403 &  0.0630 &  0.0539 &  0.0494 &
 0.0289 &  0.0165 &  0.0549 &  0.1231 &
 0.0000 &  0.0579 &  0.1177 &  0.0784 &
 0.1329 &  0.0858 &  0.0129 &  0.0843 &
z=  0.01139203 modelstat=        1.00 solvestat=        1.00
Final Matrix P Obtained by Model 2:
 0.0478 &  0.0478 &  0.0478 &  0.0478 &
 0.0335 &  0.0335 &  0.0670 &  0.1340 &
 0.0000 &  0.0678 &  0.1131 &  0.0678 &
 0.1124 &  0.0899 &  0.0225 &  0.0674 &
z=  0.02107728 modelstat=        1.00 solvestat=        1.00
Final Matrix P Obtained by Model 3:
 0.0559 &  0.0559 &  0.0559 &  0.0559 &
 0.0309 &  0.0309 &  0.0618 &  0.1237 &
 0.0000 &  0.0524 &  0.0873 &  0.0524 &
 0.1295 &  0.1036 &  0.0259 &  0.0777 &
z=  0.02502453 modelstat=        1.00 solvestat=        1.00
```

which shows the initial P, A, and B matrices and the P matrices resulting from methods 1–3. Note that the A and B matrices are incompatible, where ϵ values for methods 1, 2, and 3 are 0.01139203, 0.02107728, and 0.02502453, respectively. ■

12.4 Regression Models

The second problem we deal with is the regression problem. Least-squares methods have succeeded because of the nice mathematical behavior of the associated quadratic function, which has derivatives and lead to simple systems of equations. However, other interesting methods, such as those based on absolute values or maxima, in spite of its importance, are rarely used because of the nonexistence of derivatives of the associated functions. Fortunately, linear programming techniques allow dealing with these problems in a very satisfactory way, as it is shown in this section.

Consider the standard linear model

$$\mathbf{y} = \mathbf{Z}\boldsymbol{\beta} + \boldsymbol{\varepsilon} \tag{12.35}$$

or, equivalently

$$y_i = \mathbf{z}_i^T \boldsymbol{\beta} + \varepsilon_i, \ i = 1, \ldots, n \tag{12.36}$$

where $\mathbf{y} = (y_1, \ldots, y_n)^T$ is an $n \times 1$ vector of response variables, $\mathbf{Z}$ is an $n \times p$ matrix of predictor variables, $\mathbf{z}_i^T$ is the ith row in $\mathbf{Z}$, $\boldsymbol{\beta}$ is a $p \times 1$ vector of regression coefficients or parameters, and $\boldsymbol{\varepsilon} = (\varepsilon_1, \ldots, \varepsilon_n)^T$ is an $n \times 1$ vector of random errors.

The most popular methods for estimating the regression parameters $\boldsymbol{\beta}$ are

1. The least squares (LS) method. This method minimizes

$$\sum_{i=1}^{n} (y_i - \mathbf{z}_i^t \boldsymbol{\beta})^2 \tag{12.37}$$

This is the standard regression model, and can be easily solved using standard and well known techniques. This method penalizes large errors with respect to small errors because the errors are squared; that is, large errors are enlarged and small errors are reduced. So, this method must be used when the user is concerned about large errors but does not care about small errors.

2. The least-absolute-value (LAV) method. This method minimizes (see, for example, Arthanari and Dodge [6])

$$\sum_{i=1}^{n} |y_i - \mathbf{z}_i^t \boldsymbol{\beta}| \tag{12.38}$$

This method minimizes the sum of the distances between observed and predicted values instead of their squares. However, because of the presence of the absolute-value function, it is difficult to solve using standard regression techniques. This method treats all errors equally. Thus, this method must be used when the user is concerned about any level of error. In fact, what is important is the sum of all absolute errors, not a single error.

3. The minimax (MM) method. It minimizes

$$\max_{i} |y_i - \mathbf{z}_i^t \boldsymbol{\beta}|, \tag{12.39}$$

This method minimizes the maximum of the distances between observed and predicted values. So the user must be concerned only on the maximum error. However, because of the presence of the maximum function, it has the same difficulties as the previous method.

The estimates of $\boldsymbol{\beta}$ for the last two methods can be obtained by solving some simple linear programming problems. The LAV estimate of $\boldsymbol{\beta}$, $\hat{\boldsymbol{\beta}}$, can be obtained by solving the following linear programming problem (LPP) (see Castillo et al. [22]). Minimize

$$\sum_{i=1}^{n} \varepsilon_i \tag{12.40}$$

subject to

$$\begin{aligned} y_i - \mathbf{z}_i^t \boldsymbol{\beta} &\le \varepsilon_i, & i = 1, \ldots, n \\ \mathbf{z}_i^t \boldsymbol{\beta} - y_i &\le \varepsilon_i, & i = 1, \ldots, n \\ \varepsilon_i &\ge 0, & i = 1, \ldots, n \end{aligned} \tag{12.41}$$

Similarly, the MM estimate of $\boldsymbol{\beta}$, $\tilde{\boldsymbol{\beta}}$, can be obtained by solving the following LPP. Minimize

$$\varepsilon \tag{12.42}$$

subject to

$$\begin{aligned} y_i - \mathbf{z}_i^t \boldsymbol{\beta} &\le \varepsilon, & i = 1, \ldots, n \\ \mathbf{z}_i^t \boldsymbol{\beta} - y_i &\le \varepsilon, & i = 1, \ldots, n \\ \varepsilon &\ge 0 \end{aligned} \tag{12.43}$$

Let $\hat{\boldsymbol{\beta}}_{(i)}$ and $\tilde{\boldsymbol{\beta}}_{(i)}$ be the LAV and MM estimators of $\boldsymbol{\beta}$ when the ith observation is omitted, respectively. The influence of the ith observation on the LAV estimators can be measured by

$$d_i(LAV) = ||\hat{\mathbf{y}} - \hat{\mathbf{y}}_{(i)}|| \tag{12.44}$$

where $\hat{\mathbf{y}} = \mathbf{Z}\hat{\boldsymbol{\beta}}$ is the vector of fitted value, $\hat{\mathbf{y}}_{(i)} = \mathbf{Z}\hat{\boldsymbol{\beta}}_{(i)}$ is the vector of fitted value computed when the i observation is omitted. Thus, $d_i(LAV)$ is the norm of the difference between the vectors of predicted values based on the full and reduced data, respectively.

Similarly, the influence of the ith observation on the MM estimators can be measured by

$$d_i(MM) = ||\tilde{\mathbf{y}} - \tilde{\mathbf{y}}_{(i)}|| \tag{12.45}$$

where $\tilde{\mathbf{y}} = \mathbf{Z}\tilde{\boldsymbol{\beta}}$ and $\tilde{\mathbf{y}}_{(i)} = \mathbf{Z}\tilde{\boldsymbol{\beta}}_{(i)}$ Finally, $d_i(LS)$ can be defined in a similar way.

To compute $d_i(LAV)$ or $d_i(MM)$, one needs to solve $n+1$ linear programming problems, unless one uses some sensitivity tools, that save computational time.

The following example illustrated the different sensitivities (robustness) of the three methods with respect to outliers.

Example 12.7 (Robustness with respect to outliers). Consider the data in Table 12.4, which has been simulated with the model

$$Y_i = 0.2 + 0.1Z_1 - 0.3Z_2 + 0.4Z_3 - 0.2Z_4 + \epsilon_i$$

where $\epsilon_i \sim U(0, 0.05)$ and the data points $7, 15$ and 16 have been modified to be converted in outliers by

$$\begin{aligned} Y_7 &= 0.2 + 0.1Z_1 - 0.3Z_2 + 0.4Z_3 - 0.2Z_4 - 0.05 \times 2 \\ Y_{15} &= 0.2 + 0.1Z_1 - 0.3Z_2 + 0.4Z_3 - 0.2Z_4 + 0.05 \times 2 \\ Y_{16} &= 0.2 + 0.1Z_1 - 0.3Z_2 + 0.4Z_3 - 0.2Z_4 - 0.05 \times 2 \end{aligned}$$

The three models – LAV, MM, and LS, in (12.37), (12.38), and (12.39), respectively – have been used to fit the data points. To this end, the following GAMS code was used:

```
$ title Regression

SETS
I number of data points /1*30/
P Number of parameters/1*5/
J number of regression models/1*3/
SS(I) subset of data points used in analysis;

PARAMETER
C(P) regression coefficients used in the simulation
/1 0.2
2 0.1
3 -0.3
4 0.4
5 -0.2/;

PARAMETER
Z(I,P) observed and predicted variables;

Z(I,P)=uniform(0,1);
Z(I,'1')=C('1')+sum(P$(ord(P)>1),C(P)*Z(I,P))+uniform(0,1)*0.05;
Z('7','1')=C('1')+sum(P$(ord(P)>1),C(P)*Z('7',P))-0.05*2;
Z('15','1')=C('1')+sum(P$(ord(P)>1),C(P)*Z('15',P))+0.05*2;
Z('16','1')=C('1')+sum(P$(ord(P)>1),C(P)*Z('16',P))-0.05*2;

PARAMETER
Y(I) Observed values
Y1(I) Fitted values with all data points
Y2(I) Fitted values with one data point removed
z0 auxiliary value;

Y(I)=Z(I,'1');

POSITIVE VARIABLES
EPSILON1(I) error associated with data I
EPSILON error;

VARIABLES
z1 objective function value
z2 objective function value
z3 objective function value
BETA(P);

EQUATIONS
dz1 objective function 1 value definition
dz2 objective function 2 value definition
dz3 objective function 3 value definition
lower1(I) lower bound of error
upper1(I) upper bound of error
lower2(I) lower bound of error
upper2(I) upper bound of error;

dz1..z1=e=sum(SS,EPSILON1(SS));
```

```
dz2..z2=e=EPSILON;
dz3..z3=e=sum(SS,sqr(Y(SS)-sum(P,Z(SS,P)*BETA(P))));
lower1(SS)..Y(SS)-sum(P,Z(SS,P)*BETA(P))=l=EPSILON1(SS);
upper1(SS)..-Y(SS)+sum(P,Z(SS,P)*BETA(P))=l=EPSILON1(SS);
lower2(SS)..Y(SS)-sum(P,Z(SS,P)*BETA(P))=l=EPSILON;
upper2(SS)..-Y(SS)+sum(P,Z(SS,P)*BETA(P))=l=EPSILON;

MODEL regreslav/dz1,lower1,upper1/;
MODEL regresMM/dz2,lower2,upper2/;
MODEL regresLS/dz3/;

file out/regress1.out/;
put out;

put "Data"/;

loop(I,
put I.tl:3;
loop(P,put " & ",Z(I,P):8:4;);
put "\\"/;
);
Z(I,'1')=1.0;

loop(J,
  SS(I)=yes;
  if(ord(J) eq 1,SOLVE regreslav USING lp MINIMIZING z1;Z0=z1.l);
  if(ord(J) eq 2,SOLVE regresMM USING lp MINIMIZING z2;Z0=z2.l);
  if(ord(J) eq 3,SOLVE regresLS USING nlp MINIMIZING z3;Z0=z3.l);
  put "All data points z0=",z0:12:8 ;
    loop(P,put " & ",BETA.l(P):10:5;);
    put "\\"/;
    Y1(I)=sum(P,Z(I,P)*BETA.l(P));
  loop(I,
    SS(I)=no;
    if(ord(J) eq 1,SOLVE regreslav USING lp MINIMIZING z1;Z0=z1.l);
    if(ord(J) eq 2,SOLVE regresMM USING lp MINIMIZING z2;Z0=z2.l);
    if(ord(J) eq 3,SOLVE regresLS USING nlp MINIMIZING z3;Z0=z3.l);
    put "Removing point ",I.tl:3," z0=",z0:12:8;
    loop(P,put " & ",BETA.l(P):10:5;);
    put "\\"/;
    Y2(I)=sum(P,Z(I,P)*BETA.l(P))-Y1(I);
    SS(I)=yes;
  );
  loop(I,put I.tl:3,Y2(I):10:6/;);
);
```

The resulting values of $d_i(LAV), d_i(MM)$, and $d_i(LS)$ are shown in Table 12.5, where the outliers and the largest three values of each column are boldfaced. Note that the LS method clearly identifies the three outliers, while the MM method has difficulties in identifying the data point 16, and the LAV method is unable to identify any of them. This shows that the LAV method is more robust, against outliers, than the MM method, and thus more robust than the LS method. It is interesting to point out the zero change in the $d_i(MM)$ when removing single data that are not extreme points. This is due to the fact

Table 12.4: Simulated data points

i	Y_i	Z_{1i}	Z_{2i}	Z_{3i}	Z_{4i}
1	0.2180	0.8433	0.5504	0.3011	0.2922
2	-0.0908	0.3498	0.8563	0.0671	0.5002
3	0.2468	0.5787	0.9911	0.7623	0.1307
4	0.3431	0.1595	0.2501	0.6689	0.4354
5	0.1473	0.3514	0.1315	0.1501	0.5891
6	0.3076	0.2308	0.6657	0.7759	0.3037
7	**0.3981**	**0.5024**	**0.1602**	**0.8725**	**0.2651**
8	0.2089	0.5940	0.7227	0.6282	0.4638
9	0.0879	0.1177	0.3142	0.0466	0.3386
10	0.3795	0.6457	0.5607	0.7700	0.2978
11	0.2259	0.7558	0.6274	0.2839	0.0864
12	-0.0147	0.6413	0.5453	0.0315	0.7924
13	0.3731	0.1757	0.5256	0.7502	0.1781
14	0.1575	0.5851	0.6212	0.3894	0.3587
15	**0.5829**	**0.2464**	**0.1305**	**0.9334**	**0.3799**
16	**0.3781**	**0.3000**	**0.1255**	**0.7489**	**0.0692**
17	0.3163	0.0051	0.2696	0.4999	0.1513
18	0.0804	0.3306	0.3169	0.3221	0.9640
19	0.3913	0.3699	0.3729	0.7720	0.3967
20	-0.0961	0.1196	0.7355	0.0554	0.5763
21	0.1868	0.0060	0.4012	0.5199	0.6289
22	0.0703	0.3961	0.2760	0.1524	0.9363
23	0.2184	0.1347	0.3861	0.3746	0.2685
24	0.1059	0.1889	0.2975	0.0746	0.4013
25	0.1961	0.3839	0.3241	0.1921	0.1124
26	0.3889	0.5114	0.0451	0.7831	0.9457
27	0.2762	0.6073	0.3625	0.5941	0.6799
28	0.2523	0.1593	0.6569	0.5239	0.1244
29	0.1535	0.2281	0.6757	0.7768	0.9325
30	0.1480	0.2971	0.1972	0.2463	0.6465

that the regression line is defined by only two data extreme points (outliers or not).

Table 12.6 shows the parameter estimates for the three models, LAV, MM, and LS, estimated using all data points, and removing one of the first 5 points or the outliers. The sensitivities of each parameter with respect to each outlier can be observed. Again we can see that the LAV method is more robust to outliers (smaller changes in the parameter estimates when removing one outlier) than is the MM method.

■

Table 12.5: Values of $d_i(LAV), d_i(MM)$, and $d_i(LS)$, which allow analysis of the influence of outliers on the predictions

i	$d_i(LAV)$	$d_i(MM)$	$d_i(LS)$
1	**-0.017913**	0.000000	-0.008427
2	0.007130	0.000000	0.007199
3	0.003571	0.000000	0.006221
4	0.008645	0.000000	-0.000730
5	0.001230	0.000000	0.000278
6	-0.005584	0.000000	-0.001787
7	0.000400	**0.032893**	**0.021929**
8	0.000000	0.000000	0.001975
9	-0.001667	0.000000	-0.000668
10	-0.009012	0.000000	-0.003484
11	**-0.014035**	-0.017450	-0.010407
12	-0.011423	0.000000	-0.001843
13	-0.005477	0.000000	-0.004351
14	0.000000	0.000000	0.001429
15	-0.007005	**-0.056369**	**-0.019154**
16	0.002400	0.011714	**0.023611**
17	0.009702	0.000000	-0.003079
18	0.001470	0.017649	0.003613
19	-0.004419	0.000000	-0.001829
20	0.004550	0.000000	0.008311
21	-0.000332	0.000000	0.000371
22	-0.005133	0.000000	-0.003675
23	0.001806	0.000000	-0.000840
24	-0.002236	0.000000	-0.001722
25	0.000000	0.000000	0.001087
26	-0.005122	0.000000	-0.006582
27	0.001186	0.000000	-0.001038
28	-0.006072	0.000000	-0.004491
29	**0.014205**	**0.054685**	0.013893
30	0.000673	0.000000	0.000769

12.5 Applications to Discretization of Continuous Optimization Problems

Optimization problems in the infinite-dimensional case, where functions replace vectors in cost functions, become finite-dimensional when one discretizes such situations to compute approximations to optimal functions. Although this is a subject beyond the scope of this book, we believe that it is interesting to point out how the common link between finite and infinite-dimensional optimization

Table 12.6: Parameter estimates for the three methods using all data and removing one of the first 6 data points or the outliers

i	Y_i	β_1	β_2	β_3	β_4
Parameters estimates for the LAV method					
All	0.23134	0.11741	-0.31868	0.41337	-0.22561
1	0.22128	0.09177	-0.30947	0.43343	-0.21652
2	0.22807	0.11690	-0.30427	0.40866	-0.22850
3	0.22944	0.11313	-0.31254	0.41648	-0.22957
4	0.23360	0.09454	-0.31811	0.42908	-0.22703
5	0.23074	0.12023	-0.31984	0.41182	-0.22353
6	0.23098	0.10934	-0.32361	0.40977	-0.21668
7	0.23108	0.11721	-0.31960	0.41392	-0.22401
15	0.22807	0.11690	-0.30427	0.40866	-0.22850
16	0.23291	0.11002	-0.31565	0.41743	-0.23107
Parameters estimates for the MM method					
All	0.21723	0.00519	-0.28273	0.37877	-0.09172
1	0.21723	0.00519	-0.28273	0.37877	-0.09172
2	- 0.21723	0.00519	-0.28273	0.37877	-0.09172
3	0.21723	0.00519	-0.28273	0.37877	-0.09172
4	0.21723	0.00519	-0.28273	0.37877	-0.09172
5	0.21723	0.00519	-0.28273	0.37877	-0.09172
6	0.21723	0.00519	-0.28273	0.37877	-0.09172
7	0.09010	0.13240	-0.27174	0.48689	-0.09162
15	0.14378	0.06891	-0.18892	0.38863	-0.14452
16	0.22157	-0.03975	-0.23504	0.40342	-0.14350
Parameters estimates for the LS method					
All	0.20098	0.08080	-0.26760	0.39807	-0.17825
1	0.20066	0.06789	-0.26633	0.40153	-0.17470
2	0.20061	0.07838	-0.25826	0.39316	-0.17677
3	0.19746	0.08240	-0.26064	0.40084	-0.17975
4	0.20037	0.08183	-0.26696	0.39742	-0.17827
5	0.20131	0.08090	-0.26801	0.39779	-0.17824
6	0.20173	0.08264	-0.26996	0.39625	-0.17818
7	0.20658	0.09872	-0.29242	0.41429	-0.18902
15	0.19585	0.08594	-0.24951	0.37834	-0.17624
16	0.22322	0.08249	-0.29511	0.40603	-0.20198

problems is carried out in a few selected and typical situations. Needless to say, these examples are elementary and academic so that it will be impossible to convey through them the richness of all situations and the difficulties attached to more complex, realistic instances. From this perspective, our aim is to examine such problems to indicate how discretizations can be carried out and, in this way, broaden and enlarge the set of optimization problems that can be, at least computationally, analyzed using the techniques and tools described in this text. In particular, we shall stress the use of a computational tool like GAMS to approximate and find optimal solutions for discretized versions of infinite-dimensional optimization problems.

There are two main categories of infinite-dimensional optimization: variational problems and optimal control problems. We shall describe and make computations for two selected, well-known examples in each category: the braquistochrone and the hanging cable, for the first case; and the optimal controls of a hitting target and of an harmonic oscillator, for the second. In all these examples, the point of view we have taken is related to a basic discretization scheme. More efficient and accurate approximations would require a more specialized treatment. For some more examples, see Polak [86] and Troutman [99].

12.5.1 Variational Problems

Typically, a scalar, one-dimensional variational problem comes in the following form. Minimize

$$T(u) = \int_a^b F(t, u(t), u'(t))dt$$

subject to

$$u(a) = A, u(b) = B$$

where the unknown function u is assumed to be continuous.

Integral constraints of the type

$$k = \int_a^b G(t, u(t), u'(t))dt$$

are also common. How is a discretized version built for the original problem? To this end, we divide the interval $[a, b]$ in $n+1$ equal subintervals, and assume that feasible functions for the new, discretized optimization problem are piecewise affine, that is, affine on each subinterval

$$[a + j\Delta, a + (j + 1)\Delta]$$

where

$$\Delta = \frac{(b - a)}{(n + 1)}$$

Notice that such class of functions are uniquely determined by its values at the nodes

$$a + j\Delta, \quad j = 1, \ldots, n$$

and therefore, feasible vectors for the new optimization problem correspond to these values. We see that this process changes the original, infinite-dimensional problem to a finite-dimensional one. The point is that by letting $n + 1$, the number of subintervals becomes larger and larger, optimal solutions for these discretized optimization problems will resemble and approximate well enough, under conditions to be overlooked here, the true optimal solutions for the initial optimization problem. Let

$$x = (x_j), \quad 1 \le j \le n. \ x_0 = A, \quad x_{n+1} = B$$

be the nodal values of feasible functions. In this way the function u that we consider to optimize $T(u)$ is

$$u(t) = x_j + \frac{(x_{j+1} - x_j)}{\Delta}(t - a - j\Delta), \quad \text{if } t \in [a + j\Delta, a + (j+1)\Delta] \quad (12.46)$$

This is the continuous, piecewise affine function that takes on values x_j at nodal points $a + j\Delta$. Thus, we get

$$T(u) = \sum_{j=0}^{n} \int_{a+j\Delta}^{a+(j+1)\Delta} F\left(t, u(t), \frac{(x_{j+1} - x_j)}{\Delta}\right) dt$$

where $u(t)$ in the interval of integration is given by (12.46). If we realize that u is determined by the vector $x = (x_1, \dots, x_n)$ and interpret $T(u)$ as a function of x, we are faced with a (nonlinear, unconstrained) programming problem. By solving it, we find an approximate solution to the initial, continuous optimization problem. The form of the functional $T(u)$ in terms of x depends on each particular situation. We will see and solve from this perspective the two examples mentioned above.

The Braquistochrone

This was a famous optimization problem in the twentieth century and one of the first examples where optimal solutions were explicitly found. Assume that we are given two points, A and B, in a plane at different heights. We are asked to find the path joining those two points so that a unit mass, under the action of gravity, and without slipping, takes the least time possible in getting from the highest to the lowest point. This is one the most celebrated examples of the so-called minimum transit problems.

The first thing to do is to derive the functional measuring the time spent by the mass, in going from A to B, under the action of gravity. For convenience, we put the X axis along the vertical direction so that gravity acts along this axis, and the Y axis horizontally. Let, without loss of generality, $A = (1, 0)$, $B = (0, 1)$. Assume that $y = u(x)$ is a continuous curve joining A and B, so that $u(0) = 1$ and $u(1) = 0$. If the unit mass is supposed to travel from A to B through the path determined by the graph of u, what would the time spent in going from A to B be? We know that space is velocity times time. In our continuous situation space is measured by the elementary element's length

$$d\ell = \sqrt{1 + u'(x)^2}\, dx$$

while velocity, in terms of g, is given by

$$v = \sqrt{2gx}$$

Therefore, if $T(u)$ stands for the time spent associated with the path $y = u(x)$, we have

$$T(u) = \int_0^1 \frac{\sqrt{1+u'(x)^2}}{\sqrt{2gx}}\,dx$$

or equivalently, keeping in mind that positive multiplicative constants do not affect the solution of optimization problems, we obtain

$$T(u) = \int_0^1 \frac{\sqrt{1+u'(x)^2}}{\sqrt{x}}\,dx$$

This is the functional to be minimized with respect to all those continuous $y = u(x)$ functions satisfying the constraints $u(0) = 1$ and $u(1) = 0$.

To fully understand this situation, one has to study the convexity properties of the integrand for T and the associated Euler–Lagrange equation. Since this would take us too far, we shall restrict our initial optimization problem to a discretized version of it, and stress the use of GAMS to compute optimal solutions for the discretized version.

The resulting discretized objective function $\bar{T}(x)$, in terms of the nodal points, becomes

$$\bar{T}(x) = \sum_{j=0}^{n} \sqrt{1+\left(\frac{x_{j+1}-x_j}{\Delta}\right)^2} \int_{j/(n+1)}^{(j+1)/(n+1)} \frac{dx}{\sqrt{x}}$$

or even more explicitly, neglecting positive constants

$$\bar{T}(x) = \sum_{j=0}^{n} \left(\sqrt{\frac{j+1}{n+1}} - \sqrt{\frac{j}{n+1}}\right) \sqrt{1+\left(\frac{x_{j+1}-x_j}{\Delta}\right)^2}$$

keeping in mind that $x_0 = 1$ and $x_{n+1} = 0$. This is the function we want to minimize with the help of GAMS, for several values of the number of subintervals n. By doing so, we find a very good agreement with the arc of a cycloide, which is the optimal solution of the continuous optimization problem.

The GAMS code for this problem, together with a summary of the optimal solution, follows.

```
$title FUNCTIONAL1 n=20
SET J /0*20/;
VARIABLES z,x(J);
  x.fx(J)$(ord(J) eq 1) = 0;
  x.fx(J)$(ord(J) eq card(J)) = 1;
SCALAR n;
  n = card(J)-2;
EQUATION
  cost   objective function;
cost.. z =e= SUM(J$(ord(J) lt card(J)),
(sqrt(ord(J))-sqrt(ord(J)-1))*sqrt(1+sqr(n+1)*sqr(x(J+1)-x(J))));
MODEL funct1 /all/;
```

```
SOLVE funct1 USING nlp MINIMIZING z;

                         LOWER      LEVEL      UPPER     MARGINAL
---- VAR Z                -INF      5.776      +INF          .
---- VAR X
     LOWER      LEVEL      UPPER     MARGINAL
0      .          .          .        -2.091
1     -INF      0.005      +INF   -1.163E-6
2     -INF      0.018      +INF   1.6563E-6
3     -INF      0.036      +INF   8.1950E-7
4     -INF      0.057      +INF   -1.405E-6
5     -INF      0.082      +INF   -4.992E-7
6     -INF      0.110      +INF   -1.353E-6
7     -INF      0.141      +INF   -3.233E-6
8     -INF      0.176      +INF   3.1623E-7
9     -INF      0.215      +INF   3.0808E-6
10    -INF      0.257      +INF   1.8503E-6
11    -INF      0.303      +INF   -6.250E-7
12    -INF      0.353      +INF   -1.116E-6
13    -INF      0.408      +INF   3.2812E-6
14    -INF      0.468      +INF   -1.583E-6
15    -INF      0.534      +INF         EPS
16    -INF      0.606      +INF   7.5934E-7
17    -INF      0.687      +INF   -8.374E-7
18    -INF      0.777      +INF         EPS
19    -INF      0.880      +INF         EPS
20    1.000     1.000      1.000       2.091
```

If we plot in the same graph these computed values with the ones corresponding to the exact solution, we are unable to distinguish them apart (see Figure 12.5).

Alternatively, we can set up the discretization scheme considering the slopes on each subinterval as independent variables. This leads to a simpler form of the objective function, but we would have to enforce a (linear) constraint because the slopes in the different subintervals must be such that the value of u at 1 is given, and this imposes a constraint on the sets of possible slopes. Rather than solving the same example in this format with an integral constraint, we prefer to get into the next example.

The Hanging Cable

The hanging cable is also one of the classical problems Euler solved in the twentieth century by variational techniques. It consists of determining the shape adopted by a cable, hanging freely on its two endpoints at the same height, under the action of its own weight (see Figure 12.6).

We postulate that the optimal shape will be that corresponding to the least potential energy, and assume that the cable has constant cross sections along its length. Let us say that the two endpoints of the cable are a distance H apart, and in the same horizontal plane, and that the length of the cable is L. Obviously, we must require $L > H$ so that the problem is well posed. Assume that we place the X axis along the two endpoints of the cable, and the Y axis vertically starting at the left endpoint of the cable. In this way $(0, 0)$ and $(H, 0)$

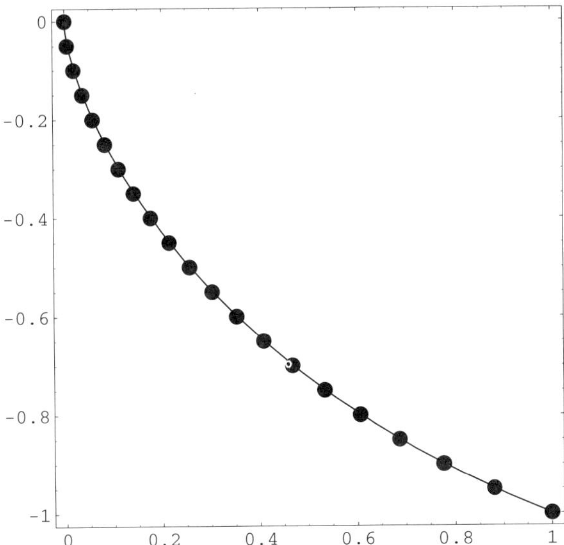

Figure 12.5: The exact (continuous) and the approximated (dotted) solutions to the braquistochrone problem.

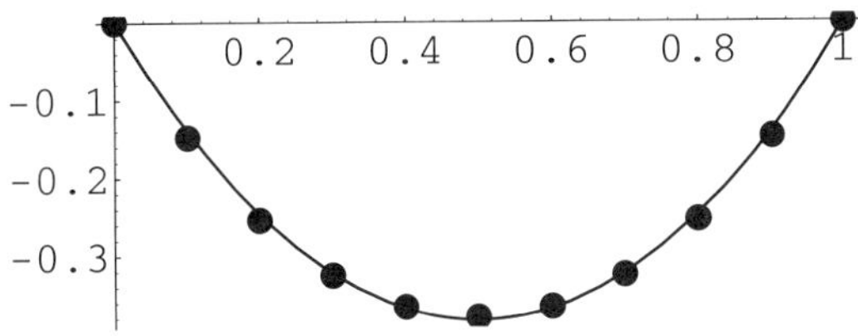

Figure 12.6: The exact (continuous) and the approximated (dotted) solutions to the hanging cable.

are the coordinates of the two, fixed endpoints. Let $y = u(x)$, $x \in (0, H)$ be any continuous function joining the two endpoints. If the cable would adopt the profile described by the graph of u, its potential energy is measured by the integral

$$k \int_0^H u(x)\sqrt{1 + u'(x)^2}\, dx$$

where $k > 0$ is a constant related to the material the cable is made of, $u(x)$ is the height of each material cross-section, and

$$\sqrt{1 + u'(x)^2}\, dx$$

is the element's length. Therefore, and discarding again positive constants, we seek the optimal profile $u(x)$ so that it minimizes the potential energy functional

$$P(u) = \int_0^H u(x)\sqrt{1 + u'(x)^2}\, dx$$

subject to $u(0) = 0$, $u(H) = 0$. This time, however, we need an extra constraint since we must enforce the cable to have total length L. Otherwise the optimization problem would not make much sense because we could lower the potential energy all the way to $-\infty$ by letting the length become larger and larger. Hence, we must enforce the constraint

$$\int_0^H \sqrt{1 + u'(x)^2}\, dx = L$$

Then, our optimization problem becomes minimization of

$$P(u) = \int_0^H u(x)\sqrt{1 + u'(x)^2}\, dx$$

subject to

$$\int_0^H \sqrt{1 + u'(x)^2}\, dx = L \quad u(0) = 0, \quad u(H) = 0$$

Again the complete analysis of this problem would compel us to use the Lagrange multipliers associated with this integral constraint, and study the Euler–Lagrange equation for the augmented Lagrangian. Alternatively, we set up a discretized version of this problem, and solve it using GAMS, as before.

The process is exactly the same as above. We restrict our optimization problem to the finite-dimensional subspace of piecewise affine functions determined by the nodal values x_j, $j = 1, 2, \ldots, n$ over a uniform partition of the interval $(0, H)$. Without loss of generality, we may take $H = 1$. As we have done before, we have

$$P(x) = \sum_{j=0}^{n} \int_{j/(n+1)}^{(j+1)/(n+1)} u(x)\sqrt{1 + \left(\frac{x_{j+1} - x_j}{\Delta}\right)^2}\, dx$$

where again $\Delta = 1/(n+1)$. Since the square roots appearing under the integral sign are constants, we can also write

$$P(x) = \sum_{j=0}^{n} \sqrt{1 + \left(\frac{x_{j+1} - x_j}{\Delta}\right)^2} \int_{j/(n+1)}^{(j+1)/(n+1)} u(x)\, dx$$

In addition we can use the trapezoidal rule to obtain a good approximation of the integrals

$$\int_{j/(n+1)}^{(j+1)/(n+1)} u(x)\, dx \simeq \frac{x_{j+1} + x_j}{2(n+1)}$$

Altogether we obtain the objective functional

$$P(x) = \sum_{j=0}^{n} \sqrt{1 + \left(\frac{x_{j+1} - x_j}{\Delta}\right)^2} \frac{x_{j+1} + x_j}{2(n+1)}$$

Finally, expressing the initial integral constraint in terms of the vector x, we get

$$L = \frac{1}{n+1} \sum_{j=0}^{n} \sqrt{1 + \left(\frac{x_{j+1} - x_j}{\Delta}\right)^2}$$

We must remember that $x_0 = 0$ and $x_{n+1} = H$. This discretized, finite-dimensional formulation is appropriate for computation using GAMS. With this package we can find good approximations to the classical catenary, which is the optimal solution of the continuous problem.

The GAMS code to approximate the optimal solution, together with part of the solution file, can be found below. L is taken to be 1.3116.

```
$title FUNCTIONAL2 n=10
SET J /0*10/;
VARIABLES z,x(J);
  x.fx(J)$(ord(J) eq 1) = 0;
  x.fx(J)$(ord(J) eq card(J)) = 0;
  x.l(J)=-0.5;
SCALAR n;
  n = card(J)-2;
EQUATION
  cost   objective function
  rest   equality constraint;
cost.. z =e= SUM(J$(ord(J) lt card(J)),
                  sqrt(1+sqr(n+1)*sqr(x(J+1)-x(J)))*(x(J+1)+x(J)));
rest.. (n+1)*1.3116 =e= SUM(J$(ord(J) lt card(J)),sqrt(1+sqr(n+1)*sqr(x(J+1)-x(J))));
MODEL funct2 /all/;
SOLVE funct2 USING nlp MINIMIZING z;
```

```
                          LOWER     LEVEL     UPPER     MARGINAL
---- VAR Z                 -INF     -6.114     +INF        .
---- VAR X
      LOWER     LEVEL     UPPER    MARGINAL
0       .         .         .       13.116
1     -INF     -0.148     +INF       EPS
2     -INF     -0.254     +INF   6.1259E-7
3     -INF     -0.325     +INF   8.3847E-7
4     -INF     -0.366     +INF   6.0564E-7
5     -INF     -0.379     +INF       .
6     -INF     -0.366     +INF   1.2127E-6
7     -INF     -0.325     +INF   1.0593E-6
8     -INF     -0.254     +INF   -5.058E-7
9     -INF     -0.148     +INF   8.8932E-7
10      .         .         .       13.116
```

Figure 12.6 shows a graphic of the exact and approximated solutions.

12.5.2 Optimal Control Problems

Optimal control problems are more complex continuous optimization problems than its variational counterpart. Indeed, variational problems are a very special class of optimal control problems. The main ingredients of an optimal control problem are

1. A vector $x(t)$ determining the dynamics of the state of a certain system under control; the number of components in x indicates the number of parameters to be uniquely determined to identify the state of the system under consideration.

2. A vector $u(t)$ designating the control that we can exercise on the system so as to modify its dynamics with some specific objective in mind; frequently, the control is restricted by requiring $u(t) \in K$ for a given set K.

3. The state equation

$$x'(t) = f(t, x(t), u(t)), \quad t \in (0, T)$$

   which governs the dynamics of the system, and expresses the interaction between states and controls.

4. Additional constraints on the initial and/or final state of the system.

5. The objective functional

$$I(u) = \int_0^T g(t, x(t), u(t))\, dt$$

   yielding a measure of optimality when the control u is exercised on the system, and the resulting dynamics, $x(t)$, is obtained by solving the state equation together with initial and/or final states, as indicated by $x(t)$.

The aim of the optimal control problem is to find the best way of acting on the system, the optimal control $u(t)$, measured in terms of the proposed cost functional $I(u)$. The relationship between the state x and the control u, through the state equation and additional constraints, is what makes optimal control problems much more complex than variational problems. Notice that variational problems correspond to the simplest, nontrivial state equation

$$x'(t) = u(t),$$

where both the initial and final states are prescribed a priori.

The computational procedure is, likewise, much more involved because it requires one to solve (often numerically by an appropriate solver) the state equation. We do not pretend to enter a full discussion of this issue. Our goal here is to provide two easy examples of optimal control problems to stress the usefulness of a tool such as GAMS, possibly in conjunction with some additional package to take care of the state equation, in approximating optimal controls.

Optimal Control of Hitting a Target

Assume that we would like to hit a target that is a distance $3 + \frac{5}{6}$ length units apart from us in 3 units of time. Then, the control we can exercise on the projectile is the magnitude of acceleration $u(t)$, so that the state of the projectile $(x(t), x'(t))$ must obey the state equation and auxiliary conditions

$$x''(t) = u(t), \quad x(0) = x'(0) = 0, \quad x(3) = 3 + \frac{5}{6}$$

The initial conditions reflect the fact that the projectile departs from rest. We must also respect a constraint in the size of $u(t)$ by imposing $0 \leq u \leq 1$. Finally, the objective is to hit the target in the cheapest possible way measured by the cost functional

$$I(u) = k \int_0^3 u(t)^2 \, dt$$

where $k > 0$ is a constant.

This is a typical optimal control problem. The peculiar form of the data allows us to obtain the optimal solution analytically, by exploiting optimality conditions through Pontryagin's maximum principle (see, for example, Polak [86] or Troutman [99]). Indeed the (unique) optimal control for this problem is

$$u(t) = \begin{cases} 1, & t \leq 1 \\ \dfrac{3-t}{2}, & t \geq 1 \end{cases}$$

A justification of this is far beyond the scope of this book. However, we can compare the exact solution with an approximated one, computed using GAMS.

To set up a discretized version of the optimal control is, as announced above, more involved because it requires to solve the state equation. In our particular example this can be done explicitly so that we will end up with a precise optimization problem suitable for GAMS. The underlying idea behind the discretization is, however, the same: we divide the time interval $(0, 3)$ in n equal subintervals, and we set the control u to be u_j, constant, in the associated subinterval $(3(j-1)/n, 3j/n)$. By doing so, and after solving the state equation for this class of piecewise constant controls, we hope to express the discretized, optimal control problem as a (nonlinearly constrained) mathematical programming problem in the u_j variables.

In the subinterval $(3(j-1)/n, 3j/n)$ the control u takes on the constant value u_j; therefore assuming, recursively, that we have solved the state equation in the previous subinterval $(3(j-2)/n, 3(j-1)/n)$ and letting a_{j-1} and b_{j-1} be the values of the velocity and position, respectively, for $t = 3(j-1)/n$, we need to find the solution of

$$x''(t) = u_j, \quad x'\left(\frac{3(j-1)}{n}\right) = a_{j-1}, \; x\left(\frac{3(j-1)}{n}\right) = b_{j-1}$$

to get

$$x(t) = \frac{u_j}{2}\left(t - \frac{3(j-1)}{n}\right)^2 + a_{j-1}\left(t - \frac{3(j-1)}{n}\right) + b_{j-1}$$

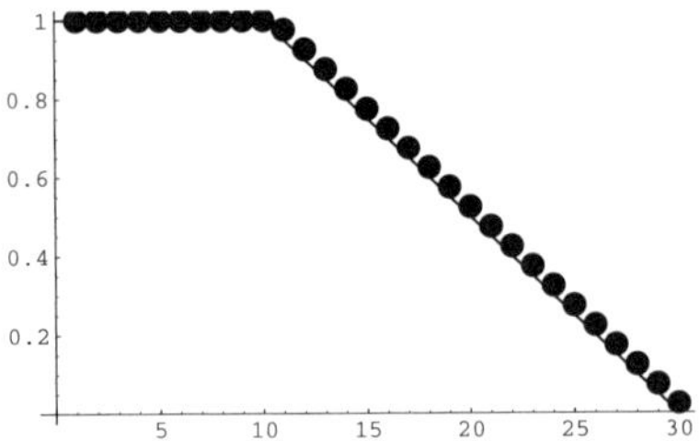

Figure 12.7: The exact (continuous) and the approximated (dotted) solution to the optimal control of a hitting target problem.

The values of the velocity and position at $t = 3j/n$ are

$$a_j = a_{j-1} + \frac{3u_j}{2n}, \quad b_j = b_{j-1} + a_{j-1}\frac{3}{n} + \frac{9u_j}{2n^2}$$

It should be noted that the initial conditions imply $a_0 = b_0 = 0$. By using this recursion formulas repeatedly, it is not hard to find

$$a_j = \frac{3}{n}\sum_{k=1}^{j} u_k, \quad b_j = \frac{9}{2n^2}\sum_{k=1}^{j}(2j - 2k + 1)u_k$$

The final target condition $x(3) = 3 + \frac{5}{6}$ will translate into the linear constraint

$$\frac{9}{2n^2}\sum_{k=1}^{n}(2n - 2k + 1)u_k = 3 + \frac{5}{6}$$

On the other hand, the cost quadratic functional, neglecting positive constants, becomes

$$I(u) = \sum_{k=1}^{n} u_k^2, \quad u = (u_k); \;\; k = 1, 2, \ldots, n$$

Thus, we must solve the following (quadratic) mathematical programming problem. Minimize

$$Z = \sum_{k=1}^{n} u_k^2$$

subject to

$$\frac{9}{2n^2}\sum_{k=1}^{n}(2n - 2k + 1)u_k = 3 + \frac{5}{6}, \quad 0 \le u_k \le 1$$

In this format, GAMS can find an approximation to the original optimal control problem, which is in close agreement with the exact solution given previously. Figure 12.7 shows the exact and the numerical solutions of this problem. The GAMS code and computed solutions follows.

```
$title FUNCTIONAL3 n=30
SET K /1*30/;
VARIABLES z,u(K);
  u.up(K) = 1;
  u.lo(K) = 0;
SCALAR n;
  n = card(K);
EQUATION
  cost   objective function
  rest   equality constraint;
cost.. z =e= SUM(K,sqr(u(K)));
rest.. 3+(5/6) =e= (9/(2*sqr(n)))*SUM(K,(2*n-2*ord(K)+1)*u(K));
MODEL funct3 /all/;
SOLVE funct3 USING nlp MINIMIZING z;
```

```
                         LOWER      LEVEL      UPPER    MARGINAL
---- VAR Z                -INF      16.671      +INF        .
---- VAR U
      LOWER      LEVEL      UPPER    MARGINAL
1       .        1.000      1.000     -0.952
2       .        1.000      1.000     -0.852
3       .        1.000      1.000     -0.752
4       .        1.000      1.000     -0.652
5       .        1.000      1.000     -0.552
6       .        1.000      1.000     -0.452
7       .        1.000      1.000     -0.351
8       .        1.000      1.000     -0.251
9       .        1.000      1.000     -0.151
10      .        1.000      1.000     -0.051
11      .        0.976      1.000        EPS
12      .        0.926      1.000 3.4120E-6
13      .        0.876      1.000 2.7012E-6
14      .        0.826      1.000 1.9903E-6
15      .        0.775      1.000 1.2794E-6
16      .        0.725      1.000 -2.347E-7
17      .        0.675      1.000 -2.185E-7
18      .        0.625      1.000         .
19      .        0.575      1.000 -1.564E-6
20      .        0.525      1.000        EPS
21      .        0.475      1.000        EPS
22      .        0.425      1.000        EPS
23      .        0.375      1.000        EPS
24      .        0.325      1.000        EPS
25      .        0.275      1.000 -1.026E-5
26      .        0.225      1.000 -8.395E-6
27      .        0.175      1.000 -6.530E-6
28      .        0.125      1.000 -4.664E-6
29      .        0.075      1.000 -2.798E-6
30      .        0.025      1.000        EPS
```

Optimal Control of an Harmonic Oscillator

There is a special class of important optimal control problems where the objective is to perform a known task in minimum time. In these situations the governing state equation

$$x'(t) = f(t, x(t), u(t)), \quad t \in (0, T), u(t) \in K$$

is completed with both initial and final states x_I and x_F, respectively. The task is to find the control u that accomplishes the final, given task

$$x(T) = x_F$$

for the smallest possible value of T.

To fix our ideas, we treat below the example of a linear harmonic oscillator with state equation

$$x''(t) + x(t) = u(t), \quad t \in (0, T), |u| \leq 1$$

The goal is to lead the oscillator from given, initial conditions

$$x(0) = a_0, \;\; x'(0) = b_0$$

to rest

$$x(T) = x'(T) = 0$$

in minimum time. Again this example can be treated analytically by exploiting the optimality conditions coming from the Pontryaguin's maximum principle. In particular, the linear (in fact, constant) dependence of the cost functional on the control implies that the optimal control will take exclusively the extremal values $+1$ and -1. But let us pretend not to have that information at our disposal, and proceed to formulate a discretized version of the optimization problem suitable for GAMS.

Since this time T is not known, we must incorporate it as one of our independent variables for the resulting mathematical programming problem, so that $u = (u_j)_{j=0,1,\ldots,n}$ is the unknown, where u_0 stands for T and the associated, piecewise constant control takes the constant value u_j on the interval $(u_0(j-1)/n, u_0 j/n)$. As before, if we let

$$a_j = x\left(\frac{u_0 j}{n}\right), \quad b_j = x'\left(\frac{u_0 j}{n}\right)$$

then we can find x by recursively solving

$$x''(t) + x(t) = u_j, \quad x\left(\frac{u_0(j-1)}{n}\right) = a_{j-1}, x'\left(\frac{u_0(j-1)}{n}\right) = b_{j-1}$$

For this example, it is much more complicated to find exact formulas for a_j and b_j. Even though we could use some symbolic computation package to find such expressions, there is no point in doing so if, after all, what we are about to compute is an approximation to the exact optimal control. Therefore it suffices to use, for instance, an Euler integrator of step size precisely u_0/n to find reasonable approximations for a_j and b_j. If we do not distinguish between the exact and approximated values for a_j and b_j, it is easy to find, in matrix notation

$$\begin{pmatrix} a_j \\ b_j \end{pmatrix} = \begin{pmatrix} a_{j-1} \\ b_{j-1} \end{pmatrix} + \frac{u_0}{n}\left[\begin{pmatrix} 0 & 1 \\ -1 & 0 \end{pmatrix}\begin{pmatrix} a_{j-1} \\ b_{j-1} \end{pmatrix} + \begin{pmatrix} 0 \\ u_j \end{pmatrix}\right]$$

for $j = 1, 2, \ldots, n$. By using this identity recursively, we obtain

$$\begin{pmatrix} a_j \\ b_j \end{pmatrix} = \begin{pmatrix} 1 & \frac{u_0}{n} \\ -\frac{u_0}{n} & 1 \end{pmatrix}^j \begin{pmatrix} a_0 \\ b_0 \end{pmatrix} + \frac{u_0}{n}\sum_{k=0}^{j-1}\begin{pmatrix} 1 & \frac{u_0}{n} \\ -\frac{u_0}{n} & 1 \end{pmatrix}^k \begin{pmatrix} 0 \\ u_{j-k} \end{pmatrix}$$

The constraints come from demanding the desired final rest conditions $a_n = b_n = 0$. Thus, the discrete (nonlinearly constrained) mathematical programming problem whose optimal solution provides a good approximation for our problem involves minimization of

$$Z = u_0$$

subject to

$$\begin{pmatrix} 0 \\ 0 \end{pmatrix} = \begin{pmatrix} 1 & \frac{u_0}{n} \\ -\frac{u_0}{n} & 1 \end{pmatrix}^n \begin{pmatrix} a_0 \\ b_0 \end{pmatrix} + \frac{u_0}{n}\sum_{j=1}^{n}\begin{pmatrix} 1 & \frac{u_0}{n} \\ -\frac{u_0}{n} & 1 \end{pmatrix}^{n-j} \begin{pmatrix} 0 \\ u_j \end{pmatrix}$$

and

$$u_0 \geq 0, \quad -1 \leq u_j \leq 1, j = 1, 2, \ldots, n$$

For specific values of the initial conditions a_0 and b_0, and a value for n, GAMS can find good approximations of the optimal control of a linear harmonic oscillator.

With the objective of simplifying the GAMS formulation of this situation, we have used the change of variables

$$\frac{u_0}{n} = \tan\alpha, \quad 0 \leq \alpha \leq \frac{\pi}{2}$$

so that

$$\begin{pmatrix} 1 & \frac{u_0}{n} \\ -\frac{u_0}{n} & 1 \end{pmatrix}^k = (1 + \tan^2\alpha)^k \begin{pmatrix} \cos(k\alpha) & \sin(k\alpha) \\ -\sin(k\alpha) & \cos(k\alpha) \end{pmatrix}$$

We have chosen several possibilities for initial conditions (a_0, b_0), namely, $(-1, 0)$, $(3, 0)$, $(-7, 0)$. What we see in the computations below, except for some inaccuracies, is the typical bang–bang control jumping from -1 to 1, as predicted by Pontryaguin maximum principle.

A GAMS input file as well as a part of the corresponding output file for the case $(a_0, b_0) = (3, 0)$ is written below:

```
$title FUNCTIONAL n=20, azero
SET J /1*20/;
SCALAR pi the pi number /3.1416/;
VARIABLES z,theta,u(J);
 u.lo(J)=-1;
```

```
 u.up(J)=1;
 theta.lo=0;
 theta.up=pi/2;
 theta.l=pi/4;
 u.l(J)=0.;
SCALARS n, azero;
  n = card(J);
  alpha=-7;
EQUATION
  cost   objective function
  const1
  const2;
cost.. z =e= SIN(theta)/COS(theta);;
const1..
alpha*COS(n*theta)*POWER((1+POWER((SIN(theta)/COS(theta)),2)),n)+(SIN(theta)
/COS(theta))*
         SUM(J,u(J)*SIN((n-ORD(J))*theta)*
           POWER((1+POWER((SIN(theta)/COS(theta)),2)),n-ORD(J))
         ) =E= 0;
const2..
-azero*SIN(n*theta)*POWER((1+POWER((SIN(theta)/COS(theta)),2)),n)+(SIN(theta
)/COS(theta))*
         SUM(J,u(J)*COS((n-ORD(J))*theta)*
         POWER((1+POWER((SIN(theta)/COS(theta)),2)),n-ORD(J))
         ) =E= 0;
MODEL funct5 /all/;
SOLVE funct5 USING nlp MINIMIZING z;

  azero=3;
                        LOWER     LEVEL     UPPER    MARGINAL
---- VAR Z               -INF      0.982     +INF        .
---- VAR THETA             .       0.776     1.571       .
---- VAR U
      LOWER     LEVEL     UPPER    MARGINAL
1     -1.000    -1.000     1.000 1.2962E+5
2     -1.000     1.000     1.000 -2.455E+5
3     -1.000     1.000     1.000 -2.118E+5
4     -1.000     1.000     1.000 -9.025E+4
5     -1.000     1.000     1.000 -1.067E+4
6     -1.000    -1.000     1.000 15633.593
7     -1.000    -1.000     1.000 14116.404
8     -1.000    -1.000     1.000  6201.667
9     -1.000    -1.000     1.000   846.919
10    -1.000     1.000     1.000  -991.476
11    -1.000     1.000     1.000  -939.412
12    -1.000     1.000     1.000  -425.373
13    -1.000     1.000     1.000   -65.560
14    -1.000    -1.000     1.000    62.581
15    -1.000    -1.000     1.000    62.430
16    -1.000    -1.000     1.000    29.126
17    -1.000    -1.000     1.000     4.979
18    -1.000     1.000     1.000    -3.929
19    -1.000     1.000     1.000    -4.143
20    -1.000     1.000     1.000    -1.991
```

12.6 Transportation Systems

In this section we deal with some network equilibrium models that are applied to transportation problems. This section has two main objectives: (1) to show some models that are routinely used in transportation planning, and (2) to illustrate how GAMS is an appropriate tool to implement these models.

We avoid the discussion of specific algorithms to solve these kinds of problems efficiently. The interested reader may consult the comprehensive reviews on this topic given in Patriksson [84].

Readers interested in mathematical models used in transportation planning are directed to the books of Ortúzar and Willumsen [82], and Sheffi [98]. A good introduction is provided in the book of Potts and Oliver [88].

12.6.1 Introduction

Traffic planning and transportation problems have motivated a large amount of mathematical models. The use of traffic planning models aids planners in predicting what effects on network performance are produced by changes in the network topology, or in their parameters. The classical transportation model the following stages:

Base Inventory. In this stage the study area is defined, and an inventory of the main transportation networks and travel patterns is obtained.

Model Analysis. The second phase of the process is the selection and calibration of the model. It has four steps:

1. *Trip generation step.* This step starts by considering a zoning and network system, and a database of each zone. These data, which include information about economic activity, social distribution, educational and recreational facilities, and shopping space are used to estimate the total number of trips generated and attracted by each zone of the study area.

2. *Distribution step.* This stage is the allocation of these trips to particular destinations, such that their distribution over space, and building an origin–destination (O–D) trip matrix.

3. *Modal split step.* The modal split step produces the allocation of trips to different transportation modes. In this phase the origin–destination matrices are obtained for every transportation mode (public, private, etc.). Their elements are the total number of trips associated with a transportation mode for each origin–destination pair Ω.

4. *Assignment step.* Finally, the last stage requires the *assignment* of these trips to the transportation network. This section deals with some assignment equilibrium models to the road network. These models predict the utilization level of the different arcs in the network. Thus, it can be used

to answer questions such as what would happen in the network service level if a new road were built or the capacity of an existing road were modified.

Travel Forecasts. In this stage, the service level and demand of the transportation network is forecast using the corresponding mathematical models for different scenarios.

Network Evaluation. In the final phase of the process, alternative future transportation systems are evaluated and the optimal one is selected.

In this section we present four assignment models for private vehicles. These models take into account the congestion effect, and use Wardrop's principle [102] as a general framework to formulate them. Wardrop's first principle states that under congested conditions drivers choose routes until no one can reduce its costs by switching to other path.

12.6.2 Elements of a Road Transportation Network

In this subsection we give a brief outline of the main elements of the road transportation network theory. The mathematical model used to represent a road transportation network is called a *directed graph*. It is defined as a pair $\mathcal{G} = (\mathcal{N}, \mathcal{A})$, where $\mathcal{N}$ is a finite set of nodes, and $\mathcal{A}$ is a set of ordered pairs (*arcs* or *links*) of elements of $\mathcal{N}$. The nodes are denoted by i or j, and the arcs or links, or more precisely the *directed links*, are denoted as (i, j). The link directions are important because they allow distinction between one-way routes and two-way routes.

Example 12.8 (Nguyen–Dupuis Network). Consider the graph $\mathcal{G}$, in Table 12.7, where $\mathcal{N} = \{1, \dots, 13\}$ and $\mathcal{A}$ has 19 links. This example is taken from Nguyen and Dupuis [80] (ND network) and is illustrated in Figure 12.8.

Table 12.7: Links of the graph $\mathcal{G}$

Links			
(1, 5)	(1, 12)	(4, 5)	(4, 9)
(5, 6)	(5, 9)	(6, 7)	(6, 10)
(7, 8)	(7, 11)	(8, 2)	(9, 10)
(9, 13)	(10, 11)	(11, 2)	(11, 3)
(12, 6)	(12, 8)	(13, 3)	

■

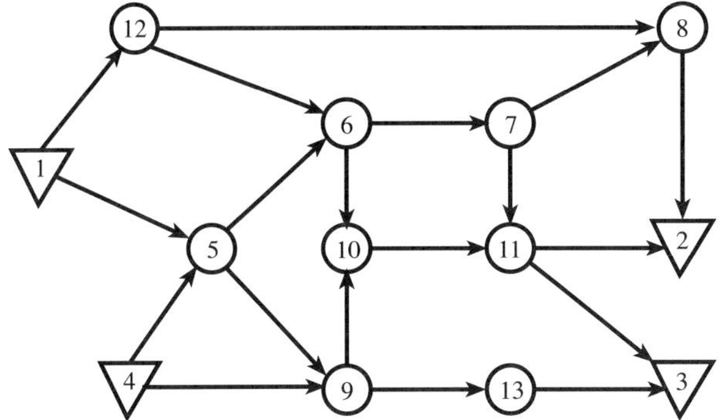

Figure 12.8: Transportation network represented by a graph. Circles refer to intermediate nodes, and triangles are the centroids (origins and destinations).

Single-Commodity Problem

In many problems, flows of vehicles, goods, or passengers can be associated with links of a graph. In this case we denominate the graph a *network* or, more specifically, a *transportation network*, if the application to transportation is to be emphasized. The term *flow* denotes quantity per unit time, such as vehicles per hour or pedestrians per minute.

Fundamental to a network, such as electrical networks, water pipe networks, or transportation networks, is the *flow conservation law*, which states that the flows are neither created nor destroyed.

The nodes of $\mathcal{N}$ are classified as *centroids* or *intermediate nodes*. The first group represents either zones where vehicle trips are *produced* by residents going on a trip elsewhere, and are called *origins*, or zones where trips are *attracted* to places of work, shopping, and so on, and are called *destinations*.

The conservation flow law states that the sum of all flows leaving a node minus the sum of the flows entering that node equals the flow generated or attracted to that node.

The main elements of this problem are

1. **Data.**

   $\mathcal{N}$: the set of nodes in the network

   $\mathcal{A}$: the set of arcs in the network

   r_i: the flow produced or attracted by node i. Note that if i is an intermediate node then $r_i = 0$. If $r_i > 0$ then i is an origin and if $r_i < 0$, i is a destination

   $A(i)$: the sets of nodes $\{j : j \in \mathcal{N}, (i,j) \in \mathcal{A}\}$ "after" node i

   $B(i)$: the sets of nodes $\{j : j \in \mathcal{N}, (j,i) \in \mathcal{A}\}$ "before" node i

   $c_{ij}(x)$: cost of the transportation associated with link $i - j$ and flow x

2. **Variables.**

   f_{ij}**:** the *link flow* on the directed link (i, j)

3. **Constraints.** The conservation equations are written in the following form:

$$\sum_{j \in A(i)} f_{ij} - \sum_{j \in B(i)} f_{ji} = r_i, \ \forall i \in \mathcal{N}$$

4. **Function to be optimized.** In this problem, we minimize

$$Z = \sum_{(i,j) \in \mathcal{A}} \int_0^{f_{ij}} c_{ij}(x) dx \tag{12.47}$$

Example 12.9 (Single-commodity problem). To illustrate the flow conservation equations, consider the network given in Figure 12.8, where circles refer to intermediate nodes, and triangles are the centroids (origins and destinations). For the intermediate node 5, we have

$$i = 5, \quad A(5) = \{6, 9\}, \quad B(5) = \{1, 4\}$$

$$\sum_{j \in A(5)} f_{5j} - \sum_{j \in B(5)} f_{j5} = f_{5,6} + f_{5,9} - f_{1,5} - f_{4,5} = r_5 = 0$$

■

Multicommodity Flow Problems

The previous discussion was based on the assumption that only one type of flow exists on the network, thus, we are in front of a *single-commodity* network.

Multicommodity flow problems arise when several commodities use the same underlying network. The commodities might be differentiated either by their physical characteristics and/or because they have different origins and/or destinations. For example, in a communication network where video and audio flows have to be considered, or in a road network, where trips are classified according to origins and destinations. In these cases, it is essential to distinguish certain O–D flows from others, to make sure travelers get their correct destinations. This imposes that each *commodity* defined by an O–D demand pair must satisfy its flow conservation equation. We denote by $\Omega = (i, j)$ an specific origin–destination (O–D) pair, and by W the set of all the O–D pairs. In traffic studies the traffic on the network is studied as a superposition of traffic between specific O–D pairs.

To formulate the multicommodity network, we consider that the flow of the commodity Ω from O^{Ω} -D^{Ω} is $g^{\Omega} > 0$. Table 12.8 defines a set of O–D pairs for the ND network of Figure 12.8.

Consequently, for the multicommodity problem the constraints described above become

Table 12.8: O–D pairs for the example network

Pair	Demand g^Ω	Pair	Demand g^Ω
$\Omega_1 = (1,2)$	400	$\Omega_2 = (1,3)$	800
$\Omega_3 = (4,2)$	600	$\Omega_4 = (4,3)$	200

(12.48)

1. The flow conservation equations for all commodities are

$$\sum_{j \in A(i)} f_{ij}^\Omega - \sum_{j \in B(i)} f_{ji}^\Omega = r_i^\Omega, \ \forall i \in \mathcal{N}, \quad \forall \Omega \in W$$

   where

$$r_i^\Omega = \begin{cases} g^\Omega & \text{if } O^\Omega = i \\ -g^\Omega & \text{if } D^\Omega = i \\ 0 & \text{if } i \text{ is an intermediate node} \\ & \text{for the commodity } \Omega. \end{cases}$$

2. The total link flow is given by superposing the link flow of all commodities:

$$\sum_{\Omega \in W} f_{ij}^\Omega = f_{ij}, \ \forall (i,j) \in \mathcal{A}$$

In matrix form the previous relations can be stated as

$$\begin{aligned} \mathbf{E}\mathbf{f}^\Omega &= \mathbf{r}^\Omega \ \forall \Omega \in W \\ \sum_{\Omega \in W} \mathbf{f}^\Omega &= \mathbf{f} \end{aligned}$$

where $\mathbf{E}$ is the $n \times l$ node–link incidence matrix whose element in the row corresponding to node i, and the column corresponding to the link (j,k) is defined as

$$e_{i(jk)} = \begin{cases} +1 & \text{if } i = j \\ -1 & \text{if } i = k \\ 0 & \text{otherwise} \end{cases}$$

Note that the conservation flow equations are equivalent to consider the total link flow, specifically, the sum of the link flows associated with all commodities. It is possible to reduce the amount of copies of the network by using a network for all the O–D pairs with the same origin, or alternatively for all O–D pairs with the same destination.

Congested Traffic

We have considered the structure of graphs and road transportation networks. It is characteristic of the road networks that the passage of flow through the network creates delays (effect of congestion) and, as a result of growing traffic volumes, the speed on a link tends to decrease. To take the congestion effect into account, we introduce the notion of the *link cost*, $c_{ij}(f_{ij})$, as the average travel time in traversing the street segment defined by the link (i,j) with flow f_{ij}. These functions are usually modeled as positive, nonlinear, and strictly increasing functions in the analysis of traffic systems. The basic parameters of a link performance function, relating travel time, c_{ij}, on link (i,j), to the flow f_{ij}, on the link, are the *free-flow travel time*, c^0_{ij}, which is a measure of the travel time at zero flow, and the practical capacity of the link, k_{ij}, which is a measure of the flow from which the travel time will increase very rapidly if the flow is further increased. The most common expression for $c_{ij}(f_{ij})$ is called the BPR function

$$c_{ij}(f_{ij}) = c^0_{ij} + b_{ij}(f_{ij}/k_{ij})^{n_{ij}}, \tag{12.49}$$

where b_{ij} and n_{ij} are parameters to be calibrated.

12.6.3 The Traffic Assignment Problem

We must introduce a principle that models user behavior in their route choice in the transportation network. Wardrop [102] was the first to formally enunciate this principle:

> "Under equilibrium conditions, traffic in congested networks arranges itself in such a way that no individual tripmaker can reduce his paths cost by switching routes."

This principle assumes perfect information of all users, and means that they would change the route if this produces a time reduction. A corollary of this principle is that if all tripmakers perceive costs in the same way, under equilibrium conditions, then traffic in congested networks arranges itself such that all used routes between an O–D pair have equal and minimum costs while all unused routes have greater or equal costs than used routes.

This principle has been used as a framework to build equilibrium assignment models. Beckman and McGuire [10] formulated the optimization problem below to express the equilibrium condition derived from Wardrop's first principle [traffic assignment problem] (TAP). Minimize

$$Z = \sum_{(i,j)\in\mathcal{A}} \int_0^{f_{ij}} c_{ij}(x)dx \tag{12.50}$$

subject to

$$\begin{aligned}
\sum_{j\in A(i)} f^{\Omega}_{ij} - \sum_{j\in B(i)} f^{\Omega}_{ji} &= r^{\Omega}_i, \ \forall i \in \mathcal{N}, \quad \forall \Omega \in W \\
\sum_{\Omega\in W} f^{\Omega}_{ij} &= f_{ij}, \ \forall (i,j) \in \mathcal{A} \\
f^{\Omega}_{ij} &\geq 0, \quad \forall (i,j) \in \mathcal{A}, \forall \Omega \in W
\end{aligned}$$

This formulation of the problem is known as the *link-node* formulation. Since the equilibrium conditions are given in terms of route flows and costs, it follows that the optimization problem is based on route flows. Next, an alternative formulation of the equilibrium conditions based on path flows, called the *linkflow formulation*, is given.

The Linkflow Formulation

The main elements of this formulation are

1. **Data.**

   $\mathcal{R}_\Omega$**:** the set of routes for the commodity Ω

   $c_a(x)$**:** the cost associated with flow x through arc a

2. **Variables.**

   h_r**:** the flow in route r

3. **Constraints.** The amount of users of a demand pair Ω is the sum of the total amount of users in different paths satisfying the demand

$$\sum_{r \in \mathcal{R}_\Omega} h_r = g_\Omega, \ \forall \Omega \in W \tag{12.51}$$

   Moreover, the flow must be nonnegative

$$h_r \geq 0, \ \forall r \in \mathcal{R}_\Omega, \ \forall \Omega \in W \tag{12.52}$$

   The relationship between linkflow and routeflow is that the flow on each link $a \in \mathcal{A}$ is the sum of the flow in all paths that use it:

$$\sum_{w \in W} \sum_{r \in \mathcal{R}_\Omega} \delta_{a,r} h_r = f_a \quad \forall a \in \mathcal{A} \tag{12.53}$$

   where

$$\delta_{a,r} = \begin{cases} 1 & \text{if } r \in \mathcal{R}_\Omega \text{ contains arc } a \\ 0 & \text{otherwise} \end{cases}$$

4. **Function to be optimized.** In this problem, we minimize

$$Z = \sum_{a \in \mathcal{A}} \int_0^{f_a} c_a(x) dx$$

Thus, the linkflow formulation of TAP can be stated as follows. Minimize

$$Z = \sum_{a \in \mathcal{A}} \int_0^{f_a} c_a(x) dx$$

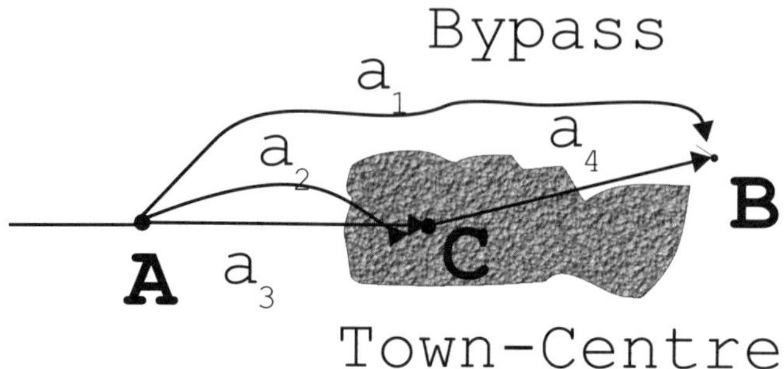

Figure 12.9: Diagram showing the routes.

subject to

$$\begin{aligned}
\sum_{r \in \mathcal{R}_\Omega} h_r &= g_\Omega, \ \forall \Omega \in W && (12.54) \\
\sum_{\Omega \in W} \sum_{r \in \mathcal{R}_\Omega} \delta_{ar} h_r &= f_a \ , \forall a \in \mathcal{A} && (12.55) \\
h_r &\geq 0, \ \forall r \in \mathcal{R}_\Omega, \ \forall \Omega \in W && (12.56)
\end{aligned}$$

Example 12.10 (Linkflow formulation). To illustrate the previous constraints, consider the problem of a town served with a bypass and several town-center routes as illustrated in Figure 12.9. Assume that there are 4000 trips from A to B, and 2500 trips from A to C. The routes available to satisfy the demand pair $\Omega_1 = (A, B)$ are $r_1 = \{a_1\}$, $r_2 = \{a_2, a_4\}$, and $r_3 = \{a_3, a_4\}$, and the routes for the pair $\Omega_2 = (A, C)$ are $r_4 = \{a_2\}$ and $r_5 = \{a_3\}$. In this example $W = \{1, 2\}$, and $\mathcal{R}_{\Omega_1} = \{r_1, r_2, r_3\}$ and $\mathcal{R}_{\Omega_2} = \{r_4, r_5\}$. The path flow variables are $h_1, \ldots, h_5$, and the line flow variables are $f_1, \ldots, f_4$.

For this example the constraints of the problem are:

- Constraints (12.51)

$$\begin{aligned}
h_1 + h_2 + h_3 &= 4000 \\
h_4 + h_5 &= 2500
\end{aligned}$$

- Constraints (12.53)

$$\begin{aligned}
h_1 &= f_1 \\
h_2 + h_4 &= f_2 \\
h_3 + h_5 &= f_3 \\
h_2 + h_3 &= f_4
\end{aligned}$$

- Constraints (12.52):

$$h_1, \dots, h_5 \;\; \geq \;\; 0$$

■

Appropriateness of the Model

We show below that this mathematical model is appropriate for describing the Wardrop's equilibrium conditions. The first observation is

$$\begin{aligned} C_a(f_a) &= \int_0^{f_a} c_a(x)dx \\ C'_a(f_a) &= c_a(f_a) \\ C''_a(f_a) &= c'_a(f_a) \end{aligned}$$

Since $c_a(f_a)$, the cost of the link as a function of the service level in this link, is a nondecreasing function, this implies that $c'_a(f_a) \geq 0$ and that $C_a(f_a)$ is a convex function. Since the objective function is the sum of convex functions, it is also a convex function. Since the constraints are linear, the optimization problem becomes a convex mathematical programming problem. This implies that the KKTCs are necessary and sufficient conditions. We shall show by means of the KKTCs that the optimal solution of the TAP meets the equilibrium conditions.

By expressing the link flows f_a in terms of the path flows in the objective function (12.54) [using Equation (12.55)], the problem can be formulated only in terms of the path flow variables $\{h_r : r \in \mathcal{R}_\Omega,\ \Omega \in W\}$ as follows. Minimize

$$Z = \sum_{a \in \mathcal{A}} \int_0^{\sum\limits_{\Omega \in W} \sum\limits_{r \in \mathcal{R}_\Omega} \delta_{ar} h_r} c_a(x)dx$$

subject to

$$\sum_{r \in \mathcal{R}_\Omega} h_r = g_\Omega, \ \forall \Omega \in W \tag{12.57}$$

$$h_r \geq 0, \ \forall r \in \mathcal{R}_\Omega, \ \forall \Omega \in W \tag{12.58}$$

The Lagrangian function of this problem is

$$\mathcal{L}(\mathbf{h}, \boldsymbol{\lambda}, \boldsymbol{\psi}) = Z + \sum_{\Omega \in W} \lambda_\Omega \left(g_\Omega - \sum_{r \in \mathcal{R}_\Omega} h_r \right) + \sum_{\Omega \in W} \sum_{r \in \mathcal{R}_\Omega} \psi_r(-h_r)$$

and the KKT condition (8.3) are

$$\frac{\partial \mathcal{L}}{\partial h_{r'}} = \sum_{a \in \mathcal{A}} c_a \left(\sum_{\Omega \in W} \sum_{r \in \mathcal{R}_\Omega} \delta_{ar} h_r \right) \delta_{ar'} - \lambda_{\Omega'} - \psi_{r'} = 0 \tag{12.59}$$

where $r' \in \mathcal{R}_{\Omega'}$. Note that $\sum_{\Omega \in W} \sum_{r' \in \mathcal{R}_{\Omega}} \delta_{ar} h_{r'}$ gives the flow on the link a, and $\sum_{a \in \mathcal{A}} c_a(f_a)\delta_{ar'}$ is the sum of the cost associated with all arcs contained in the route r', specifically, it is the length of path r'. We denote this quantity as $C^*_{r'}$. The condition (12.59) becomes

$$C^*_{r'} = \lambda_{\Omega'} + \psi_{r'}$$

The slackness condition leads to $\psi_r h_r = 0$, so $h_r = 0$ or $\psi_r = 0$. If $h_{r'} > 0$ then $\psi_{r'} = 0$ and $C^*_{r'} = \lambda_{\Omega'}$. Otherwise, using the nonnegativity of the multiplier $\psi_{r'}$ we obtain $C^*_{r'} = \lambda_{\Omega'} + \psi_{r'} \geq \lambda_{\Omega'}$. In other words, the preceding condition says that the a set of paths flows that is optimal must be positive on paths with a minimum cost length. The condition also implies that at an optimum, the paths along which the demand g_{Ω} of O–D pair Ω is split must have equal lengths (and length less than or equal to that of all other paths of Ω).

A GAMS implementation

Below, we use the GAMS package to implement the TAP formulation. The test example is that stated in Section 12.6.2 and illustrated in Figure 12.8. This example uses the BPR functions as link costs, and the objective function becomes

$$\begin{aligned} Z &= \sum_{a \in \mathcal{A}} C_a(f_a) = \sum_{a \in \mathcal{A}} \int_0^{f_a} c_a(x)dx = \sum_{a \in \mathcal{A}} \int_0^{f_a} \left[c_a^0 + b_a \left(\frac{x}{k_a} \right)^{n_a} \right] dx \\ &= \sum_{a \in \mathcal{A}} \left[c_a^0 f_a + \frac{b_a}{n_a + 1} \left(\frac{f_a}{k_a} \right)^{n_a + 1} \right] = \sum_{a \in \mathcal{A}} \left(c_a^0 f_a + d_a {f_a}^{m_a} \right) \end{aligned}$$

where

$$d_a = \frac{b_a}{(n_a + 1)k_a^{n_a+1}} \text{ and } m_a = n_a + 1$$

The parameters of the ND network are shown in Table 12.9.

A GAMS input file is provided below:

```
$title TRAFFIC ASSIGNMENT PROBLEM.

** Sets are declared in first place.
** Set N is the set of nodes of the road network.
** Set A(N,N) is the set of links.
** Set W is the set of O--D pairs

SET
N set of nodes /I1*I13/
W pairs         /W1*W4/
A(N,N) set of links
/I1.(I5,I12)
I4.(I5,I9)
I5.(I6,I9)
I6.(I7,I10)
```

Table 12.9: Parameters of the link cost functions of the Nguyen–Dupuis network

Link	c_a^0	d_a	m_a	Link	c_a^0	d_a	m_a
(1,5)	7.0	0.00625	2.0	(1,12)	9.0	0.00500	2.0
(4,5)	9.0	0.00500	2.0	(4,9)	12.0	0.00250	2.0
(5,6)	3.0	0.00375	2.0	(5,9)	9.0	0.00375	2.0
(6,7)	5.0	0.00625	2.0	(6,10)	13.0	0.00250	2.0
(7,8)	5.0	0.00625	2.0	(7,11)	9.0	0.00625	2.0
(8,2)	9.0	0.00625	2.0	(9, 10)	10.0	0.00250	2.0
(9,13)	9.0	0.00250	2.0	(10,11)	6.0	0.00125	2.0
(11,2)	9.0	0.00250	2.0	(11,3)	8.0	0.00500	2.0
(12,6)	7.0	0.00125	2.0	(12,8)	14.0	0.00500	2.0
(13,3)	11.0	0.00500	2.0				

```
I7.(I8,I11)
I8.(I2)
I9.(I10,I13)
I10.(I11)
I11.(I2,I3)
I12.(I6,I8)
I13.(I3)/

ODE(N,W) set of origins by demand
/I1.(W1,W2)
I4.(W3,W4)/

DDE(N,W) set of destinations by demand
/I2.(W1,W3)
I3.(W2,W4)/;

** The set of nodes I must be  duplicated to refer to its different element
*  in the same constraint.

ALIAS(N,I)
ALIAS(N,J)

PARAMETER
G(W) demand in the O--D pair  W
/W1 400
W2 800
W3 600
W4 200/;

TABLE  CDATA(N,N,*)  link cost parameters
          CO       D        M
 I1.I5     7.0  0.00625   2.0
 I1.I12    9.0  0.00500   2.0
 I4.I5     9.0  0.00500   2.0
 I4.I9    12.0  0.00250   2.0
 I5.I6     3.0  0.00375   2.0
```

```
I5.I9      9.0  0.00375   2.0
I6.I7      5.0  0.00625   2.0
I6.I10    13.0  0.00250   2.0
I7.I8      5.0  0.00625   2.0
I7.I11     9.0  0.00625   2.0
I8.I2      9.0  0.00625   2.0
I9.I10    10.0  0.00250   2.0
I9.I13     9.0  0.00250   2.0
I10.I11    6.0  0.00125   2.0
I11.I2     9.0  0.00250   2.0
I11.I3     8.0  0.00500   2.0
I12.I6     7.0  0.00125   2.0
I12.I8    14.0  0.00500   2.0
I13.I3    11.0  0.00500   2.0;

** Optimization variables are declared.

VARIABLES
z          objective function variable
f(I,J)     the flow at the link I-J
fc(W,I,J)  the flow of the commodity W at the link I-J
cos(I,J)   cost in the link I-J at equilibrium;

POSITIVE VARIABLE fc(W,N,N);

** Constraints are declared.

EQUATIONS
COSTZ Objective function
BALANCE(W,I) conservation of flow condition for the commodity W.
FLOW(I,J) total link flow
COST(I,J) cost of the link I-J;

** The objective function is the sum of the integrated cost
** in all links.
** The equation FLOW illustrate the use of dollar operation
** to restrict the number of constraints generated to less
** than implied by the domain of the defining sets.

COSTZ.. z=E= SUM((I,J)$A(I,J),CDATA(I,J,'CO')*f(I,J)
             +CDATA(I,J,'D')*f(I,J)**CDATA(I,J,'M'));
BALANCE(W,I) .. SUM(J$A(I,J),fc(W,I,J))-SUM(J$A(J,I),fc(W,J,I))
             =E=G(W)$ODE(I,W)-G(W)$DDE(I,W);
FLOW(I,J)$A(I,J)..SUM(W,fc(W,I,J))=E=f(I,J);
COST(I,J)$A(I,J)..cos(I,J)=E=CDATA(I,J,'CO')
             +2*CDATA(I,J,'D')*f(I,J);

** The next two sentences define the ND (Nguyen-Dupis) problem,
** considering all the above constraints, and direct GAMS to
** solve the problem using the nlp solver.

MODEL ND /ALL/;
SOLVE ND USING nlp MINIMIZING z;
```

Using the previous code we compute link flows and costs at equilibrium, which are shown in Table 12.10. The resulting information – the service level of the network transportation links – is crucial in transportation planning.

Table 12.10: Linkflows and costs at equilibrium

Link	f_a	$c_a(f_a)$	Link	f_a	$c_a(f_a)$
(1,5)	675.144	15.439	(1,12)	524.856	14.249
(4,5)	102.571	10.026	(4,9)	697.429	15.487
(5,6)	416.187	6.121	(5,9)	361.528	11.711
(6,7)	356.416	9.455	(6,10)	184.626	13.923
(7,8)	102.571	6.282	(7,11)	253.845	12.173
(8,2)	502.571	15.282	(9,10)	497.429	12.487
(9,13)	561.528	11.808	(10.I11)	682.056	7.705
(11.I2)	497.429	11.487	(11.I3)	438.472	12.385
(12.I6)	124.856	7.312	(12.I8)	400.000	18.000
(13.I3)	561.528	16.615			

We also illustrate that the equilibrium conditions are satisfied. To this end, we can compute the paths used through the link flow for every commodity. Moreover, the link costs at equilibrium are used to compute the costs of the path used. This information is shown in Table 12.11. For each origin and destination the table provides all optimal paths with their corresponding flows and costs.

It can be observed that the paths used for the same commodity have approximately the same cost (see commodities Ω_2, Ω_3). A second observation is that the linkflows at equilibrium are unique, but this is not true for pathflows. In this example paths flows satisfying

$$\begin{aligned} h_5 + h_6 &= 124.856 \\ h_3 + h_4 &= 313.616 \\ h_5 + h_3 &= 253.845 \\ h_4 + h_6 &= 184.626 \end{aligned}$$

produce linkflows at equilibrium.

12.6.4 Side-Constrained Assignment Models

For applications to transportation networks, it is sometimes important to consider a *capacitated network*, in which the following constraints are introduced:

$$s_k(\mathbf{f}) \leq 0, \quad \forall k \in \mathcal{K}$$

where the index set $\mathcal{K}$ may, for instance, consist of the index set of the links, nodes, routes or any combinations of subsets of them. A special case of these constraints are

$$0 \leq f_a \leq u_a, \quad \forall a \in \mathcal{B} \subset \mathcal{A} \tag{12.60}$$

that lead to the *capacitated traffic assignment problem*, CTAP.

Table 12.11: Paths used at the equilibrium for TAP

O–D Pair	Used paths	h_r	C_r
$\Omega_1 = (1-2)$	$r_1 = 1-12-8-2$	400	47.53
$\Omega_2 = (1-3)$	$r_2 = 1-5-9-13-3$	361.528	55.57
	$r_3 = 1-5-6-7-11-3$	h_3	55.62
	$r_4 = 1-5-6-10-11-3$	h_4	55.57
	$r_5 = 1-12-6-7-11-3$	h_5	55.62
	$r_6 = 1-12-6-10-11-3$	h_6	55.56
$\Omega_3 = (4-2)$	$r_7 = 4-5-6-7-8-2$	102.571	47.17
	$r_8 = 4-9-10-11-2$	184.626	47.16
$\Omega_4 = (4-3)$	$r_9 = 4-9-13-3$	200	43.91

In designing a future network, and depending on the service level to be provided, the traffic engineer computes the capacities of streets and intersections as a function of the road widths, number of lanes, shoulder widths, gradients, traffic signalization, and other factors.

If we use the TAP to predict the service level, then the predicted flow on some links may be lower or greater than the traffic assumed a priori by the engineer. This is due to the fact that TAP allows every road to carry an arbitrarily large traffic volume. It is clear, then, that inequalities such as (12.60) are significant for the planning process.

For the capacitated problem a simple optimality condition, similar to that of the Wardropp's first principle, can be formulated. Assume that the used routes for a demand Ω are numbered in the order of increasing costs, and that they are denoted as c_i^{Ω}; $i = 1, 2, \ldots, l$. Assume also that the number of routes is l, and among these the first m are *saturated*, containing at least one link which carries flow at its capacity limit. The equilibrium exists if and only if

$$c_1^{\Omega} \leq c_2^{\Omega}, \ldots \leq c_m^{\Omega} \leq c_{m+1}^{\Omega} = \cdots c_l^{\Omega}, \quad \forall \Omega \in W$$

Now we discuss two ways of taking into account side constraints:

1. We consider explicitly the upper bounds on the link flows (12.60). Taking into account the link capacities is equivalent to modifying the link costs in the following way

$$\hat{c}_a(f_a) = c_a(f_a) + \phi_a(f_a), \quad \forall a \in \mathcal{B}$$

where

$$\phi_a(f_a) = \begin{cases} +\infty & \text{if } f_a > k_a \\ 0 & \text{if } f_a \leq k_a \end{cases}$$

the functions ϕ_a are not continuous, and this introduces a new difficulty.

2. We use the so-called travel-time function, which tend to infinity when the link flows approach their respective capacities. This is similar to barrier

functions. In this approach, we approximate the function ϕ_a by means of a continuously differentiable function. For example, the implicit case considers approximations such as

$$\hat{c}_a(f_a) = c_a(f_a) + s_a \ln\left(1 + \frac{q_a}{k_a - f_a}\right), \forall a \in \mathcal{B}$$

where $s_a \to 0$, and k_a is the maximum capacity of link a. These functions are not defined on flow links such that their flow is over the capacity, and a special recovery procedure is needed to handle infeasible points. This fact may produce computational errors when the nonlinear solver progresses. We approximate the functions ϕ_a by a BPR formula $s_a\left(f_a/k_a\right)^{n_a}$, and the link cost functions become

$$\bar{c}_a(f_a) = c_a(f_a) + s_a\left(\frac{f_a}{k_a}\right)^{n_a}, \quad \forall a \in \mathcal{B} \tag{12.61}$$

where $n_a \to +\infty$.

Example 12.11 (capacitated network). To illustrate both approaches, we have used a test example adding to the ND network the side constraints

$$f_{10,11} \le 400$$
$$f_{12,8} \le 300$$

This example is called ND-C, and the set $\mathcal{B}$ is $\{(10,11),(12,8)\}$. In the implicit approach we have used the values $n_a = 9$, and $s_a = 20$, $\forall a \in \mathcal{B}$.

Now we list the new commands to deal with the capacitated traffic assignment problem. To compress the code, we have shown only the new commands to be added to the previously described TAP mode. The code to deal with the side constraints can be obtained by changing the links costs $c_a(f_a)$ by $\bar{c}_a(f_a)$. So, it is not listed.

```
$title CAPACITATED TRAFFIC ASSIGNMENT PROBLEM.

** Set B(N,N) is the set of bounded links. This is a new set.

SET
...

B(N,N) set of  bounded links
              /I10.I11
              I12.I8/;
PARAMETER
U(I,J) upper bound in link I-J
              /I10.I11 400
               I12.I8  300/
...

** Upper bounds for the link capacities.
```

```
VARIABLES
...
f.up(I,J)$B(I,J)=U(I,J);
```

We have obtained the link flows, the costs, and the equilibrium using the two approaches, which are shown in Table 12.12. Using the link flows for each commodity we have computed the paths used and their costs. This information is also shown in Table 12.13. It can be observed that the equilibrium conditions are satisfied. The pairs Ω_1 and Ω_3 illustrate the fact that saturated routes have less cost than nonsaturated ones. Moreover, the fact that the nonsaturated routes have the same cost is demonstrated in the pairs Ω_1 and Ω_2. In Table 12.13 we have computed the route cost using the cost $c_a(f_a)$ instead of $\bar{c}_a(f_a)$. It is relevant to note that the link cost in capacitated links is unrealistically high for the implicit approach (see Table 12.12). This indicates that the equilibrium conditions for the implicit approach, Wardrop's first principle, do not hold in the obtained approximation. This is due to the high nonlinearity of the capacitated link costs. This is a computational difficulty for the implicit approach.

From a modeling point of view, explicit upper bounds have the advantage of allowing links flows to attain the capacity values, whereas the use of travel-time functions such as (12.61) will force all link flows to be strictly greater than the capacities (see Table 12.12). Note that both approaches generate the same routes, but there exists a small difference in the load of these routes, and the cost. It can be concluded that the predictions of both approaches are not significantly different.

An important advantage of explicit modeling is the interpretation of the Lagrangian multipliers for the capacitated constraints. The Lagrangian multipliers for this example are shown in Table 12.12. They measure the time gained by the users of saturated routes compared to the fastest route still available. For example, the route r_1 is saturated because it contains the link $(12, 8)$. The Lagrangian multiplier associated with this link is 11.501, which is approximately equal to the difference between the costs of the saturated route and the nonsaturated routes, i.e., $58.70 - 47.20$.

This interpretation shows that the capacitated equilibrium link flow pattern can be found by solving the corresponding uncapacitated problem TAP with the travel time functions adjusted to $\hat{c}_a(f_a) = c_a(f_a) + \beta_a$ for all $a \in \mathcal{B}$, where β_a are the Lagrangian multipliers. The reader can prove in this example that using the link costs $\hat{c}_a$ leads used paths to satisfy Wardrop's first principle.

Finally, we can conclude that the implicit approach has the computational advantage of using the uncapacitated assignment model but has the disadvantage from a modeling point of view of overloading the capacitated routes, and difficulties in the interpretation of the Lagrangian multipliers. ■

12.6.5 The Variable-Demand Case

The TAP has been formulated as a problem with fixed demand, but it is more realistic to consider the *elastic* nature of the demand. Travelers have a number

Table 12.12: Link flows and costs at the equilibrium for CTAP, and Lagrangian multipliers

Link	Explicit approach		Implicit approach	
	f_a	$c_a(f_a)$	f_a	$c_a(f_a)$
(1,5)	689.920	15.624	690.242	15.628
(1,12)	510.080	14.101	509.758	14.098
(4,5)	200.000	11.000	178.735	10.787
(4,9)	600.000	15.000	621.265	15.106
(5,6)	400.266	6.002	393.659	5.952
(5,9)	489.655	12.672	475.318	12.565
(6,7)	610.345	12.629	586.810	12.335
(6,10)	.	13.000	.	13.000
(7,8)	267.749	8.347	253.988	8.175
(7,11)	342.597	13.282	332.822	13.160
(8,2)	567.749	16.097	570.595	16.132
(9,10)	400.000	12.000	421.265	12.106
(9,13)	689.655	12.448	675.318	12.377
(10,11)	400.000	7.000	421.265	7.053
	($\beta_a = 8.914$)			$\bar{c}_a = 342.775$
(11,2)	432.251	11.161	429.405	11.147
(11,3)	310.345	11.103	324.682	11.247
(12,6)	210.080	7.525	193.151	7.483
(12,8)	300.000	17.000	316.607	17.166
	($\beta_a = 11.501$)			$\bar{c}_a = 359.950$
(13,3)	689.655	17.897	675.318	17.753

Table 12.13: Flow and cost paths at equilibrium for CTAP

O–D Pair	Used paths	Approach			
		Explicit		Implicit	
		h_r	C_r	h_r	C_r
Ω_1 (1-2)	$r_1 = 1 - 12 - 8 - 2$	300	47.20	316.6	47.40
	$r_2 = 1 - 12 - 6 - 7 - 8 - 2$	67.75	58.70	75.25	58.23
	$r_3 = 1 - 12 - 6 - 7 - 11 - 2$	32.25	58.70	8.14	58.23
Ω_2 (1-3)	$r_4 = 1 - 5 - 6 - 7 - 11 - 3$	200.3	58.64	214.9	58.33
	$r_5 = 1 - 5 - 9 - 13 - 3$	489.7	58.64	475.3	58.33
	$r_6 = 1 - 12 - 6 - 7 - 11 - 3$	110.1	58.64	109.8	58.32
Ω_3 (4-2)	$r_9 = 4 - 9 - 10 - 11 - 2$	400	45.10	421.3	45.42
	$r_{10} = 4 - 5 - 6 - 7 - 8 - 2$	200	54.07	178.74	53.39
Ω_4 (4-3)	$r_{11} = 4 - 9 - 13 - 3$	200	45.34	200	45.24

of choices available and are motivated by economical considerations in their decisions. For example, as congestion increases, motorists may decide to use a different mode of transport (e.g. underground).

In order to take the elastic nature of the demand into account, the number of trips, g_Ω, between the pair Ω in the network can be assumed to be a function of the travel cost for that pair Ω:

$$g_\Omega = G_\Omega(c_\Omega)$$

where c_Ω is the minimum travel cost for the pair Ω, and G_Ω is the demand function. We assume that G_Ω is nonnegative, continuous, and strictly decreasing for each $\Omega \in W$. Its inverse function gives the number of trips as a function of the travel cost, i.e. $c_\Omega = G_\Omega^{-1}(g_\Omega)$.

The equilibrium condition states that the O–D trip rate satisfy the demand function, and that the travel times on all used paths between any O–D pair are equal, and are also equal to or less than the travel times on any unused paths.

The following model combines variable-demand and assignment. It is possible to demonstrate that the equilibrium conditions are those obtained by solving the following problem (TAPE). Minimize

$$Z = \sum_{(i,j)\in\mathcal{A}} \int_0^{f_{ij}} c_{ij}(x)dx - \sum_{\Omega\in W} \int_0^{g_\Omega} G_\Omega^{-1}(x)dx$$

subject to

$$\sum_{j\in A(i)} f_{ij}^\Omega - \sum_{j\in B(i)} f_{ji}^\Omega = r_i^\Omega, \ \forall i \in \mathcal{N}, \quad \forall \Omega \in W \tag{12.62}$$

$$\sum_{\Omega\in W} f_{ij}^\Omega = f_{ij}, \forall (i,j) \in \mathcal{A} \tag{12.63}$$

$$f_{ij}^\Omega \geq 0 \quad \forall \Omega \in W, \forall (i,j) \in \mathcal{A} \tag{12.64}$$

where

$$r_i^\Omega = \begin{cases} g_\Omega & \text{if } O^\Omega = i \\ -g_\Omega & \text{if } D^\Omega = i \\ 0 & \text{if i is an intermediate node for the commodity } \Omega \end{cases}$$

Note that for TAPE the terms r_i^Ω are variables, but for TAP they are constants.

To illustrate the TAPE, we deal with a particular case of TAPE. Assume that the travelers choose between several modes of transport, and a logit function (demand model) gives the number of trips taken on each alternative, g_Ω^k, by the equation

$$g_\Omega^k = G_\Omega^k(\mathbf{c}_\Omega) = \frac{\exp[-\left(\alpha^k + \beta_1 c_\Omega^k\right)]}{\sum_{k'} \exp[-\left(\alpha^{k'} + \beta_1 c_\Omega^{k'}\right)]} \bar{g}_\Omega \tag{12.65}$$

where c_Ω^k is the user's perception of the generalized cost of traveling of the pair $\Omega = (i,j)$ by mode k, that corresponds to a user optimal route choice of the network; $\{\mathbf{c}_\Omega\}$ is the vector of generalized costs for all the modes present, $\bar{g}_\Omega$ is the demand for the O–D pair Ω by all considered modes; and α^k, β_1 are parameters of the logit model. For two alternatives, such as car (a) and public transport (b), (12.65) simplifies to

$$G_\Omega^a(\mathbf{c}_\Omega) = \frac{1}{1+\exp[-\left(\alpha^{ab} + \beta_1(c_\Omega^b - c_\Omega^a)\right)]}\bar{g}_\Omega$$

where $\alpha^{ab} = \alpha^b - \alpha^a$. We assume that the travel cost c_Ω^b by public transportation is independent of traffic volumes, and by this reason it is constant. The inverse function of the demand model becomes

$$c_\Omega^a = G^{-1}(g_\Omega^a) = c_\Omega^b + \frac{1}{\beta_1}\left[\alpha^{ab} + \log(\bar{g}_\Omega - g_\Omega^a) - \log(g_\Omega^a)\right]$$

Using the relationship $g_\Omega^a + g_\Omega^b = \bar{g}_\Omega$, we obtain the equality

$$-\int_0^{g_\Omega^a} G^{-1}(x)dx = c_\Omega^b g_\Omega^b + \left(\frac{1}{\beta_1}\right)\sum_{k\in\{a,b\}} g_\Omega^k(\log g_\Omega^k - 1 + \alpha^k) + C$$

where C is a constant. The objective function then becomes

$$\sum_{(i,j)\in\mathcal{A}}\int_0^{f_{ij}} c_{ij}(x)dx + \sum_{\Omega\in W} c_\Omega^b g_\Omega^b + \left(\frac{1}{\beta_1}\right)\sum_{\Omega\in W}\sum_{k\in\{a,b\}} g_\Omega^k(\log g_\Omega^k - 1 + \alpha^k)$$

Example 12.12 (Variable-demand case). To illustrate the TAPE model consider that there exists an underground network. Assume that there exist connections to satisfy the four demands pairs, and the trip times are flow independent (Table 12.14).

Table 12.14: Inputs for TAPE

O–D pair	$\bar{g}_\Omega$	c_Ω^b	Logit parameters
Ω_1	600	41	$\alpha^a = -2.0$
Ω_2	1000	46	$\alpha^b = 0.0$
Ω_3	800	43	$\beta_1 = 0.1$
Ω_4	400	40	

The following code implements this model in GAMS.

```
$title TRAFFIC ASSIGNMENT PROBLEM WITH ELASTIC DEMAND.

** Sets are declared in first place.
** Set N is the set of nodes of the road network.
** Set A is the set of links.
** Set W is the set of O--D pairs
```

```
SET
N set of nodes /I1*I13/
W pairs          /W1*W4/
A(N,N) set of links
/I1.(I5,I12)
I4.(I5,I9)
I5.(I6,I9)
I6.(I7,I10)
I7.(I8,I11)
I8.(I2)
I9.(I10,I13)
I10.(I11)
I11.(I2,I3)
I12.(I6,I8)
I13.(I3)/

ODE(N,W) set of origins by demand
/I1.(W1,W2)
I4.(W3,W4/

DDE(N,W) set of destinations by demand
/I2.(W1,W3)
I3.(W2,W4)/;

ALIAS(N,I)
ALIAS(N,J)

PARAMETER
G(W) demand in O--D pair  W
/W1 600
W2 1000
W3 800
W4 400/

Cb(W) travel cost by public transport in O--D pair  W
/W1 41
W2 35
W3 43
W4 40/

BETA
ALPHAa
ALPHAb;
BETA=0.1;
ALPHAa=-2.;
ALPHAb=0.;

TABLE  CDATA(N,N,*)  link cost parameters
          C0      D
 I1.I5    7.0  0.00625
 I1.I12   9.0  0.005
 I4.I5    9.0  0.005
 I4.I9   12.0  0.0025
 I5.I6    3.0  0.00375
 I5.I9    9.0  0.00375
 I6.I7    5.0  0.00625
 I6.I10  13.0  0.0025
```

```
 I7.I8     5.0  0.00625
 I7.I11    9.0  0.00625
 I8.I2     9.0  0.00625
 I9.I10   10.0  0.0025
 I9.I13    9.0  0.0025
 I10.I11   6.0  0.00125
 I11.I2    9.0  0.0025
 I11.I3    8.0  0.005
 I12.I6    7.0  0.00125
 I12.I8   14.0  0.005
 I13.I3   11.0  0.005;


** Optimization variables are declared.

VARIABLES
 z          objective function variable
 f(N,N)     flow link
 fc(W,N,N)  is the flow link of the commodity W
 cos(N,N)   cost in the link at equilibrium
 ga(W)      demand by car
 gb(W)      demand by public transport;

POSITIVE VARIABLE fc(W,N,N);

ga.LO(W)=0.01;
gb.LO(W)=0.01;

** Constraints are declared.

EQUATIONS
 COST Objective function
 BALANCE(W,I) conservation of flow condition for the commodity W.
 FLOW(I,J) total link flow
 MODAL(W) Modal split of the demand
 COSTE(I,J) cost at equilibrium;

COST .. z=E= SUM((I,J)$A(I,J),CDATA(I,J,'CO')*f(I,J)
             +CDATA(I,J,'D')*f(I,J)**CDATA(I,J,'M'))
+SUM(W,Cb(W)*gb(W))+1/BETA* SUM(W,ga(W)*(-1+ALPHAa+LOG(ga(W)))
             + gb(W)*(-1+ALPHAb+LOG(gb(W))) );
BALANCE(W,I) .. SUM(J$A(I,J),fc(W,I,J))-SUM(J$A(J,I),fc(W,J,I))
                =E=ga(W)$ODE(I,W)-ga(W)$DDE(I,W);
FLOW(I,J)$A(I,J)..SUM(W,fc(W,I,J))=E=f(I,J);
COSTE(I,J)$A(I,J)..cos(I,J)=E=CDATA(I,J,'CO')
                              +2*CDATA(I,J,'D')*f(I,J);
MODAL(W).. G(W)=E=ga(W)+gb(W);
MODEL nd /ALL/;
SOLVE nd USING nlp MINIMIZING z;
```

The results obtained are shown in Table 12.15.

Some comments are in order. For example, consider that the system operator increases train frequencies to satisfy the demand pair Ω_2, and the travel time is reduced to 35. The operator wishes to compute the new demand and the congestion level on the road network for this demand. The outputs obtained with TAPE are shown in Table 12.16. Note that the demand in public trans-

Table 12.15: Cost at the equilibrium for TAPE and modal split

O–D Pair	c^a_Ω	c^b_Ω	g^a_Ω	g^b_Ω
Ω_1	49.29	41	459.821	140.179
Ω_2	56.73	46	736.288	263.712
Ω_3	48.52	43	647.730	152.270
Ω_4	45.56	40	323.124	76.876

portation for the pair Ω_2 increases. This produces a reduction of the congestion level on the road network, and the alternative car for the others pairs is slightly more attractive.

Table 12.16: Cost at equilibrium for TAPE, and modal split with intervention on the transportation system

O–D Pair	c^a_Ω	c^b_Ω	g^a_Ω	g^b_Ω
Ω_1	48.53	41	466.009	133.991
Ω_2	52.79	35	554.843	445.157
Ω_3	48.34	43	649.947	150.053
Ω_4	44.56	40	329.594	70.406

■

12.6.6 Combined Distribution and Assignment

One way of dealing with the four steps of the planning process (explained in Section 12.6.1) is to merge as many steps as possible into one, in particular, we can include assignment and distribution in the same process. We assume that the number of trips emanating from the origins O_i, and attracted to destinations D_j are known, but not the O–D trip matrix. The O–D trip matrix, g_Ω $\Omega = (i,j) \in W$, must satisfy

$$\begin{array}{rcl} \sum\limits_j g_{ij} & = & O_i, \ \forall i \\ \sum\limits_i g_{ij} & = & D_j, \ \forall j \end{array} \tag{12.66}$$

Distribution models of different kinds have been developed to assist in forecasting future trip patterns when important changes in the network take place. They consider a trip making behavior and the way that behavior is influenced by external factors such as total trip ends and traveled distance. These models consider that the number of trips from zones depends on the travel cost between the zones, and the potential of each zone. They use the expression

$$g_{ij} = p(c_{ij}) = \alpha O_i D_j f(c_{ij})$$

where α is a proportionality factor, and $f(c_{ij})$ has one or more parameters to calibrate. This function often receives the name of *deterrence function.* The most common used expressions are

$$\begin{array}{ll} p(c_{ij}) = \exp(-\beta c_{ij}) & \text{exponential function} \\ p(c_{ij}) = c_{ij}^{-n} & \text{power function} \\ p(c_{ij}) = c_{ij}^{n} \exp(-\beta c_{ij}) & \text{combined function} \end{array} \qquad (12.67)$$

Assuming that the function $p(c_{ij})$ is a decreasing function of the travel cost, the previous model is formulated as the following mathematical program (TAPD). Minimize

$$Z = \sum_{(i,j)\in\mathcal{A}} \int_0^{f_{ij}} c_{ij}(x)dx - \sum_{\Omega\in W} \int_0^{g_\Omega} p^{-1}(x)dx$$

subject to

$$\begin{aligned} \sum_j g_{ij} &= O_i, \ \forall i && (12.68) \\ \sum_i g_{ij} &= D_j, \ \forall j && (12.69) \\ \sum_{j\in A(i)} f_{ij}^{\Omega} - \sum_{j\in B(i)} f_{ji}^{\Omega} &= r_i^{\Omega}, \ \forall i \in \mathcal{N}, \ \ \forall \Omega \in W && (12.70) \\ \sum_{\Omega\in W} f_{ij}^{\Omega} &= f_{ij}, \ \forall (i,j) \in \mathcal{A} && (12.71) \\ f_{ij}^{\Omega} &\geq 0 \ \ \forall \Omega \in W, \ \ \forall (i,j) \in \mathcal{A} && (12.72) \end{aligned}$$

where

$$r_i^{\Omega} = \begin{cases} g_\Omega & \text{if } O^{\Omega} = i \\ -g_\Omega & \text{if } D^{\Omega} = i \\ 0 & \text{if} i \text{ is an intermediate nodefor the commodity } \Omega \end{cases}$$

The example below illustrates the TAPD model

Example 12.13 (gravity distribution model). We illustrate the model TAPD using the *gravity distribution model,* which is derived using an exponential function for the deterrence function. In this case we obtain

$$\begin{aligned} -\int_0^{g_\Omega} p^{-1}(x)dx &= -\int_0^{g_\Omega} \left[-\frac{1}{\beta}\log(x) + (d_i + d_j + \alpha') \right] dx \\ &= \frac{1}{\beta} g_\Omega \left(\log g_\Omega - 1\right) + (d_i + d_j + \alpha') g_\Omega \end{aligned}$$

where $d_i = \log O_i/\beta$, $d_j = \log D_j/\beta$, and $\alpha' = \alpha/\beta$. The objective function becomes

$$\sum_{(i,j)\in\mathcal{A}} \int_0^{f_{ij}} c_{ij}(\mathbf{x})d\mathbf{x} + \frac{1}{\beta} \sum_{\Omega\in W} g_\Omega \left(\log g_\Omega - 1\right) + \sum_{\Omega\in W} (d_i + d_j + \alpha') g_\Omega$$

The term $\sum_{\Omega \in W}(d_i + d_j + \alpha')g_\Omega$ is constant in the set of feasible solutions defined by (12.66), and it can be dropped. It can be shown that the previous mathematical programming problem is convex, and using the KKT conditions the optimal solution satisfies

$$g_{ij}^* = A_i O_i B_j D_j \exp(-\beta c_{ij}^*)$$

where c_{ij}^* is the equilibrium cost. The set of parameters A_i and B_j replace the proportionality factor α to enforce constraints (12.66).

The following GAMS file implements this model for the inputs given in Table 12.17. The results obtained are shown in Table 12.18

```
$title TRAFFIC DISTRIBUTION-ASSIGNMENT PROBLEM.

** Sets are declared in first place.
** Set N is the set of nodes of the road network.
** Set A is the set of links.
** Set W is the set of O--D pairs

SET
 N set of nodes /I1*I13/
 W pairs        /W1*W4/
 O origins      /I1
                 I4/
 D destinations /I2
                 I3/
OW(O,W)
/I1.(W1,W2)
I4.(W3,W4)/
DW(D,W)
/I2.(W1,W3)
I3.(W2,W4)/

A(N,N) set of links
/I1.(I5,I12)
I4.(I5,I9)
I5.(I6,I9)
I6.(I7,I10)
I7.(I8,I11)
I8.(I2)
I9.(I10,I13)
I10.(I11)
I11.(I2,I3)
I12.(I6,I8)
I13.(I3)/

ODE(N,W) set of origins by demand
/I1.W1
I1.W2
I4.W3
I4.W4/
DDE(N,W) set of destinations by demand
/I2.W1
I3.W2
I2.W3
I3.W4/;
```

```
ALIAS(N,I)
ALIAS(N,J)

PARAMETER
Oi(O) number of trips emanating from O
/I1 1200
I4 800/

Dj(D) number of trips attracted to D
/I2 1000
I3 1000/

BETA;
BETA=0.2;

TABLE  CDATA(N,N,*)  link cost parameters
          CO      D        M
 I1.I5     7.0  0.00625   2.0
 I1.I12    9.0  0.005     2.0
 I4.I5     9.0  0.005     2.0
 I4.I9    12.0  0.0025    2.0
 I5.I6     3.0  0.00375   2.0
 I5.I9     9.0  0.00375   2.0
 I6.I7     5.0  0.00625   2.0
 I6.I10   13.0  0.0025    2.0
 I7.I8     5.0  0.00625   2.0
 I7.I11    9.0  0.00625   2.0
 I8.I2     9.0  0.00625   2.0
 I9.I10   10.0  0.0025    2.0
 I9.I13    9.0  0.0025    2.0
 I10.I11   6.0  0.00125   2.0
 I11.I2    9.0  0.0025    2.0
 I11.I3    8.0  0.005     2.0
 I12.I6    7.0  0.00125   2.0
 I12.I8   14.0  0.005     2.0
 I13.I3   11.0  0.005     2.0;

** Optimization variables are declared.

VARIABLES
 z         objective function variable
 f(N,N)    flow link
 fc(W,N,N) is the flow link of the commodity W
 cos(N,N)  cost in the link at equilibrium
 g(W)      demand in the pair W;

POSITIVE VARIABLE fc(W,N,N);

g.LO(W)=0.01;

** Constraints are declared.

EQUATIONS
 COST Objective function
 BALANCE(W,I) conservation of flow condition for the commodity W.
 FLOW(I,J) total link flow
```

```
 ORIGIN(O)
 DESTIN(D)
 COSTE(I,J) cost at equilibrium;

COST .. z=E= SUM((I,J)$A(I,J),CDATA(I,J,'CO')*f(I,J)
             +CDATA(I,J,'D')*f(I,J)**CDATA(I,J,'M'))
             +(1/BETA)*SUM(W,g(W)*(LOG(g(W))-1 ));
BALANCE(W,I) .. SUM(J$A(I,J),fc(W,I,J))-SUM(J$A(J,I),fc(W,J,I))
                =E=g(W)$ODE(I,W)-g(W)$DDE(I,W);
FLOW(I,J)$A(I,J)..SUM(W,fc(W,I,J))=E=f(I,J);
COSTE(I,J)$A(I,J)..cos(I,J)=E=CDATA(I,J,'CO')
                            +2*CDATA(I,J,'D')*f(I,J);
ORIGIN(O)..Oi(O)=E= SUM(W,g(W)$OW(O,W));
DESTIN(D)..Dj(D)=E= SUM(W,g(W)$DW(D,W));
MODEL nd /ALL/;
SOLVE nd USING nlp MINIMIZING z;
```

Table 12.17: Inputs for TAPD

Origin	O_i	Destination	D_j	β
1	1200	2	1000	0.2
4	800	3	1000	–

Table 12.18: Outputs for TAPD

O–D pair	g_Ω	Equilibrium cost
Ω_1	650.312	52.21
Ω_2	549.688	53.65
Ω_3	349.688	46.77
Ω_4	450.312	45.05

■

12.7 Short-Term Hydrothermal Coordination

Short-term hydrothermal coordination (STHTC) determines the startup and shutdown of thermal plants, as well as the power output of hydro and thermal plants to meet customer demand with an appropriate level of security and so that total operating costs are minimized. Because of the high cost associated with the startup of thermal plants, selecting the plants to meet customer demand in the most economical manner can save large amounts of money. If the electric energy system under consideration does not include hydroelectric plants, the above problem is called unit commitment.

Mathematically, the STHTC problem can be formulated as a mixed-integer nonlinear optimization problem. For realistic size electric energy systems it is also a large-scale problem. Solving this large-scale nonlinear and combinatorial optimization problem is not an easy task. Lagrangian relaxation (LR) techniques are the most suitable techniques to solve this kind of problems (see Muckstadt and Koening [77], Merlin and Sandrin [74], Bertsekas et al. [12], Zhuang and Galiana [107], Yan et al. [106], Mendes et al. [73], Rakic and Marcovic [91], Wang et al. [101], Pellegrino et al. [85], Luh et al.[69], Jiménez and Conejo [60]). Dynamic programming techniques require discretization of continuous variables and drastic simplifying assumptions to make the problem computationally tractable (Hobbs et al. [52]). Mixed-integer linear programming techniques not only linearize the problem but also make important simplifications to be able to solve such a large-scale problem (see Dillon et al. [32], Brannlund et al. [16], Medina et al. [70]).

When using LR techniques to solve the STHTC problem, the resulting relaxed primal problem can be naturally decomposed into one subproblem per thermal plant and one subproblem per hydro system. Therefore, by using LR techniques, the solution of the STHTC problem (large-scale and complex optimization problem) is accomplished by the solution of many small sized and structurally homogeneous subproblems.

This decomposition property allows a very precise modeling of each generating plant as well as the possibility of applying to each subproblem the most suitable optimization technique to its structure. It also allows the natural application of parallel computing with the corresponding advantages regarding CPU time.

In addition to all these advantages, derived from the decomposition property of the relaxed primal problem, the application of LR techniques to solve the STHTC problem presents another important advantage: the dual problem variables (the Lagrange multipliers) have an economical meaning that can be very helpful in the framework of deregulated electric energy markets, and also in the traditional framework of centralized systems.

12.7.1 Problem Formulation and the LR Solution Procedure

The STHTC problem can be formulated as a nonlinear and combinatorial optimization problem in which total operating costs are minimized subject to meeting constraints modeling the technical limitations of thermal and hydro plants, and to meet load constraints. Load constraints include electric energy customer demand constraints plus spinning reserve constraints. Spinning reserve constraints ensure an appropriate level of security.

The main elements of this problem are

1. **Data.**

   I: the number of thermal plants

   J: the number of hydro systems

   $\mathbf{H}$: the vector of demands

$\mathbf{h}_i(\mathbf{x}_i)$: the contribution of thermal unit i to meet the demand

$\mathbf{h}_j(\mathbf{x}_j)$: the contribution of hydro system j to meet the demand

$\mathbf{G}$: the vector of power reserves

$\mathbf{g}_i(\mathbf{x}_i)$: the contribution of thermal unit i to meet the power reserve

$\mathbf{g}_j(\mathbf{x}_j)$: the contribution of hydro system j to meet the power reserve

$\mathbf{H}$, $\mathbf{h}_i(\mathbf{x}_i)$, $\mathbf{h}_j(\mathbf{x}_j)$, $\mathbf{G}$, $\mathbf{g}_i(\mathbf{x}_i)$ and $\mathbf{g}_j(\mathbf{x}_j)$ are vectors of dimension equal to the number of time periods in the planning horizon.

2. **Variables.**

$\mathbf{x}_i$: the vector of variables associated with thermal plant i

$\mathbf{x}_j$: the vector of variables associated with hydro system j

3. **Constraints.**

$$\begin{array}{rcl} \mathbf{s}_i(\mathbf{x}_i) & \leq & \mathbf{0}, i = 1, \ldots, I \\ \mathbf{s}_j(\mathbf{x}_j) & \leq & \mathbf{0}, j = 1, \ldots, J \\ \sum_{i=1}^{I} \mathbf{h}_i(\mathbf{x}_i) + \sum_{j=1}^{J} \mathbf{h}_j(\mathbf{x}_j) & = & \mathbf{H} \\ \sum_{i=1}^{I} \mathbf{g}_i(\mathbf{x}_i) + \sum_{j=1}^{J} \mathbf{g}_j(\mathbf{x}_j) & \leq & \mathbf{G} \end{array} \quad (12.73)$$

The first set of constraints expresses thermal plant constraints, the second one represents hydro system constraints, the third one expresses customer demand constraints, and the fourth one represents spinning reserve constraints. It should be noted that time is embedded in the formulation shown above.

4. **Function to be optimized.** The objective function represents the total operating cost (the cost of hydro power production is negligible compared to the cost of thermal power production)

$$f(\mathbf{x}) = \sum_{i=1}^{I} f_i(\mathbf{x}_i) \quad (12.74)$$

This problem, referred to as the *primal problem* (PP), is formulated as follows

$$\min_{\mathbf{x}=(\mathbf{x}_i, \mathbf{x}_j)} \; f(\mathbf{x}) = \sum_{i=1}^{I} f_i(\mathbf{x}_i) \quad (12.75)$$

subject to

$$\begin{array}{rcl} \mathbf{s}_i(\mathbf{x}_i) & \leq & \mathbf{0}, i = 1, \ldots, I \\ \mathbf{s}_j(\mathbf{x}_j) & \leq & \mathbf{0}, j = 1, \ldots, J \\ \sum_{i=1}^{I} \mathbf{h}_i(\mathbf{x}_i) + \sum_{j=1}^{J} \mathbf{h}_j(\mathbf{x}_j) & = & \mathbf{H} \\ \sum_{i=1}^{I} \mathbf{g}_i(\mathbf{x}_i) + \sum_{j=1}^{J} \mathbf{g}_j(\mathbf{x}_j) & \leq & \mathbf{G} \end{array} \quad (12.76)$$

Load constraints are the complicating or global constraints of this primal problem. Load constraints include equality constraints (demand constraints) and inequality constraints (spinning reserve constraints). They couple together decisions related to thermal and hydro plants. Because of the existence of these constraints, the preceding problem cannot be decomposed and cannot be easily solved.

By applying LR techniques, load constraints are incorporated into the objective function to form the relaxed primal problem. The vector of multipliers associated with the vector of demand constraints is called $\boldsymbol{\lambda}$ and the vector of multipliers associated to the spinning reserve constraints is called $\boldsymbol{\mu}$.

The Lagrangian function is defined as

$$\begin{aligned}\mathcal{L}(\mathbf{x}, \boldsymbol{\lambda}, \boldsymbol{\mu}) &= \sum_{i=1}^{I} f_i(\mathbf{x}_i) + \boldsymbol{\lambda}^T \left(\mathbf{H} - \sum_{i=1}^{I} \mathbf{h}_i(\mathbf{x}_i) - \sum_{j=1}^{J} \mathbf{h}_j(\mathbf{x}_j) \right) \\ &\quad + \boldsymbol{\mu}^T \left(\mathbf{G} - \sum_{i=1}^{I} \mathbf{g}_i(\mathbf{x}_i) - \sum_{j=1}^{J} \mathbf{g}_j(\mathbf{x}_j) \right)\end{aligned} \tag{12.77}$$

and the dual function is the solution of the problem

$$\theta(\boldsymbol{\lambda}, \boldsymbol{\mu}) = \min_{(\mathbf{x}_i, \mathbf{x}_j)} \mathcal{L}(\mathbf{x}_i, \mathbf{x}_j, \boldsymbol{\lambda}, \boldsymbol{\mu}) \tag{12.78}$$

subject to

$$\begin{aligned}\mathbf{s}_i(\mathbf{x}_i) &\leq \mathbf{0}; \; i = 1, 2, \ldots, I \\ \mathbf{s}_j(\mathbf{x}_j) &\leq \mathbf{0}; \; j = 1, 2, \ldots, J\end{aligned} \tag{12.79}$$

which can be expressed as

$$\theta(\boldsymbol{\lambda}, \boldsymbol{\mu}) = \boldsymbol{\lambda}^T \mathbf{H} + \boldsymbol{\mu}^T \mathbf{G} + d(\boldsymbol{\lambda}, \boldsymbol{\mu}) \tag{12.80}$$

where $d(\boldsymbol{\lambda}, \boldsymbol{\mu})$ is the solution of the following optimization problem

$$\min_{\mathbf{x}_i, \mathbf{x}_j} \left(\sum_{i=1}^{I} \left(f_i(\mathbf{x}_i) - \boldsymbol{\lambda}^T \mathbf{h}_i(\mathbf{x}_i) - \boldsymbol{\mu}^T \mathbf{g}_i(\mathbf{x}_i) \right) - \sum_{j=1}^{J} \left(\boldsymbol{\lambda}^T \mathbf{h}_j(\mathbf{x}_j) + \boldsymbol{\mu}^T \mathbf{g}_j(\mathbf{x}_j) \right) \right) \tag{12.81}$$

subject to

$$\begin{aligned}\mathbf{s}_i(\mathbf{x}_i) \leq \mathbf{0}, i = 1, \ldots, I \\ \mathbf{s}_j(\mathbf{x}_j) \leq \mathbf{0}, j = 1, \ldots, J.\end{aligned} \tag{12.82}$$

This problem can be naturally decomposed into one subproblem per thermal plant i and one subproblem per hydro system j. For fixed values of $\boldsymbol{\lambda}$ and $\boldsymbol{\mu}$, problem (12.78) is called the *relaxed primal problem*, and problem (12.81) is called the *decomposed primal problem.*

The subproblem associated with thermal plant i is

$$\min_{\mathbf{x}_i} f_i(\mathbf{x}_i) - \boldsymbol{\lambda}^T \mathbf{h}_i(\mathbf{x}_i) - \boldsymbol{\mu}^T \mathbf{g}_i(\mathbf{x}_i) \tag{12.83}$$

subject to

$$\mathbf{s}_i(\mathbf{x}_i) \leq \mathbf{0} \tag{12.84}$$

and the subproblem associated with hydro system j is

$$\max_{\mathbf{x}_j} \boldsymbol{\lambda}^T \mathbf{h}_j(\mathbf{x}_j) + \boldsymbol{\mu}^T \mathbf{g}_j(\mathbf{x}_j) \tag{12.85}$$

subject to

$$\mathbf{s}_j(\mathbf{x}_j) \leq \mathbf{0} \tag{12.86}$$

LR techniques are based on the solution of the following dual problem [as problem (9.64)]

$$\max_{\boldsymbol{\lambda},\boldsymbol{\mu}} \theta(\boldsymbol{\lambda}, \boldsymbol{\mu}) \tag{12.87}$$

subject to

$$\boldsymbol{\mu} \geq \mathbf{0} \tag{12.88}$$

Because of the existence of integer variables (e.g., thermal plant unit commitment variables) in the formulation of the primal problem, the STHTC is a non-convex problem. Therefore, the optimal solution of the dual problem is not the optimal solution of the primal problem but it is a lower bound. Nevertheless, as the size of the problem increases the per unit duality gap (see Section 1.1) decreases (see Ferreira [37], Bertsekas et al. [12], Everett [34]), therefore the optimal solution of the dual problem becomes closer to the optimal solution of the primal problem. Once the optimal solution of the dual problem is found, heuristic procedures can be easily applied to derive a near-optimal primal problem solution (Zhuang and Galiana [107]).

In the STHTC problem, inequality complicating constraints are closely related to the integer variables (the thermal plant unit commitment variables) and the equality constraints are closely related to the continuous variables (thermal and hydro plant power output variables). This motivates the decomposition of phase 2 into two consecutive phases. In the first one, called *phase 2A*, the solution of the dual problem (or phase 1) is slightly modified to find values of the integer variables that meet inequality global constraints (spinning reserve constraints). In the second one, called *phase 2B*, the solution of phase 2A is modified by adjusting the values of the continuous variables to meet equality global constraints (demand constraints).

Therefore, the LR procedure to solve the STHTC problem consists of

Phase 1. Solution of the dual problem.

Phase 2A. Search for a spinning reserve primal feasible solution.

Phase 2B. Search for a load balanced primal feasible solution: multiperiod economic dispatch.

The solution of the dual problem (phase 1) is the key element to solve the STHTC problem by LR. The efficiency of a STHTC algorithm relies on the efficiency of the solution of this phase.

Phase 2A is an iterative procedure in which the $\boldsymbol{\mu}$ multipliers are updated in those periods where spinning reserve constraints are not satisfied. In these periods, the corresponding $\boldsymbol{\mu}$ multipliers are increased proportionally to the mismatches in spinning reserve constraints (subgradient type updating) until these constraints are met in all the time periods of the planning horizon. At the end of phase 2A a primal set of feasible commitment decisions is found. This phase requires typically little CPU time to achieve a solution very close to the solution of phase 1. phase 2B is a multiperiod economic dispatch procedure (Wood and Wollenberg [104]) in which, once unit commitment variables are set to the solution of phase 2A, power output is adjusted in order to meet the demand constraints in all time periods. This is a traditional problem routinely solved by electric energy system operators.

12.7.2 Dual-Problem Solution: Multiplier Updating Techniques

The procedure to solve the dual problem of the STHTC problem was described earlier. Specifically, at each iteration the relaxed primal problem is solved and, with the information obtained, the multiplier vector is updated. The information derived from the resolution of the relaxed primal problem is

- The value of the dual function

  $$\theta(\boldsymbol{\lambda}^{(t)}, \boldsymbol{\mu}^{(t)}) = \boldsymbol{\lambda}^{(t)T}\mathbf{H} + \boldsymbol{\mu}^{(t)T}\mathbf{G} + d(\boldsymbol{\lambda}^{(t)}, \boldsymbol{\mu}^{(t)}) \tag{12.89}$$

  which is

  $$\begin{aligned} d(\boldsymbol{\lambda}^{(t)}, \boldsymbol{\mu}^{(t)}) = & \sum_{i=1}^{I} \left(f_i(\mathbf{x}_i^{*(t)}) - \boldsymbol{\lambda}^{(t)T}\mathbf{h}_i(\mathbf{x}_i^{*(t)}) - \boldsymbol{\mu}^{(t)T}\mathbf{g}_i(\mathbf{x}_i^{*(t)}) \right) \\ & - \sum_{j=1}^{J} \left(\boldsymbol{\lambda}^{(t)T}\mathbf{h}_j(\mathbf{x}_j^{*(t)}) + \boldsymbol{\mu}^{(t)T}\mathbf{g}_j(\mathbf{x}_j^{*(t)}) \right) \end{aligned} \tag{12.90}$$

  where $\mathbf{x}_i^{*(t)}$ is the vector of optimal values for the variables associated with thermal plant i and $\mathbf{x}_j^{*(t)}$ is the vector of optimal values for the variables associated with hydro system j obtained from the solution of the relaxed primal problem

- A subgradient $\mathbf{s}^{(t)}$ of the dual function at the optimal solution of the relaxed primal problem: $\mathbf{x}_i^*, \forall i$, $\mathbf{x}_j^*, \forall j$

A subgradient can be easily computed as the vector of mismatches in demand constraints and the vector of mismatches in spinning reserve constraints:

$$\mathbf{s}^{(t)} = \text{column}[\mathbf{h}^{(t)}, \mathbf{g}^{(t)}] \tag{12.91}$$

where

$$\begin{aligned}\mathbf{h}^{(t)} &= \mathbf{H} - \sum_{i=1}^{I} \mathbf{h}_i(\mathbf{x}_i^{*(t)}) - \sum_{j=1}^{J} \mathbf{h}_j(\mathbf{x}_j^{*(t)}) \\ \mathbf{g}^{(t)} &= \mathbf{G} - \sum_{i=1}^{I} \mathbf{g}_i(\mathbf{x}_i^{*(t)}) - \sum_{j=1}^{J} \mathbf{g}_j(\mathbf{x}_j^{*(t)})\end{aligned} \tag{12.92}$$

Although the four methods previously described can be applied to update the multipliers, the most suitable methods for the STHTC problem are the bundle method Pellegrino et al. [85]) and the dynamically constrained cutting plane method (Jiménez and Conejo [60]).

12.7.3 Economical Meaning of the Multipliers

As it has been indicated above, one of the advantages of the use of LR techniques is the availability of the useful economical information provided by the variables of the dual problem, the Lagrange multipliers.

Multiplier $\boldsymbol{\lambda}$ at a given time represents, from the point of view of the system, the cost of producing one extra unit of electric energy [megawatt-hour (MWh)], that is, the electric energy marginal cost. Equivalently, from the point of view of a generating company, multiplier $\boldsymbol{\lambda}$ at a given time represents an indicator of the price a plant should be paid for each MWh of energy. It also represents an indicator of the price that a generating company should bid to get its plant online.

Analogously, multiplier $\boldsymbol{\mu}$ at a given time represents the cost of keeping an incremental unit (MW) of power reserve or equivalently, an indicator of the price a generator should be paid for each MW of reserve.

This economical interpretation is useful in the traditional framework of centralized electric energy systems to elaborate electric tariffs, but also in the framework of modern deregulated electric energy markets.

In the framework of deregulated electric energy markets, the LR procedure to solve the STHTC problem can be interpreted as the actual functioning of a free market. In other words, it can be thought of as a mechanism to meet customer demand with an appropriate level of security (measured in terms of the spinning reserve) by choosing the cheapest generator offers.

Hourly energy prices proposals (Lagrange multipliers) are specified by the market operator for the planning horizon. Each generator (or generating company) schedules its production independently along the planning horizon to maximize its benefit (i.e., each generator solves an optimization problem). Analogously, each hydro system is scheduled so that its benefit is maximum (i.e., each hydro system solves a problem). After the submissions of production proposals by all generators the demand equation is evaluated in each hour of the planning horizon. Hourly prices are updated by the market operator with any of the techniques stated above, and the previous procedure is repeated until the demand is satisfied. This mechanism constitutes a competitive energy market. Similarly, a spinning reserve market can be established.

It should be noted that by applying the LR technique to solve the STHTC problem, each generator schedules its production attending only to prices of energy and prices of reserve. The interchange of information between the market operator and the generators is clear and concise. The resulting market is therefore economically efficient and transparent.

The Augmented Lagrangian decomposition technique has also been applied to the STHTC problem. Relevant information can be found in Batut et al. [7], Batut and Renaud [8], Renaud [94], and Wang et al. [101].

Chapter 13

Some Useful Modeling Tricks

13.1 Introduction

This chapter has two parts. In the first part we deal with general tricks used in mathematical programming problems no matter which software is used to solve it. In the second part we deal with specific GAMS tricks.

13.2 Some General Tricks

In this section we include some useful tricks that facilitate the formulation and solution of many problems and allow stating as LPP some problems that otherwise would be NLPP. In particular, we describe some useful tricks for solving the following problems:

1. Dealing with unrestricted variables when the computer software requires non-negative variables

2. Converting a set of linear inequality constraints into an equivalent set of linear equality constraints

3. Converting a set of linear equality constraints into an equivalent set of linear inequality constraints

4. Converting a maximization problem into a minimization problem

5. Converting a nonlinear objective function to a linear one

6. Treating some nonlinear functions as linear ones

7. Treating a linear space as a cone

8. Dealing with alternative sets of constraints

9. Dealing with conditional constraints
10. Dealing with discontinuous functions
11. Dealing with piecewise nonconvex functions

13.2.1 Dealing with Unrestricted Variables

A set of unrestricted variables $\{x_1, x_2, \ldots, x_r\}$ can be replaced by a set

$$\{y_1 - z_1, y_2 - z_2, \ldots, y_r - z_r | y_1, y_2, \ldots, y_r \geq 0; z_1, z_2, \ldots, z_r \geq 0\}$$

of differences between two nonnegative variables. This implies duplicating the number of variables in the initial set.

This is possible because we can introduce the positive and negative parts of a variable x_i, defined by

$$\begin{array}{rclcl} y_i & = & x_i^+ & = & \max\{0, x_i\} \\ z_i & = & x_i^- & = & \max\{0, -x_i\} \end{array} \tag{13.1}$$

respectively. Then, it is elementary to check that $x = x_i^+ - x_i^-$, where both x_i^+ and x_i^- are nonnegative. However, since we can also write $x = (x_i^+ + k) - (x_i^- + k)$ for any $k > 0$, there are an infinite set of possible decompositions.

An alternative and better than the previous approach consists of replacing the set of r unrestricted variables $\{x_1, \ldots, x_r\}$ by the set or $r+1$ nonnegative variables $\{x_1^*, \ldots, x_r^*, x^*\}$, where

$$x_i = x_i^* - x^*; \quad i = 1, 2, \ldots, r \tag{13.2}$$

In this way we add a single new variable instead of r new variables.

The reader can easily verify that if we choose

$$x^* = -\min\{x_1, x_2, \ldots, x_r\}; \quad x_i^* = x_i + x^*$$

then the variables $\{x_1^*, \ldots, x_r^*, x^*\}$ are nonnegative. Note that if $\{x_1^*, \ldots, x_r^*, x^*\}$ satisfies (13.2), then $\{x_1^* + k, \ldots, x_r^* + k, x^* + k\}$ also satisfies (13.2), for any $k > 0$.

Example 13.1 (Conversion to nonnegative variables). Assume that we are given the set of constraints

$$\begin{array}{cccrr} x_1 & +x_2 & +x_3 & = & 1 \\ x_1 & -x_2 & -x_3 & \leq & 1 \\ x_1 & & +x_3 & \geq & -1 \\ x_1 & & & \geq & 0 \end{array} \tag{13.3}$$

where only x_1 is restricted to be nonnegative. If we want to transform this set into an equivalent set with only nonnegative variables, we make the transformation

$$\begin{array}{rcl} x_2 & = & y_2 - z_2 \\ x_3 & = & y_3 - z_3 \\ y_2, y_3, z_2, z_3 & \geq & 0 \end{array} \tag{13.4}$$

where $y_2 = x_2^+$, $y_3 = x_3^+$, and $z_2 = x_2^-$, $z_3 = x_3^-$. Thus, the initial set of constraints transforms to

$$
\begin{array}{cccccccc}
x_1 & +y_2 & -z_2 & +y_3 & -z_3 & = & 1 \\
x_1 & -y_2 & +z_2 & -y_3 & +z_3 & \leq & 1 \\
x_1 & & & +y_3 & -z_3 & \geq & -1 \\
x_1, & y_2, & z_2, & y_3, & z_3 & \geq & 0
\end{array}
\tag{13.5}
$$

■

Example 13.2 (Conversion to nonnegative variables: alternative solution). A better alternative to the solution given in the previous page consists of using the change of variables:

$$
\begin{array}{ccc}
x_2 & = & y_2 - y \\
x_3 & = & y_3 - y
\end{array}
\tag{13.6}
$$

and then, the initial set of constraints becomes:

$$
\begin{array}{ccccccc}
x_1 & +y_2 & +y_3 & -2y & = & 1 \\
x_1 & -y_2 & -y_3 & +2y & \leq & 1 \\
x_1 & & +y_3 & -y & \geq & -1 \\
x_1, & y_2, & y_3, & y, & \geq & 0
\end{array}
\tag{13.7}
$$

which has one less variable than the set (13.5). ■

13.2.2 Converting Inequalities into Equalities

Inequality constraints can be converted to equality constraints by adding new variables known as *slack variables*:

- If

$$a_{i1}x_1 + a_{i2}x_2 + \cdots + a_{in}x_n \leq b_i$$

then there exists a variable $x_{n+1} \geq 0$ such that

$$a_{i1}x_1 + a_{i2}x_2 + \cdots + a_{in}x_n + x_{n+1} = b_i$$

- If

$$a_{i1}x_1 + a_{i2}x_2 + \cdots + a_{in}x_n \geq b_i$$

then there exists a variable $x_{n+1} \geq 0$ such that

$$a_{i1}x_1 + a_{i2}x_2 + \cdots + a_{in}x_n - x_{n+1} = b_i$$

Example 13.3 (Converting inequalities to equality constraints). Introducing the slack variables, u_1 and u_2, the set of inequalities (13.5) can be transformed into the set of equalities

$$\begin{array}{cccccccc} x_1 & -y_2 & +z_2 & -y_3 & +z_3 & +u_1 & & = & 1 \\ x_1 & & & +y_3 & -z_3 & & -u_2 & = & -1 \end{array} \tag{13.8}$$

where now

$$x_1, y_2, y_3, z_2, z_3, u_1, u_2 \geq 0$$

■

13.2.3 Converting Equalities into Inequalities

A set of m equalities can always be transformed into an equivalent set of $m+1$ inequalities, as the following proposition shows.

Proposition 13.1 (Converting equalities into inequalities). *The set of equalities*

$$\mathbf{a}_i^T \mathbf{x} = b_i \tag{13.9}$$

is equivalent to the set of inequalities

$$\begin{array}{rcl} \mathbf{a}_i^T \mathbf{x} & \leq & b_i, \ i = 1, \ldots, m \\ \left(\sum\limits_{i=1}^{m} \mathbf{a}_i^T \right) \mathbf{x} & \geq & \sum\limits_{i=1}^{m} b_i \end{array} \tag{13.10}$$

■

Proof. That (13.9) implies (13.10) is obvious. Conversely, consider $k \in \{1, \ldots, m\}$. Then

$$\mathbf{a}_k^T \mathbf{x} = \left(\sum_{i=1}^{m} \mathbf{a}_i^T \right) \mathbf{x} - \left(\sum_{i \neq k} \mathbf{a}_i^T \right) \mathbf{x} \geq \sum_{i=1}^{m} b_i - \sum_{i \neq k} b_i = b_k$$

and, taking into account that $\mathbf{a}_k^T \mathbf{x} \leq b_k$, then

$$\mathbf{a}_k^T \mathbf{x} = b_k$$

■

Example 13.4 (Converting equalities into inequalities). The set of equalities

$$\begin{array}{cccccc} x_1 & +x_2 & +x_3 & = & 0 \\ x_1 & -x_2 & -x_3 & = & 2 \\ x_1 & & +x_3 & = & -1 \end{array}$$

is equivalent to the set of inequalities

$$\begin{array}{cccccc} x_1 & +x_2 & +x_3 & \leq & 0 \\ x_1 & -x_2 & -x_3 & \leq & 2 \\ x_1 & & +x_3 & \leq & -1 \\ 3x_1 & & +x_3 & \geq & 1 \end{array}$$

■

13.2.4 Converting Maximization into Minimization Problems

A maximization problem is equivalent to a minimization one by a change of sign in the objective function. Specifically, what we mean is that maximization of

$$Z_{max} = \mathbf{c}^T \mathbf{x}$$

is equivalent to minimization of

$$Z_{min} = -\mathbf{c}^T \mathbf{x}$$

when both problems are subject to the same constraints. Note that the optimal values are attained at the same points, but $Z_{max} = -Z_{min}$.

13.2.5 Converting Nonlinear Objective Functions into Linear

The problem involving minimization of

$$Z = f(\mathbf{x})$$

subject to

$$\begin{array}{rcl} \mathbf{h}(\mathbf{x}) & = & \mathbf{0} \\ \mathbf{g}(\mathbf{x}) & \leq & \mathbf{0} \end{array} \tag{13.11}$$

is equivalent to the problem involving minimization of

$$Z = y$$

subject to

$$\begin{array}{rcl} \mathbf{h}(\mathbf{x}) & = & \mathbf{0} \\ \mathbf{g}(\mathbf{x}) & \leq & \mathbf{0} \\ f(\mathbf{x}) & \leq & y \end{array} \tag{13.12}$$

The proof of this equivalence is straightforward.

This trick is useful when the solution technique requires a linear objective function.

13.2.6 Nonlinear Functions Treated as Linear Functions

The LPP is restricted to linear objective functions and linear constraints. When one of these two conditions is violated, we are in front of a NLPP. However, in some cases a NLPP can be transformed into an equivalent LPP.

Dealing with Absolute Values

A very frequently appearing function is the absolute value, which is not only a nonlinear but also a nondifferentiable function. We show how to deal with some interesting cases where this function appears.

The NLPP involving minimization of

$$Z = |\mathbf{c}^T\mathbf{x}|$$

subject to

$$\mathbf{A}\mathbf{x} = \mathbf{b}$$

is equivalent to the LPP involving minimization of

$$Z = y$$

subject to

$$\begin{array}{rcl} \mathbf{A}\mathbf{x} & = & \mathbf{b} \\ \mathbf{c}^T\mathbf{x} & \leq & y \\ -\mathbf{c}^T\mathbf{x} & \leq & y \\ y & \geq & 0 \end{array}$$

More generally, the NLPP involving minimization of

$$Z = \sum_{i=1}^{m} |\mathbf{c}_i^T\mathbf{x}|$$

subject to

$$\mathbf{A}\mathbf{x} = \mathbf{b}$$

is equivalent to the LPP involving minimization of

$$Z = \sum_{i=1}^{m} y_i$$

subject to

$$\begin{array}{rcll} \mathbf{A}\mathbf{x} & = & \mathbf{b} & \\ \mathbf{c}_i^T\mathbf{x} & \leq & y_i, & i = 1, 2, \ldots, m \\ -\mathbf{c}_i^T\mathbf{x} & \leq & y_i, & i = 1, 2, \ldots, m \\ y_i & \geq & 0, & i = 1, 2, \ldots, m \end{array}$$

Dealing with the Max Function

Another interesting function, which is also nonlinear and nondifferentiable is the max function. Next, we give an interesting trick to deal with this function.

The NLPP involving minimization of

$$Z = \max_{i=1,2,\ldots,m} |\mathbf{c}_i^T\mathbf{x}|$$

subject to

$$\mathbf{Ax} = \mathbf{b}$$

is equivalent to the LPP involving minimization of

$$Z = y$$

subject to

$$\begin{array}{rcll} \mathbf{Ax} & = & \mathbf{b} & \\ \mathbf{c}_i^T\mathbf{x} & \leq & y, & i = 1, 2, \ldots, m \\ -\mathbf{c}_i^T\mathbf{x} & \leq & y, & i = 1, 2, \ldots, m \\ y & \geq & 0 & \end{array}$$

The Fractional Programming Problem

A particular case is the *linear fractional programming problem*, which consists of optimizing the ratio or two linear functions subject to linear constraints. This problem can be stated as follows. Minimize

$$\frac{\mathbf{p}^\mathbf{T}\mathbf{x} + \alpha}{\mathbf{q}^\mathbf{T}\mathbf{x} + \beta} \tag{13.13}$$

subject to

$$\begin{array}{rcl} \mathbf{Ax} & \leq & \mathbf{b} \\ \mathbf{x} & \geq & \mathbf{0} \end{array} \tag{13.14}$$

where $\mathbf{p}$ and $\mathbf{q}$ are n vectors, $\mathbf{b}$ is an m vector, $\mathbf{A}$ is an $m \times n$ matrix, and α and β are scalars.

This problem can be converted to an equivalent LPP, using the following theorem, due to Charnes and Cooper [25].

Theorem 13.1 (Fractional programming). *If the feasible set*

$$\mathbf{X} = \{\mathbf{x} | \mathbf{Ax} \leq \mathbf{b} \text{ and } \mathbf{x} \geq \mathbf{0}\}$$

is nonempty and bounded, and $\mathbf{q}^T\mathbf{x} + \beta > 0$ for each $\mathbf{x} \in \mathbf{X}$, the problem (13.13)–(13.14) is equivalent to the LPP. Minimize

$$\mathbf{p}^T\mathbf{y} + \alpha z \tag{13.15}$$

subject to

$$\begin{array}{rcll} \mathbf{Ay} - \mathbf{b}z & \leq & \mathbf{0} & (13.16) \\ \mathbf{q}^T\mathbf{y} + \beta z & = & 1 & (13.17) \\ \mathbf{y} & \geq & \mathbf{0} & (13.18) \\ z & \geq & 0 & (13.19) \end{array}$$

that has one additional variable and one additional constraint. ∎

This theorem is based on the following transformation:

$$z = \frac{1}{\mathbf{q}^T\mathbf{x} + \beta} \quad \text{and} \quad \mathbf{y} = z\mathbf{x}, \tag{13.20}$$

Example 13.5 (Fractional programming). Assume that we are given the following linear fractional program. Minimize

$$\frac{x_1 + 1}{x_2 + 2}$$

subject to

$$\begin{array}{rcl} x_1 + x_2 & \leq & 1 \\ x_1, x_2 & \geq & 0 \end{array}$$

To transform this linear fractional problem to a linear one, we make the transformation (13.20)

$$z = \frac{1}{x_2 + 1}; \ \mathbf{y} = z\mathbf{x},$$

and we obtain the following LPP. Minimize

$$y_1 + z$$

subject to

$$\begin{array}{rcl} y_1 + y_2 - z & \leq & 0 \\ y_2 + 2z & = & 1 \\ y_1, y_2, z & \geq & 0 \end{array}$$

■

13.2.7 Linear Space as a Cone

Sometimes one can be interested in working only with nonnegative components of vectors. In these cases we can use the following proposition.

Proposition 13.2 (Linear space as a cone). *A linear space $\mathbf{A}_\rho$ is a particular case of a cone:*

$$\mathbf{A}_\rho \equiv (\mathbf{A} : -\mathbf{A})_\pi, \tag{13.21}$$

where $-\mathbf{A}$ is the negative of $\mathbf{A}$. In other words, a linear space is the cone generated by its generators and their opposite vectors. ■

Note that this is equivalent to duplicating the number of generators. However, there is a much better solution, as shown by the following theorem.

Theorem 13.2 (Linear space as a cone). *Given a linear space $\mathbf{A}_\rho$ and a vector $\mathbf{x} \in \mathbf{A}_\rho$ with all its components with respect to the set of generators being positive, then $\mathbf{A}_\rho \equiv (\mathbf{A} : -\mathbf{x})_\pi$.* ■

Proof. Since $(\mathbf{A} : -\mathbf{x})_\pi \subseteq \mathbf{A}_\rho$, we only need to prove that $\mathbf{A}_\rho \subseteq (\mathbf{A} : -\mathbf{x})_\pi$. Since $\mathbf{x}$ can be written as

$$\mathbf{x} = \sum_{i=1}^{m} \sigma_i \mathbf{a}_i; \quad \sigma_i > 0; i = 1, \ldots, m \tag{13.22}$$

if $\mathbf{y} \in \mathbf{A}_\rho$, for an arbitrary real π, we have

$$\mathbf{y} = \sum_{i=1}^{m} \rho_i \mathbf{a}_i = \sum_{i=1}^{m} \rho_i \mathbf{a}_i + \pi \mathbf{x} - \pi \mathbf{x} = \sum_{i=1}^{m} (\rho_i + \pi \sigma_i) \mathbf{a}_i + \pi(-\mathbf{x}) \tag{13.23}$$

Choosing

$$\pi = \max_{i=1,\ldots,m} \left| \frac{\rho_i}{\sigma_i} \right| \geq 0 \tag{13.24}$$

we have $\rho_i + \pi \sigma_i \geq 0; i = 1, \ldots, m$, and then $\mathbf{y} \in (\mathbf{A} : -\mathbf{x})_\pi$. ■

Remark 13.1 *The important practical implication of Theorem 13.2 is that any linear space generated by m vectors can be considered as a cone generated by $m+1$ vectors. This improves the weak result in Proposition 13.2, Expression (13.21), which required $2m$ generators.*

If one is interested in writing the linear space as a cone, there are many possible selections of $\mathbf{x}$, however, $\mathbf{x} = \sum_{i=1}^{m} \mathbf{a}_i$, which leads to $\pi = \max_{i=1,\ldots,m} |\rho_i|$, is an advantageous selection of $\mathbf{x}$. ■

Example 13.6 (Linear space as a cone). Consider the linear space $\mathbf{A}_\rho$ with

$$\mathbf{A} = \begin{pmatrix} 1 & -1 \\ 2 & 0 \\ 3 & 1 \end{pmatrix}$$

According to Remark 13.1, we can use Theorem 13.2, with

$$\mathbf{x} = \sum_{i=1,2} \mathbf{a}_i = (1, 2, 3)^T + (-1, 0, 1)^T = (0, 2, 4)^T$$

and get $\mathbf{A}_\rho \equiv (\mathbf{A} : -\mathbf{x})_\pi \equiv \mathbf{B}_\pi$, with

$$\mathbf{B} = \begin{pmatrix} 1 & -1 & 0 \\ 2 & 0 & -2 \\ 3 & 1 & -4 \end{pmatrix}$$

Now, given a vector $\mathbf{y}$ of $\mathbf{A}_\rho$, for example

$$\mathbf{y} = -(1, 2, 3)^T - 2(-1, 0, 1)^T = (1, -2, -5)^T$$

we can choose

$$\pi = \max_{i=1,2} |\rho_i| = \max(|-1|, |-2|) = 2$$

and then

$$\begin{aligned} \mathbf{y} &= (1,-2,-5)^T = (-1+2)(1,2,3)^T + (-2+2)(-1,0,1)^T + 2(0,-2,-4)^T \\ &= (1,2,3)^T + 2(0,-2,-4)^T = \mathbf{a}_1 + 2(-\mathbf{x}) \end{aligned} \tag{13.25}$$

which is a nonnegative linear combination of the vector $\mathbf{a}_1$ and the vector $-\mathbf{x}$. ■

13.2.8 Alternative Sets of Constraints

Binary variables constitute a powerful tool to model nonlinearities of diverse nature, which quite often occur in engineering. Some examples are provided below. One interesting application of binary variables allows us to deal with at least one of two sets of constraints. This can be done using the proposition below.

Proposition 13.3 (Alternative sets of constraints). *Consider two sets of constraints*

$$\mathbf{A}_1^T\mathbf{x} \le \mathbf{b}_1 \tag{13.26}$$

$$\mathbf{A}_2^T\mathbf{x} \le \mathbf{b}_2 \tag{13.27}$$

A set of constraints stating that at least one of the two above sets of constraints must be satisfied can be written as

$$\mathbf{A}_1^T\mathbf{x} - y_1\mathbf{d}_1 \le \mathbf{b}_1 \tag{13.28}$$

$$\mathbf{A}_1^T\mathbf{x} - y_2\mathbf{d}_2 \le \mathbf{b}_2 \tag{13.29}$$

$$y_1 + y_2 \le 1 \tag{13.30}$$

$$y_1, y_2 \in \{0,1\} \tag{13.31}$$

Column matrices $\mathbf{d}_1$ and $\mathbf{d}_2$ must satisfy for all $\mathbf{x}$:

$$\mathbf{A}_1^T\mathbf{x} \le \mathbf{b}_1 + \mathbf{d}_1 \Leftrightarrow \mathbf{d}_1 \ge \mathbf{A}_1^T\mathbf{x} - \mathbf{b}_1 \tag{13.32}$$

$$\mathbf{A}_2^T\mathbf{x} \le \mathbf{b}_2 + \mathbf{d}_2 \Leftrightarrow \mathbf{d}_2 \ge \mathbf{A}_2^T\mathbf{x} - \mathbf{b}_2. \tag{13.33}$$

If only one constraint must be satisfied from the original set, constraint $y_1 + y_2 \le 1$ must be replaced by $y_1 + y_2 = 1$. ■

Proof. Since (13.30) and (13.31) hold, we have three possible cases.

Case 1. $y_1 = y_2 = 0$. With this, (13.28) and (13.29) respectively become (13.26) and (13.27); thus both constraints are satisfied.

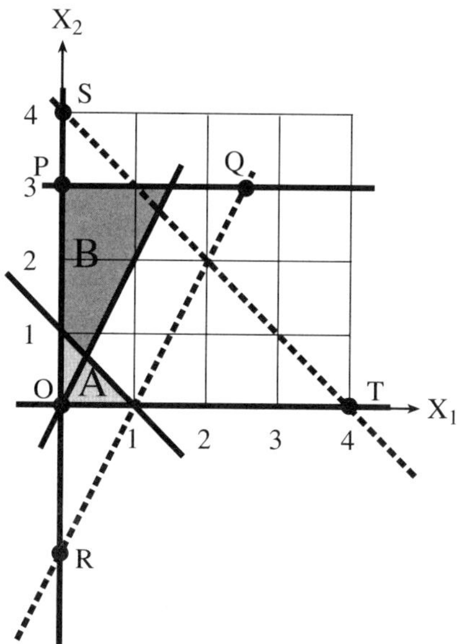

Figure 13.1: Graphical illustration of alternative sets of constraints in Example 13.7.

Case 2. $y_1 = 0$; $y_2 = 1$. Since (13.33) holds for every x, and (13.29) becomes (13.33), (13.29) is not active, that is, it can be removed.

Case 3. $y_1 = 1$; $y_2 = 0$. Since (13.32) holds for every x, and (13.28) becomes (13.32), (13.28) is not active, that is, it can be removed.

■

Example 13.7 (Alternative set of constraints). Consider the following sets of constraints:

$$\begin{array}{rcl} -x_1 & \le & 0 \\ x_2 & \le & 2 \\ 2x_1 - x_2 & \le & 0 \end{array} \Leftrightarrow \begin{pmatrix} -1 & 0 \\ 0 & 1 \\ 2 & -1 \end{pmatrix} \begin{pmatrix} x_1 \\ x_2 \end{pmatrix} \le \begin{pmatrix} 0 \\ 2 \\ 0 \end{pmatrix} \tag{13.34}$$

$$\begin{array}{rcl} -x_1 & \le & 0 \\ -x_2 & \le & 0 \\ x_1 + x_2 & \le & 1 \end{array} \Leftrightarrow \begin{pmatrix} -1 & 0 \\ 0 & 1 \\ 1 & 1 \end{pmatrix} \begin{pmatrix} x_1 \\ x_2 \end{pmatrix} \le \begin{pmatrix} 0 \\ 0 \\ 1 \end{pmatrix} \tag{13.35}$$

which lead to the two feasible sets A and B, light and dark shadowed sets, in Figure 13.1, respectively.

One way to calculate the bounds in (13.32) and (13.33) consists of finding the smallest modified region with bounds paralel to those of A containing B, this is region OST. Similarly, the smallest modified region with bounds parallel to those of B containing A is region PQR.

These two new regions can be defined by the sets of constraints

$$\begin{array}{rcl} -x_1 & \leq & 0 \\ x_2 & \leq & 2 \\ 2x_1 - x_2 & \leq & 2 \end{array} \Leftrightarrow \begin{pmatrix} -1 & 0 \\ 0 & 1 \\ 2 & -1 \end{pmatrix} \begin{pmatrix} x_1 \\ x_2 \end{pmatrix} - y_1 \begin{pmatrix} 0 \\ 0 \\ 2 \end{pmatrix} \leq \begin{pmatrix} 0 \\ 2 \\ 0 \end{pmatrix} \tag{13.36}$$

$$\begin{array}{rcl} -x_1 & \leq & 0 \\ -x_2 & \leq & 0 \\ x_1 + x_2 & \leq & 3 \end{array} \Leftrightarrow \begin{pmatrix} -1 & 0 \\ 0 & 1 \\ 1 & 1 \end{pmatrix} \begin{pmatrix} x_1 \\ x_2 \end{pmatrix} - y_2 \begin{pmatrix} 0 \\ 0 \\ 2 \end{pmatrix} \leq \begin{pmatrix} 0 \\ 0 \\ 1 \end{pmatrix} \tag{13.37}$$

from which we find that

$$\mathbf{d}_1 = \mathbf{d}_2 = \begin{pmatrix} 0 \\ 0 \\ 2 \end{pmatrix} \tag{13.38}$$

However, it is sufficient to select $\mathbf{d}_1$ and $\mathbf{d}_2$ large enough. ■

13.2.9 Dealing with Conditional Constraints

A conditional constraint of the form

$$f_1(x_1, \ldots, x_n) > b_1 \qquad \text{implies} \qquad f_2(x_1, \ldots, x_n) \leq b_2 \tag{13.39}$$

which is equivalent to the alternative set of constraints

$$f_1(x_1, \ldots, x_n) \leq b_1 \qquad \text{and/or} \qquad f_2(x_1, \ldots, x_n) \leq b_2 \tag{13.40}$$

Alternative sets of constrains have been analyzed in Section 13.2.8.

This equivalence is proved below. The original conditional constraint is not satisfied only when

$$f_1(x_1, \ldots, x_n) > b_1 \qquad \text{and} \qquad f_2(x_1, \ldots, x_n) > b_2 \tag{13.41}$$

and therefore it is satisfied when

$$f_1(x_1, \ldots, x_n) \leq b_1 \qquad \text{and/or} \qquad f_2(x_1, \ldots, x_n) \leq b_2 \tag{13.42}$$

13.2.10 Dealing with Discontinuous Functions

Binary variables also allow dealing with discontinuous functions, as it is shown by the following proposition.

Proposition 13.4 (Discontinuity). *The discontinuous function to be minimized*

$$f(x) = \begin{cases} 0 & x = 0 \\ k + cx & 0 < x \leq b; \; k > 0 \end{cases} \tag{13.43}$$

can be written as

$$\begin{array}{rcl} f(x) & = & ky + cx \\ x & \leq & by \\ -x & \leq & 0 \\ y & \in & \{0, 1\} \end{array} \tag{13.44}$$

■

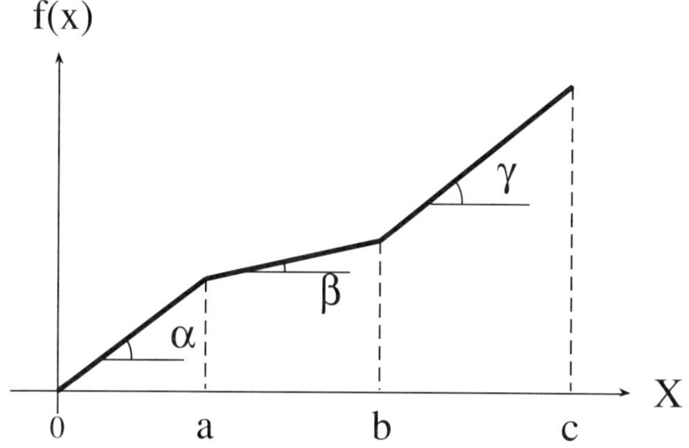

Figure 13.2: Piecewise nonconvex function.

Proof. We analyze the two possible options:

Option 1. If $y = 0$ from the second and third equations in (13.44), we get $0 \le x \le 0$; that is, $x = 0$ and replacing $x = y = 0$ into the first we get $f(x) = 0$.

Option 2. If, on the contrary, $y = 1$, we get $0 \le x \le b$ and $f(x) = k + cx$. Note that the discontinuity in zero ($f(0) = 0, f(0^+) = k$) is resolved by the condition that $f(x)$ has to be minimized and therefore $f(0) = 0$.

■

13.2.11 Dealing with Piecewise Nonconvex Functions

In this section we show how to use binary variables for dealing with piecewise nonconvex functions.

Proposition 13.5 (Piecewise nonConvex function). *The piecewise nonconvex function (see Figure 13.2)*

$$f(x) = \begin{cases} \alpha x & 0 \le x \le a \\ \alpha a + \beta(x-a) & a < x \le b \\ \alpha a + \beta(b-a) + \gamma(x-b) & b < x \le c \end{cases}$$

where

$$\beta < \alpha < \gamma; \;\; \alpha, \beta, \gamma > 0; \;\; a < b < c; \;\; a, b, c > 0$$

can be written as

$$\begin{aligned} f(x) &= \alpha x_1 + \beta x_2 + \gamma x_3 && (13.45) \\ x &= x_1 + x_2 + x_3 && (13.46) \\ a w_1 &\le x_1 \le a && (13.47) \end{aligned}$$

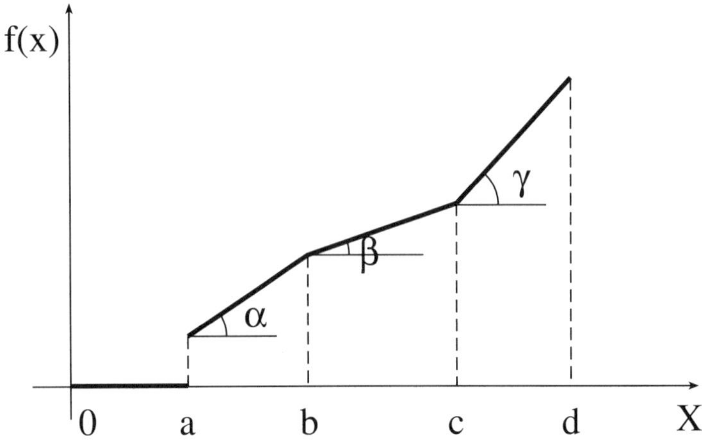

Figure 13.3: Piecewise nonconvex function with initial discontinuity.

$$
\begin{aligned}
w_2(b-a) &\le x_2 \le (b-a)w_1 & (13.48)\\
0 &\le x_3 \le (c-b)w_2 & (13.49)\\
w_2 &\le w_1 & (13.50)\\
w_1, w_2 &\in \{0,1\} & (13.51)
\end{aligned}
$$

■

Proof. As a result of the last two constraints (13.50) and (13.51), we have three cases:

Case 1. $w_2 = w_1 = 0$. This, with (13.47)–(13.49), implies $x_2 = x_3 = 0$ and $0 \le x_1 \le a$, and then (13.46) leads to $x = x_1$ and $0 \le x \le a$, and $f(x) = \alpha x$.

Case 2. $w_2 = 0$; $w_1 = 1$. This, with (13.54)–(13.56), implies $x_3 = 0$ and $0 \le x_2 \le b - a$ and $x_1 = a$, and then (13.53) leads to $x = a + x_2 \Rightarrow a \le x \le b$ and $f(x) = \alpha a + \beta(x - a)$.

Case 3. $w_2 = w_1 = 1$. This, with (13.54)–(13.56), implies $0 \le x_3 \le c - b$, $x_2 = b - a$ and $x_1 = a$, and then (13.53) leads to $x = b + x_3 \Rightarrow b \le x \le c$ and $f(x) = \alpha a + \beta(b - a) + \gamma(x - b)$.

■

Proposition 13.6 (Piecewise nonconvex function with initial discontinuity). *The piecewise nonconvex function with initial discontinuity (see Figure*

13.3)

$$f(x) = \begin{cases} 0 & 0 \leq x < a \\ f_0 + \alpha(x-a) & a < x \leq b \\ f_0 + \alpha(b-a) + \beta(x-b) & b < x \leq c \\ f_0 + \alpha(b-a) + \beta(c-b) + \gamma(x-c) & c < x \leq d \end{cases}$$

where

$$0 < \beta < \alpha < \gamma; \;\; 0 < a < b < c < d$$

can be written as

$$f(x) = vf_0 + \alpha x_1 + \beta x_2 + \gamma x_3 \tag{13.52}$$

$$x = va + x_1 + x_2 + x_3 \tag{13.53}$$

$$w_1(b-a) \leq x_1 \leq (b-a)v \tag{13.54}$$

$$w_2(c-b) \leq x_2 \leq (c-b)w_1 \tag{13.55}$$

$$0 \leq x_3 \leq (d-c)w_2 \tag{13.56}$$

$$w_2 \leq w_1 \leq v \tag{13.57}$$

$$v, w_1, w_2 \in \{0, 1\} \tag{13.58}$$

■

Proof. As a result of the last two constraints (13.57) and (13.58), we have four cases:

Case 1. $w_2 = w_1 = v = 0$. This, with (13.54)–(13.56), implies $x_1 = x_2 = x_3 = 0$, and then (13.53) leads to $x = 0$ and $f(x) = 0$.

Case 2. $w_2 = w_1 = 0;\;\; v = 1$. This, with (13.54)–(13.56), implies $x_2 = x_3 = 0$ and $0 \leq x_1 \leq b - a$, and then (13.53) leads to $x = a + x_1 \Rightarrow a \leq x \leq b$ and $f(x) = f_0 + \alpha(x - a)$.

Case 3. $w_2 = 0;\;\; w_1 = v = 1$. This, with (13.54)–(13.56), implies $x_3 = 0$ and $0 \leq x_2 \leq c - b$ and $x_1 = b - a$, and then (13.53) leads to $x = b + x_2 \Rightarrow b \leq x \leq c$ and $f(x) = f_0 + \alpha(b-a) + \beta(x-b)$.

Case 4. $w_2 = w_1 = v = 1$. This, with (13.54)–(13.56), implies $0 \leq x_3 \leq d - c$, $x_2 = c - b$ and $x_1 = b - a$, and then (13.53) leads to $x = c + x_3 \Rightarrow c \leq x \leq d$ and $f(x) = f_0 + \alpha(b-a) + \beta(c-b) + \gamma(x-c)$.

■

13.3 Some GAMS Tricks

In this section, we introduce the following GAMS tricks:

- Assigning values to a matrix
- Defining a symmetric matrix
- Defining a sparse matrix
- Splitting a problem into subproblems
- Adding constraints iteratively
- Dealing with initial and final states
- Performing a sensitivity analysis
- Making the program flow dependent on previous problem solutions

13.3.1 Assigning Values to a Matrix

Most programming languages require loop sentences to assign values to a matrix. In GAMS, the use of loops is not required for this purpose. The user can assign values to a matrix for all its indexes without loops because the assignment is made in 'parallel'. To illustrate this feature, the next example is given. Consider the following matrices:

$$\mathbf{m}_1 = \begin{pmatrix} 1 & 1 & 1 \\ 1 & 1 & 1 \end{pmatrix}, \quad \mathbf{m}_2 = \begin{pmatrix} 2 & 3 & 4 \\ 3 & 4 & 5 \end{pmatrix}$$

A possible code for defining both matrices is

```
SETS   I  row index    /1*2/
       J  column index /1*3/;

PARAMETER m1(I,J),m2(i,J);
** Paralell assignment for all indices
 m1(I,J)=1;
 m2(I,J)=ord(I)+ord(J);

DISPLAY m1,m2;
```

The above `DISPLAY` statement produces:

```
----        9 PARAMETER M1
               1            2            3
1          1.000        1.000        1.000
2          1.000        1.000        1.000

----        9 PARAMETER M2
               1            2            3
1          2.000        3.000        4.000
2          3.000        4.000        5.000
```

13.3.2 Defining a Symmetric Matrix

A simple way to represent a symmetric matrix in GAMS consists of defining just its upper triangular submatrix. The remaining submatrix is assigned by using a conditional expression on the assignment statement. To illustrate this simple trick consider the following symmetric matrix:

$$\mathbf{a} = \begin{pmatrix} 1 & 0 & 3 \\ 0 & 2 & 4 \\ 3 & 4 & 5 \end{pmatrix}$$

The corresponding GAMS code to define this matrix is

```
SET   I  row and column index    /1*3/

ALIAS(I,J);

PARAMETER a(I,J) data matrix
          /1.1  1
           2.2  2
           3.3  5
           1.3  3
           2.3  4
          /;
** Firstly the upper triangular part is displayed
DISPLAY a;

** The lower triangular part is assigned
 a(I,J)$(ord(I) gt ord(J))=a(J,I);

** The whole matrix is displayed
DISPLAY a;
```

The results of the DISPLAY statements are shown below:

```
----       13 PARAMETER A                 data matrix
              1            2            3
1        1.000                      3.000
2                      2.000        4.000
3                                   5.000

----       19 PARAMETER A                 data matrix
              1            2            3
1        1.000                      3.000
2                      2.000        4.000
3        3.000         4.000        5.000
```

A different example is presented in Section 11.4.10.

13.3.3 Defining a Sparse Matrix

There are different ways of defining a sparse matrix in GAMS. The most simple one is by enumeration of the nonzero matrix elements. This way is appropriate if the position of nonzero elements does not follow any pattern. Large sparse matrices usually show a special distribution of their nonzero elements. In these cases, the individual enumeration of the elements may be an inefficient procedure, and one is recommended to take advantage of the element distribution. If the indices of nonzero elements follow a pattern, the assignments can be made using conditions on the indices. Otherwise, a different set should be defined to map the indices of the nonzero elements with their corresponding values. Given the following matrix:

$$\mathbf{s} = \begin{pmatrix} 0 & 0 & 1 & 2 & 0 & 0 \\ 0 & 1 & 2 & 0 & 0 & 1 \\ 1 & 2 & 0 & 0 & 1 & 0 \\ 2 & 0 & 0 & 1 & 0 & 2 \\ 0 & 0 & 1 & 0 & 2 & 0 \\ 0 & 1 & 0 & 2 & 0 & 1 \end{pmatrix}$$

a possible code to take advantage of the distribution of nonzero elements is

```
SET  I  row and column index /I1*I6/;

ALIAS(I,J);

PARAMETER s(I,J) sparse matrix;

 s(I,J)$(mod((ord(I)+ord(J)),4) eq 0)=1;
 s(I,J)$(mod((ord(I)+ord(J)),5) eq 0)=2;

DISPLAY s;
```

The following code defines the matrix **s** giving the nonzero indices by enumeration:

```
SETS  I  row and column index /I1*I6/
      MAP1(I,I)     /I1.I3,I2.I2,I2.I6,I3.I5,I4.I4,I6.I6/
      MAP2(I,I)     /I1.I4,I2.I3,I4.I6,I5.I5 /;

ALIAS(I,J);

PARAMETER s(I,J) sparse matrix;

 s(I,J)$MAP1(I,J)=1;
 s(I,J)$MAP2(I,J)=2;

s(I,J)$(ord(I) gt ord(J))=s(J,I);

DISPLAY s;
```

Both codes produces the same matrix:

```
----      10 PARAMETER S                 sparse matrix
             I1          I2          I3          I4          I5          I6
I1                                1.000       2.000
I2                    1.000       2.000                               1.000
I3        1.000       2.000                               1.000
I4        2.000                               1.000                   2.000
I5                                1.000                   2.000
I6                    1.000                   2.000                   1.000
```

13.3.4 Splitting a Separable Problem

Occasionally, it may be useful from a computational perspective to decompose a problem into similar subproblems. In these cases, GAMS allows us to define the corresponding equations only once, therefore it is not necessary to define as many equation blocks as the number of subproblems. For example, given the following optimization problem. Minimize

$$Z = \sum_{j=1}^{n} \sum_{i=1}^{n} c_{ij} x_{ij}$$

subject to

$$\sum_{i=1}^{n} a_i x_{ij} \geq b_j \quad j = 1, \ldots, n$$

This problem can be split into as many subproblems as the range of j index (a total of n problems) by minimizing

$$Z = \sum_{i=1}^{n} c_{ij} x_{ij}$$

subject to

$$\sum_{i=1}^{n} a_i x_{ij} \geq b_j$$

To see how this problem is decomposed in GAMS, we suggest that the reader check the output file associated with the following code:

```
SETS   I  row index    /1*2/
       J  column index /1*3/
       DIN(J) dynamic set;

ALIAS(J,J1);

PARAMETERS c(I,J),a(I),b(J);
 c(I,J)=1; a(I)=1; b(J)=1;

POSITIVE VARIABLE x(I,J);
VARIABLE z;

EQUATIONS
 Cost(J)  objective function
 Rest(J)  restriction;

Cost(J)$DIN(J)..    SUM(I,c(I,J)*x(I,J)) =e= z;
Rest(J)$DIN(J)..    SUM(I,a(I)*x(I,J))   =g=  b(J);

Model split /all/;

DIN(J)=NO;
loop(J1,
     DIN(J1)=YES;
     Solve split using lp minimizing z;
     DIN(J1)=NO;
);
```

13.3.5 Adding Constraints Iteratively to a Problem

This trick may be useful for those algorithms that solve a problem in an iterative way. The solution of the original problem is improved by adding a new constraint at every iteration. The added constraints are structurally identical but differ in the input data. Loops and dynamic sets are needed for programming these kinds of algorithms in GAMS. First, the dynamic set is initialized to be the empty set. Then, this set is updated adding a new index element at every loop iteration. The constraints to be added should depend on the dynamic set. Consider the following sequence of problems. Minimize

$$Z = cx$$

subject to

$$ax \geq b^{(i)}, \quad i = 1, \ldots, t$$

Suppose that the maximum number of iterations required by the algorithm is 5, ($t = 5$), and also that parameter $b^{(i)}$ is increased by one unit at every iteration; then the following code can solve the preceding problem within 5 iterations:

```
SETS   I iteration counter  /1*5/
       DIN(I) dynamic set;

ALIAS(I,I1);

SCALARS c,a;
PARAMETER b(I);
 c=1; a=1; b(I)=1;

POSITIVE VARIABLE x;
VARIABLE z;

EQUATIONS
 Cost     objective function
 Rest(I)  restriction;

Cost..                c*x =e= z;
Rest(I)$DIN(I)..      a*x =g= b(I);

Model add /all/;

DIN(I)=NO;
loop(I1,
     DIN(I1)=YES;
     Solve add using lp minimizing z;
     b(I+1)=b(I)+1;
);
```

The reader should check the output list to fully comprehend this code.

13.3.6 Dealing with Initial and Final States

In many problems it is important to consider initial and final states with fixed values of their corresponding variables. In GAMS, these states may be modeled using conditional statements that affect the generic constraints. The programmer should define an ordered set containing all the indices including those referred to the initial and the final states. In GAMS, it is always possible to obtain the first and the last indices of an ordered set. For instance, given the set K referring to all the time periods, possible conditions on the constraints are

- `$(ord(K) EQ 1)` to satisfy the affected constraint for the initial period
- `$(ord(K) EQ card(K))` to satisfy the affected constraint for the final period
- `$((ord(K) GT 1) AND (ord(K) LT card(K)))` to satisfy the affected constraint for every period except the initial and final ones

For more details, see the production-scheduling and the unit commitment examples in Sections 11.2.2 and 11.3.6, respectively.

13.3.7 Performing a Sensitivity Analysis

In some applications, it may be useful to show how the solutions depend on specific parameters. The model is solved repeatedly changing some data or bounds on its variables.

For instance, the example below shows the dependence of the problem solution on the constraint coefficients. This value is changed at every iteration. Minimize

$$Z = cx$$

subject to

$$a^{(t)}x \geq b.$$

A simple change may consist of increasing the coefficient value by one unit, that is, $a^{(t+1)} = a^{(t)} + 1$. The next code solves the problem for each updated parameter, this process is repeated 5 times.

```
SET   I  iteration counter /1*5/

SCALARS a,b,c;
 a=1; b=1; c=1;

POSITIVE VARIABLE x;
VARIABLE z;

EQUATIONS
 Cost      objective function
 Rest      restriction;

Cost..    c*x =e= z;
Rest..    a*x =g= b;
```

```
Model analysis /all/;

loop(I,
     Solve analysis using lp minimizing z;
     a=a+1;
);
```

13.3.8 Making the Model Dependent on Problem States

In some cases, modelers may change the structure of their models subject to the results of previously solved problems. For instance, if the problem is unbounded, a new constraint can be added to the model and solve it again. The following problem is a simple example. Minimize

$$Z = cx$$

If the solution of the problem above is unbounded, a new constraint is added, and the next problem to be solved consists of minimization of

$$Z = cx$$

subject to

$$ax \geq b$$

The code below deals with the explained conditional modeling. Some comments are included to clarify the code.

```
SCALARS c,a,b,flag;
 c=1; a=1; b=1;

VARIABLE x,z;

EQUATIONS
 COST      objective function
 REST      restriction to be added;

COST..                  c*x =e= z;
REST$(flag eq 1)..      a*x =g= b;

MODEL analysis /ALL/;

** If the flag is 0 the constraint REST is not included in the model
   flag=0;

** The model is solved without REST constraint
   SOLVE analysis USING lp MINIMIZING z;

** If the problem is unbounded a new constraint is added
   if(analysis.modelstat eq 3,

**   The constraint REST is included in the model
     flag=1;
```

```
**   the model is solved again with the new constraint
     SOLVE analysis USING lp MINIMIZING z;
  );
```

The reader should check the output listing for a better understanding of this example.

Exercises

13.1 An estimation problem can be formulated as

$$\text{Minimize}_{\Theta \in C} \ F(\hat{\mathbf{y}}, \mathbf{y}(\Theta))$$

where $\hat{\mathbf{y}}$ is the observed data, and $\mathbf{y}(\Theta)$ is the predicted values by a model in function of the vector of parameters that we wish to estimate. F may be any suitable metric between the observed and predicted data. If F is the metric derived from the euclidean norm ($\|\cdot\|_2$), the method is called the least-squares method (LSM), and F becomes $\sum_{i=1}^{n} (\hat{y}_i - y_i)^2$. If F is derived from ℓ_1 norm ($\|\cdot\|_1$), that is, $F(\hat{\mathbf{y}}, \mathbf{y}) = \sum_{i=1}^{n} |\hat{y}_i - y_i|$, the method is called the *least-absolute-value* (LAV) *method*. Other example is derived from the supremum norm ($\|\cdot\|_\infty$) that leads to $F(\hat{\mathbf{y}}, \mathbf{y}) = \max\{|\hat{y}_i - y_i \; i = 1, \cdots, n\}$. This method is called the *min-max estimation method* (MM). Consider two theoretical models to describe the variable y_i as a function of the parameters a and b

$$\begin{array}{lrcl} \text{Linear model} & y_i & = & a + bx_i \\ \text{Nonlinear model} & y_i & = & (x_i - a)^b \end{array}$$

Estimate the parameters a and b assuming that $(a, b) \in \mathbb{R}^2$ of both models for the inputs shown in Table 13.1 using the methods LSM, LAV, and MM.

Table 13.1: Inputs for the estimation problem

x_i	$\hat{y}_i$
0	1.4
1	3.8
1.5	6.7
2	9.2
3	15.2
4	24.3
5	30.2

13.2 Consider the following sequence of problems. Minimize

$$Z = cx$$

subject to

$$a^{(i)}x \geq b^{(i)} \quad i = 1, \ldots, t$$

Suppose that the maximum number of iterations required by the algorithm is 5, ($t = 5$), and also that parameters $(a^{(i)}, b^{(i)})$ are shown in Table 13.2. Write a GAMS code to the preceding sequence of problems.

Table 13.2: Parameters of the sequence of problems

$a^{(i)}$	$b^{(i)}$
1	1.7
2	2.4
3	3.8
4	4.2
5	5.7

13.3 Consider a power circuit of 3 buses and 3 lines. Line admitances and capacity transmission limits are given in the table below:

Line	Admitance	Limit
1–2	2.5	0.3
1–3	3.5	0.7
2–3	3.0	0.7

Power demand for bus 3 is 0.85 MW. Two generators are located in buses 1 and 2, respectively. Their production limits are given in the table below:

Generator	Upper limit	Lower limit
1	0.9	0
2	0.9	0

The production cost of any of the 2 generators is expressed as

$$c_i = \begin{cases} 0 & \text{if } p_i = 0 \\ F_i + V_i p_i & \text{if } p_i > 0 \end{cases}$$

where F_i is the fixed cost, V_i the variable cost, and p_i the output power. The table below gives data for the 2 generators:

Generator	F_i	V_i
1	10	6
2	5	7

Determine the production of each generator so that the total cost of supplying the demand is minimized. Repeat the problem using $F_2 = 10$.

13.4 Consider the following linear programming problem. Minimize

$$Z = x_1$$

subject to

$$\begin{array}{rrcr} -2x_1 & + x_2 & \leq & 0 \\ x_1 & - x_2 & \leq & 2 \\ x_1 & + x_2 & \leq & 6 \end{array}$$

(a) Solve this problem graphically

(b) Formulate this problem in standard form using the following two means:

(i) Dealing with unrestricted variables as difference between two nonnegative variables

(ii) Dealing with unrestricted variables adding a single new variable

(c) Use the GAMS package to solve both formulations and compare the results

13.5 Consider the following linear fractional programming problem. Minimize

$$Z = \frac{-2x_1 - 2x_3 + 3}{x_1 + x_2 + x_3 + 1}$$

subject to

$$\begin{array}{rrcr} -2x_1 & + x_2 & \leq & 0 \\ & - x_2 & \leq & 0 \\ x_1 & - x_2 & \leq & 2 \\ x_1 & + x_2 & \leq & 6 \end{array}$$

(a) Use the GAMS package to solve the linear fractional programming problem

(b) Formulate this problem as a linear programming problem

(c) Use the GAMS package to solve the linear formulation and compare with the results obtained from (a).

13.6 Consider the two sets of constraints

$$A = \left\{ \begin{array}{rrcr} -2x_1 & + x_2 & \leq & 0 \\ x_1 & - x_2 & \leq & 2 \\ x_1 & + x_2 & \leq & 6 \end{array} \right.$$

$$B = \left\{ \begin{array}{rrcr} -x_1 & + x_2 & \leq & -6 \\ -x_1 & + x_2 & \leq & -2 \\ 3x_1 & - x_2 & \leq & 18 \end{array} \right.$$

Table 13.3: Temperature distribution within the wall

Distance	0	0.2	0.4	0.6	0.8	1.0
Temperature	400	350	250	175	100	50

(a) Formulate mathematically the following problem. Minimize

$$Z = x_1 - 3x_2$$

subject to

$$(x_1, x_2) \quad \in \quad A \cup B$$

(b) Solve this problem graphically

(c) Use the GAMS package to solve this problem. Modify the GAMS code to obtain the other optimal solution

13.7 Consider a linear programming problem with equality and inequality constraints. Write an equivalent one containing only inequality constraints.

13.8 Consider a mathematical programming problem with a nonlinear objective function. Write an equivalent one with a linear objective function.

13.9 Table 13.3 provides the temperature at various points within a wall, heated in one of its sides, where distances are given in proportions with respect to total width. We would like to adjust this data set to a linear model yielding temperature in terms of distance

$$T = a + bx$$

where a and b are constants. Determine these coefficients, by using the GAMS package and the tricks described in this chapter, when the parameters are estimated by minimizing the errors:

(a)

$$\sum_i \|a + bx_i - T_i\|$$

(b)

$$\max_i \|a + bx_i - T_i\|$$

where the pairs (x_i, T_i) are given in Table 13.3.

13.10 Given a matrix $T(i, j)$, write just one sentence in GAMS to count the elements that belong to the main diagonal and differ from 10.

13.11 Once a variable is fixed to a value, it remains constant throughout the program execution. Suppose that it is needed to solve a problem iteratively and the variable to be fixed is different at every iteration. Write a generic GAMS code to deal with this situation.

Appendix A

Compatibility and Set of All Feasible Solutions

In Chapter 5 we have seen that the set of all feasible solutions of a system of linear equations and inequalities is a polyhedron, which can degenerate to an empty set, a linear space, a cone, a polytope or to a sum of several of these three structures. The aim of this appendix consists of studying the set of all feasible solutions and solving the following two problems:

1. Determining whether a feasible solution exists, that is, if the feasible set is not empty

2. Obtaining the set of all feasible solutions, that is, identifying all elements in this set

3. Interpreting the set of feasible solutions for many of the practical examples introduced in previous chapters

Solving these problems requires adequate tools that are not easily available in the existing literature. Thus, we have included in this appendix the most important concepts and tools required for working with linear equations and inequalities.

The appendix is structured as follows. In Section A.1 we introduce the concept of dual cone, which plays an important role in discussing the compatibility and in solving systems of linear constraints. In Section A.2 we show how the cone associated with a polyhedron can be obtained. In Section A.3 the Γ algorithm that allows us to obtain the dual cone and find the set of all feasible solutions is given. In Section A.4 we deal with the problem of compatibility of a set of constraints. In Section A.5 we find the set of all feasible solutions. Finally, in Section A.6 we discuss the compatibility problem and find the set of all feasible solutions of some of the problems discussed in previous chapters, and give a physical interpretation of all results.

As mentioned previously, although nonlinear constraints are important, this chapter is devoted only to the case of linear constraints.

Most of the theorem proofs in this chapter have been omitted because they are considered classical results. The interested reader can consult classical textbooks in convex analysis, as Rockafellar [95], for example. On the other hand, detailed proofs are only given for material relevant to this chapter.

A.1 The Dual Cone

Definition A.1 (Nonpositive dual or polar cone). *Let $\mathbf{A}_\pi$ be a cone in $\mathbb{R}^n$ with generators $\mathbf{a}_1, \ldots, \mathbf{a}_k$. The nonpositive dual $\mathbf{A}^*_\pi$ of $\mathbf{A}_\pi$ is defined as the set*

$$\mathbf{A}^*_\pi \equiv \left\{\mathbf{u} \in \mathbb{R}^n \mid \mathbf{A}^T\mathbf{u} \le \mathbf{0}\right\} \equiv \left\{\mathbf{u} \in \mathbb{R}^n \mid \mathbf{a}_i^T\mathbf{u} \le 0;\ i = 1, \ldots, k\right\}$$

that is, the set of all vectors such that their dot products by all vectors in $\mathbf{A}_\pi$ are nonpositive. ∎

Using Theorem 5.4 it can be shown that $\mathbf{A}^*_\pi$ is a cone too, which is known as the *dual cone.* Using the general form of a cone (see Definition 5.9), we can write the dual cone $\mathbf{A}^*_\pi$ of $\mathbf{A}_\pi$ in the form $\mathbf{A}^*_\pi \equiv \mathbf{V}_\rho + \mathbf{W}_\pi$, where $\mathbf{V}$ is the set of generators of a linear subspace and $\mathbf{W}$ is the set of generators of a proper cone.

Remark A.1 *The nonpositive dual of a linear space coincides with its orthogonal:*

$$\begin{aligned} \mathbf{A}^*_\rho &\equiv (\mathbf{A}, -\mathbf{A})^*_\pi \equiv \{\mathbf{u} \in \mathbb{R}^n | \mathbf{a}_i^T\mathbf{u} \le 0 \text{ and } -\mathbf{a}_i^T\mathbf{u} \le 0; i = 1, \ldots, k\} \\ &\equiv \{\mathbf{u} \in \mathbb{R}^n | \mathbf{a}_i^T\mathbf{u} = 0; i = 1, \ldots, k\} \end{aligned}$$

∎

Remark A.2 *The cone $\mathbf{Ax} \le \mathbf{0}$, that is, the feasible set of the linear system of inequalities $\mathbf{Ax} \le \mathbf{0}$, coincides with the dual of the cone generated by the rows of $\mathbf{A}$. Then, obtaining the generators of $\mathbf{Ax} \le \mathbf{0}$ is the same as obtaining the generators of the dual cone of the cone generated by the rows of $\mathbf{A}$. Thus, solving the linear set of equations and inequalities $\mathbf{Ax} \le \mathbf{0}$ is equivalent to obtaining the generators of the dual cone of the cone generated by the rows of $\mathbf{A}$.* ∎

Note that this remark is an immediate consequence of the definition of a dual cone.

Example A.1 (Dual cone). Figure A.1 shows the dual cone, $\mathbf{A}^*_\pi \equiv (\mathbf{u}_1, \mathbf{u}_2)_\pi$, associated with the cone $(\mathbf{a}_1, \mathbf{a}_2)_\pi$.

∎

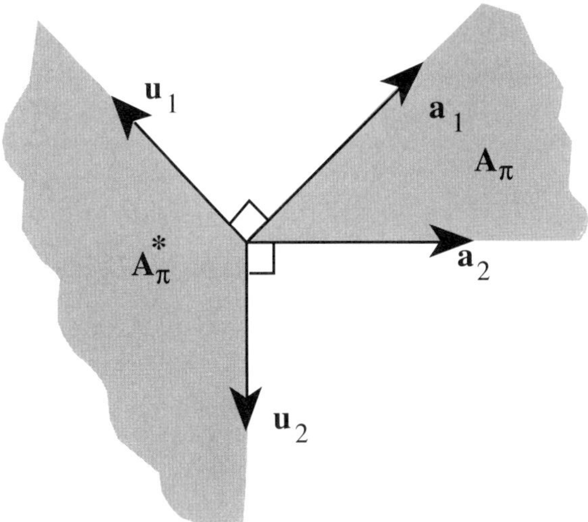

Figure A.1: Example of a cone $\mathbf{A}_\pi$ and its dual cone, $\mathbf{A}_\pi^*$,

The dual cone satisfies the following properties:

1. The dual is unique: $\mathbf{A}_\pi \equiv \mathbf{B}_\pi \iff \mathbf{A}_\pi^* \equiv \mathbf{B}_\pi^*$.
2. The dual of a subset contains the dual of the superset:

$$\mathbf{E}_\pi \subseteq \mathbf{A}_\pi \iff \mathbf{A}_\pi^* \subseteq \mathbf{E}_\pi^*.$$

The following lemma is due to Farkas (1902) and is the basis for the duality theory.

Lemma A.1 (Farkas). *If $\mathbf{A}_\pi^*$ is the nonpositive dual of $\mathbf{A}_\pi$, then*

$$\mathbf{x} \in \mathbf{A}_\pi \Leftrightarrow \boldsymbol{\mu}^T \mathbf{x} \leq 0; \forall \boldsymbol{\mu} \in \mathbf{A}_\pi^*. \tag{A.1}$$

In other words, the dual of the dual of $\mathbf{A}_\pi$ is $\mathbf{A}_\pi$ itself: $\mathbf{A}_\pi^{**} \equiv \mathbf{A}_\pi$.

This lemma leads to the following theorem, that allows determining whether a vector belongs to a cone.

Theorem A.1 (Vector in a cone). *The vector $\mathbf{x}$ belongs to the cone $\mathbf{A}_\pi$ if and only if $\mathbf{V}^T\mathbf{x} = \mathbf{0}$ and $\mathbf{W}^T\mathbf{x} \leq \mathbf{0}$:*

$$\mathbf{x} \in \mathbf{A}_\pi \Leftrightarrow \mathbf{V}^T \mathbf{x} = \mathbf{0} \text{ and } \mathbf{W}^T \mathbf{x} \leq \mathbf{0}, \tag{A.2}$$

where $\mathbf{V}_\rho + \mathbf{W}_\pi$ is the general form of the dual cone $\mathbf{A}_\pi^*$ ∎

Proof. The dual cone $\mathbf{A}_\pi^*$ can be written as

$$\mathbf{A}_\pi^* \equiv \mathbf{U}_\pi \equiv \mathbf{V}_\rho + \mathbf{W}_\pi \equiv (\mathbf{V} : -\mathbf{V} : \mathbf{W})_\pi \tag{A.3}$$

According to the Farkas lemma, we have

$$\begin{array}{rcl} \mathbf{x} \in \mathbf{A}_\pi & \Leftrightarrow & \mathbf{U}^T\mathbf{x} \le \mathbf{0} \Leftrightarrow \mathbf{V}^T\mathbf{x} \le \mathbf{0} \text{ and } -\mathbf{V}^T\mathbf{x} \le \mathbf{0} \text{ and } \mathbf{W}^T\mathbf{x} \le \mathbf{0} \\ & \Leftrightarrow & \mathbf{V}^T\mathbf{x} = \mathbf{0} \text{ and } \mathbf{W}^T\mathbf{x} \le \mathbf{0} \end{array} \tag{A.4}$$

■

A.2 Cone Associated with a Polyhedron

Dual cones are very useful for obtaining minimal sets of generators, the linear space, the cone, and the polytope, defining the polyhedron of feasible solutions. To this aim, we replace our $\mathbb{R}^n$ Euclidean space by another $\mathbb{R}^{n+1}$ Euclidean space (see Castillo et al. [21]). Using this trick, we convert polyhedra in $\mathbb{R}^n$ into cones in $\mathbb{R}^{n+1}$. Thus, we need to deal only with cone structures. However, at the end, we need to go back to our initial Euclidean space $\mathbb{R}^n$.

More precisely, the polyhedron

$$S = \{\mathbf{x} \in \mathbb{R}^n | \mathbf{Hx} \le \mathbf{a}\}$$

is replaced by the cone

$$\mathbf{C}_S = \left\{ \begin{pmatrix} \mathbf{x} \\ x_{n+1} \end{pmatrix} \in \mathbb{R}^{n+1} | \mathbf{Hx} - x_{n+1}\mathbf{a} \le \mathbf{0};\ -x_{n+1} \le 0 \right\}$$

This process is illustrated in Figure A.2, where we start from a pentagon (polyhedron) in $\mathbb{R}^2$ and we move to $\mathbb{R}^3$ by adding one more unit component to its vertices (extreme points). Next, we consider the cone generated by these extreme directions. Note that extreme points in $\mathbb{R}^2$ transform to extreme directions in $\mathbb{R}^3$. Once we are in $\mathbb{R}^3$, we look for the cone generators, using the Γ algorithm to be described below. Finally, we recover our initial constraints in $\mathbb{R}^2$ by imposing the constraint $x_n = 1$, thus, obtaining the intersection of the cone with the plane (hyperplane) $x_n = 1$.

It is important to note that if the polyhedron has directions, these would generate directions in the associated cone. The following theorem relates the polyhedron and cone sets of generators.

Theorem A.2 (Polyhedron and associated cone generators). *Let* $\mathbf{H}$ *be a matrix and consider the polyhedron* $\mathbf{Hx} \le \mathbf{a}$. *Then the system of inequalities* $\mathbf{Hx} \le \mathbf{a}$ *is equivalent to*

$$\begin{array}{rcl} \mathbf{C}\begin{pmatrix} \mathbf{x} \\ x_{n+1} \end{pmatrix} & \le & \mathbf{0} \\ x_{n+1} & = & 1 \end{array}$$

where

$$\mathbf{C} = \begin{pmatrix} \mathbf{H} & | & -\mathbf{a} \\ -- & + & -- \\ \mathbf{0} & | & -1 \end{pmatrix}$$

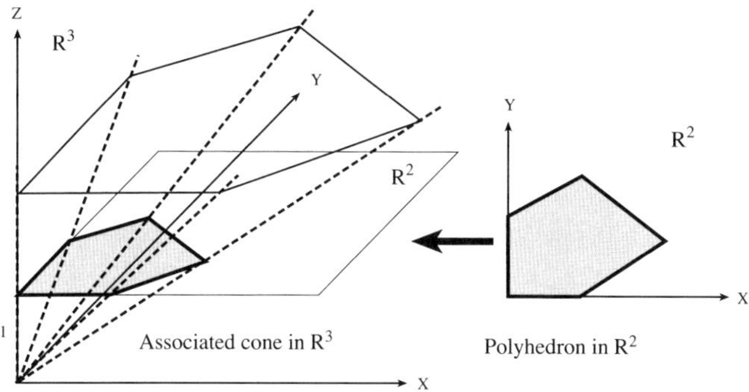

Figure A.2: Illustration of how the cone associated with a polyhedron is obtained.

If $\mathbf{C}_\pi^{T^*} = \mathbf{V}_\rho^0 + \mathbf{W}_\pi^0$ *is the general form of* $\mathbf{C}_\pi^{T^*}$, *where* $\mathbf{W}^0$ *has all last components equal to zero and one, then the general form of the polyhedron* $\mathbf{Hx} \le \mathbf{a}$ *is*

$$S = \mathbf{V}_\rho + \mathbf{W}_\pi + \mathbf{Q}_\lambda$$

where $\mathbf{V}$ *is* $\mathbf{V}^0$ *with their last components removed, and* $\mathbf{W}$ *and* $\mathbf{Q}$ *contain all column vectors in* $\mathbf{W}^0$ *such that their last components are null and nonnull, respectively, with their last components removed.* ■

Proof. Assume that $\mathbf{C}_\pi^{T^*} = \mathbf{V}_\rho^0 + \mathbf{W}_\pi^0$. Then, all the vectors in $\mathbf{V}^0$ must have a null last component, because of the redundant constraint $-x_{n+1} \le 0$. In addition, and without loss of generality, we can assume that w_{n+1k}^0, the last component of any vectors in $\mathbf{W}^0$, is 0 or 1, because if it is not, we can divide the whole vector by w_{n+1k}^0.

Finally, condition $x_{n+1} = 1$ leads to

$$\sum_{k \in K_{W^0}} \pi_k = 1 \tag{A.5}$$

where

$$K_{W^0} \equiv \{k | w_{n+1k}^0 \ne 0\} \tag{A.6}$$

is the set of all vectors in $\mathbf{W}^0$ with its last component nonnull.

We have the following two cases:

1. $K_{W^0} \equiv \emptyset$: In this case the polyhedron is empty, since this implies $x_{n+1} = 0 \ne 1$.

2. $K_{W^0} \ne \emptyset$: Letting $\lambda_k = \pi_k$, (A.5) becomes

$$\sum_{k \in K_{W^0}} \lambda_k = 1 \tag{A.7}$$

and then, returning to our initial $\mathbb{R}^n$ space, the cone $\mathbf{V}_\rho^0 + \mathbf{W}_\pi^0$ transforms to the polyhedron $\mathbf{V}_\rho + \mathbf{W}_\pi + \mathbf{Q}_\lambda$, where $\mathbf{W}$ and $\mathbf{Q}$ are the matrices whose columns have null and unit last component, respectively, and its last components have been removed.

■

Thus, if we have a procedure to identify the cone generators, we immediately have a procedure to identify the polyhedron generators. One of these methods has been developed by Castillo et al. [21], and is known as the "Γ-algorithm", to be described in the following section.

Example A.2 (Polyhedron and associated cone generators). Consider the polyhedron S:

$$\begin{array}{rrcl} -x_2 & & \leq & 0 \\ & -x_3 & \leq & 0 \\ & x_3 & \leq & 1 \end{array} \tag{A.8}$$

that can be written as

$$\begin{array}{rrrcl} -x_2 & & & \leq & 0 \\ & -x_3 & & \leq & 0 \\ & x_3 & -x_4 & \leq & 0 \\ & & -x_4 & \leq & 0 \\ & & x_4 & = & 1 \end{array} \tag{A.9}$$

or as

$$\mathbf{C}\begin{pmatrix} \mathbf{x} \\ x_4 \end{pmatrix} = \begin{pmatrix} 0 & -1 & 0 & 0 \\ 0 & 0 & -1 & 0 \\ 0 & 0 & 1 & -1 \\ 0 & 0 & 0 & -1 \end{pmatrix} \begin{pmatrix} x_1 \\ x_2 \\ x_3 \\ x_4 \end{pmatrix} \leq \mathbf{0}; \;\; x_4 = 1 \tag{A.10}$$

The dual $\mathbf{C}^{T^*}_\pi$ is the set of vectors of the form

$$\begin{pmatrix} x_1 \\ x_2 \\ x_3 \\ x_4 \end{pmatrix} = \rho_1 \begin{pmatrix} 1 \\ 0 \\ 0 \\ 0 \end{pmatrix} + \pi_1 \begin{pmatrix} 0 \\ 1 \\ 0 \\ 0 \end{pmatrix} + \pi_2 \begin{pmatrix} 0 \\ 0 \\ 1 \\ 1 \end{pmatrix} \Rightarrow \mathbf{V}^0 = \begin{pmatrix} 1 \\ 0 \\ 0 \\ 0 \end{pmatrix}, \;\; \mathbf{W}^0 = \begin{pmatrix} 0 & 0 \\ 1 & 0 \\ 0 & 1 \\ 0 & 1 \end{pmatrix} \tag{A.11}$$

Note that all last components are already unit or null.

Then, according to Theorem A.2, the polyhedron is

$$S = \mathbf{V}_\rho + \mathbf{W}_\pi + \mathbf{Q}_\lambda = \begin{pmatrix} 1 \\ 0 \\ 0 \end{pmatrix}_\rho + \begin{pmatrix} 0 \\ 1 \\ 0 \end{pmatrix}_\pi + \begin{pmatrix} 0 \\ 0 \\ 1 \end{pmatrix}_\lambda, \tag{A.12}$$

that is, the vectors in the polyhedron are of the form [see Figure 5.1(c)]

$$\begin{pmatrix} x_1 \\ x_2 \\ x_3 \end{pmatrix} = \rho_1 \begin{pmatrix} 1 \\ 0 \\ 0 \end{pmatrix} + \pi_1 \begin{pmatrix} 0 \\ 1 \\ 0 \end{pmatrix} + \lambda_1 \begin{pmatrix} 0 \\ 0 \\ 1 \end{pmatrix} \tag{A.13}$$

■

A.3 The Γ Procedure

In this section we give the Γ algorithm, which allows us to solve systems of linear equations and inequalities, by calculating, with a minimal sets of generators, the dual cone $\mathbf{A}_\pi^*$ of $\mathbf{A}_\pi$ in its general form.

The computational process associated with what has been named the Γ algorithm has been treated by Jubete [61, 62]. This procedure allows us to obtain the dual $\mathbf{A}_\pi^* \equiv \mathbf{V}_\rho + \mathbf{W}_\pi$ of $\mathbf{A}_\pi \equiv \mathbf{B}_\rho + \mathbf{C}_\pi$:

$$\begin{array}{rcl} \mathbf{A}_\pi^* & \equiv & \{\boldsymbol{\mu} \in \mathbb{R}^n | \boldsymbol{\mu}^T \mathbf{a}_i \leq 0; i = 1, \ldots, k\} \\ & \equiv & \{\boldsymbol{\mu} \in \mathbb{R}^n | \boldsymbol{\mu}^T \mathbf{b}_i = 0, \boldsymbol{\mu}^T \mathbf{c}_j \leq 0, \forall i, j\} \end{array} \quad (A.14)$$

with a minimum number of generators. A description of this algorithm is given below. More details and examples can be seen in Castillo et al. [21].

The reader should keep in mind that the ultimate goal consists of finding the general solution of systems of inequalities. Duals are only an intermediate step.

Algorithm A.1 (The Γ Algorithm).

- **Input.** A cone defined by the set of its generators $\mathbf{A} = \{\mathbf{a}_1, \ldots, \mathbf{a}_m\}$ in the Euclidean space $\mathbb{R}^n$.

- **Output.** The generators of the dual cone.

Initialization.

- Let $\mathbf{U} = \mathbf{I}_n$, where $\mathbf{I}_n$ is the identity matrix of dimension n, and let V equal to the list (ordered set) from 1 to n.

- Initialize one list $A_0(\mathbf{u}_j)$ (the list of $\mathbf{A}$ vectors (vectors in $\mathbf{A}$) orthogonal to $\mathbf{u}_j$ (column of $\mathbf{U}$) for each vector $\mathbf{u}_j; j = 1, 2, \ldots, n$ to the empty set, and let $h = 1$.

Step 1. (Calculate the dot products). Calculate $\mathbf{t} = \mathbf{a}_h^T \mathbf{U}$.

Step 2. (Look for the pivot). Find a pivot column ($t_{pivot} \neq 0$) among those columns with indices in V.

Step 3. (Test for processes 1 and 2). If no pivot has been found, go to Process 2 (Step 5). Otherwise go to Process 1 (step 4).

Step 4. (Process 1). Normalize the pivot column by dividing it by $-t_{pivot}$ (note that the pivot is an index), and perform the pivoting operation by letting $u_{ij} = u_{ij} + t_j u_{i\,pivot}$ for all $j \neq pivot$. Append the index h to the $A_0(\mathbf{u}_j)$ lists for all $j \neq pivot$. Remove index $pivot$ from V and change vector $\mathbf{u}_{pivot}$ from $\mathbf{v}$ to $\mathbf{w}$. Then, go to step 6.

Table A.1: Added constraint, **T** matrix, and associated dot products

Phase 1					Phase 2				
$\mathbf{a}_1$	$\mathbf{v}_1^1$	$\mathbf{v}_2^1$	$\mathbf{v}_3^1$	$\mathbf{v}_4^1$	$\mathbf{a}_2$	$\mathbf{w}_1^2$	$\mathbf{v}_1^2$	$\mathbf{v}_2^2$	$\mathbf{v}_3^2$
1	**1**	0	0	0	2	–1	**0**	0	–1
0	**0**	1	0	0	1	0	**1**	0	0
0	**0**	0	1	0	0	0	**0**	1	0
1	**0**	0	0	1	0	0	**0**	0	1
$\mathbf{t}^1$	**1**	0	0	1	$\mathbf{t}^2$	–2	**1**	0	–2
		1	1		A_0		1	1 2	1

Step 5. (Process 2). Determine the sets

$$\begin{array}{lcl} \mathbf{I}^- & \equiv & \{i \mid \mathbf{w}_i^T \mathbf{a}_h < 0; i \notin V\} \\ \mathbf{I}^+ & \equiv & \{i \mid \mathbf{w}_i^T \mathbf{a}_h > 0; i \notin V\} \\ \mathbf{I}^0 & \equiv & \{i \mid \mathbf{w}_i^T \mathbf{a}_h = 0; i \notin V\} \end{array} \tag{A.15}$$

and append to $A_0(\mathbf{u}_j)$ the index h for all $j \in \mathbf{I}_0$. If $\mathbf{I}^+ \equiv \emptyset$, then go to step 6. Otherwise, if $\mathbf{I}^- \equiv \emptyset$, remove from $\mathbf{U}$ all vectors $\mathbf{u}_j$; $j \in \mathbf{I}^+$. Otherwise, for each $i \in \mathbf{I}^-$ and $j \in \mathbf{I}^+$ build the set of vectors

$$\mathbf{W}^{\pm} \equiv \{\mathbf{w} \mid \mathbf{w} = t_j \mathbf{w}_i - t_i \mathbf{w}_j\}$$

and initialize for each set $\mathbf{w} \in \mathbf{W}^{\pm}$ a list $\mathbf{A}_0(\mathbf{w})$ to the set $(A_0(\mathbf{u}_i) \cap A_0(\mathbf{u}_j)) \cup \{h\}$. Remove from $\mathbf{U}$ all vectors $\mathbf{u}_j; j \in \mathbf{I}^+$, and append to $\mathbf{U}$ the vectors $\mathbf{w} \in \mathbf{W}^{\pm}$ such that

$$\mathbf{w} \neq \mathbf{0};\ A_0(\mathbf{w}) \not\subset A_0(\mathbf{w}^*), \mathbf{w} \neq \mathbf{w}^* \in \mathbf{W}^{\pm};\ A_0(\mathbf{w}) \not\subset A_0(\mathbf{w}_j), j \in \mathbf{I}^0$$

Step 6. (Process 2). If $h < m$, let $h = h + 1$ and go to step 1; otherwise, return matrix $\mathbf{U}$, and exit.

■

Example A.3 (Dual cone). Consider the cone

$$\mathbf{A}_\pi = \begin{pmatrix} 1 & 2 & 0 & 0 & 1 & 0 \\ 0 & 1 & -1 & -1 & 1 & 0 \\ 0 & 0 & 1 & 1 & 1 & 0 \\ 1 & 0 & 2 & 1 & 1 & -1 \end{pmatrix} \tag{A.16}$$

Initialization. We let $\mathbf{U} = \mathbf{I}_n$ as indicated in Table A.1 (phase 1) and let $V = \{1, 2, 3, 4\}$; that is, in this initial step all vectors are of type $\mathbf{v}$, that is, generators of a linear space. We also initialize 4 lists, one per each vector $\mathbf{v}$ to empty, and let $h = 1$.

Table A.2: Added constraint, **T** matrix, and associated dot products

Phase 3					Phase 4				
$\mathbf{a}_3$	$\mathbf{w}_1^3$	$\mathbf{w}_2^3$	$\mathbf{v}_1^3$	$\mathbf{v}_2^3$	$\mathbf{a}_4$	$\mathbf{w}_1^4$	$\mathbf{w}_2^4$	$\mathbf{w}_3^4$	$\mathbf{v}_1^4$
0	–1	0	**0**	–1	0	–1	0	0	**–1**
–1	2	–1	**0**	2	–1	2	–1	0	**2**
1	0	0	**1**	0	1	2	–1	–1	**0**
2	0	0	**0**	1	1	0	0	0	**1**
$\mathbf{t}^3$	–2	1	**1**	0	$\mathbf{t}^4$	0	0	–1	**–1**
A_0	2	1	1	1	A_0	2	1	1	1
			2	2		3	3	2	2
				3		4	4		3

Step 1. We incorporate the first column of **A** in the first column of Table A.1 (phase 1), calculate the dot products $\mathbf{t} = \mathbf{a}_1^T\mathbf{U}$, and write them in the last row of Table A.1 (phase 1).

Step 2. We look for the pivot and find pivot=1.

Step 3. Since we have found a pivot, we go to process 1.

Step 4. We normalize the pivot column, perform the pivoting process, append index 1 to the $A_0(\mathbf{u}_j)$ lists for all $j \neq 1$, and change vector $\mathbf{u}_1$ from **v** to **w**, obtaining the matrix in Table A.1 (phase 2).

Step 6. Since $h < 6$, we let $h = 2$ and go to step 1.

Step 1. We incorporate the second column of **A** in the first column of Table A.1 (phase 2), calculate the dot products $\mathbf{t} = \mathbf{a}_2^T\mathbf{U}$, and write them in the last row of Table A.1 (phase 2).

Step 2. We look for the pivot and find pivot=2.

Step 3. Since we have found a pivot, we go to the process 1.

Step 4. We normalize the pivot column, perform the pivoting process, append the index 2 to the $A_0(\mathbf{u}_j)$ lists for all $j \neq 2$, and change vector $\mathbf{u}_2$ from **v** to **w**, obtaining the matrix in Table A.2 (phase 3).

Step 6. Since $h < 6$, we let $h = 3$ and go to step 1.

Step 1. We incorporate the third column of **A** in the first column of Table A.2 (phase 3), calculate the dot products $\mathbf{t} = \mathbf{a}_3^T\mathbf{U}$, and write them in the last row

Table A.3: Added constraint, **T** matrix, and associated dot products

Phase 5					Phase 6						
$\mathbf{a}_5$	$\mathbf{w}_1^5$	$\mathbf{w}_2^5$	$\mathbf{w}_3^5$	$\mathbf{w}_4^5$	$\mathbf{a}_6$	$\mathbf{w}_1^6$	$\mathbf{w}_2^6$	$\mathbf{w}_3^6$	$\mathbf{w}_4^6$	$\mathbf{w}_5^6$	$\mathbf{w}_6^6$
1	–1	0	1	–1	0	0	1	–2	–2	0	–1
1	2	–1	–2	2	0	–1	–2	1	2	0	2
1	2	–1	–1	0	0	–1	–1	1	–2	3	–2
1	0	0	–1	1	–1	0	–1	0	2	–3	1
$\mathbf{t}^5$	3	–2	–3	2	$\mathbf{t}^6$	0	1	0	–2	3	–1
A_0	2	1	1	1	A_0	1	1	3	1	2	1
	3	3	2	2		3	2	4	3	4	2
	4	4	4	3		4	4	5	5	5	5
						6		6			

of Table A.2 (phase 3).

Step 2. We look for the pivot and find pivot=3.

Step 3. Since we have found a pivot, we go to process 1.

Step 4. We normalize the pivot column, perform the pivoting process, append the index 3 to the $A_0(\mathbf{u}_j)$ lists for all $j \neq 3$, and change vector $\mathbf{u}_3$ from $\mathbf{v}$ to $\mathbf{w}$, obtaining the matrix in Table A.2 (phase 4).

Step 6. Since $h < 6$, we let $h = 4$ and go to step 1.

Step 1. We incorporate the fourth column of **A** in the first column of Table A.2 (phase 4), calculate the dot products $\mathbf{t} = \mathbf{a}_4^T\mathbf{U}$, and write them in the last row of Table A.2 (phase 4).

Step 2. We look for the pivot and find pivot=4.

Step 3. Since we have found a pivot, we go to process 1.

Step 4. We normalize the pivot column, perform the pivoting process, append the index 4 to the $A_0(\mathbf{u}_j)$ lists for all $j \neq 4$, and change vector $\mathbf{u}_4$ from $\mathbf{v}$ to $\mathbf{w}$, obtaining the matrix in Table A.3 (phase 5).

Step 6. Since $h < 6$, we let $h = 5$ and go to step 1.

Step 1. We incorporate the fifth column of **A** in the first column of Table A.3 (phase 5), calculate the dot products $\mathbf{t} = \mathbf{a}_5^T\mathbf{U}$, and write them in the last row of Table A.3 (phase 5).

Step 2. We look for the pivot and find no pivot (no $\mathbf{v}$ vectors exist).

Step 3. Since we have found no pivot, we go to process 2.

Step 5. We determine the sets

$$\mathbf{I}^- \equiv \{2,3\};\ \ \mathbf{I}^+ \equiv \{1,4\};\ \ \mathbf{I}^0 \equiv \{\}$$

We build the set of vectors

$$\mathbf{W}^\pm \equiv \{(-2,1,1,0)^T, (-2,2,-2,2)^T, (0,0,3,-3)^T, (-1,2,-2,1)^T\}$$

and initialize the lists A_0 for those vectors as $(A_0(\mathbf{u}_i) \cap A_0(\mathbf{u}_j)) \cup \{h\}$ [see the A_0 sets in Table A.3 (phase 6)]. We remove from $\mathbf{U}$ all vectors $\mathbf{u}_j; j \in \mathbf{I}^+$, and since none of the $A_0(\mathbf{w})$ sets is contained in other $A_0(\mathbf{w}^*)$ set for $\mathbf{w}, \mathbf{w}^* \in \mathbf{W}^\pm$, and there are no sets $A_0(\mathbf{w}_j; j \in \mathbf{I}^0)$, we append the four vectors in $\mathbf{W}^\pm$ to $\mathbf{U}$.

Step 6. Since $h < 6$, we let $h = 6$ and go to step 1.

Step 1. We incorporate the sixth column of $\mathbf{A}$ in the first column of Table A.4 (phase 6), calculate the dot products $\mathbf{t} = \mathbf{a}_6^T\mathbf{U}$, and write them in the last row of Table A.4 (phase 6).

Step 2. We look for the pivot and find no pivot (no $\mathbf{v}$ vectors exist).

Step 3. Since we have found no pivot, we go to process 2.

Step 5. We determine the sets

$$\mathbf{I}^- \equiv \{4,6\};\ \ \mathbf{I}^+ \equiv \{2,5\};\ \ \mathbf{I}^0 \equiv \{1,3\}$$

We build the set of vectors

$$\begin{array}{rcl}\mathbf{W}^\pm & \equiv & \{(-2,1,1,0)^T, (-2,2,-2,2)^T, (0,0,3,-3)^T \\ & & (-1,2,-2,1)^T, (0,-2,-4,0)^T, (-6,6,0,0)^T\}\end{array}$$

and initialize the lists A_0 for those vectors as $(A_0(\mathbf{u}_i) \cap A_0(\mathbf{u}_j)) \cup \{h\}$ (see the A_0 sets in Table A.4). Since $A_0(\mathbf{w}_7) \subset A_0(\mathbf{w}_6)$ and $A_0(\mathbf{w}_8) \subset A_0(\mathbf{w}_6)$, we only add to $\mathbf{U}$ the vectors $\mathbf{w}_5$ and $\mathbf{w}_6$. Finally, if we normalized all vectors, that is, if we divide those vectors with its last component nonnull by it, we get the normalized dual in Table A.5.

Thus, the dual cone of $\mathbf{A}_\pi$ in (A.16) is

$$\pi_1 \begin{pmatrix} 0 \\ -1 \\ -1 \\ 0 \end{pmatrix} + \pi_2 \begin{pmatrix} -2 \\ 1 \\ 1 \\ 0 \end{pmatrix} + \pi_3 \begin{pmatrix} -1 \\ 1 \\ -1 \\ 1 \end{pmatrix} + \pi_4 \begin{pmatrix} -1 \\ 2 \\ -2 \\ 1 \end{pmatrix} + \pi_5 \begin{pmatrix} 0 \\ 0 \\ -1 \\ 0 \end{pmatrix} + \pi_6 \begin{pmatrix} -1 \\ 2 \\ -1 \\ 0 \end{pmatrix}$$

■

Table A.4: Unnormalized dual

	$\mathbf{w}_1$	$\mathbf{w}_2$	$\mathbf{w}_3$	$\mathbf{w}_4$	$\mathbf{w}_5$	$\mathbf{w}_6$	$\mathbf{w}_7$	$\mathbf{w}_8$
A_0	0	–2	–2	–1	0	–3	0	–6
	–1	1	2	2	0	6	–2	6
	–1	1	–2	–2	–3	–3	–4	0
	0	0	2	1	0	0	0	0
A_0	1	3	1	1	1	6	1	5
	3	4	3	2	2		6	6
	4	5	5	5	6			
	6	6						

Table A.5: Normalized dual

$\mathbf{w}_1$	$\mathbf{w}_2$	$\mathbf{w}_3$	$\mathbf{w}_4$	$\mathbf{w}_5$	$\mathbf{w}_6$
0	–2	–1	–1	0	–1
–1	1	1	2	0	2
–1	1	–1	–2	–1	–1
0	0	1	1	0	0

Once we know how the dual cone $\mathbf{A}_\pi^*$ of a given cone $\mathbf{A}_\pi$ can be obtained, we return to our initial two problems described at the beginning of this appendix:

1. Determining whether a feasible solution of the system of inequalities $\mathbf{Hx} \leq \mathbf{a}$ exists, that is, whether the feasible set is not empty

2. Obtaining the set of all feasible solutions of $\mathbf{Hx} \leq \mathbf{a}$, that is, identifying all elements in this set.

To this aim, we use the transformation from $\mathbb{R}^n$ to $\mathbb{R}^{n+1}$ described at the beginning of Section A.3 and illustrated in Figure A.2 to transform the system $\mathbf{Hx} \leq \mathbf{a}$ into the system

$$\mathbf{Hx} - \mathbf{a}x_{n+1} = \mathbf{H}^*\mathbf{x}^* \leq \mathbf{0}; \;\; x_{n+1} = 1$$

where $\mathbf{H}^*$ and $\mathbf{x}^*$ are the new matrix of coefficients and unknowns.

Next, the first problem is solved in the following section using the Γ algorithm with the cone generated by the columns of $\mathbf{H}^*$, and the second problem is solved using the Γ algorithm with the cone generated by the rows of $\mathbf{H}^*$.

A.4 Compatibility of Linear Systems

In this section we analyze whether the set of feasible solutions of the system $\mathbf{Hx} \leq \mathbf{a}$ is empty.

To analyze the compatibility problem, we need to transform the set of constraints into a set of equalities in nonnegative variables. This is not a problem because we can always find this transformation. For example

- In the case of inequalities, we use nonnegative slack variables to reduce the inequalities to equalities (we add one new variable per inequality).
- In the mixed case of equalities and inequalities, we reduce the inequalities as in the previous case.
- In the mixed case of unrestricted and restricted variables, we replace the unrestricted variables by differences between two nonnegative variables with a common subtracting variable (we add only one more variable).

Thus, we consider the system of linear equations

$$\begin{aligned} \mathbf{C}\mathbf{x}_\pi &= \mathbf{a} \\ \mathbf{x}_\pi &\geq \mathbf{0} \end{aligned} \tag{A.17}$$

where $\mathbf{C}$ is the coefficient matrix, and the subindex π is used to refer to nonnegative variables.

Since the system (A.17) can be written as

$$x_1 \begin{pmatrix} c_{11} \\ c_{21} \\ \vdots \\ c_{n1} \end{pmatrix} + x_2 \begin{pmatrix} c_{12} \\ c_{22} \\ \vdots \\ c_{n2} \end{pmatrix} + \ldots + x_n \begin{pmatrix} c_{1n} \\ c_{2n} \\ \vdots \\ c_{nn} \end{pmatrix} = \begin{pmatrix} a_1 \\ a_2 \\ \vdots \\ a_n \end{pmatrix}; \; x_i \geq 0, \; \forall i \tag{A.18}$$

it has solution if and only if the vector $\mathbf{a}$ is a linear nonnegative combination of the columns of $\mathbf{C}$, that is, if $\mathbf{a}$ belongs to the cone generated by the columns of $\mathbf{C}$ ($\mathbf{a} \in \mathbf{C}_\pi$). Using Theorem A.1, we obtain

$$\mathbf{C}\mathbf{x}_\pi = \mathbf{a} \text{ has a solution } \Leftrightarrow \mathbf{V}^T\mathbf{a} = \mathbf{0} \text{ and } \mathbf{W}^T\mathbf{a} \leq \mathbf{0} \tag{A.19}$$

where $\mathbf{C}_\pi^* \equiv \mathbf{U}_\pi \equiv \mathbf{V}_\rho + \mathbf{W}_\pi$. Thus, checking the compatibility reduces to obtaining the dual $\mathbf{C}_\pi^* \equiv \mathbf{U}_\pi \equiv \mathbf{V}_\rho + \mathbf{W}_\pi$ and checking (A.19).

Example A.4 (Compatibility). Consider the set of constraints

$$\begin{array}{cccc} x_1 & & & \leq p \\ 2x_1 & +x_2 & & \leq q \\ & -x_2 & +x_3 & \leq r \\ x_1, & x_2, & x_3 & \geq 0 \end{array} \tag{A.20}$$

which, using the slack variables x_4, x_5 and x_6, can be written as

$$\begin{array}{ccccccc} x_1 & & & +x_4 & & & = p \\ 2x_1 & +x_2 & & & +x_5 & & = q \\ & -x_2 & +x_3 & & & +x_6 & = r \\ x_1, & x_2, & x_3, & x_4, & x_5, & x_6 & \geq 0 \end{array} \tag{A.21}$$

Table A.6: Illustration of the application of the Γ algorithm to Example A.4

Phase 1				Phase 2				Phase 3			
$\mathbf{a}_1$	$\mathbf{v}_1^1$	$\mathbf{v}_2^1$	$\mathbf{v}_3^1$	$\mathbf{a}_2$	$\mathbf{w}_1^2$	$\mathbf{v}_1^2$	$\mathbf{v}_2^2$	$\mathbf{a}_3$	$\mathbf{w}_1^3$	$\mathbf{w}_2^3$	$\mathbf{v}_1^3$
1	**1**	0	0	0	–1	**–2**	0	0	–1	2	**–2**
2	**0**	1	0	1	0	**1**	0	0	0	–1	**1**
0	**0**	0	1	–1	0	**0**	1	1	0	0	**1**
$\mathbf{t}^1$	**1**	2	0	$\mathbf{t}^2$	0	**1**	–1	$\mathbf{t}^3$	0	0	**1**
A_0			1	A_0	2	1	1	A_0	2	1	1
									3	3	2

Phase 4				Phase 5				Phase 6			
$\mathbf{a}_4$	$\mathbf{w}_1^4$	$\mathbf{w}_2^4$	$\mathbf{w}_3^4$	$\mathbf{a}_5$	$\mathbf{w}_1^5$	$\mathbf{w}_2^5$	$\mathbf{w}_3^5$	$\mathbf{a}_6$	$\mathbf{w}_1^6$	$\mathbf{w}_2^6$	$\mathbf{w}_3^6$
1	–1	2	2	0	–1	0	0	0	–1	0	0
0	0	–1	–1	1	0	–1	–1	0	0	–1	–1
0	0	0	–1	0	0	0	–1	1	0	0	–1
$\mathbf{t}^4$	–1	2	2	$\mathbf{t}^5$	0	–1	–1	$\mathbf{t}^6$	0	0	–1
A_0	2	1	1	A_0	2	3	2	A_0	2	3	2
	3	3	2		3	4	4		3	4	4
					5				5	6	
									6		

Thus, the system A.20 has solution if and only if

$$\begin{pmatrix} p \\ q \\ r \end{pmatrix} \in \mathbf{A}_\pi = \begin{pmatrix} 1 & 0 & 0 & 1 & 0 & 0 \\ 2 & 1 & 0 & 0 & 1 & 0 \\ 0 & -1 & 1 & 0 & 0 & 1 \end{pmatrix}_\pi$$

that is, if the vector of independent terms belongs to the cone generated by the columns of the matrix of coefficients. To know this, we obtain the dual of the cone $\mathbf{A}_\pi$ using the Γ algorithm. Table A.6 gives the resulting tableaux, and shows that the dual is

$$\mathbf{A}_\pi^* = \mathbf{W}_\pi = \begin{pmatrix} -1 & 0 & 0 \\ 0 & -1 & -1 \\ 0 & 0 & -1 \end{pmatrix}$$

According to (A.19), the system is compatible if and only if

$$\begin{pmatrix} -1 & 0 & 0 \\ 0 & -1 & -1 \\ 0 & 0 & -1 \end{pmatrix}^T \begin{pmatrix} p \\ q \\ r \end{pmatrix} = \begin{pmatrix} -1 & 0 & 0 \\ 0 & -1 & 0 \\ 0 & -1 & -1 \end{pmatrix} \begin{pmatrix} p \\ q \\ r \end{pmatrix} = \begin{pmatrix} -p \\ -q \\ -q-r \end{pmatrix} \leq \begin{pmatrix} 0 \\ 0 \\ 0 \end{pmatrix}$$

which results in $p \geq 0$, $q \geq 0$ and $r \geq -q$.

■

A.5 Solving Linear Systems

In this section we use Theorem A.2 and Algorithm A.1 to solve linear systems of inequalities of the form

$$\mathbf{Hx} \le \mathbf{a}$$

More precisely, given the polyhedron

$$S = \{\mathbf{x}|\mathbf{Hx} \le \mathbf{a}\}$$

we first build the cone

$$C_S = \mathbf{C}_\pi^* = \left\{ \begin{pmatrix} \mathbf{x} \\ x_{n+1} \end{pmatrix} |\mathbf{Hx} - x_{n+1}\mathbf{a} \le \mathbf{0};\ -x_{n+1} \le 0 \right\}$$

which is the dual cone of the cone generated by the rows of the matrix

$$\mathbf{C} = \begin{bmatrix} \mathbf{H} & | & -\mathbf{a} \\ -- & + & -- \\ 0 & | & -1 \end{bmatrix}$$

In other words, we start by writing the system in the more convenient form

$$\begin{aligned} \mathbf{c}_i^T \mathbf{x} \le \mathbf{0}, \forall i \\ x_n = 1 \end{aligned} \tag{A.22}$$

where it is assumed that now $\mathbf{x}$ includes x_{n+1} and that the redundant constraint $-x_n \le 0$ has been added to $\mathbf{c}_i^T \mathbf{x} \le \mathbf{0}$. This facilitates to obtain the solution later.

It is worth noting that it is always possible to write a given system in this form just by transforming

- All relations of the type $\mathbf{c}_i^T \mathbf{x} \le \mathbf{a}_i$ by means of

$$\mathbf{c}_i^T \mathbf{x} - x_n \mathbf{a}_i \le \mathbf{0} \text{ and } x_n \ge 0 \text{ and } x_n = 1$$

- The variables $x_i \ge 0$ by means of $\mathbf{c}_i^T \mathbf{x} \le 0$ with $\mathbf{c}_i = (0, \dots, -1, \dots, 0)^T$ with the -1 in the ith position.

If $\mathbf{X}$ is the set of solutions of (A.22), the definition of the dual cone leads to

$$\mathbf{X} \equiv \{\mathbf{x} \in \mathbf{C}_\pi^* | x_n = 1\} \tag{A.23}$$

This implies that we can find first the dual cone $\mathbf{C}_\pi^*$ associated with the first set of constraints of (A.22) and then add the last condition.

Using the Γ algorithm, we find the C_S generators and write $C_S = \mathbf{V}_\rho + \mathbf{W}_\pi$. Next, we identify those generators in $\mathbf{W}$ with last null component $\mathbf{W}$, and the rest $\mathbf{Q}$. Finally, Theorem A.2 allows us to write

$$S = \mathbf{V}_\rho + \mathbf{W}_\pi + \mathbf{Q}_\lambda$$

that is, the solution of $\mathbf{Hx} \le \mathbf{a}$.

In summary, the system of linear equations and inequalities (A.22) can be solved using the following steps:

1. Obtaining the dual cone $(\mathbf{B}_\rho + \mathbf{C})^*_\pi = \mathbf{V}_\rho + \mathbf{W}^1_\pi$
2. Normalizing the vectors of $\mathbf{W}^1$ with nonnull last component w_n by dividing it by w_n
3. Writing $\mathbf{C}^*_\pi$ as $\mathbf{V}_\rho + \mathbf{W}_\pi + \mathbf{Q}_\lambda$, where $\mathbf{W}$ and $\mathbf{Q}$ are the vectors in $\mathbf{W}^*$ with null and unit last component, respectively

Example A.5 (System of inequalities). Consider the system in (A.20) with $p = q = r = 1$:

$$\begin{array}{rrrcl} x_1 & & & \leq & 1 \\ 2x_1 & +x_2 & & \leq & 1 \\ & -x_2 & +x_3 & \leq & 1 \\ x_1, & x_2, & x_3 & \geq & 0 \end{array} \tag{A.24}$$

To solve this system, we write it as

$$\begin{array}{rrrrcl} x_1 & & & -x_4 & \leq & 0 \\ 2x_1 & +x_2 & & -x_4 & \leq & 0 \\ & -x_2 & +x_3 & -x_4 & \leq & 0 \\ -x_1 & & & & \leq & 0 \\ & -x_2 & & & \leq & 0 \\ & & -x_3 & & \leq & 0 \\ & & & -x_4 & \leq & 0 \\ & & & x_4 & = & 1 \end{array} \tag{A.25}$$

Next, we solve the system (A.25) using the three steps above, as follows.

1. We obtain the dual cone associated with the homogeneous system in (A.25). Using the Γ algorithm, we obtain Table A.7, which in its last table gives the desired dual:

$$\pi_1 \begin{pmatrix} 0 \\ 1 \\ 2 \\ 1 \end{pmatrix} + \pi_2 \begin{pmatrix} 0 \\ 0 \\ 1 \\ 1 \end{pmatrix} + \pi_3 \begin{pmatrix} 1 \\ 0 \\ 2 \\ 2 \end{pmatrix} + \pi_4 \begin{pmatrix} 0 \\ 1 \\ 0 \\ 1 \end{pmatrix} + \pi_5 \begin{pmatrix} 0 \\ 0 \\ 0 \\ 1 \end{pmatrix} + \pi_6 \begin{pmatrix} 1 \\ 0 \\ 0 \\ 2 \end{pmatrix}$$

2. We normalize these vectors (dividing by its last component) and get

$$\pi_1 \begin{pmatrix} 0 \\ 1 \\ 2 \\ 1 \end{pmatrix} + \pi_2 \begin{pmatrix} 0 \\ 0 \\ 1 \\ 1 \end{pmatrix} + \pi_3 \begin{pmatrix} \frac{1}{2} \\ 0 \\ 1 \\ 1 \end{pmatrix} + \pi_4 \begin{pmatrix} 0 \\ 1 \\ 0 \\ 1 \end{pmatrix} + \pi_5 \begin{pmatrix} 0 \\ 0 \\ 0 \\ 1 \end{pmatrix} + \pi_6 \begin{pmatrix} \frac{1}{2} \\ 0 \\ 0 \\ 1 \end{pmatrix}$$

3. We write the polyhedron, which in this case is the polytope

$$\begin{pmatrix} x_1 \\ x_2 \\ x_3 \end{pmatrix} = \lambda_1 \begin{pmatrix} 0 \\ 1 \\ 2 \end{pmatrix} + \lambda_2 \begin{pmatrix} 0 \\ 0 \\ 1 \end{pmatrix} + \lambda_3 \begin{pmatrix} \frac{1}{2} \\ 0 \\ 1 \end{pmatrix} + \lambda_4 \begin{pmatrix} 0 \\ 1 \\ 0 \end{pmatrix} + \lambda_5 \begin{pmatrix} 0 \\ 0 \\ 0 \end{pmatrix} + \lambda_6 \begin{pmatrix} \frac{1}{2} \\ 0 \\ 0 \end{pmatrix}$$

with $\sum_{i=0}^{6} \lambda_i = 1$.

■

Table A.7: Tableaux corresponding to the Γ algorithm of Example A.5

Phase 1

$\mathbf{a}_1$	$\mathbf{v}_1^1$	$\mathbf{v}_2^1$	$\mathbf{v}_3^1$	$\mathbf{v}_4^1$
1	**1**	0	0	0
0	**0**	1	0	0
0	**0**	0	1	0
−1	**0**	0	0	1
$\mathbf{t}^1$	**1**	0	0	−1
A_0		1	1	

Phase 2

$\mathbf{a}_2$	$\mathbf{w}_1^2$	$\mathbf{v}_1^2$	$\mathbf{v}_2^2$	$\mathbf{v}_3^2$
2	−1	**0**	0	1
1	0	**1**	0	0
0	0	**0**	1	0
−1	0	**0**	0	1
$\mathbf{t}^2$	−2	**1**	0	1
A_0		1	1	1
			2	

Phase 3

$\mathbf{a}_3$	$\mathbf{w}_1^3$	$\mathbf{w}_2^3$	$\mathbf{v}_1^3$	$\mathbf{v}_2^3$
0	−1	0	**0**	1
−1	2	−1	**0**	−1
1	0	0	**1**	0
−1	0	0	**0**	1
$\mathbf{t}^3$	−2	1	**1**	0
A_0	2	1	1	1
			2	2
				3

Phase 4

$\mathbf{a}_4$	$\mathbf{w}_1^4$	$\mathbf{w}_2^4$	$\mathbf{w}_3^4$	$\mathbf{v}_1^4$
−1	−1	0	0	**1**
0	2	−1	0	**−1**
0	2	−1	−1	**0**
0	0	0	0	**1**
$\mathbf{t}^4$	1	0	0	**−1**
A_0	2	1	1	1
	3	3	2	2
		4	4	3

Phase 5

$\mathbf{a}_5$	$\mathbf{w}_1^5$	$\mathbf{w}_2^5$	$\mathbf{w}_3^5$	$\mathbf{w}_4^5$
0	0	0	0	1
−1	1	−1	0	−1
0	2	−1	−1	0
0	1	0	0	1
$\mathbf{t}^5$	−1	1	0	1
A_0	2	1	1	1
	3	3	2	2
	4	4	4	3
			5	

Phase 6

$\mathbf{a}_6$	$\mathbf{w}_1^6$	$\mathbf{w}_2^6$	$\mathbf{w}_3^6$	$\mathbf{w}_4^6$
0	0	0	0	1
0	0	1	0	0
−1	−1	2	1	2
0	0	1	1	2
$\mathbf{t}^6$	1	−2	−1	−2
A_0	1	2	3	2
	2	3	4	3
	4	4	5	5
	5			

Phase 7 and Dual

$\mathbf{a}_7$	$\mathbf{w}_1^7$	$\mathbf{w}_2^7$	$\mathbf{w}_3^7$	$\mathbf{w}_4^7$	$\mathbf{w}_5^7$	$\mathbf{w}_6^7$
0	0	0	1	0	0	1
0	1	0	0	1	0	0
0	2	1	2	0	0	0
−1	1	1	2	1	1	2
$\mathbf{t}^7$	−1	−1	−2	−1	−1	−2
A_0	2	3	2	2	4	2
	3	4	3	4	5	5
	4	5	5	6	6	6

Example A.6 (System of inequalities). Consider the following system of inequalities:

$$\begin{array}{rrrcr} x_1 & & & \le & -1 \\ 2x_1 & +x_2 & & \le & 0 \\ & -x_2 & +x_3 & \le & -2 \\ & -x_2 & +x_3 & \le & -1 \\ x_1 & +x_2 & +x_3 & \le & -1 \end{array} \tag{A.26}$$

which using an auxiliary unknown can be written as

$$\begin{array}{rrrrcr} x_1 & & & +x_4 & \le & 0 \\ 2x_1 & +x_2 & & & \le & 0 \\ & -x_2 & +x_3 & +2x_4 & \le & 0 \\ & -x_2 & +x_3 & +x_4 & \le & 0 \\ x_1 & +x_2 & +x_3 & +x_4 & \le & 0 \\ & & & -x_4 & \le & 0 \\ & & & x_4 & = & 1 \end{array} \tag{A.27}$$

Since the vectors associated with the rows of this system coincide with those in matrix **A** in Example A.3, we can use the results obtained there. Thus, from Table A.5 we conclude that the solution is:

$$\begin{pmatrix} x_1 \\ x_2 \\ x_3 \end{pmatrix} = \lambda_1 \begin{pmatrix} -1 \\ 1 \\ -1 \end{pmatrix} + \lambda_2 \begin{pmatrix} -1 \\ 2 \\ -2 \end{pmatrix} + \pi_1 \begin{pmatrix} 0 \\ -1 \\ -1 \end{pmatrix} + \pi_2 \begin{pmatrix} -2 \\ 1 \\ 1 \end{pmatrix} + \pi_3 \begin{pmatrix} 0 \\ 0 \\ -3 \end{pmatrix} + \pi_4 \begin{pmatrix} -3 \\ 6 \\ -3 \end{pmatrix}$$

■

A.6 Applications to Several Examples

In this section we apply the methods developed in previous sections to several examples discussed in previous chapters. In particular, we discuss the compatibility problem and obtain the associated sets of all feasible solutions.

A.6.1 The Transportation Problem

In this section we analyze the compatibility and the set of possible solutions of the transportation problem discussed in Section 1.2. This problem consists of minimization of

$$Z = \sum_{i=1}^{m} \sum_{j=1}^{n} c_{ij} x_{ij}$$

subject to

$$\begin{array}{rcl} \sum\limits_{j=1}^{n} x_{ij} & = & u_i;\ i = 1, \dots, m \\ \sum\limits_{i=1}^{m} x_{ij} & = & v_j;\ j = 1, \dots, n \\ x_{ij} & \ge & 0;\ i = 1, \dots, m;\ j = 1, \dots, n \end{array} \tag{A.28}$$

Compatibility

To analyze the compatibility problem, we need to write, as in (A.17), the set of constraints as a set of equalities

$$\begin{array}{rcl} \mathbf{Cx} & = & \mathbf{a} \\ \mathbf{x} & \geq & \mathbf{0} \end{array} \tag{A.29}$$

that is, in the form of (A.28). Next, we need to calculate the dual $\mathbf{C}_\pi^{T*} = \mathbf{V}_\rho + \mathbf{W}_\pi$ of the cone $\mathbf{C}_\pi^T$ and next checking the conditions $\mathbf{V}^T\mathbf{a} = \mathbf{0}$ and $\mathbf{W}^T\mathbf{a} \leq \mathbf{0}$, where $\mathbf{a} = \begin{pmatrix} \mathbf{u} \\ \mathbf{v} \end{pmatrix}$. Using this process, it can be shown that the compatibility conditions of the system (1.2) are

$$\sum_{i=1}^{m} u_i = \sum_{j=1}^{n} v_j; \quad u_i, v_j \geq 0, \quad i = 1, 2, \ldots, m; \quad j = 1, 2, \ldots, n \tag{A.30}$$

which states that the total amount shipped from all origins must be equal to the total amount received at all destinations, and that these totals must be nonnegative.

The Set of All Feasible Solutions

To obtain the set of possible solutions, we need to write the system (A.28) in the following equivalent homogeneous form:

$$\begin{array}{rcl}
c_{11}x_{11} + c_{12}x_{12} + \ldots + c_{1n}x_{1n} - a_1 y & = & 0 \\
c_{21}x_{21} + c_{22}x_{22} + \ldots + c_{2n}x_{2n} - a_2 y & = & 0 \\
\vdots \quad\quad \vdots \quad\quad \vdots \quad\quad \vdots \quad\quad \vdots & = & \vdots \\
c_{m1}x_{m1} + c_{m2}x_{m2} + \ldots + c_{mn}x_{mn} - a_n y & = & 0 \\
-x_{11} & \leq & 0 \\
-x_{12} & \leq & 0 \\
\vdots & \vdots & \vdots \\
-x_{mn} & \leq & 0 \\
-y & \leq & 0 \\
y & = & 1
\end{array} \tag{A.31}$$

From (A.28), we see that the solution of (A.31) can be obtained by selecting the vectors that satisfy $y = 1$ and belong to the dual of the cone $\mathbf{V}_\rho + \mathbf{W}_\pi$, where the columns of $\mathbf{V}$ are the coefficients in the first m rows of (A.31), and the columns of $\mathbf{W}$ are the coefficients in rows $m + 1$ to $m + mn + 1$ of (A.31).

Example A.7 (Transportation problem). Now we discuss the problem in (1.4)

$$
\mathbf{Cx} = \begin{pmatrix} 1 & 1 & 1 & 0 & 0 & 0 & 0 & 0 & 0 \\ 0 & 0 & 0 & 1 & 1 & 1 & 0 & 0 & 0 \\ 0 & 0 & 0 & 0 & 0 & 0 & 1 & 1 & 1 \\ 1 & 0 & 0 & 1 & 0 & 0 & 1 & 0 & 0 \\ 0 & 1 & 0 & 0 & 1 & 0 & 0 & 1 & 0 \\ 0 & 0 & 1 & 0 & 0 & 1 & 0 & 0 & 1 \end{pmatrix} \begin{pmatrix} x_{11} \\ x_{12} \\ x_{13} \\ x_{21} \\ x_{22} \\ x_{23} \\ x_{31} \\ x_{32} \\ x_{33} \end{pmatrix} = \begin{pmatrix} u_1 \\ u_2 \\ u_3 \\ v_1 \\ v_2 \\ v_3 \end{pmatrix} \tag{A.32}
$$

$$
x_{ij} \geq 0; \quad i, j = 1, 2, 3
$$

Compatibility

To derive the compatibility conditions associated with this problem, we need to obtain the dual cone of the cone generated by the columns of $\mathbf{C}$. This cone is

$$
\begin{aligned}
\mathbf{C}^*_\pi &= \mathbf{V}_\rho + \mathbf{W}_\pi \\
&= \rho \begin{pmatrix} -1 \\ -1 \\ -1 \\ 1 \\ 1 \\ 1 \end{pmatrix} + \pi_1 \begin{pmatrix} -1 \\ 0 \\ 0 \\ 0 \\ 0 \\ 0 \end{pmatrix} + \pi_2 \begin{pmatrix} 0 \\ -1 \\ 0 \\ 0 \\ 0 \\ 0 \end{pmatrix} + \pi_3 \begin{pmatrix} 0 \\ 0 \\ -1 \\ 0 \\ 0 \\ 0 \end{pmatrix} \\
&\quad + \pi_4 \begin{pmatrix} 0 \\ 0 \\ 0 \\ -1 \\ 0 \\ 0 \end{pmatrix} + \pi_5 \begin{pmatrix} 0 \\ 0 \\ 0 \\ 0 \\ -1 \\ 0 \end{pmatrix} + \pi_6 \begin{pmatrix} 0 \\ 0 \\ 0 \\ 0 \\ 0 \\ -1 \end{pmatrix}; \quad \rho \in \mathbb{R}; \quad \pi_i \geq 0; \quad \forall i
\end{aligned} \tag{A.33}
$$

which, according to (A.19) leads to the compatibility conditions

$$
\begin{aligned}
\mathbf{V}^T\mathbf{a} = 0 \quad &\equiv \quad -u_1 - u_2 - u_3 + v_1 + v_2 + v_3 = 0 \\
\mathbf{W}^T\mathbf{a} \leq 0 \quad &\equiv \quad u_1, u_2, u_3, v_1, v_2, v_3 \geq 0
\end{aligned} \tag{A.34}
$$

which coincides with (A.30).

The Set of All Feasible Solutions

To find the set of all feasible solutions of (A.32) with $u_1 = 2, u_2 = 3, u_3 =$

$4; v_1 = 5, v_2 = 2, v_3 = 2$, we write the set of constraints (A.32) as follows:

$$\begin{pmatrix} 1 & 1 & 1 & 0 & 0 & 0 & 0 & 0 & 0 & -2 \\ 0 & 0 & 0 & 1 & 1 & 1 & 0 & 0 & 0 & -3 \\ 0 & 0 & 0 & 0 & 0 & 0 & 1 & 1 & 1 & -4 \\ 1 & 0 & 0 & 1 & 0 & 0 & 1 & 0 & 0 & -5 \\ 0 & 1 & 0 & 0 & 1 & 0 & 0 & 1 & 0 & -2 \\ 0 & 0 & 1 & 0 & 0 & 1 & 0 & 0 & 1 & -2 \end{pmatrix} \begin{pmatrix} x_{11} \\ x_{12} \\ x_{13} \\ x_{21} \\ x_{22} \\ x_{23} \\ x_{31} \\ x_{32} \\ x_{33} \\ y \end{pmatrix} = \begin{pmatrix} 0 \\ 0 \\ 0 \\ 0 \\ 0 \\ 0 \end{pmatrix}$$

$$\begin{array}{rcl} -x_{ij} & \le & 0; \;\; i, j = 1, 2, 3 \\ -y & \le & 0 \\ y & = & 1 \end{array} \tag{A.35}$$

Thus, we need to calculate the dual of the cone $\mathbf{V}_\rho + \mathbf{W}_\pi$, where the rows of $\mathbf{V}$ are the coefficients in the first 6 equations of (A.35), and the columns of $\mathbf{W}$ are the coefficients in equations 7–16 of (A.35).

This dual, which has been obtained using the Γ algorithm, is

$$\begin{pmatrix} x_{11} \\ x_{12} \\ x_{13} \\ x_{21} \\ x_{22} \\ x_{23} \\ x_{31} \\ x_{32} \\ x_{33} \\ y \end{pmatrix} = \pi_1 \begin{pmatrix} 0 \\ 0 \\ 2 \\ 3 \\ 0 \\ 0 \\ 2 \\ 2 \\ 0 \\ 1 \end{pmatrix} + \pi_2 \begin{pmatrix} 0 \\ 0 \\ 2 \\ 1 \\ 2 \\ 0 \\ 4 \\ 0 \\ 0 \\ 1 \end{pmatrix} + \pi_3 \begin{pmatrix} 0 \\ 2 \\ 0 \\ 3 \\ 0 \\ 0 \\ 2 \\ 0 \\ 2 \\ 1 \end{pmatrix} + \pi_4 \begin{pmatrix} 2 \\ 0 \\ 0 \\ 3 \\ 0 \\ 0 \\ 0 \\ 2 \\ 2 \\ 1 \end{pmatrix} + \pi_5 \begin{pmatrix} 2 \\ 0 \\ 0 \\ 0 \\ 2 \\ 1 \\ 3 \\ 0 \\ 1 \\ 1 \end{pmatrix} + \pi_6 \begin{pmatrix} 2 \\ 0 \\ 0 \\ 1 \\ 2 \\ 0 \\ 2 \\ 0 \\ 2 \\ 1 \end{pmatrix}$$

$$+\pi_7 \begin{pmatrix} 0 \\ 2 \\ 0 \\ 1 \\ 0 \\ 2 \\ 4 \\ 0 \\ 0 \\ 1 \end{pmatrix} + \pi_8 \begin{pmatrix} 2 \\ 0 \\ 0 \\ 1 \\ 0 \\ 2 \\ 2 \\ 2 \\ 0 \\ 1 \end{pmatrix} + \pi_9 \begin{pmatrix} 1 \\ 1 \\ 0 \\ 0 \\ 1 \\ 2 \\ 4 \\ 0 \\ 0 \\ 1 \end{pmatrix} + \pi_{10} \begin{pmatrix} 2 \\ 0 \\ 0 \\ 0 \\ 1 \\ 2 \\ 3 \\ 1 \\ 0 \\ 1 \end{pmatrix} + \pi_{11} \begin{pmatrix} 1 \\ 0 \\ 1 \\ 0 \\ 2 \\ 1 \\ 4 \\ 0 \\ 0 \\ 1 \end{pmatrix} \tag{A.36}$$

where $\pi_i \ge 0; \;\; i = 1, 2, \ldots, 11$. Finally, forcing $y = 1$ and eliminating the last component (associated with the artificial variable y) of all vectors leads to the polytope:

$$\begin{array}{rcl}
\begin{pmatrix} x_{11} \\ x_{12} \\ x_{13} \\ x_{21} \\ x_{22} \\ x_{23} \\ x_{31} \\ x_{32} \\ x_{33} \end{pmatrix} & = & \lambda_1 \begin{pmatrix} 0 \\ 0 \\ 2 \\ 3 \\ 0 \\ 0 \\ 2 \\ 2 \\ 0 \end{pmatrix} + \lambda_2 \begin{pmatrix} 0 \\ 0 \\ 2 \\ 1 \\ 2 \\ 0 \\ 4 \\ 0 \\ 0 \end{pmatrix} + \lambda_3 \begin{pmatrix} 0 \\ 2 \\ 0 \\ 3 \\ 0 \\ 0 \\ 2 \\ 0 \\ 2 \end{pmatrix} + \lambda_4 \begin{pmatrix} 2 \\ 0 \\ 0 \\ 3 \\ 0 \\ 0 \\ 0 \\ 2 \\ 2 \end{pmatrix} + \lambda_5 \begin{pmatrix} 2 \\ 0 \\ 0 \\ 0 \\ 2 \\ 1 \\ 3 \\ 0 \\ 1 \end{pmatrix} + \lambda_6 \begin{pmatrix} 2 \\ 0 \\ 0 \\ 1 \\ 2 \\ 0 \\ 2 \\ 0 \\ 2 \end{pmatrix} \\
& & +\lambda_7 \begin{pmatrix} 0 \\ 2 \\ 0 \\ 1 \\ 0 \\ 2 \\ 4 \\ 0 \\ 0 \end{pmatrix} + \lambda_8 \begin{pmatrix} 2 \\ 0 \\ 0 \\ 1 \\ 0 \\ 2 \\ 2 \\ 2 \\ 0 \end{pmatrix} + \lambda_9 \begin{pmatrix} 1 \\ 1 \\ 0 \\ 0 \\ 1 \\ 2 \\ 4 \\ 0 \\ 0 \end{pmatrix} + \lambda_{10} \begin{pmatrix} 2 \\ 0 \\ 0 \\ 0 \\ 1 \\ 2 \\ 3 \\ 1 \\ 0 \end{pmatrix} + \lambda_{11} \begin{pmatrix} 1 \\ 0 \\ 1 \\ 0 \\ 2 \\ 1 \\ 4 \\ 0 \\ 0 \end{pmatrix} \\
\lambda_i & \geq & 0 \\
\sum_{i=1}^{11} \lambda_i & = & 1
\end{array} \tag{A.37}$$

which is the set of all possible (feasible) solutions. Note that each column is a solution corresponding to an extreme point of the feasible set.

In fact, the optimum solution obtained in (1.6) can be obtained from (A.37). To this end we calculate the cost function value (1.5) for the general solution in (A.37) and get

$$\begin{pmatrix} 1 & 2 & 3 & 2 & 1 & 2 & 3 & 2 & 1 \end{pmatrix} \begin{pmatrix} x_{11} \\ x_{12} \\ x_{13} \\ x_{21} \\ x_{22} \\ x_{23} \\ x_{31} \\ x_{32} \\ x_{33} \end{pmatrix}$$

$$= 22\lambda_1 + 22\lambda_2 + 18\lambda_3 + 14\lambda_4 + 16\lambda_5 + 14\lambda_6 + 22\lambda_7 + 18\lambda_8 + 20\lambda_9 + 18\lambda_{10} + 20\lambda_{11}. \tag{A.38}$$

Expression (A.38) contains all possible values of the cost function. To obtain a possible value, all we need is to distribute one unit among the 11 lambdas. Since there are many possible options for this distribution, and we need to minimize this value, we look for the lambdas with the minimum coefficient.

These are λ_4 and λ_6, that share a value of 14. Then, we distribute the one unit between these two lambdas and make zeros the remaining lambdas. Thus, from (A.38) we conclude that the minimum value of the cost function is 14, which corresponds to any linear convex combination of the points associated with λ_4 and λ_6, because both have a coefficient of 14. Thus, the set of all solutions, namely, those leading to a value 14 of the cost function, is

$$\begin{pmatrix} x_{11} \\ x_{12} \\ x_{13} \\ x_{21} \\ x_{22} \\ x_{23} \\ x_{31} \\ x_{32} \\ x_{33} \end{pmatrix} = \lambda \begin{pmatrix} 2 \\ 0 \\ 0 \\ 3 \\ 0 \\ 0 \\ 0 \\ 2 \\ 2 \end{pmatrix} + (1-\lambda) \begin{pmatrix} 2 \\ 0 \\ 0 \\ 1 \\ 2 \\ 0 \\ 2 \\ 0 \\ 2 \end{pmatrix} ; \quad 0 \leq \lambda \leq 1 \tag{A.39}$$

which, of course, includes the solution $(2, 0, 0, 1, 2, 0, 2, 0, 2)^T$, which was given in (1.6), in Section 1.6. ■

A.6.2 Production Scheduling Problem

In this section we analyze the compatibility and the set of possible solutions of the production scheduling problems discussed in Section 1.3. This problem consists of minimizing

$$\sum_{t=1}^{n} (a_t y_t - b_t x_t - c_t s_t)$$

subject to

$$\begin{array}{rcl} s_{t-1} + x_t - s_t & = & y_t, \; t = 1, 2, \ldots, n \\ s_t, x_t, y_t & \geq & 0 \end{array} \tag{A.40}$$

Compatibility

Since (A.40) is a system of equations in nonnegative variables, we can use the results in Section A.4 directly to discuss the compatibility problem. Our system becomes

$$\begin{array}{rcl} x_1 - s_1 & = & y_1 - s_0 \\ s_1 + x_2 - s_2 & = & y_2 \\ \vdots & \vdots & \vdots \\ s_{n-1} + x_n - s_n & = & y_n \\ s_i, x_i & \geq & 0; \quad i = 1, 2, \ldots, n \end{array} \tag{A.41}$$

To analyze its compatibility, we need to calculate the dual $\mathbf{C}_\pi^* = \mathbf{V}_\rho + \mathbf{W}_\pi$ of the corresponding cone $\mathbf{C}_\pi$ and next checking the conditions $\mathbf{V}^T \mathbf{a} = \mathbf{0}$ and $\mathbf{W}^T \mathbf{a} \leq \mathbf{0}$, where $\mathbf{a} = (y_1 - s_0, y_2, \ldots, y_n)^T$.

It can be shown that the dual cone is empty and thus, the problem is always compatible.

The Set of All Feasible Solutions

To obtain the set of all possible solutions we need to write the system (A.41) in the following equivalent homogeneous form:

$$\begin{array}{rcl} x_1 - s_1 - (y_1 - s_0)x_{n+1} & = & 0 \\ s_1 + x_2 - s_2 - y_2 x_{n+1} & = & 0 \\ \vdots & \vdots & \vdots \\ s_{n-1} + x_n - s_n - y_n x_{n+1} & = & 0 \\ -s_i & \leq & 0; \quad i = 1, 2, \ldots, n \\ -x_i & \leq & 0; \quad i = 1, 2, \ldots, n \\ -x_{n+1} & \leq & 0 \\ x_{n+1} & = & 1 \end{array} \tag{A.42}$$

Thus, the solution of (A.41) can be obtained by selecting the vectors that satisfy $x_{n+1} = 1$ and belong to the dual of the cone $\mathbf{V}_\rho + \mathbf{W}_\pi$, where the columns of $\mathbf{V}$ are the coefficients in the first n equations of (A.42), and the columns of $\mathbf{W}$ are the coefficients in equations $n+1$ to $3n+1$ of (A.42).

Example A.8 (Production scheduling problem). To derive the compatibility conditions of the system (A.40), we need to obtain the dual cone of the cone generated by its columns. This cone is empty. Thus, the problem is always feasible for any demand function and initial storage value s_0.

Now assume that we have a storage limit s_t^{max} at time t. Then, introducing four auxiliary nonnegative slack variables z_1, z_2, z_3, z_4, system (1.10) becomes

$$\mathbf{Cx} = \begin{pmatrix} -1 & 0 & 0 & 0 & 1 & 0 & 0 & 0 & 0 & 0 & 0 & 0 \\ 1 & -1 & 0 & 0 & 0 & 1 & 0 & 0 & 0 & 0 & 0 & 0 \\ 0 & 1 & -1 & 0 & 0 & 0 & 1 & 0 & 0 & 0 & 0 & 0 \\ 0 & 0 & 1 & -1 & 0 & 0 & 0 & 1 & 0 & 0 & 0 & 0 \\ 1 & 0 & 0 & 0 & 0 & 0 & 0 & 0 & 1 & 0 & 0 & 0 \\ 0 & 1 & 0 & 0 & 0 & 0 & 0 & 0 & 0 & 1 & 0 & 0 \\ 0 & 0 & 1 & 0 & 0 & 0 & 0 & 0 & 0 & 0 & 1 & 0 \\ 0 & 0 & 0 & 1 & 0 & 0 & 0 & 0 & 0 & 0 & 0 & 1 \end{pmatrix} \begin{pmatrix} s_1 \\ s_2 \\ s_3 \\ s_4 \\ x_1 \\ x_2 \\ x_3 \\ x_4 \\ z_1 \\ z_2 \\ z_3 \\ z_4 \end{pmatrix} = \begin{pmatrix} 0 \\ 3 \\ 6 \\ 1 \\ s_1^{max} \\ s_2^{max} \\ s_3^{max} \\ s_4^{max} \end{pmatrix}$$

$$s_t, x_t, z_t \geq 0 \tag{A.43}$$

In this case, the dual of the cone generated by the columns of $\mathbf{C}$ is the cone generated by the columns of the matrix $\mathbf{W}$:

$$\begin{pmatrix} 0 & 0 & -1 & 0 & 0 & -1 & 0 & 0 & 0 & -1 & 0 & 0 & 0 & -1 \\ 0 & 0 & 0 & 0 & -1 & -1 & 0 & 0 & -1 & -1 & 0 & 0 & -1 & -1 \\ 0 & 0 & 0 & 0 & 0 & 0 & 0 & -1 & -1 & -1 & 0 & -1 & -1 & -1 \\ 0 & 0 & 0 & 0 & 0 & 0 & 0 & 0 & 0 & 0 & -1 & -1 & -1 & -1 \\ 0 & -1 & -1 & 0 & 0 & 0 & 0 & 0 & 0 & 0 & 0 & 0 & 0 & 0 \\ 0 & 0 & 0 & -1 & -1 & -1 & 0 & 0 & 0 & 0 & 0 & 0 & 0 & 0 \\ 0 & 0 & 0 & 0 & 0 & 0 & -1 & -1 & -1 & -1 & 0 & 0 & 0 & 0 \\ -1 & 0 & 0 & 0 & 0 & 0 & 0 & 0 & 0 & 0 & -1 & -1 & -1 & -1 \end{pmatrix} \tag{A.44}$$

which leads to the following conditions:

$$\begin{array}{rcl} s_4^{max} & \geq & 0 \\ s_1^{max} & \geq & 0 \\ -s_0 + s_1^{max} + y_1 & \geq & 0 \\ s_2^{max} & \geq & 0 \\ s_2^{max} + y_2 & \geq & 0 \\ -s_0 + s_2^{max} + y_1 + y_2 & \geq & 0 \\ s_3^{max} & \geq & 0 \\ s_3^{max} + y_3 & \geq & 0 \\ s_3^{max} + y_2 + y_3 & \geq & 0 \\ -s_0 + s_3^{max} + y_1 + y_2 + y_3 & \geq & 0 \\ s_4^{max} + y_4 & \geq & 0 \\ s_4^{max} + y_3 + y_4 & \geq & 0 \\ s_4^{max} + y_2 + y_3 + y_4 & \geq & 0 \\ -s_0 + s_4^{max} + y_1 + y_2 + y_3 + y_4 & \geq & 0 \end{array}$$

where s_t^{max} is the storage capacity at time t.

The Set of All Feasible Solutions

To find the set of all possible solutions for (1.10), we write the set of constraints as

$$\begin{pmatrix} -1 & 0 & 0 & 0 & 1 & 0 & 0 & 0 & 0 \\ 1 & -1 & 0 & 0 & 0 & 1 & 0 & 0 & -3 \\ 0 & 1 & -1 & 0 & 0 & 0 & 1 & 0 & -6 \\ 0 & 0 & 1 & -1 & 0 & 0 & 0 & 1 & -1 \end{pmatrix} \begin{pmatrix} s_1 \\ s_2 \\ s_3 \\ s_4 \\ x_1 \\ x_2 \\ x_3 \\ x_4 \\ y \end{pmatrix} = \begin{pmatrix} 0 \\ 0 \\ 0 \\ 0 \end{pmatrix} \tag{A.45}$$

Table A.8: Set of extreme points associated with the system (1.10).

$\mathbf{w}_1$	$\mathbf{w}_2$	$\mathbf{w}_3$	$\mathbf{w}_4$	$\mathbf{w}_5$	$\mathbf{w}_6$	$\mathbf{w}_7$	$\mathbf{w}_8$	$\mathbf{w}_9$	$\mathbf{w}_{10}$	$\mathbf{w}_{11}$	$\mathbf{w}_{12}$
0	0	$\frac{10}{3}$	0	3	0	9	0	3	0	10	0
0	0	$\frac{10}{3}$	$\frac{10}{3}$	0	0	6	6	0	0	7	7
0	1	$\frac{10}{3}$	$\frac{10}{3}$	0	0	0	0	1	1	1	1
1	1	$\frac{10}{3}$	$\frac{10}{3}$	0	0	0	0	0	0	0	0
0	0	$\frac{10}{3}$	0	3	0	9	0	3	0	10	0
0	0	0	$\frac{10}{3}$	0	3	0	9	0	3	0	10
0	1	0	0	6	6	0	0	7	7	0	0
1	0	0	0	1	1	1	1	0	0	0	0
0	0	0	0	1	1	1	1	1	1	1	1

and

$$\begin{array}{rcl} -s_i & \leq & 0; \;\; i = 1, 2, 3, 4 \\ -x_i & \leq & 0; \;\; i = 1, 2, 3, 4 \\ -y & \leq & 0 \\ y & = & 1 \end{array} \tag{A.46}$$

Thus, we need to calculate the dual of the cone $\mathbf{V}_\rho + \mathbf{W}_\pi$, where the columns of $\mathbf{V}$ are the coefficients in the first four rows of (A.45)–(A.46), and the columns of $\mathbf{W}$ are the coefficients in the rows 1 to 9 of (A.46).

This dual is $\mathbf{C}_\pi^* = \mathbf{W}_\pi$, where the matrix $\mathbf{W}$ is as given in Table A.8. Now, forcing the last components of the vectors in $\mathbf{C}_\pi^*$ to be unity ($y = 1$), we get that the general solution $\mathbf{W}_\pi^1 + \mathbf{W}_\lambda^2$ of (1.10) is a nonnegative linear combination of the vectors in $\mathbf{W}^1$ plus a linear convex combination of the vectors in $\mathbf{W}^2$, where $\mathbf{W}^1$ is the matrix including the first four columns, and $\mathbf{W}^2$, the remaining columns in Table A.8, namely, a cone plus a polytope.

Note that

1. Each vector in $\mathbf{W}^2$ is a solution of (A.43). These solutions represent the minimum production required to satisfy the demand. They lead to a zero storage in at least one time step.

2. Each vector in $\mathbf{W}^1$ is a solution of the associated homogeneous system. These solutions represent the production excesses over demand and therefore only represent storage increases, making the production policy more robust against possible demand increments.

If, alternatively, we find the set of possible solutions of (A.43), with $s_t^{max} =$

4; $\forall t$, we write the set of constraints as

$$\begin{pmatrix} -1 & 0 & 0 & 0 & 1 & 0 & 0 & 0 & 0 & 0 & 0 & 0 & 0 \\ 1 & -1 & 0 & 0 & 0 & 1 & 0 & 0 & 0 & 0 & 0 & 0 & -3 \\ 0 & 1 & -1 & 0 & 0 & 0 & 1 & 0 & 0 & 0 & 0 & 0 & -6 \\ 0 & 0 & 1 & -1 & 0 & 0 & 0 & 1 & 0 & 0 & 0 & 0 & -1 \\ 1 & 0 & 0 & 0 & 0 & 0 & 0 & 0 & 1 & 0 & 0 & 0 & -4 \\ 0 & 1 & 0 & 0 & 0 & 0 & 0 & 0 & 0 & 1 & 0 & 0 & -4 \\ 0 & 0 & 1 & 0 & 0 & 0 & 0 & 0 & 0 & 0 & 1 & 0 & -4 \\ 0 & 0 & 0 & 1 & 0 & 0 & 0 & 0 & 0 & 0 & 0 & 1 & -4 \end{pmatrix} \begin{pmatrix} s_1 \\ s_2 \\ s_3 \\ s_4 \\ x_1 \\ x_2 \\ x_3 \\ x_4 \\ z_1 \\ z_2 \\ z_3 \\ z_4 \\ y \end{pmatrix} = \begin{pmatrix} 0 \\ 0 \\ 0 \\ 0 \\ 0 \\ 0 \\ 0 \\ 0 \end{pmatrix}$$

$$\begin{array}{rcl} -s_t, -x_t, -z_t & \leq & 0; \ \forall t \\ -y & \leq & 0 \\ y & = & 1 \end{array} \tag{A.47}$$

and get the dual in Table A.9. Thus, the general solution of (A.43) is a linear convex combination of the vectors in Table A.9, with their last components removed, namely, a polytope.

Replacing the general solution in with Table A.8 in the function to be maximized

$$Z = \sum_{i=1}^{n} (a_t d_t - b_t x_t - c_t s_t)$$

with $a_t = 3, b_t = 1, c_t = 1$, and the demand function in Table 1.1, we get

$$\begin{array}{rcl} Z & = & 36 - 2\pi_1 - 3\pi_2 - \frac{50}{3}\pi_3 - \frac{54}{3}\pi_4 - 13\lambda_1 - 10\lambda_2 - 25\lambda_3 - 16\lambda_4 \\ & & -14\lambda_5 - 11\lambda_6 - 28\lambda_7 - 18\lambda_8. \end{array} \tag{A.48}$$

Since all the π coefficients are negative, and the minimum λ coefficient is -10, and corresponds to λ_2, the optimal solution is unique and corresponds to $\pi_1 = \pi_2 = \pi_3 = \pi_4 = 0$ and $\lambda_i = 0; i \neq 2; \lambda_2 = 1$, which leads to $Z = 36 - 10 = 26$.

One alternative option consists of minimizing the storage cost, as in (1.9):

$$Z = s_1 + s_2 + s_3 + s_4. \tag{A.49}$$

Calculating the value of the function to be minimized using the general solution associated with Table A.8, we get

$$Z = \pi_1 + 2\pi_2 + \frac{40}{3}\pi_3 + 10\pi_4 + 3\lambda_1 + 15\lambda_3 + 6\lambda_4 + 4\lambda_5 + 1\lambda_6 + 18\lambda_7 + 8\lambda_8 \tag{A.50}$$

Since all π coefficients are positive, and the minimum λ coefficient is 0, for λ_2, we obtain a minimum value of $Z = 0$, which corresponds to $\pi_1 = \pi_2 = \pi_3 = \pi_4 = 0$ and $\lambda_i = 0; i \neq 2; \lambda_2 = 1$.

Table A.9: Set of extreme points associated with the system (A.43)

$\mathbf{w}_1$	$\mathbf{w}_2$	$\mathbf{w}_3$	$\mathbf{w}_4$	$\mathbf{w}_5$	$\mathbf{w}_6$	$\mathbf{w}_7$	$\mathbf{w}_8$	$\mathbf{w}_9$	$\mathbf{w}_{10}$	$\mathbf{w}_{11}$	$\mathbf{w}_{12}$	$\mathbf{w}_{13}$
3	0	3	0	4	4	0	0	4	4	3	0	4
0	0	0	0	1	1	4	4	4	4	0	0	1
0	0	1	1	0	1	0	1	0	1	4	4	4
0	0	0	0	0	0	0	0	0	0	3	3	3
3	0	3	0	4	4	0	0	4	4	3	0	4
0	3	0	3	0	0	7	7	3	3	0	3	0
6	6	7	7	5	6	2	3	2	3	10	10	9
1	1	0	0	1	0	1	0	1	0	0	0	0
1	4	1	4	0	0	4	4	0	0	1	4	0
4	4	4	4	3	3	0	0	0	0	4	4	3
4	4	3	3	4	3	4	3	4	3	0	0	0
4	4	4	4	4	4	4	4	4	4	1	1	1
1	1	1	1	1	1	1	1	1	1	1	1	1

$\mathbf{w}_{14}$	$\mathbf{w}_{15}$	$\mathbf{w}_{16}$	$\mathbf{w}_{17}$	$\mathbf{w}_{18}$	$\mathbf{w}_{19}$	$\mathbf{w}_{20}$	$\mathbf{w}_{21}$	$\mathbf{w}_{22}$	$\mathbf{w}_{23}$	$\mathbf{w}_{24}$	$\mathbf{w}_{25}$
0	4	3	0	4	0	4	3	0	4	0	4
4	4	0	0	1	4	4	0	0	1	4	4
4	4	0	0	0	0	0	4	4	4	4	4
3	3	4	4	4	4	4	4	4	4	4	4
0	4	3	0	4	0	4	3	0	4	0	4
7	3	0	3	0	7	3	0	3	0	7	3
6	6	6	6	5	2	2	10	10	9	6	6
0	0	5	5	5	5	5	1	1	1	1	1
4	0	1	4	0	4	0	1	4	0	4	0
0	0	4	4	3	0	0	4	4	3	0	0
0	0	4	4	4	4	4	0	0	0	0	0
1	1	0	0	0	0	0	0	0	0	0	0
1	1	1	1	1	1	1	1	1	1	1	1

If, instead of constraints (1.10), we use (A.43), we obtain the dual in Table A.9 and (A.48) becomes

$$\begin{aligned} Z &= 36 - 13\lambda_1 - 10\lambda_2 - 14\lambda_3 - 11\lambda_4 - 15\lambda_5 - 16\lambda_6 - 14\lambda_7 \\ &= -15\lambda_8 - 18\lambda_9 - 19\lambda_{10} - 23\lambda_{11} - 20\lambda_{12} - 25\lambda_{13} - 24\lambda_{14} \\ &= -28\lambda_{15} - 21\lambda_{16} - 18\lambda_{17} - 23\lambda_{18} - 22\lambda_{19} - 26\lambda_{20} - 25\lambda_{21} - 22\lambda_{22} \\ & \quad -27\lambda_{23} - 26\lambda_{24} - 30\lambda_{25} \end{aligned} \tag{A.51}$$

which leads to a maximum value of 26 for $\lambda_2 = 1; \lambda_i = 0; \forall i \neq 2$. This value, according to Table A.9, is attained at

$$(s_1, s_2, s_3, s_4, x_1, x_2, x_3, x_4) = (0, 0, 0, 0, 0, 3, 6, 1)$$

where only the first eight components of $\mathbf{w}_2$ have been used (we have removed the values of the auxiliary variables $z_i = 4; i = 1, 2, 3, 4$).

Table A.10: Example of an input–output table

Sales	Railroad	Steel	Coal	Other	Final demand	Total
Railroad	x_{11}	x_{12}	x_{13}	x_{14}	y_1	x_1
Steel	x_{21}	x_{22}	x_{23}	x_{24}	y_2	x_2
Coal	x_{31}	x_{32}	x_{33}	x_{34}	y_3	x_3
Other	x_{41}	x_{42}	x_{43}	x_{44}	y_4	x_4

Note that now production is limited by the limited storage. Note also that all solutions in Table A.8 with their first four components smaller than 4 (the storage limit) are included in Table A.9. All other solutions in Table A.9 have at least one of their first four components equal to 4 (the storage limit).

Similarly, with (A.43), (A.50) becomes

$$\begin{array}{rcl} M & = & 3\lambda_1 + 4\lambda_3 + 1\lambda_4 + 5\lambda_5 + 6\lambda_6 + 4\lambda_7 + 5\lambda_8 + 8\lambda_9 \\ & & +9\lambda_{10} + 10\lambda_{11} + 7\lambda_{12} + 12\lambda_{13} + 11\lambda_{14} + 15\lambda_{15} + 7\lambda_{16} + 4\lambda_{17} \\ & & +9\lambda_{18} + 8\lambda_{19} + 12\lambda_{20} + 11\lambda_{21} + 8\lambda_{22} + 13\lambda_{23} + 12\lambda_{24} + 16\lambda_{25} \end{array} \tag{A.52}$$

which leads to a minimum value of 0 for $\lambda_2 = 1; \lambda_i = 0; i \neq 2$. This value, according to Table A.9, is also attained at

$$(s_1, s_2, s_3, s_4, x_1, x_2, x_3, x_4) = (0, 0, 0, 0, 0, 3, 6, 1)$$

■

A.6.3 The Input–Output Tables

As an introduction to the input–output tables, consider an economy that consists of only three basic industries – railroad, steel, and coal – and a fourth category of all other industries. The aim of the problem consists of analyzing the interrelationships of these industries in terms of sales to each other and to other elements of the economy, during a given period of time. This analysis can be done by means of an input–output table. This problem does not fall into the category of optimization problems.

Each element in Table A.10 represents a total sale activities of each industry during the considered time period. For example, the first row of the table describes the sales of the railroad industry to each of the other industries.

Thus, the three main elements in the input–output tables problem are:

1. **Data.**

   x_{ij}: the total sales of industry i to industry j during the base period

   y_i: the amount of final demand for industry i

2. **Variables.**

   x_i: the total output of industry i during the base period

Table A.11: Example of an input–output table

Sales	Railroad	Steel	Coal	Other	Final demand	Total
Railroad	2	2	6	4	6	20
Steel	5	1	2	4	8	20
Coal	3	4	2	5	6	20
Other	6	6	3	6	9	30

Table A.12: Example of an input–output coefficients table

Sales	Railroad	Steel	Coal	Other	Final demand
Railroad	0.10	0.10	0.30	0.20	0.30
Steel	0.25	0.05	0.10	0.20	0.40
Coal	0.15	0.20	0.10	0.25	0.30
Other	0.20	0.20	0.10	0.20	0.30

3. **Constraints.** The interrelationships of the industries in the considered economy can be written as

$$x_i - \sum_{j=1,\ldots,n} x_{ij} = y_i; \quad i = 1, \ldots, n \tag{A.53}$$

In the 1930s Leontieff assumed that the linear model of the economy for a base period could be used to predict the structure of the economy for a future period, and called $a_{ij} = x_{ij}/x_j$, the *input–output coefficient*, which is the amount of industry i necessary to produce one unit of commodity j. Thus system (A.53) becomes

$$\begin{array}{rcl} (1 - a_{ii})x_i - \sum_{j \neq i} a_{ij}x_j & = & y_i; \quad i = 1, \ldots, n \\ y_i & \geq & 0; \quad i = 1, \ldots, n \\ x_{ij} & \geq & 0 \end{array} \tag{A.54}$$

Example A.9 (Input–output table). Consider Table A.11, an input–output table, and its corresponding coefficients in Table A.12. In this case, system (A.54) becomes

$$\begin{array}{c} \begin{pmatrix} 0.90 & -0.10 & -0.30 & -0.20 \\ -0.25 & 0.95 & -0.10 & -0.20 \\ -0.15 & -0.20 & 0.90 & -0.25 \\ -0.20 & -0.20 & -0.10 & 0.80 \end{pmatrix} \begin{pmatrix} x_1 \\ x_2 \\ x_3 \\ x_4 \end{pmatrix} = \begin{pmatrix} 6 \\ 8 \\ 6 \\ 9 \end{pmatrix} \\ x_i \geq 0; \quad i = 1, 2, 3, 4. \end{array} \tag{A.55}$$

Table A.13: Dual of the cone generated by the columns of matrix in (A.55)

$\mathbf{w}_1$	$\mathbf{w}_2$	$\mathbf{w}_3$	$\mathbf{w}_4$
–1.481	–0.568	–0.534	–0.579
–0.426	–1.323	–0.504	–0.500
–0.615	–0.403	–1.423	–0.432
–0.669	–0.599	–0.704	–1.655

Compatibility

The compatibility conditions of system (A.55) are obtained by finding the dual of the cone generated by its associated columns, which is given in Table A.13. Thus, it is compatible if and only if

$$\begin{pmatrix} -1.481 & -0.426 & -0.615 & -0.669 \\ -0.568 & -1.323 & -0.403 & -0.599 \\ -0.534 & -0.504 & -1.423 & -0.704 \\ -0.579 & -0.500 & -0.432 & -1.655 \end{pmatrix} \begin{pmatrix} d_1 \\ d_2 \\ d_3 \\ d_4 \end{pmatrix} \leq \begin{pmatrix} 0 \\ 0 \\ 0 \\ 0 \end{pmatrix} \tag{A.56}$$

where d_1, d_2, d_3, d_4 are the final demands of the four industries.

Note that for nonnegative final demands the system always has a solution. If the input–output tables are correctly defined, this problem always has a solution for positive final demands.

The Set of All Feasible Solutions

The general solution of the system (A.55) is

$$\begin{pmatrix} x_1 & x_2 & x_3 & x_4 \end{pmatrix}^T = \begin{pmatrix} \frac{3125}{142} & \frac{3095}{142} & \frac{1570}{71} & \frac{3545}{142} \end{pmatrix}^T, \tag{A.57}$$

which gives a unique solution.

■

A.6.4 The Diet Problem

Example A.10 (The diet problem). Let us study the compatibility conditions and the set of all feasible solution in the diet problem. We will interpret the physical meaning of each of the components appearing in the general solution, that is, the linear convex and the nonnegative linear combinations; beatiful results obtained from the mathematical analysis.

Compatibility

Using auxiliary slack variables $z_1, \ldots, z_4$, the system (1.14) becomes

$$\begin{pmatrix} 78.6 & 70.1 & 80.1 & 67.2 & 77.0 & -1 & 0 & 0 & 0 \\ 6.50 & 9.40 & 8.80 & 13.7 & 30.4 & 0 & -1 & 0 & 0 \\ 0.02 & 0.09 & 0.03 & 0.14 & 0.41 & 0 & 0 & -1 & 0 \\ 0.27 & 0.34 & 0.30 & 1.29 & 0.86 & 0 & 0 & 0 & -1 \end{pmatrix} \begin{pmatrix} x_1 \\ x_2 \\ x_3 \\ x_4 \\ x_5 \\ z_1 \\ z_2 \\ z_3 \\ z_4 \end{pmatrix} = \begin{pmatrix} 74.2 \\ 14.7 \\ 0.14 \\ 0.55 \end{pmatrix}$$

$$x_1, x_2, x_3, x_4, x_5, z_1, z_2, z_3, z_4 \geq 0. \tag{A.58}$$

This system is compatible because the dual cone associated with the columns of the matrix in (A.58) is empty.

The Set of All Feasible Solutions

The feasible solution is obtained from the homogeneous system associated with (1.14). This solution is $\mathbf{W}_\pi + \mathbf{Q}_\lambda$, where $\mathbf{W}$ and $\mathbf{Q}$ are as shown in Table A.14. Note that this general solution consists of a nonnegative linear combination of the vectors in $\mathbf{W}$ plus a convex linear combination of the vectors in $\mathbf{Q}$, namely, a cone plus a polytope. The meanings of both parts are clear:

- $\mathbf{Q}$ contains solutions with the minimum amounts of nutrients required to satisfy the minimum requirements.

- $\mathbf{W}$ contains any amount of nutrient leading to an excess of requirements.

As in previous examples, the minimum can be directly obtained from Table A.14. ∎

A.6.5 The Network Flow Problem

Since system (1.19) is not in the form of (A.17), that is, in terms of nonnegative variables, we define the new variables

$$y_{ij} = x_{ij} + m_{ij}; \forall i, j$$

Table A.14: Vector generating the solution of the diet problem example

	Corn	Oats	Milo	Bran maize	Linseed meal
			W		
$\mathbf{w}_1$	1	0	0	0	0
$\mathbf{w}_2$	0	1	0	0	0
$\mathbf{w}_3$	0	0	1	0	0
$\mathbf{w}_4$	0	0	0	1	0
$\mathbf{w}_5$	0	0	0	0	1
			Q		
$\mathbf{q}_1$	0	0	0	1.10417	0
$\mathbf{q}_2$	0	0	0	0	0.963636
$\mathbf{q}_3$	0.0448423	0	0	1.05172	0
$\mathbf{q}_4$	0	0.0873163	0	1.01308	0
$\mathbf{q}_5$	0	0	0.0567182	1.03656	0
$\mathbf{q}_6$	7.	0	0	0	0
$\mathbf{q}_7$	0	0	4.66667	0	0
$\mathbf{q}_8$	0.220126	0	0	0.968553	0
$\mathbf{q}_9$	0	0	0.170524	0.963459	0
$\mathbf{q}_{10}$	0	1.61765	0	0	0
$\mathbf{q}_{11}$	0.458531	0	0	0	0.495577
$\mathbf{q}_{12}$	0	0.629273	0	0	0.390752
$\mathbf{q}_{13}$	0	0	0.468746	0	0.476019
$\mathbf{q}_{14}$	0.544348	0	0	0.096701	0.323583
$\mathbf{q}_{15}$	0	1.53026	0	0.023032	0
$\mathbf{q}_{16}$	0	0.736664	0	0.088176	0.216031
$\mathbf{q}_{17}$	0	0	0.574136	0.116168	0.265004
$\mathbf{q}_{18}$	0.10857	1.53143	0	0	0
$\mathbf{q}_{19}$	1.12406	0	0	0	0.286631
$\mathbf{q}_{20}$	0	1.51786	0.11309	0	0
$\mathbf{q}_{21}$	0	0	1.08128	0	0.262346
$\mathbf{q}_{22}$	0.022243	1.51631	0	0.022055	0
$\mathbf{q}_{23}$	0.814007	0	0	0.070968	0.277523
$\mathbf{q}_{24}$	0	1.51325	0.017539	0.023436	0
$\mathbf{q}_{25}$	0	0	0.598964	0.114763	0.258449

and then system (1.19) transforms to

$$\begin{pmatrix} 1 & 1 & 1 & 0 & 0 & 0 & 0 & 0 & 0 & 0 \\ -1 & 0 & 0 & 1 & 0 & 0 & 0 & 0 & 0 & 0 \\ 0 & -1 & 0 & 0 & 1 & 0 & 0 & 0 & 0 & 0 \\ 0 & 0 & -1 & -1 & -1 & 0 & 0 & 0 & 0 & 0 \\ 1 & 0 & 0 & 0 & 0 & 1 & 0 & 0 & 0 & 0 \\ 0 & 1 & 0 & 0 & 0 & 0 & 1 & 0 & 0 & 0 \\ 0 & 0 & 1 & 0 & 0 & 0 & 0 & 1 & 0 & 0 \\ 0 & 0 & 0 & 1 & 0 & 0 & 0 & 0 & 1 & 0 \\ 0 & 0 & 0 & 0 & 1 & 0 & 0 & 0 & 0 & 1 \end{pmatrix} \begin{pmatrix} y_{12} \\ y_{13} \\ y_{14} \\ y_{24} \\ y_{34} \\ z_{12} \\ z_{13} \\ z_{14} \\ z_{24} \\ z_{34} \end{pmatrix} = \begin{pmatrix} f_1^* \\ f_2^* \\ f_3^* \\ f_4^* \\ 2m_{12} \\ 2m_{13} \\ 2m_{14} \\ 2m_{24} \\ 2m_{34} \end{pmatrix},$$

$$\begin{aligned} y_{ij} &\geq 0; \forall i,j \\ z_{ij} &\geq 0; \forall i,j \end{aligned} \tag{A.59}$$

where $z_{ij} \geq 0;\ \forall i,j$ are auxiliary slack variables used to convert the system of inequalities (1.19) into the system of equalities (A.59), and

$$\begin{pmatrix} f_1^* \\ f_2^* \\ f_3^* \\ f_4^* \end{pmatrix} = \begin{pmatrix} f_1 + m_{12} + m_{13} + m_{14} \\ f_2 - m_{12} + m_{24} \\ f_3 - m_{13} + m_{34} \\ f_4 - m_{14} - m_{24} - m_{34} \end{pmatrix} \tag{A.60}$$

Compatibility

An analysis of the compatibility of the system requires that we obtain the dual $\mathbf{C}_\pi^*$ of the cone generated by the columns of the matrix $\mathbf{C}$ in (A.59), whose generators are given in the first column of Table A.15. It can be seen that there is one single $\mathbf{V}$ vector and seventeen $\mathbf{W}$ vectors, which lead to one compatibility equation (the first in Table A.15) and 17 compatibility inequalities (the remaining rows in Table A.15).

The compatibility equation

$$f_1 + f_2 + f_3 + f_4 = 0 \tag{A.61}$$

says only that the conservation of flow must be satisfied globally, that is, input flow to the system and output flow from the system must be equal (lack of losses is assumed).

The inequality conditions can be divided into two groups:

Nonnegative inequality constraints, which indicate only that the capacities must be nonnegative.

Table A.15: Compatibility conditions.

Vector	Constraint	Section
	Global flow balance equality constraints	
$(1,1,1,1,0,0,0,0,0)^T$	$f_1+f_2+f_3+f_4=0$	
	Nonnegative inequality constraints	
$(0,0,0,0,-1,0,0,0,0)^T$	$m_{12}\geq 0$	
$(0,0,0,0,0,-1,0,0,0)^T$	$m_{13}\geq 0$	
$(0,0,0,0,0,0,-1,0,0)^T$	$m_{14}\geq 0$	
$(0,0,0,0,0,0,0,-1,0)^T$	$m_{24}\geq 0$	
$(0,0,0,0,0,0,0,0,-1)^T$	$m_{34}\geq 0$	
	Maximum flow inequality constraints	
$(-1,0,0,0,0,0,0,0,0)^T$	$-f_1\leq m_{12}+m_{13}+m_{14}$	S_1
$(1,0,0,0,-1,-1,-1,0,0)^T$	$f_1\leq m_{12}+m_{13}+m_{14}$	S_1
$(0,1,0,0,0,0,0,-1,0)^T$	$f_2\leq m_{12}+m_{24}$	S_2
$(0,-1,0,0,-1,0,0,0,0)^T$	$-f_2\leq m_{12}+m_{24}$	S_2
$(0,0,1,0,0,0,0,0,-1)^T$	$f_3\leq m_{13}+m_{34}$	S_3
$(0,0,-1,0,0,-1,0,0,0)^T$	$-f_3\leq m_{13}+m_{34}$	S_3
$(1,1,1,0,0,0,-1,-1,-1)^T$	$f_1+f_2+f_3\leq m_{14}+m_{24}+m_{34}$	S_4
$(-1,-1,-1,0,0,0,0,0,0)^T$	$-f_1-f_2-f_3\leq m_{14}+m_{24}+m_{34}$	S_4
$(1,0,1,0,-1,0,-1,0,-1)^T$	$f_1+f_3\leq m_{12}+m_{14}+m_{34}$	S_5
$(-1,0,-1,0,0,0,0,0,0)^T$	$-f_1-f_3\leq m_{12}+m_{14}+m_{34}$	S_5
$(-1,-1,0,0,0,0,0,0,0)^T$	$-f_1-f_2\leq m_{13}+m_{14}+m_{24}$	S_6
$(1,1,0,0,0,-1,-1,-1,0)^T$	$f_1+f_2\leq m_{13}+m_{14}+m_{24}$	S_6

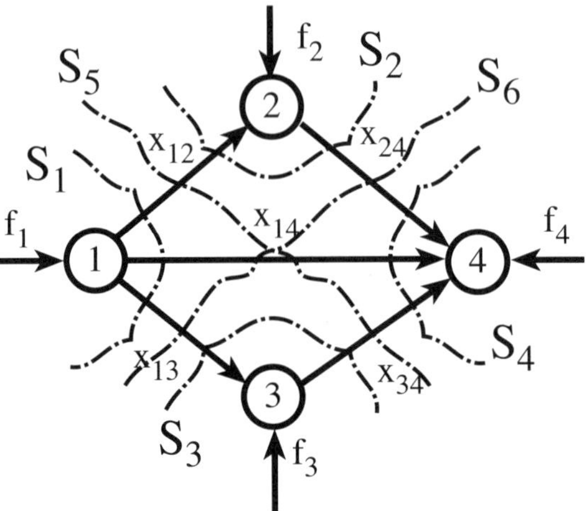

Figure A.3: Sections associated with the inequality constraints of Example 1.4.

Maximum flow inequality constraints, which require that the input or output flow cannot exceed the maximum capacity associated with given sections (indicated in Figure A.3). For example, in section S_1, the maximum flow is given by $m_{12} + m_{13} + m_{14}$ (the maximum flow of the links associated with node 1), which is an upper bound for either f_1 or $-f_1$ (flow can go in either direction). Similar arguments explain the physical meaning of the remaining compatibility conditions, which are flow balances in different parts of the network assuming maximum allowed flow to occur in their links.

The Set of All Feasible Solutions

Now system (1.19) is solved for the particular case

$$(f_1, f_2, f_3, f_4)^T = (7, -4, -1, -2)^T$$

and $m_{ij} = 4;\ \forall i, j$. Using an auxiliary unknown $y = 1$ the system (1.19) can be written as

$$\begin{pmatrix} 1 & 1 & 1 & 0 & 0 & 0 & 0 & 0 & 0 & 0 & -19 \\ -1 & 0 & 0 & 1 & 0 & 0 & 0 & 0 & 0 & 0 & 4 \\ 0 & -1 & 0 & 0 & 1 & 0 & 0 & 0 & 0 & 0 & 1 \\ 0 & 0 & -1 & -1 & -1 & 0 & 0 & 0 & 0 & 0 & 14 \\ 1 & 0 & 0 & 0 & 0 & 1 & 0 & 0 & 0 & 0 & -8 \\ 0 & 1 & 0 & 0 & 0 & 0 & 1 & 0 & 0 & 0 & -8 \\ 0 & 0 & 1 & 0 & 0 & 0 & 0 & 1 & 0 & 0 & -8 \\ 0 & 0 & 0 & 1 & 0 & 0 & 0 & 0 & 1 & 0 & -8 \\ 0 & 0 & 0 & 0 & 1 & 0 & 0 & 0 & 0 & 1 & -8 \end{pmatrix} \begin{pmatrix} y_{12} \\ y_{13} \\ y_{14} \\ y_{24} \\ y_{34} \\ z_{12} \\ z_{13} \\ z_{14} \\ z_{24} \\ z_{34} \\ y \end{pmatrix} = \begin{pmatrix} 0 \\ 0 \\ 0 \\ 0 \\ 0 \\ 0 \\ 0 \\ 0 \\ 0 \end{pmatrix} \tag{A.62}$$

Table A.16: Dual cone associated with the system (A.62)–(A.63)

$\mathbf{w}_1$	$\mathbf{w}_2$	$\mathbf{w}_3$	$\mathbf{w}_4$
4	8	4	8
8	8	7	3
7	3	8	8
0	4	0	4
7	7	6	2
4	0	4	0
0	0	1	5
1	5	0	0
8	4	8	4
1	1	2	6
1	1	1	1

and

$$\begin{pmatrix} -1 & 0 & 0 & 0 & 0 & 0 & 0 & 0 & 0 & 0 & 0 \\ 0 & -1 & 0 & 0 & 0 & 0 & 0 & 0 & 0 & 0 & 0 \\ 0 & 0 & -1 & 0 & 0 & 0 & 0 & 0 & 0 & 0 & 0 \\ 0 & 0 & 0 & -1 & 0 & 0 & 0 & 0 & 0 & 0 & 0 \\ 0 & 0 & 0 & 0 & -1 & 0 & 0 & 0 & 0 & 0 & 0 \\ 0 & 0 & 0 & 0 & 0 & -1 & 0 & 0 & 0 & 0 & 0 \\ 0 & 0 & 0 & 0 & 0 & 0 & -1 & 0 & 0 & 0 & 0 \\ 0 & 0 & 0 & 0 & 0 & 0 & 0 & -1 & 0 & 0 & 0 \\ 0 & 0 & 0 & 0 & 0 & 0 & 0 & 0 & -1 & 0 & 0 \\ 0 & 0 & 0 & 0 & 0 & 0 & 0 & 0 & 0 & -1 & 0 \\ 0 & 0 & 0 & 0 & 0 & 0 & 0 & 0 & 0 & 0 & -1 \end{pmatrix} \begin{pmatrix} y_{12} \\ y_{13} \\ y_{14} \\ y_{24} \\ y_{34} \\ z_{12} \\ z_{13} \\ z_{14} \\ z_{24} \\ z_{34} \\ y \end{pmatrix} \leq \begin{pmatrix} 0 \\ 0 \\ 0 \\ 0 \\ 0 \\ 0 \\ 0 \\ 0 \\ 0 \\ 0 \\ 0 \end{pmatrix} \tag{A.63}$$

and we obtain the dual of the cone generated by the rows of the matrices in (A.62)–(A.63), which is given in Table A.16. Now, forcing $y = 1$ and returning to the initial x variables we obtain as the final solution. the polytope:

$$\begin{array}{rcl} \mathbf{x} & = & \lambda_1(0,4,3,-4,3)^T + \lambda_2(4,4,-1,0,3)^T \\ & & +\lambda_3(0,3,4,-4,2)^T + \lambda_4(4,-1,4,0,-2)^T\} \\ 1 & = & \lambda_1 + \lambda_2 + \lambda_3 + \lambda_4 \end{array} \tag{A.64}$$

Note that the four vectors in (A.64) satisfy the initial system (1.19).

Exercises

A.1 Consider the water supply network system in Figure A.4, where $\{q_1, q_2\}$ are the supply flows, $\{q_3, \ldots, q_{10}\}$ are the consumption flows at the nodes of the network, and $\{x_1, \ldots, x_{14}\}$ are the corresponding flows for the required supply. By convention, we consider $q_1, q_2 < 0$ and $q_i \geq 0$ for $i = 3, \ldots, 14$.

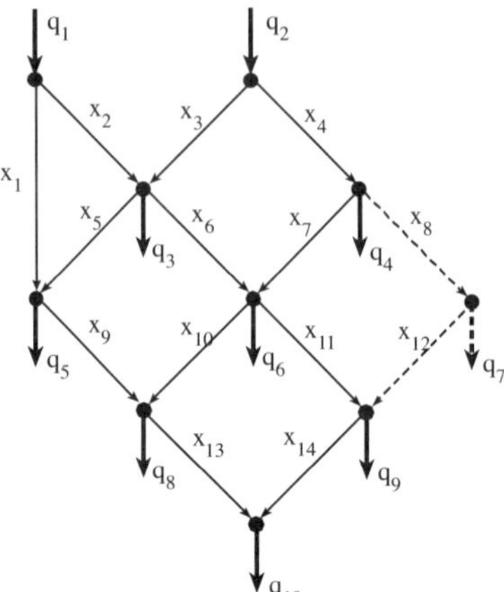

Figure A.4: Water supply system showing the flows in all conduits.

(a) Write the balance of flow in each node to obtain the system of equations for the water supply system.

(b) Obtain the compatibility conditions and give them a physical interpretation.

(c) Assume that the supplies satisfy $q_1 = q_2$ and the consumptions are $q_3 = q_4 = q_5 = q_6 = q_7 = q_8 = q_9 = q_{10} = 1\ \text{m}^3/\text{s}$ (cubic meter per second). What is the common value of q_1 and q_2?

(d) Obtain the general solution of the system of equations in this case.

(e) Consider the following two situations:

(i) Consumer q_7 has a normal supply, but its consumption is null.

(ii) We cut off the water supply to consumer q_7.

Obtain the new solutions based on the old ones.

A.2 Consider the network flow problem in Figure A.5. Show that for this case, Equation (1.16) becomes

$$\begin{pmatrix} 1 & 1 & 0 & 0 & 0 & 0 & 0 & 0 \\ -1 & 0 & 1 & 1 & 0 & 0 & 0 & 0 \\ 0 & -1 & 0 & 0 & 1 & 1 & 0 & 0 \\ 0 & 0 & -1 & 0 & -1 & 0 & 1 & 0 \\ 0 & 0 & 0 & 0 & 0 & -1 & 0 & 1 \\ 0 & 0 & 0 & -1 & 0 & 0 & -1 & -1 \end{pmatrix} \begin{pmatrix} x_{12} \\ x_{13} \\ x_{24} \\ x_{26} \\ x_{34} \\ x_{35} \\ x_{46} \\ x_{56} \end{pmatrix} = \begin{pmatrix} f_1 \\ f_2 \\ f_3 \\ f_4 \\ f_5 \\ f_6 \end{pmatrix}$$

$$\begin{aligned} x_{ij} &\le m_{ij}; \forall i,j \\ -x_{ij} &\le m_{ij}; \forall i,j \end{aligned} \tag{A.65}$$

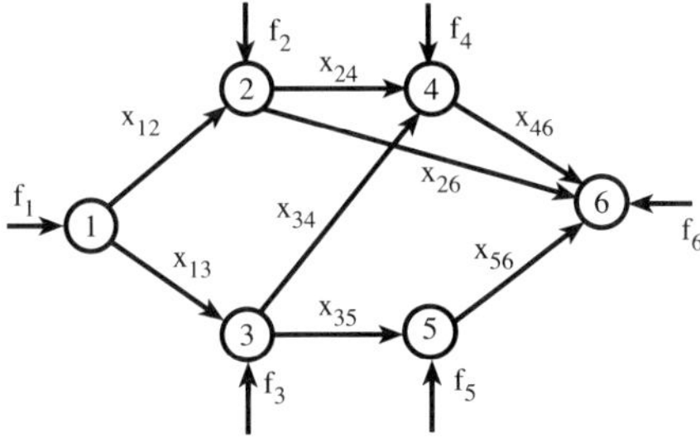

Figure A.5: A network flow problem.

Defining the new variables $y_{ij} = x_{ij} + m_{ij}; \forall i, j$, show that system (A.65) transforms to

$$\begin{pmatrix} 1 & 1 & 0 & 0 & 0 & 0 & 0 & 0 \\ -1 & 0 & 1 & 1 & 0 & 0 & 0 & 0 \\ 0 & -1 & 0 & 0 & 1 & 1 & 0 & 0 \\ 0 & 0 & -1 & 0 & -1 & 0 & 1 & 0 \\ 0 & 0 & 0 & 0 & 0 & -1 & 0 & 1 \\ 0 & 0 & 0 & -1 & 0 & 0 & -1 & -1 \end{pmatrix} \begin{pmatrix} y_{12} \\ y_{13} \\ y_{24} \\ y_{26} \\ y_{34} \\ y_{35} \\ y_{46} \\ y_{56} \end{pmatrix} = \begin{pmatrix} f_1^* \\ f_2^* \\ f_3^* \\ f_4^* \\ f_5^* \\ f_6^* \end{pmatrix}$$

$$\begin{array}{rcl} y_{ij} + z_{ij} & = & 2m_{ij}; \forall i, j \\ y_{ij} & \geq & 0; \forall i, j \\ z_{ij} & \geq & 0; \forall i, j \end{array} \tag{A.66}$$

where $z_{ij} \geq 0; \forall i, j$ are auxiliary slack variables used to convert the system of inequalities into the system of equalities (A.66), and

$$\begin{pmatrix} f_1^* \\ f_2^* \\ f_3^* \\ f_4^* \\ f_5^* \\ f_6^* \end{pmatrix} = \begin{pmatrix} f_1 - m_{12} - m_{13} \\ f_2 + m_{12} - m_{24} - m_{26} \\ f_3 + m_{13} - m_{34} - m_{35} \\ f_4 + m_{24} + m_{34} - m_{46} \\ f_5 + m_{35} - m_{56} \\ f_6 + m_{26} + m_{46} + m_{56} \end{pmatrix} \tag{A.67}$$

Obtain the compatibility conditions and the general solution of the problem.

A.3 Consider the network flow problem in Figure A.6.

(a) Write the system of equations associated with this problem.

(b) Obtain the compatibility conditions and give them a physical interpretation.

(c) Obtain the set of feasible solutions and explain the result.

(d) Obtain the optimal solution for unit costs and $f_i = 10; \forall i$.

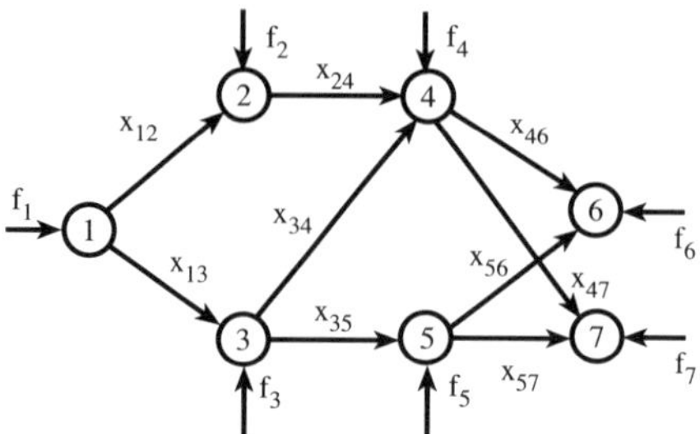

Figure A.6: Network flow problem.

Table A.17: Demand table

Time	1	2	3	4	5	6	7	8	9	10	11	12
Demand	3	4	5	4	3	2	3	1	2	1	3	4

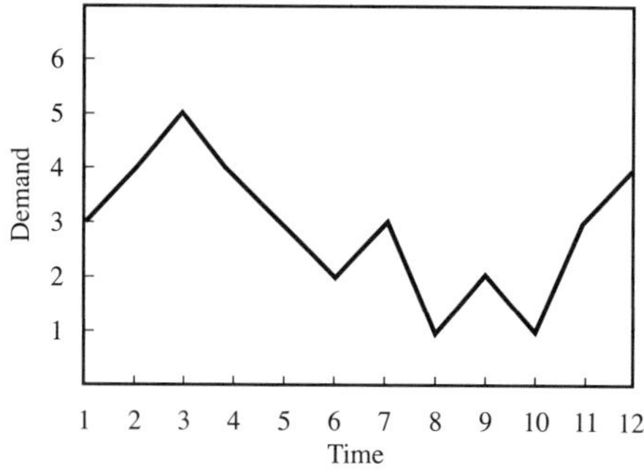

Figure A.7: Production-scheduling problem.

(e) Assuming that the shipping associated with x_{47} fails; redo parts (a)–(d) without starting from scratch.

A.4 In the following production scheduling problem, consider the demand graph in Figure A.7 (see also Table A.17) and assume an initial stock of two units.

(a) Write the system of equations associated with this problem.

(b) Obtain the compatibility conditions and give them a physical interpretation.

(c) Obtain the set of feasible solutions and explain the result.

(d) Fix the required costs.

(e) Obtain the optimal solution for these costs.

Appendix B

Notation

In this appendix we list the main notation used in this book. We have attempted to keep the notation as consistent as possible throughout the book.

Chapter 1

a_{ij}	Amount of nutrient i in one unit of food j
a_t	Selling price of one unit of good at time t
b_i	Current number of shares of stock i
	Minimum amount of nutrient i required
b_t	Production cost at time t
B	Set of beams
B_{ij}	Susceptance of line i–j
c_j	Cost of one unit of food j
c_t	Storage cost at time t
c_{ij}	Cost of sending a unit of product from origin i to destination j
C_i	Cost of producing electric power from generator i
d_i	Dividend to be paid at the end of the year for stock i
d_t	Number of units (demand) required in month t
dl_i	Distance of load i to the left endpoint of the beam where it acts
dr_s	Distance of supporting rope s to the left endpoint of the beam b it supports
D_i	Demand at bus i
f_i	Input flow at node i
$\mathcal{G}$	Graph $\mathcal{G} = (\mathcal{N}, \mathcal{A})$ where $\mathcal{N}$ is the set of nodes and $\mathcal{A}$ is the set of links
I	Set of loads
m	Number of origins
	Number of nutrients

	Number of stocks
m_{ij}	Maximum flow capacity of link going from node i to node j
n	Number of different foods
	Number of months to be considered
	Number of electric power generators
	Number of destinations
	Number of nodes in the network
$\underline{P}_i$	Minimum output electric power of generator i
p_i	Actual power produced by generator i
$\overline{P}_i$	Maximum output electric power of generator i
P_{ij}^{max}	Maximum transmission capacity of line i–j
s_0	Amount of an item available in storage at the beginning of the considered period
s_t	Number of units in storage in month t
s_t^{max}	Storage capacity
S	Set of ropes
t_s	Tension generated on rope s under the action of the set of loads
T_s	Maximum admissible load on rope s
u_i	Amount of product to be shipped from origin i
v_i	Current value of stock i
v_j	Amount of product to be received in destination j
w_i	New value of stock i
x_i	Load i
	Change in the number of shares of stock i
x_{ij}	Amount to be shipped from origin i to destination j
	Flow going from node i to node j
x_j	Amount of food j to be purchased
x_t	Number of units produced in month t
$\mathbf{z}^T$	Transpose of a vector $\mathbf{z}$
Z	Value of the objective function
δ_i	Angle of bus i
Ω_b	Set of loads applied in the mid point of beam b
Ω_i	Set of buses connected through lines to bus i
Ψ_b	Set of ropes supporting beam b
Θ_b	Set of ropes supported in beam b

Chapter 2

a	Minimum required discrepancy level
a_j	Weight of the object j of the knapsack problem
A_j	Fixed cost of the thermal unit j
b	Capacity of the knapsack
b_i	Demand of the ith client
B_j	Variable cost of the thermal unit j

c_{ij}	Level of symptom j associated with disease i
	Profit per unit of sale of the goods originated at facility j to client i
c_j	Utility of object j
C_j	Total score associated with candidate j
	Startup cost of thermal unit j
$d(x)$	A discrepancy measure
$\mathcal{D}$	Set of diseases
D_k	Demand in period k
E_j	Shutdown cost of unit j
f_j	Fixed cost of the opening facility placed at j
I	Actual number of members in the Academy of Engineering
	Set $\{1, \ldots, n\}$ of n clients
J	The number of candidates
	Number of power units
	Set $\{1, \ldots, m\}$ of m sites where facilities can be located
K	Number of time interval in the unit commitment problem
m	Number of symptoms
n	Number of diseases
	Number of objects of the knapsack problem
n_b	Number of academic blocks
n_c	Number of classrooms
n_h	Number of available teaching hours
n_i	Number of subjects taught by instructor i
n_s	Number of subjects
p_{jk}	Output power of unit j during period k
$\underline{P}_j$	Minimum output power of unit j
p_s	The sth score
$\overline{P}_j$	Maximum output power of unit j
P_j^0	Output power of unit j just before the first period of the planning horizon
r	Number of new members to be incorporated
R_k	Amount of required reserve (over demand) in period k
S	Number of different scores that can be assigned
S_j	Maximum rampup power increment of unit j
$\mathcal{S}$	Set of symptoms
$\mathcal{S}_a$	A subset of the set of symptoms $\mathcal{S}$
T_j	Maximum rampdown power decrement of unit j
u_j	Capacity of the facility located at j
$v(s, c, h)$	Binary variable that takes value 1 if subject s is taught in classroom c at hour h and 0 otherwise
v_{jk}	Binary variable that is equal to 1, if unit j is online during period k and 0 otherwise
V_j^0	Binary constant which is equal to 1, if unit j is

	online during the period preceding the first period of the planning horizon, and 0 otherwise
x_{ij}	Amount of commodities sent from facility j to client i
x_{ijs}	Binary variable equal to 1 if member i assigns score p_s to candidate j; otherwise, it takes value 0
y_j	Binary variable to model the choice of opening a facility at the site j
y_{jk}	Binary variable equal to 1 if unit j is started up at the beginning of period k and 0 otherwise
z_{jk}	Binary variable equal to 1 if unit j is shut down at the beginning of period k and 0 otherwise
Z	The value of the objective function
Δ_b	Set of subjects belonging to academic block b
Ω	Set of all subjects to be taught
Ω_i	Set of subjects taught by instructor i

Chapter 3

a	Dimension of the square base of a tent
$(\mathcal{A}, \mathcal{N})$	A directed graph $(\mathcal{A}, \mathcal{N})$ where $\mathcal{A}$ is the set of links and $\mathcal{N}$ is the set of nodes
b	Height of the vertical walls of a tent
B	Set of beams
c_j	Number of trips attracted by zone j
	Sum of the elements in column j of a matrix
$C_a(f_a)$	Cost function associated with the link a
C_i	Cost of producing one unit of active power using generator i
d_ω	Number of trips by car from the origin i to destination j, for any origin–destination pair ω
D	Joint displacement
	Unit density of the material to be used
D_0	Maximum admissible joint displacement
$D(x, y, z)$	Distance function
E	Young's modulus of the material to be used
f_a	Flow on link a
F	Load at the tip of of a cantilever beam
	Applied load at the fixed joint
g	Acceleration of gravity
h	Height of the pyramid
	Height of the bulb
	Height of the trust
h_r	Flow on route r
H	Total height of the tent
	Column height
I	Intensity of the light

	Moment of inertia
l_B	Total length of beam b
L	Length of a cantilever beam
k	Proportionality constant
M	Mass to be supported by the column
n	Number of buses in the electrical network
$\hat{p}_{kl}$	measurement of the active power leaving bus k toward bus I of line k–I toward bus l of line k–l
p_{Gi}	Active power generation in bus i
P_{Di}	Active power demand in bus i
$\underline{P}_{Gi}$	Minimum active output power of generator i
$\overline{P}_{Gi}$	Maximum active output power of generator i
$\hat{q}_{kl}$	the measurement of the reactive power leaving bus k toward bus I of line k–I
q_{Gi}	Reactive power generation in bus i
$\underline{Q}_{Gi}$	Minimum reactive output power of generator i
$\overline{Q}_{Gi}$	Maximum reactive output power of generator i
Q_{Di}	Reactive power demand in bus i
r_i	Number of trips generated at zone i Sum of the row i of a matrix
$\mathcal{R}_\omega$	Set of simple routes for the pair of demand ω
S	Maximum allowable deflection of the cantilever beam Surface of the tent Actual maximum stress (force per unit surface)
S_0	Maximum admissible stress
S^i	Stress at joint i
t_{ij}	Observed cell i–j content Observed trips from i to j
T_{ij}	Predicted trips from i to j
v_i	Voltage magnitude at bus i
$\hat{v}_i$	The measured voltage magnitude at bus i
V	Total volume of a tent Total volume of a postal package
$\underline{V}_i$	Lower bound of the voltage magnitude in bus i
$\overline{V}_i$	Upper bound of the voltage magnitude at bus i
W	Total weight of the bar truss Set of origin–destinations pairs
x	Distance from the fixed joints to the Y axis
xl_i	Distance of load i to the left endpoint of the beam where it acts
(y_{ik}, θ_{ik})	A complex constant given as its magnitude y_{ik}, and argument θ_{ik}
z	Area of the cross section of the truss arms
$z_{k\ell}$	Impedance magnitude associated with line k–l

δ_i	Voltage angle at bus i
γ or ρ	Unit weight of the bar material
Ω	Set of buses of the electrical network
Ω_k	Set of buses connected to bus k
σ_i^v	Voltage measurement quality level
σ_{kl}^p	Degree of precision of $\hat{p}_{kl}$
σ_{kl}^q	Degree of precision of $\hat{q}_{kl}$
$\theta_{k\ell}$	Impedance angle associated with line k–l

Chapter 4

$\mathbf{A}$	Matrix determining the constraints in a LPP
$\mathbf{A} = (\mathbf{B} \quad \mathbf{N})$	Decomposition on basic and nonbasic columns
$\mathbf{b}$	Vector defining the constraints in a LPP
$\mathbf{B}$	A feasible basic matrix
$\mathbf{B}^*$	Optimal basic matrix
$\mathbf{c}$	Vector defining the cost function in a LPP
$\mathbf{c}_B^T$	Basic cost vector
D	Dual problem
P	Primal problem
x^+, x^-	Positive and negative parts of a number
$\mathbf{x}_B$	The vector of all basic variables
$\mathbf{x}_B^*$	Basic optimal solution
$\tilde{\mathbf{x}}$	A generic feasible vector for the primal
$\mathbf{y}$	Vector of dual variables
$\tilde{\mathbf{y}}$	A generic feasible vector for the dual
z^*	Optimal cost value
$Z = f(\mathbf{x})$	Cost or objective function
λ_k	Coefficients of a linear convex combination
$\boldsymbol{\lambda}^{*T}$	Vector of sensitivity parameters or dual variables
π_j	Coefficients used in nonnegative linear combinations
ρ_i	Real coefficients in linear combinations of vectors
$\Delta\mathbf{b}$	Marginal change in vector $\mathbf{b}$
$\Delta\mathbf{x}_B$	Change in the optimal solution
Δz	Change in the optimal objective function value

Chapter 5

$\mathbf{a}_k$	kth column vector of matrix $\mathbf{A}$
$\mathbf{A}$	A typical matrix
$\mathbf{A}_\lambda$	Set of all linear convex combinations of the column vectors of $\mathbf{A}$
$\mathbf{A}_\pi$	Set of all nonnegative linear combinations of the column vectors of $\mathbf{A}$
$\mathbf{A}_\pi \equiv \mathbf{B}_\rho + \mathbf{C}_\pi$	General form of a cone as the sum of a linear space plus a proper cone

$\mathbf{A}_\rho$	Set of all linear combinations of the column vectors of $\mathbf{A}$
$\mathbf{B}$	Set of generators whose opposite ones belong to the cone
C	A generic convex cone
$\mathbf{C}$	Set of generators whose opposite ones do not belong to the cone
$\mathbf{d}$	A generic direction of a convex set
D	The set of directions
D_E	The set of extreme directions
E	The set of extreme points of a convex set
$\mathbf{H}$	Matrix determining a linear space, a cone, a polytope, or a polyhedron
$\mathbf{p}$	A nonzero vector in $\mathbb{R}^n$
$\mathbf{q}_k$	kth generator of a polytope
$\mathbf{Q}$	Set of extreme points
S	A generic convex set
S^+, S^-	Half spaces associated with S
$S_1 + S_2$	The sum of two convex sets
$\{S_i\}_{i \in I}$	A family of convex sets
$\mathbf{v}_i$	ith generator of a linear space
$\mathbf{V}_\rho + \mathbf{W}_\pi + \mathbf{Q}_\lambda$	General form of a polyhedron
$\mathbf{w}_j$	jth generator of a cone
$\mathbf{W}$	Set of extreme directions
α	A typical scalar
λ_k	Coefficients used in linear convex combinations
$\lambda\mathbf{x} + (1-\lambda)\mathbf{y}$	A typical convex combination
ρ_i	Coefficients used in linear combinations
π_j	Coefficients used in nonnegative linear combinations of vectors
$\mathcal{V}$	A linear space
$\mathcal{W}$	A linear subspace

Chapter 6

$\mathbf{A}$	$m \times n$ matrix containing constraints coefficients
$(\mathbf{B} \quad \mathbf{N})$	Partition of matrix $\mathbf{A}$ in basic and nonbasic columns
$\mathbf{c} = (c_1, c_2, \ldots, c_n)^T$	Column matrix of cost coefficients
$(\mathbf{c}_\mathbf{B}^T \quad \mathbf{c}_\mathbf{N}^T)$	Partition of objective function coefficients in basic and nonbasic components
(t)	Iteration number
$\mathbf{x} = (x_1, x_2, \ldots, x_n)^T$	Column vector of initial variables
$(\mathbf{x}_\mathbf{B} \quad \mathbf{x}_\mathbf{N})$	Partition of vector $\mathbf{x}$ in basic and nonbasic coordinates
$z_{\alpha\beta}^{(t)}$	A pivot
Z	Value of the objective function
$\mathbf{Z}^{(t)}$	Matrix iterates in the SM to which the elemental pivoting operation is applied

α	Row index of the pivot element
β	Column index of the pivot element
$\Lambda_\alpha^{(t)}$	Function-pivot ratio

Chapter 7

$\mathbf{B}\quad\mathbf{N}$	Partition of matrix $\mathbf{A}$ in basic $\mathbf{B}$ and nonbasic $\mathbf{N}$ components
$\mathbf{I}$	Identity matrix
$\mathbf{U}=\mathbf{B}^{-1}\mathbf{N}$	Central matrix in the Gomory algorithm
$\mathbf{x}_B\cup\mathbf{x}_N$	Partition of the set of variables $\{x_1,x_2,\ldots,x_n\}$ in basic $\mathbf{x}_B$ and nonbasic $\mathbf{x}_N$ variables
x_{B_i}	Basic noninteger but integer-to-be variable
i_{ij}	Integer part of u_{ij}
f_{ij}	Fractional part of u_{ij}
Z	Objective linear function

Chapter 8

$\mathbf{A}^{-1}$	Inverse of matrix $\mathbf{A}$
$c_{ti}(p_{ti})$	Production cost of producer i in period t to produce p_{ti}
d	A generic direction
d_t	Demand in period t
$\det(\mathbf{A})$	Determinant of matrix $\mathbf{A}$
$f(\mathbf{x})$	Objective function $f:\mathbb{R}^n\to\mathbb{R}$
$\mathbf{F}(z)$	KKT condition for an equality constrained problem
$\mathbf{g}(\mathbf{x})$	Inequality constraint mapping $\mathbf{g}:\mathbb{R}^n\to\mathbb{R}^m$
$\mathbf{h}(\mathbf{x})$	Equality constraint mapping $\mathbf{h}:\mathbb{R}^n\to\mathbb{R}^\ell$
I	Number of producers
$I(\bar{\mathbf{x}})$	Set of active indices at $\mathbf{x}$ for the inequality constraints $I(\bar{\mathbf{x}})=\{j\vert g_j(\bar{\mathbf{x}})=0\}$
$\text{Infimum}_{\,\mathbf{x}\in S}\;q(\mathbf{x})$	Infimum of the function q on S
$\text{Inf}_{\,\mathbf{x}}\;q(\mathbf{x},\mathbf{y})$	Infimum of the function q with respect to the variable $\mathbf{x}$
$\mathcal{L}(\mathbf{x},\boldsymbol{\mu},\boldsymbol{\lambda})$	Lagrangian function for the NPP
$\lim_{\mathbf{y}\to\mathbf{x}}q(\mathbf{y})$	Limit of the mapping q at $\mathbf{x}$
$\max_{\mathbf{x}\in S}q(\mathbf{x})$	Maximum of the function q on S
$\min_{\mathbf{x}\in S}q(\mathbf{x})$	Minimum of the function q on S
p_{ti}	Amount of commodity produced by producer i during perioc
$\mathbb{R}$	Real line
$\mathbb{R}^n$	Real n-dimensional space
S	Set of feasible solutions
$\text{Sup}_{\,\mathbf{x}}\;q(\mathbf{x},\mathbf{y})$	Supremum of the function q with respect to the variable $\mathbf{x}$
T	Number of time periods

$\mathbf{x}$	Vector of the decision variables
x_i	Coordinate ith of the vector of the decision variables
$\bar{\mathbf{x}}$	Local minimum of an optimization problem
$\mathbf{x}^*$	Global minimum of an optimization problem
$\mathbf{x}(\boldsymbol{\lambda}, \boldsymbol{\mu})$	Minimum of the Lagrangian for the multipliers $(\boldsymbol{\lambda}, \boldsymbol{\mu})$
$\mathbf{z}^T$	Transpose of vector $\mathbf{z}$
$\mathbf{z}^T\mathbf{y}$	Standard inner product of vectors $\mathbf{z}$ and $\mathbf{y}$ in $\mathbb{R}^n$
Z	Value of the objective function
$\boldsymbol{\Delta}\mathbf{z}$	Increment of variable $\mathbf{z}$
$\lambda_t^{(k)}$	Selling price for period t
$\boldsymbol{\lambda}$	Vector of Lagrangian multipliers associated with the equality constraints
$\boldsymbol{\mu}$	Vector of Lagrangian multiplier associated with the inequality constraints
$(\boldsymbol{\lambda}, \boldsymbol{\mu})$	Dual variables
$\sigma(t)$	A curve in $\mathbb{R}^n$ $(\sigma : (-\epsilon, \epsilon) \to \mathbb{R}^n)$
$\theta(\boldsymbol{\lambda}, \boldsymbol{\mu})$	Dual function
Π_i	Producer i operation feasible region
$\dfrac{\partial q}{\partial x_i}$	Partial derivative of q with respect to x_i
$q'(x)$	First derivative of function q
∇q	Gradient of function q
	$m \times n$ Jacobian of $q : \mathbb{R}^n \to \mathbb{R}^m$ $(m \geq 2)$
$\nabla_{\mathbf{y}} q(\mathbf{x}, \mathbf{y})$	Partial Jacobian matrix of q with respect to $\mathbf{y}$
$\nabla^2 q$	Hessian matrix of q
$\nabla_{\mathbf{yy}} q(\mathbf{x}, \mathbf{y})$	Partial Hessian matrix of q with respect to $\mathbf{y}$
$\|\cdot\|$	Euclidean vector norm
[a,b]	Closed interval

Chapter 9

$\arg\text{minimize}_{\mathbf{x}} q(\mathbf{x})$	Value of $\mathbf{x}$ leading to the the minimum of function q
$\mathbf{A}^{-1}$	Inverse of matrix $\mathbf{A}$
$\mathbf{B}^{(t)}$	Approximation of the Hessian matrix $[\nabla^2 f(\mathbf{x}^{(t)})]$
D	Set of feasible solutions of the dual problem
$\mathbf{d}$	A search direction
$\mathbf{d}^{(t)}$	Search direction at iteration t
$f(\mathbf{x})$	Objective function
$\mathbf{g}(\mathbf{x})$	Inequality constraint $\mathbf{g} : \mathbb{R}^n \to \mathbb{R}^m$
$\mathbf{h}(\mathbf{x})$	Equality constraint $\mathbf{h} : \mathbb{R}^n \to \mathbb{R}^\ell$
$\mathbf{H}^{(t)}$	Approximation of the inverse of the Hessian matrix of f at $\mathbf{x}^{(t)}$
$\text{Infimum}_{\ \mathbf{x} \in S}\ q(\mathbf{x})$	Infimum of the function q on S
$\text{Inf}_{\ \mathbf{x}}\ q(\mathbf{x}, \mathbf{y})$	Infimum of the function q with respect to the variable $\mathbf{x}$

ℓ	Lower bound
$\mathcal{L}(\mathbf{x}, \boldsymbol{\mu}, \boldsymbol{\lambda})$	Lagrangian function
$\lim_{t\to+\infty} x_t$ or x_∞	Limit of the sequence of real numbers $\{x_t\}$
$\lim_{\mathbf{y}\to\mathbf{x}} q(\mathbf{y})$	Limit of the q function at $\mathbf{x}$
$\max_{\mathbf{x}\in S} q(\mathbf{x})$	Maximum of the function q on S
$\min_{\mathbf{x}\in S} q(\mathbf{x})$	Minimum of the function q on S
$\mathbf{P}$	Projection matrix
$P(\mathbf{x}; r)$	A generic exterior penalty function
$P_1(\mathbf{x}; r)$	Absolute value penalty function
$P_2(\mathbf{x}; r)$	Quadratic penalty function
r	Penalty parameter
r_d	Dual residual for the interior point method
r_p	Primal residual for the interior point method
r_t	Penalty parameter at iteration t
$\mathbb{R}$	Real line
$\mathbb{R}^n$	Real n-dimensional space
S	Set of feasible solutions of the primal problem
S^-	A subset of the feasible region S defined by $S^- = \{\mathbf{x} \mid \mathbf{g}(\mathbf{x}) < \mathbf{0}\}$
$\text{Sup}_{\mathbf{x}}\ q(\mathbf{x}, \mathbf{y})$	Supremum of the function q with respect to the variable $\mathbf{x}$
t	Iteration counter
u	Upper bound
$\mathbf{x}$	Vector of the primal variables
$\mathbf{x}^{(t)}$	$\mathbf{x}$-vector iterates
x_t	scalar iterates
$\{\mathbf{x}^{(t)}\}$	Sequence of vectors
$\{x_t\}$	Sequence of scalars
$\mathbf{x}(r)$	Minimum of the penalty function $P(\mathbf{x}, r)$
$\mathbf{x}(\boldsymbol{\lambda}, \boldsymbol{\mu})$	Minimum of the $(\boldsymbol{\lambda}, \boldsymbol{\mu})$ Lagrangian function
$X(\boldsymbol{\lambda}, \boldsymbol{\mu})$	Set of points $\mathbf{x}(\boldsymbol{\lambda}, \boldsymbol{\mu})$
$\mathbf{z}^T$	Transpose of vector $\mathbf{z}$
$\mathbf{z}^T\mathbf{y}$	Standard inner product of vectors in $\mathbb{R}^n$
$Z,\ Z_P$	Value of the objective function of the primal problem
Z_D	Value of the objective function or the dual problem
Z_D^*	Optimal value of the dual problem
Z_P^*	Optimal value of the primal problem
Z_{LS}	Value of the line search problem objective function
$\boldsymbol{\Delta}\mathbf{z}$	Increment of variable $\mathbf{z}$
$\boldsymbol{\alpha}$	Step length
$\boldsymbol{\alpha}_d$	Dual step length for the interior point method
$\boldsymbol{\alpha}_p$	Primal step length for the interior point method
$\boldsymbol{\alpha}_t$	Step length of the tth line search
β	A parameter of the conjugate gradient method
δ	Parameter used in Armijo's rule tolerance for the interior point method

$\boldsymbol{\lambda}$	Vector of the Lagrangian multipliers associated with the equality constraints
$\boldsymbol{\lambda}^c$	Lagrangian multiplier associated with the columns in the matrix balancing problem
$\boldsymbol{\lambda}^r$	Lagrangian multiplier associated with the rows in the matrix balancing problem
$\boldsymbol{\mu}$	Vector of the Lagrangian multipliers associated with the inequality constraints
η	Parameter used to increase the value of the penalty parameter of the exterior penalty method
ε	Tolerance parameter to test convergence
$(\boldsymbol{\lambda}, \boldsymbol{\mu})$	Dual variables
$\psi(\mathbf{z}, \mathbf{y})$	Exterior penalty function defined as $\psi : \mathbb{R}^{\ell+m} \to \mathbb{R}$
$\phi(\mathbf{x})$	Interior penalty function
$\theta(\boldsymbol{\lambda}, \boldsymbol{\mu})$	Dual function
$q'(x)$	First derivate of the function $q : \mathbb{R} \to \mathbb{R}$
$q'(\mathbf{x}; \mathbf{d})$	Directional derivative of q in the direction $\mathbf{d}$ at the point $\mathbf{x}$
$q''(x)$	Second derivate of the function $q : \mathbb{R} \to \mathbb{R}$
∇q	Gradient of $q : \mathbb{R}^n \to \mathbb{R}$ $m \times n$ Jacobian of $q : \mathbb{R}^n \to \mathbb{R}^m$ $(m \geq 2)$
$\nabla_{\mathbf{y}} q(\mathbf{x}, \mathbf{y})$	Partial Jacobian matrix of q with respect to $\mathbf{y}$
$\dfrac{\partial q}{\partial x_i}$	Partial derivate of function q with respect to x_i
$\left.\dfrac{\partial q}{\partial x_i}\right\vert_{\mathbf{x}}$	Partial derivate of q with respect to x_i at $\mathbf{x}$
$\nabla^2 q$	Hessian matrix of $q : \mathbb{R}^n \to \mathbb{R}$
$\Vert \cdot \Vert$	Euclidean vector norm

Chapter 12

$\overline{a}$	Mean of the (mesh) area a
a	Index referring to transportation by car
$a(i)$	Area of the (mesh) triangle i
a_{ij}, b_{ij}	Component ijth of matrices $\mathbf{A}$, $\mathbf{B}$, respectively
a_j	Projectile velocity at time j
A	Constant
$\mathbf{A}$	Conditional probability matrix
$\mathcal{A}$	Ordered set of directed links or arcs
A_i	Proportionallity factor of the function $p(c_{ij})$
$A(i)$	Sets of nodes "after" node i
b	Transportation by public transport
b_j	Projectile position at time j
B	Constant
$\mathbf{B}$	Conditional probability matrix
$\mathcal{B}$	Subset of bounded links

B_i	Proportionallity factor of the function $p(c_{ij})$
$B(i)$	Sets of nodes "before" node i
$c_a(f_a)$	Cost of link a as a function of its service level
$\hat{c}_a(f_a)$	Cost of link a for its maximum link capacity
c_Ω	Minimum travel cost for the pair Ω
c_Ω^a	Cost (variable) of car traveling associated with pair Ω
c_Ω^b	Cost (constant) of public transport traveling associated with pair Ω
c_i^Ω	Cost of the used route i for a commodity Ω
$c_{ij}(x)$	Link cost function for the transportation problem
c_{ij}^0	Free-flow travel time
$\mathbf{C}_\pi^*$	Dual or polar cone of $\mathbf{C}_\pi$
$C_{r'}^*$	Length of route r'
CPU	Central processing unit
$d_i(LAV)$	ith observation influence on the LAV estimators
$d_i(LS)$	ith observation influence on the LS estimators
$d_i(MM)$	ith observation influence on the MM estimators
$d(\boldsymbol{\lambda}, \boldsymbol{\mu})$	Solution of the relaxed primal problem given the values of $\boldsymbol{\lambda}$, $\boldsymbol{\mu}$ (Lagrangian relaxation algorithm)
D	Set of input and output data of a neuron Space dimension of the mesh
D_j	Number of trips attracted to destination j
$e_{i(jk)}$	Element in the row associated with node i, and the column associated with link (j, k)
$\mathbf{E}$	Node-link incidence matrix
$\mathbf{f}$	Link flow of all commodities
f_{ij}	Link flow on the directed link (i, j)
f_{ij}^Ω	Link flow on the directed link (i, j) associated with commodity Ω
$f(\mathbf{x})$	Primal objective function of the STHTC problem
g_Ω	Demand (number of trips) of commodity Ω
$\overline{g}_\Omega$	Sum of demands for commodity Ω considering all the possible modes of transportation
g^Ω	Flow of commodity Ω from O^Ω–D^Ω
$\mathbf{g}_i(\mathbf{x}_i)$	Contribution of thermal unit i to meet power reserve
$\mathbf{g}_j(\mathbf{x}_j)$	Contribution of hydro system j to meet power reserve
g_{ij}^*, c_{ij}^*	Optimal values of the functions g_{ij}, c_{ij}, respectively
$\mathbf{G}$	Power reserve vector of the STHTC problem
$\mathcal{G}$	Directed graph
$G_\Omega(c_\Omega)$	Variable demand as a function of c_Ω
$\mathbf{h}$	Flow vector
h_r	Flow in route r
$\mathbf{h}_i(\mathbf{x}_i)$	Contribution of thermal unit i to meet the demand
$\mathbf{h}_j(\mathbf{x}_j)$	Contribution of hydro system j to meet the demand
H	Distance between the two endpoints of the cable

$\mathbf{H}$	Demand vector of the STHTC problem
i	Index of elements of the mesh
(i, j), a	Elements of the set $\mathcal{A}$
I	Dimension of the data input vector
	Number of thermal plants
$I(u)$	Cost functional
k	Constant
k_a	Maximum capacity of link a
$\mathcal{K}$, k	Set of links, nodes, routes, or any combination of them
l	Number of used routes
L	Length of the hanging cable
$\mathcal{L}(.)$	Lagrangian function
LAV	Least-absolute-value method
LR	Lagrangian relaxation
m	Number of saturated routes
MM	Minimax method
n	Number of discretization intervals
	Number of mesh nodes
N	Domain set of the (X, Y) variable
$\|N\|$	Cardinality of the domain set N
$\mathcal{N}$	Unordered finite set of nodes of the graph $\mathcal{G}$
O-D	Origin–destination pair
O_i	Number of trips emanating from origin i
$p_{i.}$	Marginal probability of X
$p_{.j}$	Marginal probability of Y
p_{ij}	Joint probability of (X, Y)
$p(c_{ij})$	Deterence function
$\mathbf{P}$	Joint probability matrix
PP	Primal problem
Q	Objective function value
Q_j	Objective function value associated with the neural learning option j
r, r'	Indices of routes
r_i	Flow produced or attracted by node i
r_i^{Ω}	Variable flow produced or attracted by node i
r	Index of invertible basic neural functions
R	Number of invertible basic neural functions
$\mathcal{R}_{\Omega}$	Set of routes for commodity Ω
$\mathbf{s}^{(t)}$	A subgradient of the dual function at iteration t
$s_k(\mathbf{f})$	Generic kth function dependent on the link flow
S	Number of measurements in the neural network
STHTC	Short-term hydro-thermal coordination
t	Time index
(t)	Iteration index (Lagrangian relaxation algorithm)
$u(t)$	Function that establishes the control to modify the

	dynamics of a system given some specific objective
$u'(t)$, $u''(t)$	First and second derivative of the function $u(t)$
u_a	Maximum capacity of link a
$\mathbf{V}$, $\mathbf{W}$	Generators of the dual cone $\mathbf{C}_\pi^*$
W	Set of all the O-D pairs
w_i	Neuron weight for the component i of the input
w_0	Neuron threshold value
$\mathbf{x}_i$	Vector of (primal) variables associated with thermal plant i
$\mathbf{x}_j$	Vector of (primal) variables associated with hydro system j
x_{is}	Data input vector of a neuron
x_{max}	Maximum coordinate x value in the mesh
x_{min}	Minimum coordinate x value in the mesh
$\mathbf{x}_i^{*(t)}$	Optimal value of vector $\mathbf{x}_i$ at iteration t
$x(t)$	Function determining the dynamics of the state of the system under control
(X, Y)	Bidimensional discrete random variable
$\mathbf{y}$	Response variables in the regression model
$\hat{\mathbf{y}}$	Vector of fitted values of the response variable $\mathbf{y}$ with all data points
$\hat{\mathbf{y}}_{(i)}$	Vector of fitted values of the response variable $\mathbf{y}$ when the ith observation is omitted
y_i	Component ith of vector $\mathbf{y}$
y_{max}	Maximum value of coordinate y in the mesh
y_{min}	Minimum value of coordinate y in the mesh
y_s	Data output vector of a neuron
y_{s_1}, y_{s_2}	Consecutive components of vector y_s
$\mathbf{Z}$	Matrix of predictor variables
$\mathbf{z}_i^T$	ith row in $\mathbf{Z}$
α	Proportionallity factor of the function $p(c_{ij})$
α', β, d_i, d_j	Parameters of the traffic distribution problem
α^k	Parameter of the logit model
α_r	rth coefficient of the neural function
$\boldsymbol{\beta}$	Vector of regression coefficients
$\hat{\boldsymbol{\beta}}$	Least-absolute-value estimator of $\boldsymbol{\beta}$
$\tilde{\boldsymbol{\beta}}$	Minimax estimator of $\boldsymbol{\beta}$
β_a	Lagrangian multiplier associated with the a maximum link capacity constraint
$\hat{\boldsymbol{\beta}}_{(i)}$	Least-absolute-value estimate of $\boldsymbol{\beta}$ when the ith observation is omitted
$\tilde{\boldsymbol{\beta}}_{(i)}$	Minimax $\boldsymbol{\beta}$ estimate when the ith observation is omitted
β_1	Parameter of the logit model
$\delta_{a,r}$	Binary data indicating whether the arc a belongs to the route r

δ_s	Error associated with the output y_s
Δ	Discretization step
ϵ	Nonnegative variable which represents the maximum error for learning option 3
	Global degree of incompatibility for two conditional probability matrices
$\phi_a(f_a)$	ath function to model the maximum link capacity
$\phi_r(x)$	Invertible basic neural functions
γ_{ij}	Relative importance of the error for p_{ij}
$\boldsymbol{\lambda}$	Demand constraints Lagrangian multipliers
$\boldsymbol{\lambda}$, $\boldsymbol{\psi}$	Lagrangian multipliers
$\boldsymbol{\mu}$	Reserve constraints Lagrangian multipliers
Ω	An specific origin–destination (O–D) pair
$\boldsymbol{\tau}$, $\boldsymbol{\eta}$	Marginal probability vectors
τ_i, η_j	Component ith and jth of vectors $\boldsymbol{\tau}, \boldsymbol{\eta}$, respectively
$\theta(.)$	Dual function
$\boldsymbol{\varepsilon}$	Vector of random errors

Chapter 13

$\mathbf{A}_\pi$	Polyhedral convex cone or cone generated by $\mathbf{A}$
$\mathbf{A}_\rho$	Linear space
x^*	Nonnegative variable
x^+,x^-	Positive and negative parts of variable x, respectively
x_{n+1}	Slack variables

Appendix A

$\mathbf{A}_\pi$	Polyhedral convex cone or cone generated by $\mathbf{A}$
$\mathbf{A}_\pi^*$	Nonpositive dual cone of the cone $\mathbf{A}_\pi$
$\mathbf{A}_\rho$	Linear space
$\mathbf{A}_\rho^*$	Nonpositive dual of the linear space $\mathbf{A}_\rho$
$\mathbf{B}$	Linear space generators of the cone $\mathbf{A}_\pi$
$\mathbf{C}$	Proper cone generators of the cone $\mathbf{A}_\pi$
$\mathbf{C}_S$	Cone associated with polyhedron S
h	Iteration counter for the Γ algorithm
$\mathbf{I}^-$	Indices associated with negative dot products
$\mathbf{I}^+$	Indices associated with positive dot products
$\mathbf{I}^0$	Indices associated with null dot products
$\mathbf{I}_n$	Identity matrix of dimension n
K_{W^0}	Set of all vectors in $\mathbf{W}^0$ with its last component nonnull
$pivot$	Index of the pivot column
$\mathbf{Q}_\lambda$	Polytope generated by the column vectors in $\mathbf{Q}$
ρ	Unrestricted real number
$\mathbf{V}_\rho^0$	Linear space component of $\mathbf{C}^{T^*}$
$\mathbf{V}$, $\mathbf{W}$	Matrix of generators of the dual cone $\mathbf{A}_\pi^*$

$\mathbf{W}_{\pi}^{0}$	Proper cone component of $\mathbf{C}^{T^*}$
$\mathbf{W}^{\pm}$	Set of vectors $\{\mathbf{w} \mid \mathbf{w} = t_j \mathbf{w}_i - t_i \mathbf{w}_j\}$
λ	Real number in the interval [0,1]
π	Nonnegative real number

Bibliography

[1] Arnold, B., Castillo, E., and Sarabia, J. M., *Conditional Specification of Statistical Models*, Springer-Verlag, New York, 1999.

[2] Arnold, B., Castillo, E., and Sarabia, J. M., *An Alternative Definition of Near Compatibility*, Technical Report 99-1, Univ. Cantabria, DMACC, 1999.

[3] Arnold, B., Castillo, E., and Sarabia, J. M., *Compatibility and Near Compatibility with Given Marginal and Conditional Information*, Technical Report 99-2, Univ. Cantabria, DMACC, 1999.

[4] Arnold, B., Castillo, E., and Sarabia, J. M., *Compatibility and Near Compatibility of Bayesian Networks*, Technical Report 99-3, Univ. Cantabria, DMACC, 1999.

[5] Arnold, B., Castillo, E., and Sarabia, J. M., *Exact and Near Compatibility of Discrete Conditional Distributions*, Technical Report, 00-1, Univ. of Cantabria, DMACC, 1999.

[6] Arthanari, T. S., and Dodge, Y., *Mathematical Programming in Statistics*, WIley, New York, 1993.

[7] Batut, J., Renaud, A., and Sandrin, P., "New Software for the Generation Rescheduling in the Future EDF National Control Center," *Proc. 10th Power Systems Computacional Conf.*, PSCC'90, Graz, Austria, 1990.

[8] Batut, J., and Renaud, A., "Daily Generation Scheduling Optimization with Transmission Constraints: a New Class of Algorithms," *IEEE Trans. Power Syst.* **7**(3), 982–989 (1992).

[9] Bazaraa M. S., Sherali, H. D., and Shetty C. M., *Nonlinear Programming, Theory and Algorithms*, 2nd ed., Wiley, New York, 1993.

[10] Beckman, M. J., and McGuire, C. B., *Studies in the Economics of Transportation*, Yale Univ. Press, New Haven, CT, 1956.

[11] Bertsekas, D. P., *Constrained Optimization and Lagrange Multiplier Methods*, Academic Press, New York, 1982.

[12] Bertsekas, D. P., Lauer, G. S., Sandell, N. R., and Posbergh, T. A., "Optimal Short-Term Scheduling of Large-Scale Power Systems," *IEEE Trans. Automatic Control*, **28**(1), 1–11 (1983).

[13] Bisschop, J., and Roelofs, M., *AIMMS, The User's Guide*, Paragon Decision Technology, 1999.

[14] Boyer, C. B., *A History of Mathematics*, Wiley, New York, 1968.

[15] Bradley, S. P., Hax, A. C., and Magnanti, T. L., *Applied Mathematical Programming*, Addison-Wesley, Reading, MA, 1977.

[16] Brannlund, H., Sjelvgren, D., and Bubenko, J., "Short Term Generation Scheduling with Security Constraints," *IEEE Trans. Power Syst.*, **1**(3), 310–316 (1988).

[17] Bronson, R., *Operations Research*, McGraw-Hill, New York, 1982.

[18] Canann, S. A., Muthukrishnan, S. N., and Phillips, R. K., "Topological Refinement Procedures for Triangular Finite Element Meshes," *Eng. Comput.*, **12**, 243–255 (1996).

[19] Canann, S. A., Tristano, J. R., and Staten, M. L., "An Approach to Combined Laplacian and Optimization-Based Smoothing for Triangular, Quadrilateral, and Quad-Dominant Meshes," *Proc. 7th International Meshing Roundtable*, 1998.

[20] Castillo, E., Fontenla, O., Guijarro, B., and Alonso, A., "A Global Optimum Approach for One-Layer Neural Networks," *Neural Computation*, (in press).

[21] Castillo, E., Cobo, A., Jubete, F., and Pruneda, E., *Orthogonal Sets and Polar Methods in Linear Algebra: Applications to Matrix Calculations, Systems of Equations and Inequalities, and Linear Programming*, Wiley, New York, 1999.

[22] Castillo, E., Hadi, A. S., and Lacruz, B., "Regression Diagnostics for the Least Absolute Value and the Minimax Methods," *Commun. Statist. Theor. Meth.* (in press).

[23] Castillo, E., and Mínguez, R., *An Optimization-Based Method for Mesh Generation and Smoothing*, Technical Report, 01-1, Univ. of Cantabria, DMACC, 2001.

[24] Chattopadhyay, D., "Application of General Algebraic Modeling System to Power System Optimization," *IEEE Trans. Power Systems* **14**(1), 15–22 (Feb. 1999).

[25] Charnes, A., and Cooper, W.,"Programming with Linear Fractional Programming," *Naval Research Logistics Quart.* **9**, 181–186 (1962).

[26] Clarke, F. H., *Optimization and Nonsmooth Analysis*, SIAM, Philadelphia, PA, 1990.

[27] Conn, A. R., Gould, N. I. M., and Toint, P. L., *LANCELOT: A FORTRAN Package for Large-Scale Nonlinear Optimization* (*Relase A*), No. 17 in Springer Series in Computational Mathematics, Springer-Verlag, New York, 1992.

[28] Cook, R. D., "Detection of Influential Observations in Linear Regression," *Technometrics* **19**, 15–18 (1977).

[29] Cuilliere, J. C., "An adaptive method for the Automatic Triangulation of 3D Parametric Surfaces," *Computer-Aided Design* **30**(2), 139–149 (1998).

[30] Dantzig, G. B., *Linear Programming and Extensions*, Princeton Univ. Press, Princeton, NJ, 1963.

[31] Davidon, W. C., *Variable Metric Method for Mininimization*, EC Research Development Report ANL-5990, Argonne National Laboratory, Chicago, IL, 1959.

[32] Dillon, T.S., Edwin, K.W., Kochs, H.D., and Tand, R.J., "Integer Programming Approach to the Problem of Optimal Unit Commitment with Probabilistic Reserve Determination," *IEEE Trans. Power Apparatus and Systems* **PAS-97**(6), 2154–2166 (1978).

[33] Du, D. Z., and Zhang, X. S., "Global convergence of Rosen's Gradient Projection Method," *Mathematical Programming* **44**, 357–366 (1989).

[34] Everett, H., "Generalized Lagrange Multiplier Method for Solving Problems of Optimum Allocation of Resources," *Operations Research* **11**, 399–417 (1963).

[35] Farkas, J., "Uber der einfachen Ungleichungen," *J. Reine und Angewandte Mathematik* **124**, 1–27 (1902).

[36] Farouki, R. T., "Optimal Paramaterizations," *Computer Aided Geometric Design*, **14**, 153-168 (1997).

[37] Ferreira, L. A. F. M., "On the Duality Gap for Thermal Unit Commitment Problems," *Proc ISCAS'93*, 2204–2207 (1993).

[38] Fiacco, A. V., and McCormick, G. P., *Nonlinear Programming: Sequential Unconstrained Minimization Techniques*, Wiley, New York, 1968.

[39] Fletcher, F., and Powell, "A Rapidly Convergent Descent Method for Minimization," *Comput. J.*, **6**, 163–168 (1963).

[40] Forrest, J. J. H., and Tomlin, J. A., "Implementing Interior Point Linear Programming Methods for the Optimization Subroutine Library," *IBM Syst. J.* **31**(1), 26–38 (1992).

[41] Fourer, R., Gay, D. M., and Kernighan, B. W., *A Modeling Language for Mathematical Programming*, The Scientific Press Series, 1993.

[42] Freitag, L. A., Jones, M., and Plassmann, P., "An Efficient Parallel Algorithm for Mesh Smoothing," *Proc. 4th Int. Meshing Roundtable*, 47–58 (1995).

[43] Freitag, L. A., "On Combining Laplacian and Optimization-Based Mesh Smoothing Techniques," **AMD-220**, Trends in Unstructured Mesh Generation, 37–43 (1997).

[44] Freitag, L. A., and Gooch, C. O., "Tetrahedral Mesh Improvement Using Swapping and Smoothing," *Int. J. Num. Meth. Eng.*, **40**, 3979–4002 (1997).

[45] Frisch, K. R., *The Logarithmic Potential Method for Convex Programming*, unpublished manuscript, Institute of Economics, Univ. Oslo, Norway, May 1955.

[46] García, R., López M. L., and Verastegui, D., "Extensions of Dinkelbach's Algorithm for Solving Nonlinear Fractional Programming Problems," *Top* **7**, 33–70 (1999).

[47] Garfinkel, R. S., and Nemhauser, G. L., *Integer Programming*, Wiley, New York, 1984.

[48] Gill, P. E., Murray, W., Saunders, M. A., Tomlin, J. A., and Wright, M. H., "On Projected Newton Barrier Methods for Linear Programming and an Equivalence to Karmarkar's Projective Method," *Math. Program.*, **36**, 183–209 (1986).

[49] Goldstein, A. A. "Convex programmingin Hilbert space," *Bull. Am. Math. Soc.* **70**, 709–710 (1964).

[50] Gomory, R. E., *An Algorithm for the Mixed Integer Problem*, Rand Report RM–25797., Rand, Santa Monica, CA, July 1960.

[51] Hiriart-Urruty, J. B., and Lemaréchal C., *Convex Analysis and Minimization Algorithms* Vols. I and II, Springer-Verlag, Berlin, 1996.

[52] Hobbs, W. J., Hermon, G., Warner, S., and Sheble, G. B., "An Enhanced Dynamic Programming Approach for Unit Commitment," *IEEE Trans. Power Systems* **3**(3), 1201–1205 (1988).

[53] *http://www.gams.com.*

[54] *http://www.ampl.com.*

[55] *http://www.aimms.com.*

[56] *IBM Optimization Subroutine Library (OSL)*, Release 2, Publication No. SC23-0519-03, 1992.

[57] Karmarkar, N. K., "A New Polynomial-Time Algorithm for Linear Programming," *Combinatorica*, **4**, 373–395 (1984).

[58] Kort, B. W., "Rate of Convergence of the Method of Multipliers with Inexact Minimization," in *Nonlinear Programming*, **2**, Academic Press, New York, 193–214, 1975.

[59] Jahn, J., *Introduction to the Theory of Nonlinear Optimization.* Springer-Verlag, Berlin, 1996.

[60] Jiménez Redondo, N., and Conejo, A. J., "Short-Term Hydro-Thermal Coordination by Lagrangian Relaxation: Solution of the Dual Problem," *IEEE Trans. Power Syst.*, **14**(1), 266–273 (1999).

[61] Jubete, F., *Programación no Lineal Convexa. El Método Polar*, CIS, Santander, Spain, 1987.

[62] Jubete, F., *El Cono Poliédrico Convexo. Su Incidencia en el Álgebra Lineal y la Programación no Lineal*, CIS, Santander, Spain, 1991.

[63] Khachiyan, L. G., "A Polynomial Algorithm in Linear Programming," *Doklady Akademii Nauk SSSR* No. 244, 1093–1096, [translated into English in Soviet Mathematics Doklady No. 20, 191–194 (1979)].

[64] Kuenzi, H. P., Tzchach, H. G., and Zehnder, C. A.,*Numerical Methods of Mathematical Optimization*, Academic Press, Orlando, FL, 1971.

[65] Lemaréchal, C., *Nondifferentiable Optimization*, in Optimization (G. L. Nemhauser, A. H. G., Rinnooy and M. J., Tood, eds., Vol. 1 in *Handbooks in Operations Research and Management Science*, Elsevier, Amsterdam, 1989.

[66] Levitin, E. S., and Polyak, B. T., "Constrained Minimization Methods," *USSR Comput. Math. and Math. Phys.* **6**, 1–50 (1966).

[67] Li, T. S., McKeag, R. M., and Armstrong, C. G., "Hexahedral Meshing Using Midpoint Subdivision and Integer Programming," *Comput. Meth. Appl. Mech. Eng.*, **124**, 171–193 (1995).

[68] Luenberger, D. G., *Linear and Nonlinear Programming*, 2nd ed., Addison-Wesley, Reading, MA, 1989.

[69] Luh, P. B., Zhang, D., and Tomastik, R. N., "An Algorithm for Solving the Dual Problem of Hydrothermal Scheduling," *IEEE Trans. Power Syst.* **13**(2), 593–600 (1998).

[70] Medina, J., Quintana, V.H., Conejo, A.J., and Pérez Thoden, F., "A Comparison of Interior-Point Codes for Medium-Term Hydro-Thermal Coordination," *IEEE Trans. Power Syst.* **13**(3), 836–843 (1998).

[71] Medina, J., Quintana, V. H., and Conejo, A. J., "A Clipping-Off Interior-Point Technique for Medium-Term Hydro-Thermal Coordination," *IEEE Trans. Power Syst.* **14**(1), 266–273, (Feb. 1999).

[72] Meggido, N., "Pathway to the Optimal Set in Linear Programming," in *Progress in Mathematical Programming: Interior Point and Related Methods*, Springer Verlag, New York, 1989, pp. 131–158.

[73] Mendes, V. M., Ferreira, L. A. F. M., Roldao, P., and Pestana, R., "Optimal Short-Term Scheduling in Large Hydrothermal Power Systems," *Proc. 11th Power Systems Computation Conf. PSCC'93.* Vol. II, Avignon, France, 1993, pp. 1297–1303.

[74] Merlin, A., and Sandrin, P., "A New Method for Unit Commitment at Electricité de France," *IEEE Trans. Power Apparatus Syst.*, **PAS-102**(5), 1218–1225 (1983).

[75] Mehrotra, S., *On the Implementation of a (Primal Dual) Interior Point Method*, Technical Report 90–03, Dep. Industrial Engineering and Management Sciences, Northwestern Univ., Evanston, IL, 1990.

[76] Migdalas, A., "A Regularization of the Frank–Wolfe Method and Unification of Certain Nonlinear Programming Methods," *Math. Program.* **65**, 331–345 (1994).

[77] Muckstadt, J. A., and Koening, S. A., "An Application of Lagrangian Relaxation to Scheduling in Power Generation Systems," *Operation Research* **25**(3), 387–403 (1977).

[78] Murtagh, B. A., and Saunders, M. A., *MINOS 5.0 Users's Guide*, Report SOL 83–20, Dep. Operations Research, Stanford Univ.

[79] Nemhauser, G. L., and Wolsey, L. A., *Integer and Combinatorial Optimization.* Wiley, New York, 1988.

[80] Nguyen S., and Dupis, C., "An Efficient Method for Computing Traffic Equilibria in Networks with Asymmetric Transportation Costs," *Transportation Science*, **18**, 185–202 (1984).

[81] Nocedal, A. J., and Wright, J., *Numerical Optimization*, Springer, New York, 1999.

[82] Ortúzar, J. de D., and Willumsen, L. G., *Modelling Transport*, Wiley, New York, 1994.

[83] Patrikson, M., "Partial linearization methods in nonlinear programming," *J. Optim. Theor. and Appl.* **78**, 227–246 (1993).

[84] Patriksson, M., *The Traffic Assignment Problem. Models and Methods*, VSP, Utrecht, The Netherlands, 1994.

[85] Pellegrino, F., Renaud, A., and Socroun, T., "Bundle. Augmented Lagrangian Methods for Short-Term Unit Commitment," *Proc. 12th Power Systems Computation Conf. PSCC'96*, Vol. II, 730–739. Dresden, Germany, 1996.

[86] Polak, E., *Optimization: Algorithms and Consistent Approximations*, Springer, New York, 1997.

[87] Polyak B. T., *Introduction to Optimization*, Optimization Software, New York, 1987.

[88] Potts, R., and Oliver, R., *Flows in Transportation Networks*, Academic Press, New York, 1972.

[89] Press, W. H., Teukolsky, S. A., Vetterling, W. T., and Flannery, B. P., *Numerical Recipes in C. The Art of Scientific Computing*, Cambridge Univ. Press, New York, 1985.

[90] Torres, G. L., and Quintana, V. H., *Power System Optimization via Successive Linear Programming with Interior Point Methods*, Univ. of Waterloo. Dep. Electrical and Computer Engineering, Technical Report UW E&CE 96–02, Jan. 1996.

[91] Rakic, M. V., and Marcovic, Z. M., "Short Term Operation and Power Exchange Planning of Hydro-Thermal Power Systems," *IEEE Trans. Power Syst.*, **9**(1), 359–365 (1994).

[92] Rao, S. S., *Optimization Theory and Applications.* 2nd ed., Wiley Eastern Limited, New Delhi, 1984.

[93] Rao, S. S., *Engineering Optimization. Theory and Practice*, 3rd ed., Wiley, New York, 1996.

[94] Renaud, A., "Daily Generation Management at Electricité de France: from Planning towards Real Time," *IEEE Trans. Automatic Control* **38**(7), 1080–1093 (1993).

[95] Rockafellar, R. T., *Convex Analysis*, Princeton Univ. Press, Princeton. NJ, 1970.

[96] Rosen, J. B., "The gradient projection method for nonlinear programming," Parts I and II, *J. Soc. Ind. Appl. Math.*, **8**, 181–217 (1960); **9**, 514–532 (1961).

[97] Shimizu, K., Ishizuka, Y., and Bard, J. F., *Nondifferentiable and Two-Level Mathematical Programming*, Kluwer Academic Publishers, Boston MA, 1997.

[98] Sheffi, Y., *Urban Transportation Networks: Equilibrium Analysis with Mathematical Programming Methods*, Prentice-Hall, Englewood Cliffs, NJ, 1984.

[99] Troutman, J. L., *Variational Calculus and Optimal Control: Optimization with Elementary Convexity*, 2nd ed., Springer, New York, 1996.

[100] Vanderbei, R. J., "Interior Point Methods: Algorithms and Formulations," *ORSA J. Comput.*, **6**(1), 32–34 (1994).

[101] Wang, S. J., Shahidehpour, S. M., Kirschen, D. S., Mokhtari, S., and Irisarri, G. S., "Short-Term Generation Scheduling with Transmission and Environmental Constraints using an Augmented Lagrangian Relaxation," *IEEE Trans. Power Syst.*, **10**(3), 1294–1301, 1995.

[102] Wardrop, J. G., "Some Theoretical Aspects of Road Traffic Research," *Proc. Institute of Civil Engineers Part II*, 325–378 (1952).

[103] Wolfram, S., *The Mathematica Book*, Wolfram Media and Cambridge Univ. Press, Cambridge, UK, 1996.

[104] Wood, A. J., and Wollenberg, B. F., *Power Generation Operation and Control*, 2nd ed., Wiley, New York, 1996.

[105] Wright, S. J., *Primal-Dual Interior-Point Methods*, SIAM, Philadelphia, 1997.

[106] Yan, H., Luh, P. B., Guan, X., and Rogan, P. M., "Scheduling of Hydrothermal Power Systems," *IEEE Trans. Power Syst.* **8**(3), 1135–1365 (1993).

[107] Zhuang, F., and Galiana, F. D., "Toward a More Rigorous and Practical Unit Commitment by Lagrangian Relaxation," *IEEE Trans. Power Syst.* **3**(2), 763–773 (1988).

Index

PURE AND APPLIED MATHEMATICS

A Wiley-Interscience Series of Texts, Monographs, and Tracts

ADÁMEK, HERRLICH, and STRECKER—Abstract and Concrete Catetories
ADAMOWICZ and ZBIERSKI—Logic of Mathematics
AINSWORTH and ODEN—A Posteriori Error Estimation in Finite Element Analysis
AKIVIS and GOLDBERG—Conformal Differential Geometry and Its Generalizations
ALLEN and ISAACSON—Numerical Analysis for Applied Science
*ARTIN—Geometric Algebra
AUBIN—Applied Functional Analysis, Second Edition
AZIZOV and IOKHVIDOV—Linear Operators in Spaces with an Indefinite Metric
BERG—The Fourier-Analytic Proof of Quadratic Reciprocity
BERMAN, NEUMANN, and STERN—Nonnegative Matrices in Dynamic Systems
BOYARINTSEV—Methods of Solving Singular Systems of Ordinary Differential Equations
BURK—Lebesgue Measure and Integration: An Introduction
*CARTER—Finite Groups of Lie Type
CASTILLO, COBO, JUBETE, and PRUNEDA—Orthogonal Sets and Polar Methods in Linear Algebra: Applications to Matrix Calculations, Systems of Equations, Inequalities, and Linear Programming
CASTILLO, CONEJO, PEDREGAL, GARCIÁ, and ALGUACIL—Building and Solving Mathematical Programming Models in Engineering and Science
CHATELIN—Eigenvalues of Matrices
CLARK—Mathematical Bioeconomics: The Optimal Management of Renewable Resources, Second Edition
COX—Primes of the Form $x^2 + ny^2$: Fermat, Class Field Theory, and Complex Multiplication
*CURTIS and REINER—Representation Theory of Finite Groups and Associative Algebras
*CURTIS and REINER—Methods of Representation Theory: With Applications to Finite Groups and Orders, Volume I
CURTIS and REINER—Methods of Representation Theory: With Applications to Finite Groups and Orders, Volume II
DINCULEANU—Vector Integration and Stochastic Integration in Banach Spaces
*DUNFORD and SCHWARTZ—Linear Operators
Part 1—General Theory
Part 2—Spectral Theory, Self Adjoint Operators in Hilbert Space
Part 3—Spectral Operators
FARINA and RINALDI—Positive Linear Systems: Theory and Applications
FOLLAND—Real Analysis: Modern Techniques and Their Applications
FRÖLICHER and KRIEGL—Linear Spaces and Differentiation Theory
GARDINER—Teichmüller Theory and Quadratic Differentials
GREENE and KRANTZ—Function Theory of One Complex Variable
*GRIFFITHS and HARRIS—Principles of Algebraic Geometry

*Now available in a lower priced paperback edition in the Wiley Classics Library.
†Now available in paperback.

GRILLET—Algebra
GROVE—Groups and Characters
GUSTAFSSON, KREISS and OLIGER—Time Dependent Problems and Difference Methods
HANNA and ROWLAND—Fourier Series, Transforms, and Boundary Value Problems, Second Edition
*HENRICI—Applied and Computational Complex Analysis
Volume 1, Power Series—Integration—Conformal Mapping—Location of Zeros
Volume 2, Special Functions—Integral Transforms—Asymptotics—Continued Fractions
Volume 3, Discrete Fourier Analysis, Cauchy Integrals, Construction of Conformal Maps, Univalent Functions
*HILTON and WU—A Course in Modern Algebra
*HOCHSTADT—Integral Equations
JOST—Two-Dimensional Geometric Variational Procedures
KHAMSI and KIRK—An Introduction to Metric Spaces and Fixed Point Theory
*KOBAYASHI and NOMIZU—Foundations of Differential Geometry, Volume I
*KOBAYASHI and NOMIZU—Foundations of Differential Geometry, Volume II
KOSHY—Fibonacci and Lucas Numbers with Applications
LAX—Linear Algebra
LOGAN—An Introduction to Nonlinear Partial Differential Equations
McCONNELL and ROBSON—Noncommutative Noetherian Rings
MORRISON—Functional Analysis: An Introduction to Banach Space Theory
NAYFEH—Perturbation Methods
NAYFEH and MOOK—Nonlinear Oscillations
PANDEY—The Hilbert Transform of Schwartz Distributions and Applications
PETKOV—Geometry of Reflecting Rays and Inverse Spectral Problems
*PRENTER—Splines and Variational Methods
RAO—Measure Theory and Integration
RASSIAS and SIMSA—Finite Sums Decompositions in Mathematical Analysis
RENELT—Elliptic Systems and Quasiconformal Mappings
RIVLIN—Chebyshev Polynomials: From Approximation Theory to Algebra and Number Theory, Second Edition
ROCKAFELLAR—Network Flows and Monotropic Optimization
ROITMAN—Introduction to Modern Set Theory
*RUDIN—Fourier Analysis on Groups
SENDOV—The Averaged Moduli of Smoothness: Applications in Numerical Methods and Approximations
SENDOV and POPOV—The Averaged Moduli of Smoothness
*SIEGEL—Topics in Complex Function Theory
Volume l—Elliptic Functions and Uniformization Theory
Volume 2—Automorphic Functions and Abelian Integrals
Volume 3—Abelian Functions and Modular Functions of Several Variables
SMITH and ROMANOWSKA—Post-Modern Algebra
STAKGOLD—Green's Functions and Boundary Value Problems, Second Editon
*STOKER—Differential Geometry
*STOKER—Nonlinear Vibrations in Mechanical and Electrical Systems
*STOKER—Water Waves: The Mathematical Theory with Applications
WESSELING—An Introduction to Multigrid Methods
†WHITHAM—Linear and Nonlinear Waves
†ZAUDERER—Partial Differential Equations of Applied Mathematics, Second Edition

*Now available in a lower priced paperback edition in the Wiley Classics Library.
†Now available in paperback.